TS7.8 BER

Convex Analysis and Optimization

Dimitri P. Bertsekas

with

Angelia Nedić and Asuman E. Ozdaglar

Massachusetts Institute of Technology

WWW site for book information and orders

http://www.athenasc.com

Athena Scientific, Belmont, Massachusetts

Athena Scientific
Post Office Box 391
Belmont, Mass. 02478-9998
U.S.A.

Email: info@athenasc.com
WWW: http://www.athenasc.com

Publisher's Cataloging-in-Publication Data

Bertsekas, Dimitri P., Nedić, Angelia, Ozdaglar Asuman E.
Convex Analysis and Optimization
Includes bibliographical references and index
1. Nonlinear Programming 2. Mathematical Optimization. I. Title.
T57.8.B475 2003   519.703
Library of Congress Control Number: 2002092168

ISBN 1-886529-45-0

ATHENA SCIENTIFIC
OPTIMIZATION AND COMPUTATION SERIES

1. Convex Analysis and Optimization, by Dimitri P. Bertsekas, with Angelia Nedić and Asuman E. Ozdaglar, 2003, ISBN 1-886529-45-0, 560 pages

2. Introduction to Probability, by Dimitri P. Bertsekas and John N. Tsitsiklis, 2002, ISBN 1-886529-40-X, 430 pages

3. Dynamic Programming and Optimal Control, Two-Volume Set (2nd Edition), by Dimitri P. Bertsekas, 2001, ISBN 1-886529-08-6, 840 pages

4. Nonlinear Programming, 2nd Edition, by Dimitri P. Bertsekas, 1999, ISBN 1-886529-00-0, 791 pages

5. Network Optimization: Continuous and Discrete Models by Dimitri P. Bertsekas, 1998, ISBN 1-886529-02-7, 608 pages

6. Network Flows and Monotropic Optimization by R. Tyrrell Rockafellar, 1998, ISBN 1-886529-06-X, 634 pages

7. Introduction to Linear Optimization by Dimitris Bertsimas and John N. Tsitsiklis, 1997, ISBN 1-886529-19-1, 608 pages

8. Parallel and Distributed Computation: Numerical Methods by Dimitri P. Bertsekas and John N. Tsitsiklis, 1997, ISBN 1-886529-01-9, 718 pages

9. Neuro-Dynamic Programming, by Dimitri P. Bertsekas and John N. Tsitsiklis, 1996, ISBN 1-886529-10-8, 512 pages

10. Constrained Optimization and Lagrange Multiplier Methods, by Dimitri P. Bertsekas, 1996, ISBN 1-886529-04-3, 410 pages

11. Stochastic Optimal Control: The Discrete-Time Case by Dimitri P. Bertsekas and Steven E. Shreve, 1996, ISBN 1-886529-03-5, 330 pages

Contents

1. Basic Convexity Concepts . . . . **p. 1**

1.1. Linear Algebra and Real Analysis . . . . p. 3
    1.1.1. Vectors and Matrices . . . . p. 5
    1.1.2. Topological Properties . . . . p. 8
    1.1.3. Square Matrices . . . . p. 15
    1.1.4. Derivatives . . . . p. 16
1.2. Convex Sets and Functions . . . . p. 20
1.3. Convex and Affine Hulls . . . . p. 35
1.4. Relative Interior, Closure, and Continuity . . . . p. 39
1.5. Recession Cones . . . . p. 49
    1.5.1. Nonemptiness of Intersections of Closed Sets . . . . . p. 56
    1.5.2. Closedness Under Linear Transformations . . . . . . p. 64
1.6. Notes, Sources, and Exercises . . . . p. 68

2. Convexity and Optimization . . . . **p. 83**

2.1. Global and Local Minima . . . . p. 84
2.2. The Projection Theorem . . . . p. 88
2.3. Directions of Recession and Existence of Optimal Solutions . . p. 92
    2.3.1. Existence of Solutions of Convex Programs . . . . . . p. 94
    2.3.2. Unbounded Optimal Solution Sets . . . . p. 97
    2.3.3. Partial Minimization of Convex Functions . . . . . . p. 101
2.4. Hyperplanes . . . . p. 107
2.5. An Elementary Form of Duality . . . . p. 117
    2.5.1. Nonvertical Hyperplanes . . . . p. 117
    2.5.2. Min Common/Max Crossing Duality . . . . p. 120
2.6. Saddle Point and Minimax Theory . . . . p. 128
    2.6.1. Min Common/Max Crossing Framework for Minimax . p. 133
    2.6.2. Minimax Theorems . . . . p. 139
    2.6.3. Saddle Point Theorems . . . . p. 143
2.7. Notes, Sources, and Exercises . . . . p. 151

3. Polyhedral Convexity . . . . **p. 165**

3.1. Polar Cones . . . . p. 166
3.2. Polyhedral Cones and Polyhedral Sets . . . . p. 168
    3.2.1. Farkas' Lemma and Minkowski-Weyl Theorem . . . . p. 170
    3.2.2. Polyhedral Sets . . . . p. 175
    3.2.3. Polyhedral Functions . . . . p. 178
3.3. Extreme Points . . . . p. 180
    3.3.1. Extreme Points of Polyhedral Sets . . . . p. 183
3.4. Polyhedral Aspects of Optimization . . . . p. 186
    3.4.1. Linear Programming . . . . p. 188
    3.4.2. Integer Programming . . . . p. 189
3.5. Polyhedral Aspects of Duality . . . . p. 192
    3.5.1. Polyhedral Proper Separation . . . . p. 192
    3.5.2. Min Common/Max Crossing Duality . . . . p. 196
    3.5.3. Minimax Theory Under Polyhedral Assumptions . . . p. 199
    3.5.4. A Nonlinear Version of Farkas' Lemma . . . . p. 203
    3.5.5. Convex Programming . . . . p. 208
3.6. Notes, Sources, and Exercises . . . . p. 210

4. Subgradients and Constrained Optimization . . . . . **p. 221**

4.1. Directional Derivatives . . . . p. 222
4.2. Subgradients and Subdifferentials . . . . p. 227
4.3. ϵ-Subgradients . . . . p. 235
4.4. Subgradients of Extended Real-Valued Functions . . . . . p. 241
4.5. Directional Derivative of the Max Function . . . . p. 245
4.6. Conical Approximations . . . . p. 248
4.7. Optimality Conditions . . . . p. 255
4.8. Notes, Sources, and Exercises . . . . p. 261

5. Lagrange Multipliers . . . . **p. 269**

5.1. Introduction to Lagrange Multipliers . . . . p. 270
5.2. Enhanced Fritz John Optimality Conditions . . . . p. 281
5.3. Informative Lagrange Multipliers . . . . p. 288
    5.3.1. Sensitivity . . . . p. 297
    5.3.2. Alternative Lagrange Multipliers . . . . p. 299
5.4. Pseudonormality and Constraint Qualifications . . . . . . p. 302
5.5. Exact Penalty Functions . . . . p. 313
5.6. Using the Extended Representation . . . . p. 319
5.7. Extensions Under Convexity Assumptions . . . . p. 324
5.8. Notes, Sources, and Exercises . . . . p. 335

6. Lagrangian Duality . . . . **p. 345**

    6.1. Geometric Multipliers . . . . p. 346
    6.2. Duality Theory . . . . p. 355
    6.3. Linear and Quadratic Programming Duality . . . . p. 362
    6.4. Existence of Geometric Multipliers . . . . p. 367
        6.4.1. Convex Cost – Linear Constraints . . . . p. 368
        6.4.2. Convex Cost – Convex Constraints . . . . p. 371
    6.5. Strong Duality and the Primal Function . . . . p. 374
        6.5.1. Duality Gap and the Primal Function . . . . p. 374
        6.5.2. Conditions for No Duality Gap . . . . p. 377
        6.5.3. Subgradients of the Primal Function . . . . p. 382
        6.5.4. Sensitivity Analysis . . . . p. 383
    6.6. Fritz John Conditions when there is no Optimal Solution . p. 384
        6.6.1. Enhanced Fritz John Conditions . . . . p. 390
        6.6.2. Informative Geometric Multipliers . . . . p. 406
    6.7. Notes, Sources, and Exercises . . . . p. 413

7. Conjugate Duality . . . . **p. 421**

    7.1. Conjugate Functions . . . . p. 424
    7.2. Fenchel Duality Theorems . . . . p. 434
        7.2.1. Connection of Fenchel Duality and Minimax Theory . . p. 437
        7.2.2. Conic Duality . . . . p. 439
    7.3. Exact Penalty Functions . . . . p. 441
    7.4. Notes, Sources, and Exercises . . . . p. 446

8. Dual Computational Methods . . . . **p. 455**

    8.1. Dual Derivatives and Subgradients . . . . p. 457
    8.2. Subgradient Methods . . . . p. 460
        8.2.1. Analysis of Subgradient Methods . . . . p. 470
        8.2.2. Subgradient Methods with Randomization . . . . . . p. 488
    8.3. Cutting Plane Methods . . . . p. 504
    8.4. Ascent Methods . . . . p. 509
    8.5. Notes, Sources, and Exercises . . . . p. 512

References . . . . **p. 517**

Index . . . . **p. 529**

Preface

The knowledge at which geometry aims is the knowledge of the eternal

(Plato, Republic, VII, 52)

This book focuses on the theory of convex sets and functions, and its connections with a number of topics that span a broad range from continuous to discrete optimization. These topics include Lagrange multiplier theory, Lagrangian and conjugate/Fenchel duality, minimax theory, and nondifferentiable optimization.

The book evolved from a set of lecture notes for a graduate course at M.I.T. It is widely recognized that, aside from being an eminently useful subject in engineering, operations research, and economics, convexity is an excellent vehicle for assimilating some of the basic concepts of real analysis within an intuitive geometrical setting. Unfortunately, the subject's coverage in academic curricula is scant and incidental. We believe that at least part of the reason is the shortage of textbooks that are suitable for classroom instruction, particularly for nonmathematics majors. We have therefore tried to make convex analysis accessible to a broader audience by emphasizing its geometrical character, while maintaining mathematical rigor. We have included as many insightful illustrations as possible, and we have used geometric visualization as a principal tool for maintaining the students' interest in mathematical proofs.

Our treatment of convexity theory is quite comprehensive, with all major aspects of the subject receiving substantial treatment. The mathematical prerequisites are a course in linear algebra and a course in real analysis in finite dimensional spaces (which is the exclusive setting of the book). A summary of this material, without proofs, is provided in Section 1.1.

The coverage of the theory has been significantly extended in the exercises, which represent a major component of the book. Detailed solutions of all the exercises (nearly 200 pages) are internet-posted in the book's www page

http://www.athenasc.com/convexity.html

Some of the exercises may be attempted by the reader without looking at the solutions, while others are challenging but may be solved by the advanced reader with the assistance of hints. Still other exercises represent substantial theoretical results, and in some cases include new and unpublished research. Readers and instructors should decide for themselves how to make best use of the internet-posted solutions.

An important part of our approach has been to maintain a close link between the theoretical treatment of convexity and its application to optimization. For example, in Chapter 2, after the development of some of the basic facts about convexity, we discuss some of their applications to optimization and saddle point theory; in Chapter 3, after the discussion of polyhedral convexity, we discuss its application in linear and integer programming; and in Chapter 4, after the discussion of subgradients, we discuss their use in optimality conditions. We follow this style in the remaining chapters, although having developed in Chapters 1-4 most of the needed convexity theory, the discussion in the subsequent chapters is more heavily weighted towards optimization.

The chart of the opposite page illustrates the main topics covered in the book, and their interrelations. At the top level, we have the most basic concepts of convexity theory, which are covered in Chapter 1. At the middle level, we have fundamental topics of optimization, such as existence and characterization of solutions, and minimax theory, together with some supporting convexity concepts such as hyperplane separation, polyhedral sets, and subdifferentiability (Chapters 2-4). At the lowest level, we have the core issues of convex optimization: Lagrange multipliers, Lagrange and Fenchel duality, and numerical dual optimization (Chapters 5-8).

An instructor who wishes to teach a course from the book has a choice between several different plans. One possibility is to cover in detail just the first four chapters, perhaps augmented with some selected sections from the remainder of the book, such as the first section of Chapter 7, which deals with conjugate convex functions. The idea here is to concentrate on convex analysis and illustrate its application to minimax theory through the minimax theorems of Chapters 2 and 3, and to constrained optimization theory through the Nonlinear Farkas' Lemma of Chapter 3 and the optimality conditions of Chapter 4. An alternative plan is to cover Chapters 1-4 in less detail in order to allow some time for Lagrange multiplier theory and computational methods. Other plans may also be devised, possibly including some applications or some additional theoretical topics of the instructor's choice.

While the subject of the book is classical, the treatment of several of its important topics is new and in some cases relies on new research. In

Chapter 1

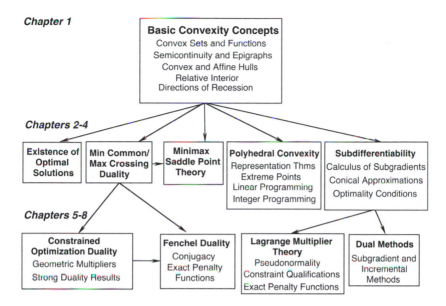

particular, our new lines of analysis include:

(a) A unified development of minimax theory and constrained optimization duality as special cases of the duality between two simple geometrical problems: the min common point problem and the max crossing point problem. Here, by minimax theory, we mean the analysis relating to the minimax equality

$$\inf_{x \in X} \sup_{z \in Z} \phi(x, z) = \sup_{z \in Z} \inf_{x \in X} \phi(x, z),$$

and the attainment of the "inf" and the "sup." By constrained optimization theory, we mean the analysis of problems such as

$$\text{minimize} \quad f(x)$$
$$\text{subject to} \quad x \in X, \qquad g_j(x) \leq 0, \quad j = 1, \dots, r,$$

and issues such as the existence of optimal solutions and Lagrange multipliers, and the absence of a duality gap [equality of the optimal value of the above problem and the optimal value of an associated dual problem, obtained by assigning multipliers to the inequality constraints $g_j(x) \leq 0$].

(b) A unification of conditions for existence of solutions of convex optimization problems, conditions for the minimax equality to hold, and conditions for the absence of a duality gap in constrained optimization. This unification is based on conditions guaranteeing that a nested family of closed convex sets has a nonempty intersection.

(c) A unification of the major constraint qualifications that guarantee the existence of Lagrange multipliers for nonconvex constrained optimization. This unification is achieved through the notion of constraint pseudonormality, which is motivated by an enhanced form of the Fritz John necessary optimality conditions.

(d) The development of incremental subgradient methods for dual optimization, and the analysis of their advantages over classical subgradient methods.

We provide some orientation by informally summarizing the main ideas of each of the above topics.

Min Common/Max Crossing Duality

In this book, duality theory is captured in two easily visualized problems: the min common point problem and the max crossing point problem, introduced in Chapter 2. Fundamentally, these problems revolve around the existence of nonvertical supporting hyperplanes to convex sets that are unbounded from above along the vertical axis. When properly specialized, this turns out to be the critical issue in constrained optimization duality and saddle point/minimax theory, under standard convexity and/or concavity assumptions.

The salient feature of the min common/max crossing framework is its simple geometry, in the context of which the fundamental constraint qualifications needed for strong duality theorems are visually apparent, and admit straightforward proofs. This allows the development of duality theory in a unified way: first within the min common/max crossing framework in Chapters 2 and 3, and then by specialization, to saddle point and minimax theory in Chapters 2 and 3, and to optimization duality in Chapter 6. All of the major duality theorems discussed in this book are derived in this way, including the principal Lagrange multiplier and Fenchel duality theorems for convex programming, and the von Neuman Theorem for zero sum games.

From an instructional point of view, it is particularly desirable to unify constrained optimization duality and saddle point/minimax theory (under convexity/concavity assumptions). Their connection is well known, but it is hard to understand beyond a superficial level, because there is not enough overlap between the two theories to develop one in terms of the other. In our approach, rather than trying to build a closer connection between constrained optimization duality and saddle point/minimax theory, we show how they both stem from a common geometrical root: the min common/max crossing duality.

We note that the constructions involved in the min common and max crossing problems arise in the theories of subgradients, conjugate convex functions, and duality. As such they are implicit in several earlier analyses; in fact they have been employed for visualization purposes in the first author's nonlinear programming textbook [Ber99]. However, the two problems have not been used as a unifying theoretical framework for constrained optimization duality, saddle point theory, or other contexts, except implicitly through the theory of conjugate convex functions, and the complicated and specialized machinery of conjugate saddle functions. Pedagogically, it appears desirable to postpone the introduction of conjugacy theory until it is needed for the limited purposes of Fenchel duality (Chapter 7), and to bypass altogether conjugate saddle function theory, which is what we have done.

Existence of Solutions and Strong Duality

We show that under convexity assumptions, several fundamental issues in optimization are intimately related. In particular, we give a unified analysis of conditions for optimal solutions to exist, for the minimax equality to hold, and for the absence of a duality gap in constrained optimization.

To provide a sense of the main idea, we note that given a constrained optimization problem, lower semicontinuity of the cost function and compactness of the constraint set guarantee the existence of an optimal solution (the Weierstrass Theorem). On the other hand, the same conditions plus convexity of the cost and constraint functions guarantee not only the existence of an optimal solution, but also the absence of a duality gap. This is not a coincidence, because as it turns out, the conditions for both cases critically rely on the same fundamental properties of compact sets, namely that the intersection of a nested family of nonempty compact sets is nonempty and compact, and that the projections of compact sets on any subspace are compact.

In our analysis, we extend this line of reasoning under a variety of assumptions relating to convexity, directions of recession, polyhedral sets, and special types of sets specified by quadratic and other types of inequalities. The assumptions are used to establish results asserting that the intersection of a nested family of closed convex sets is nonempty, and that the function $f(x) = \inf_z F(x, z)$, obtained by partial minimization of a convex function F, is lower semicontinuous. These results are translated in turn to a broad variety of conditions that guarantee the existence of optimal solutions, the minimax equality, and the absence of a duality gap.

Pseudonormality and Lagrange Multipliers

In Chapter 5, we discuss Lagrange multiplier theory in the context of optimization of a smooth cost function, subject to smooth equality and inequality constraints, as well as an additional set constraint. Our treatment of Lagrange multipliers is new, and aims to generalize, unify, and streamline the theory of constraint qualifications.

The starting point for our development is an enhanced set of necessary conditions of the Fritz John type, that are sharper than the classical Karush-Kuhn-Tucker conditions (they include extra conditions, which may narrow down the field of candidate local minima). They are also more general in that they apply when there is an abstract (possibly nonconvex) set constraint, in addition to the equality and inequality constraints. To achieve this level of generality, we bring to bear notions of nonsmooth analysis, and we find that the notion of regularity of the abstract constraint set provides the critical distinction between problems that do and do not admit a satisfactory theory.

Fundamentally, Lagrange multiplier theory should aim to identify the essential constraint structure that guarantees the existence of Lagrange multipliers. For smooth problems with equality and inequality constraints, but no abstract set constraint, this essential structure is captured by the classical notion of quasiregularity (the tangent cone at a given feasible point is equal to the cone of first order feasible variations). However, in the presence of an additional set constraint, the notion of quasiregularity breaks down as a viable unification vehicle. Our development introduces the notion of pseudonormality as a substitute for quasiregularity for the case of an abstract set constraint. Pseudonormality unifies and expands the major constraint qualifications, and simplifies the proofs of Lagrange multiplier theorems. In the case of equality constraints only, pseudonormality is implied by either one of two alternative constraint qualifications: the linear independendence of the constraint gradients and the linearity of the constraint functions. In fact, in this case, pseudonormality is not much different than the union of these two constraint qualifications. However, pseudonormality is a meaningful unifying property even in the case of an additional set constraint, where the classical proof arguments based on quasiregularity fail. Pseudonormality also provides the connecting link between constraint qualifications and the theory of exact penalty functions.

An interesting byproduct of our analysis is a taxonomy of different types of Lagrange multipliers for problems with nonunique Lagrange multipliers. Under some convexity assumptions, we show that if there exists at least one Lagrange multiplier vector, there exists at least one of a special type, called informative, which has nice sensitivity properties. The nonzero components of such a multiplier vector identify the constraints that need to be violated in order to improve the optimal cost function value. Furthermore, a particular informative Lagrange multiplier vector characterizes the direction of steepest rate of improvement of the cost function for a given level of the norm of the constraint violation. Along that direction, the equality and inequality constraints are violated consistently with the signs of the corresponding multipliers.

The theory of enhanced Fritz John conditions and pseudonormality are extended in Chapter 6 to the case of a convex programming problem, without assuming the existence of an optimal solution or the absence of a duality gap. They form the basis for a new line of analysis for asserting the existence of informative multipliers under the standard constraint qualifications.

Incremental Subgradient Methods

In Chapter 8, we discuss one of the most important uses of duality: the numerical solution of dual problems, often in the context of discrete optimization and the method of branch-and-bound. These dual problems are often nondifferentiable and have special structure. Subgradient methods have been among the most popular for the solution of these problems, but they often suffer from slow convergence.

We introduce incremental subgradient methods, which aim to accelerate the convergence by exploiting the additive structure that a dual problem often inherits from properties of its primal problem, such as separability. In particular, for the common case where the dual function is the sum of a large number of component functions, incremental methods consist of a sequence of incremental steps, each involving a single component of the dual function, rather than the sum of all components.

Our analysis aims to identify effective variants of incremental methods, and to quantify their advantages over the standard subgradient methods. An important question is the selection of the order in which the components are selected for iteration. A particularly interesting variant uses randomization of the order to resolve a worst-case complexity bottleneck associated with the natural deterministic order. According to both analysis and experiment, this randomized variant performs substantially better than the standard subgradient methods for large scale problems that typically arise in the context of duality. The randomized variant is also particularly well-suited for parallel, possibly asynchronous, implementation, and is the only available method, to our knowledge, that can be used efficiently within this context.

We are thankful to a few persons for their contributions to the book. Several colleagues contributed information, suggestions, and insights. We would like to single out Paul Tseng, who was extraordinarily helpful by proofreading portions of the book, and collaborating with us on several research topics, including the Fritz John theory of Sections 5.7 and 6.6. We would also like to thank Xin Chen and Janey Yu, who gave us valuable feedback and some specific suggestions. Finally, we wish to express our appreciation for the stimulating environment at M.I.T., which provided an excellent setting for this work.

Dimitri P. Bertsekas, dimitrib@mit.edu
Angelia Nedić, angelia.nedich@alphatech.com
Asuman E. Ozdaglar, asuman@mit.edu

1

Basic Convexity Concepts

Contents

1.1. Linear Algebra and Real Analysis . . . . p. 3
   1.1.1. Vectors and Matrices . . . . p. 5
   1.1.2. Topological Properties . . . . p. 8
   1.1.3. Square Matrices . . . . p. 15
   1.1.4. Derivatives . . . . p. 16
1.2. Convex Sets and Functions . . . . p. 20
1.3. Convex and Affine Hulls . . . . p. 35
1.4. Relative Interior, Closure, and Continuity . . . . p. 39
1.5. Recession Cones . . . . p. 49
   1.5.1. Nonemptiness of Intersections of Closed Sets . . . . p. 56
   1.5.2. Closedness Under Linear Transformations . . . . . p. 64
1.6. Notes, Sources, and Exercises . . . . p. 68

In this chapter and the following three, we develop the theory of convex sets, which is the mathematical foundation for minimax theory, Lagrange multiplier theory, and duality. We assume no prior knowledge of the subject, and we give a detailed development. As we embark on the study of convexity, it is worth listing some of the properties of convex sets and functions that make them so special in optimization.

(a) *A convex function has no local minima that are not global.* Thus the difficulties associated with multiple disconnected local minima, whose global optimality is hard to verify in practice, are avoided (see Section 2.1).

(b) *A convex set has a nonempty relative interior.* In other words, relative to the smallest affine set containing it, a convex set has a nonempty interior (see Section 1.4). Thus convex sets avoid the analytical and computational optimization difficulties associated with "thin" and "curved" constraint surfaces.

(c) *A convex set is connected and has feasible directions at any point* (assuming it consists of more than one point). By this we mean that given any point x in a convex set X, it is possible to move from x along some directions y and stay within X for at least a nontrivial interval, i.e., $x + \alpha y \in X$ for all sufficiently small but positive stepsizes α (see Section 4.6). In fact a stronger property holds: given any two distinct points x and $\bar{x}$ in X, the direction $\bar{x} - x$ is a feasible direction at x, and all feasible directions can be characterized this way. For optimization purposes, this is important because it allows a calculus-based comparison of the cost of x with the cost of its close neighbors, and forms the basis for some important algorithms. Furthermore, much of the difficulty commonly associated with discrete constraint sets (arising for example in combinatorial optimization), is not encountered under convexity.

(d) *A nonconvex function can be "convexified" while maintaining the optimality of its global minima*, by forming the convex hull of the epigraph of the function (see Exercise 1.20).

(e) *The existence of a global minimum of a convex function over a convex set is conveniently characterized in terms of directions of recession* (see Section 2.3).

(f) *A polyhedral convex set (one that is specified by linear equality and inequality constraints) is characterized in terms of a finite set of extreme points and extreme directions.* This is the basis for finitely terminating methods for linear programming, including the celebrated simplex method (see Sections 3.3 and 3.4).

(g) *A convex function is continuous within the interior of its domain, and has nice differentiability properties.* In particular, a real-valued

convex function is directionally differentiable at any point. Furthermore, while a convex function need not be differentiable, it possesses
subgradients, which are nice and geometrically intuitive substitutes
for a gradient (see Chapter 4). Just like gradients, subgradients figure
prominently in optimality conditions and computational algorithms.

(h) *Convex functions are central in duality theory.* Indeed, the dual problem of a given optimization problem (discussed in Chapter 6) consists
of minimization of a convex function over a convex set, even if the
original problem is not convex.

(i) *Closed convex cones are self-dual with respect to polarity.* In words,
we have $C = (C^*)^*$ for any closed and convex cone C, where C^* is
the polar cone of C (the set of vectors that form a nonpositive inner
product with all vectors in C), and $(C^*)^*$ is the polar cone of C^*. This
simple and geometrically intuitive property (discussed in Section 3.1)
underlies important aspects of Lagrange multiplier theory.

(j) *Convex lower semicontinuous functions are self-dual with respect to
conjugacy.* It will be seen in Chapter 7 that a certain geometrically
motivated conjugacy operation on a convex, lower semicontinuous
function generates another convex, lower semicontinuous function,
and when applied for the second time regenerates the original function. The conjugacy operation relies on a fundamental dual characterization of a closed convex set: as the union of the closures of all line
segments connecting its points, and as the intersection of the closed
halfspaces within which the set is contained. Conjugacy is central in
duality theory, and has a nice interpretation that can be used to visualize and understand some of the most interesting aspects of convex
optimization.

In this first chapter, after an introductory first section, we focus on
the basic concepts of convex analysis: characterizations of convex sets and
functions, convex and affine hulls, topological concepts such as closure,
continuity, and relative interior, and the important notion of the recession
cone.

1.1 LINEAR ALGEBRA AND REAL ANALYSIS

In this section, we list some basic definitions, notational conventions, and
results from linear algebra and real analysis. We assume that the reader is
familiar with this material, so no proofs are given. For related and additional material, we recommend the books by Hoffman and Kunze [HoK71],
Lancaster and Tismenetsky [LaT85], and Strang [Str76] (linear algebra),

and the books by Ash [Ash72], Ortega and Rheinboldt [OrR70], and Rudin [Rud76] (real analysis).

Set Notation

If X is a set and x is an element of X, we write $x \in X$. A set can be specified in the form $X = \{x \mid x$ satisfies $P\}$, as the set of all elements satisfying property P. The union of two sets X_1 and X_2 is denoted by $X_1 \cup X_2$ and their intersection by $X_1 \cap X_2$. The symbols $\exists$ and $\forall$ have the meanings "there exists" and "for all," respectively. The empty set is denoted by $\emptyset$.

The set of real numbers (also referred to as scalars) is denoted by $\Re$. The set $\Re$ augmented with $+\infty$ and $-\infty$ is called the *set of extended real numbers*. We write $-\infty < x < \infty$ for all real numbers x, and $-\infty \leq x \leq \infty$ for all extended real numbers x. We denote by $[a, b]$ the set of (possibly extended) real numbers x satisfying $a \leq x \leq b$. A rounded, instead of square, bracket denotes strict inequality in the definition. Thus $(a, b]$, $[a, b)$, and (a, b) denote the set of all x satisfying $a < x \leq b$, $a \leq x < b$, and $a < x < b$, respectively. Furthermore, we use the natural extensions of the rules of arithmetic: $x \cdot 0 = 0$ for every extended real number x, $x \cdot \infty = \infty$ if $x > 0$, $x \cdot \infty = -\infty$ if $x < 0$, and $x + \infty = \infty$ and $x - \infty = -\infty$ for every scalar x. The expression $\infty - \infty$ is meaningless and is never allowed to occur.

Inf and Sup Notation

The *supremum* of a nonempty set X of scalars, denoted by $\sup X$, is defined as the smallest scalar y such that $y \geq x$ for all $x \in X$. If no such scalar exists, we say that the supremum of X is ∞. Similarly, the *infimum* of X, denoted by $\inf X$, is defined as the largest scalar y such that $y \leq x$ for all $x \in X$, and is equal to $-\infty$ if no such scalar exists. For the empty set, we use the convention

$$\sup \emptyset = -\infty, \qquad \inf \emptyset = \infty.$$

If $\sup X$ is equal to a scalar $\bar{x}$ that belongs to the set X, we say that $\bar{x}$ is the *maximum point* of X and we write $\bar{x} = \max X$. Similarly, if $\inf X$ is equal to a scalar $\bar{x}$ that belongs to the set X, we say that $\bar{x}$ is the *minimum point* of X and we write $\bar{x} = \min X$. Thus, when we write $\max X$ (or $\min X$) in place of $\sup X$ (or $\inf X$, respectively), we do so just for emphasis: we indicate that it is either evident, or it is known through earlier analysis, or it is about to be shown that the maximum (or minimum, respectively) of the set X is attained at one of its points.

Function Notation

If f is a function, we use the notation $f : X \mapsto Y$ to indicate the fact that f is defined on a nonempty set X (its *domain*) and takes values in a set Y (its *range*). Thus when using the notation $f : X \mapsto Y$, we implicitly assume that X is nonempty. If $f : X \mapsto Y$ is a function, and U and V are subsets of X and Y, respectively, the set $\{f(x) \mid x \in U\}$ is called the *image* or *forward image of* U *under* f, and the set $\{x \in X \mid f(x) \in V\}$ is called the *inverse image of* V *under* f.

1.1.1   Vectors and Matrices

We denote by $\Re^n$ the set of n-dimensional real vectors. For any $x \in \Re^n$, we use x_i to indicate its ith *coordinate*, also called its ith *component*.

Vectors in $\Re^n$ will be viewed as column vectors, unless the contrary is explicitly stated. For any $x \in \Re^n$, x' denotes the transpose of x, which is an n-dimensional row vector. The *inner product* of two vectors $x, y \in \Re^n$ is defined by $x'y = \sum_{i=1}^{n} x_i y_i$. Two vectors $x, y \in \Re^n$ satisfying $x'y = 0$ are called *orthogonal*.

If x is a vector in $\Re^n$, the notations $x > 0$ and $x \geq 0$ indicate that all components of x are positive and nonnegative, respectively. For any two vectors x and y, the notation $x > y$ means that $x - y > 0$. The notations $x \geq y$, $x < y$, etc., are to be interpreted accordingly.

If X is a set and λ is a scalar, we denote by λX the set $\{\lambda x \mid x \in X\}$. If X_1 and X_2 are two subsets of $\Re^n$, we denote by $X_1 + X_2$ the set

$$\{x_1 + x_2 \mid x_1 \in X_1, \, x_2 \in X_2\},$$

which is referred to as the *vector sum of* X_1 *and* X_2. We use a similar notation for the sum of any finite number of subsets. In the case where one of the subsets consists of a single vector $\bar{x}$, we simplify this notation as follows:

$$\bar{x} + X = \{\bar{x} + x \mid x \in X\}.$$

We also denote by $X_1 - X_2$ the set

$$\{x_1 - x_2 \mid x_1 \in X_1, \, x_2 \in X_2\}.$$

Given sets $X_i \subset \Re^{n_i}$, $i = 1, \ldots, m$, the *Cartesian product* of the X_i, denoted by $X_1 \times \cdots \times X_m$, is the set

$$\big\{(x_1, \ldots, x_m) \mid x_i \in X_i, \, i = 1, \ldots, m\big\},$$

which is a subset of $\Re^{n_1 + \cdots + n_m}$.

Subspaces and Linear Independence

A nonempty subset S of $\Re^n$ is called a *subspace* if $ax + by \in S$ for every $x, y \in S$ and every $a, b \in \Re$. An *affine set* in $\Re^n$ is a translated subspace, i.e., a set X of the form $X = \bar{x} + S = \{\bar{x} + x \mid x \in S\}$, where $\bar{x}$ is a vector in $\Re^n$ and S is a subspace of $\Re^n$, called the *subspace parallel to* X. Note that there can be only one subspace S associated with an affine set in this manner. [To see this, let $X = x + S$ and $X = \bar{x} + \bar{S}$ be two representations of the affine set X. Then, we must have $x = \bar{x} + \bar{s}$ for some $\bar{s} \in \bar{S}$ (since $x \in X$), so that $X = \bar{x} + \bar{s} + S$. Since we also have $X = \bar{x} + \bar{S}$, it follows that $S = \bar{S} - \bar{s} = \bar{S}$.] The *span* of a finite collection $\{x_1, \ldots, x_m\}$ of elements of $\Re^n$ is the subspace consisting of all vectors y of the form $y = \sum_{k=1}^{m} \alpha_k x_k$, where each α_k is a scalar.

The vectors $x_1, \ldots, x_m \in \Re^n$ are called *linearly independent* if there exists no set of scalars $\alpha_1, \ldots, \alpha_m$, at least one of which is nonzero, such that $\sum_{k=1}^{m} \alpha_k x_k = 0$. An equivalent definition is that $x_1 \neq 0$, and for every $k > 1$, the vector x_k does not belong to the span of $x_1, \ldots, x_{k-1}$.

If S is a subspace of $\Re^n$ containing at least one nonzero vector, a *basis* for S is a collection of vectors that are linearly independent and whose span is equal to S. Every basis of a given subspace has the same number of vectors. This number is called the *dimension* of S. By convention, the subspace $\{0\}$ is said to have dimension zero. The *dimension of an affine set* $\bar{x} + S$ is the dimension of the corresponding subspace S. Every subspace of nonzero dimension has a basis that is orthogonal (i.e., any pair of distinct vectors from the basis is orthogonal).

Given any set X, the set of vectors that are orthogonal to all elements of X is a subspace denoted by $X^\perp$:

$$X^\perp = \{y \mid y'x = 0, \ \forall \ x \in X\}.$$

If S is a subspace, $S^\perp$ is called the *orthogonal complement* of S. Any vector x can be uniquely decomposed as the sum of a vector from S and a vector from $S^\perp$. Furthermore, we have $(S^\perp)^\perp = S$.

Matrices

For any matrix A, we use A_{ij}, $[A]_{ij}$, or a_{ij} to denote its ijth element. The *transpose* of A, denoted by A', is defined by $[A']_{ij} = a_{ji}$. For any two matrices A and B of compatible dimensions, the transpose of the product matrix AB satisfies $(AB)' = B'A'$.

If X is a subset of $\Re^n$ and A is an $m \times n$ matrix, then the *image of* X *under* A is denoted by AX (or $A \cdot X$ if this enhances notational clarity):

$$AX = \{Ax \mid x \in X\}.$$

If Y is a subset of $\Re^m$, the *inverse image of Y under A* is denoted by $A^{-1}Y$ or $A^{-1} \cdot Y$:

$$A^{-1}Y = \{x \mid Ax \in Y\}.$$

If X and Y are subspaces, then AX and $A^{-1}Y$ are also subspaces.

Let A be a square matrix. We say that A is *symmetric* if $A' = A$. We say that A is *diagonal* if $[A]_{ij} = 0$ whenever $i \neq j$. We use I to denote the identity matrix (the diagonal matrix whose diagonal elements are equal to 1). We denote the *determinant* of A by $\det(A)$.

Let A be an $m \times n$ matrix. The *range space* of A, denoted by $R(A)$, is the set of all vectors $y \in \Re^m$ such that $y = Ax$ for some $x \in \Re^n$. The *nullspace* of A, denoted by $N(A)$, is the set of all vectors $x \in \Re^n$ such that $Ax = 0$. It is seen that the range space and the null space of A are subspaces. The *rank* of A is the dimension of the range space of A. The rank of A is equal to the maximal number of linearly independent columns of A, and is also equal to the maximal number of linearly independent rows of A. The matrix A and its transpose A' have the same rank. We say that A has *full rank*, if its rank is equal to $\min\{m, n\}$. This is true if and only if either all the rows of A are linearly independent, or all the columns of A are linearly independent.

The range space of an $m \times n$ matrix A is equal to the orthogonal complement of the nullspace of its transpose, i.e.,

$$R(A) = N(A')^{\perp}.$$

Another way to state this result is that given vectors $a_1, \ldots, a_n \in \Re^m$ (the columns of A) and a vector $x \in \Re^m$, we have $x'y = 0$ for all y such that $a_i'y = 0$ for all i if and only if $x = \lambda_1 a_1 + \cdots + \lambda_n a_n$ for some scalars $\lambda_1, \ldots, \lambda_n$. This is a special case of Farkas' Lemma, an important result for constrained optimization, which will be discussed in Section 3.2. A useful application of this result is that if S_1 and S_2 are two subspaces of $\Re^n$, then

$$S_1^{\perp} + S_2^{\perp} = (S_1 \cap S_2)^{\perp}.$$

This follows by introducing matrices B_1 and B_2 such that $S_1 = \{x \mid B_1 x = 0\} = N(B_1)$ and $S_2 = \{x \mid B_2 x = 0\} = N(B_2)$, and writing

$$S_1^{\perp} + S_2^{\perp} = R([\, B_1' \quad B_2' \,]) = N\left(\begin{bmatrix} B_1 \\ B_2 \end{bmatrix}\right)^{\perp} = \big(N(B_1) \cap N(B_2)\big)^{\perp} = (S_1 \cap S_2)^{\perp}$$

A function $f : \Re^n \mapsto \Re$ is said to be *affine* if it has the form $f(x) = a'x + b$ for some $a \in \Re^n$ and $b \in \Re$. Similarly, a function $f : \Re^n \mapsto \Re^m$ is said to be *affine* if it has the form $f(x) = Ax + b$ for some $m \times n$ matrix A and some $b \in \Re^m$. If $b = 0$, f is said to be a *linear function* or *linear transformation*. Sometimes, with slight abuse of terminology, an equation or inequality involving a linear function, such as $a'x = b$ or $a'x \leq b$, is referred to as a *linear equation or inequality*, respectively.

1.1.2 Topological Properties

Definition 1.1.1: A *norm* $\| \cdot \|$ on $\Re^n$ is a function that assigns a scalar $\|x\|$ to every $x \in \Re^n$ and that has the following properties:

(a) $\|x\| \geq 0$ for all $x \in \Re^n$.

(b) $\|\alpha x\| = |\alpha| \cdot \|x\|$ for every scalar α and every $x \in \Re^n$.

(c) $\|x\| = 0$ if and only if $x = 0$.

(d) $\|x + y\| \leq \|x\| + \|y\|$ for all $x, y \in \Re^n$ (this is referred to as the *triangle inequality*).

The *Euclidean norm* of a vector $x = (x_1, \ldots, x_n)$ is defined by

$$\|x\| = (x'x)^{1/2} = \left(\sum_{i=1}^{n} |x_i|^2 \right)^{1/2}.$$

We will use the Euclidean norm almost exclusively in this book. In particular, *in the absence of a clear indication to the contrary, $\| \cdot \|$ will denote the Euclidean norm*. Two important results for the Euclidean norm are:

Proposition 1.1.1: (Pythagorean Theorem) For any two vectors x and y that are orthogonal, we have

$$\|x + y\|^2 = \|x\|^2 + \|y\|^2.$$

Proposition 1.1.2: (Schwarz Inequality) For any two vectors x and y, we have

$$|x'y| \leq \|x\| \cdot \|y\|,$$

with equality holding if and only if $x = \alpha y$ for some scalar α.

Two other important norms are the *maximum norm* $\| \cdot \|_\infty$ (also called *sup-norm* or ℓ_∞-*norm*), defined by

$$\|x\|_\infty = \max_{i=1,\ldots,n} |x_i|,$$

and the ℓ_1-*norm* $\|\cdot\|_1$, defined by

$$\|x\|_1 = \sum_{i=1}^{n} |x_i|.$$

Sequences

We use both subscripts and superscripts in sequence notation. Generally, we prefer subscripts, but we use superscripts whenever we need to reserve the subscript notation for indexing components of vectors and functions. The meaning of the subscripts and superscripts should be clear from the context in which they are used.

A sequence $\{x_k \mid k = 1, 2, \ldots\}$ (or $\{x_k\}$ for short) of scalars is said to *converge* if there exists a scalar x such that for every $\epsilon > 0$ we have $|x_k - x| < \epsilon$ for every k greater than some integer K (that depends on ϵ). The scalar x is said to be the *limit* of $\{x_k\}$, and the sequence $\{x_k\}$ is said to *converge to* x; symbolically, $x_k \to x$ or $\lim_{k\to\infty} x_k = x$. If for every scalar b there exists some K (that depends on b) such that $x_k \geq b$ for all $k \geq K$, we write $x_k \to \infty$ and $\lim_{k\to\infty} x_k = \infty$. Similarly, if for every scalar b there exists some integer K such that $x_k \leq b$ for all $k \geq K$, we write $x_k \to -\infty$ and $\lim_{k\to\infty} x_k = -\infty$. Note, however, that implicit in any of the statements "$\{x_k\}$ converges" or "the limit of $\{x_k\}$ exists" or "$\{x_k\}$ has a limit" is that the limit of $\{x_k\}$ is a scalar.

A scalar sequence $\{x_k\}$ is said to be *bounded above* (respectively, *below*) if there exists some scalar b such that $x_k \leq b$ (respectively, $x_k \geq b$) for all k. It is said to be *bounded* if it is bounded above and bounded below. The sequence $\{x_k\}$ is said to be monotonically *nonincreasing* (respectively, *nondecreasing*) if $x_{k+1} \leq x_k$ (respectively, $x_{k+1} \geq x_k$) for all k. If $x_k \to x$ and $\{x_k\}$ is monotonically nonincreasing (nondecreasing), we also use the notation $x_k \downarrow x$ ($x_k \uparrow x$, respectively).

Proposition 1.1.3: Every bounded and monotonically nonincreasing or nondecreasing scalar sequence converges.

Note that a monotonically nondecreasing sequence $\{x_k\}$ is either bounded, in which case it converges to some scalar x by the above proposition, or else it is unbounded, in which case $x_k \to \infty$. Similarly, a monotonically nonincreasing sequence $\{x_k\}$ is either bounded and converges, or it is unbounded, in which case $x_k \to -\infty$.

Given a scalar sequence $\{x_k\}$, let

$$y_m = \sup\{x_k \mid k \geq m\}, \qquad z_m = \inf\{x_k \mid k \geq m\}.$$

The sequences $\{y_m\}$ and $\{z_m\}$ are nonincreasing and nondecreasing, respectively, and therefore have a limit whenever $\{x_k\}$ is bounded above or is bounded below, respectively (Prop. 1.1.3). The limit of y_m is denoted by $\limsup_{k\to\infty} x_k$, and is referred to as the *upper limit* of $\{x_k\}$. The limit of z_m is denoted by $\liminf_{k\to\infty} x_k$, and is referred to as the *lower limit* of $\{x_k\}$. If $\{x_k\}$ is unbounded above, we write $\limsup_{k\to\infty} x_k = \infty$, and if it is unbounded below, we write $\liminf_{k\to\infty} x_k = -\infty$.

Proposition 1.1.4: Let $\{x_k\}$ and $\{y_k\}$ be scalar sequences.

(a) We have

$$\inf\{x_k \mid k \geq 0\} \leq \liminf_{k\to\infty} x_k \leq \limsup_{k\to\infty} x_k \leq \sup\{x_k \mid k \geq 0\}.$$

(b) $\{x_k\}$ converges if and only if

$$-\infty < \liminf_{k\to\infty} x_k = \limsup_{k\to\infty} x_k < \infty.$$

Furthermore, if $\{x_k\}$ converges, its limit is equal to the common scalar value of $\liminf_{k\to\infty} x_k$ and $\limsup_{k\to\infty} x_k$.

(c) If $x_k \leq y_k$ for all k, then

$$\liminf_{k\to\infty} x_k \leq \liminf_{k\to\infty} y_k, \qquad \limsup_{k\to\infty} x_k \leq \limsup_{k\to\infty} y_k.$$

(d) We have

$$\liminf_{k\to\infty} x_k + \liminf_{k\to\infty} y_k \leq \liminf_{k\to\infty}(x_k + y_k),$$

$$\limsup_{k\to\infty} x_k + \limsup_{k\to\infty} y_k \geq \limsup_{k\to\infty}(x_k + y_k).$$

A sequence $\{x_k\}$ of vectors in $\Re^n$ is said to converge to some $x \in \Re^n$ if the ith component of x_k converges to the ith component of x for every i. We use the notations $x_k \to x$ and $\lim_{k\to\infty} x_k = x$ to indicate convergence for vector sequences as well. The sequence $\{x_k\}$ is called bounded if each of its corresponding component sequences is bounded. It can be seen that $\{x_k\}$ is bounded if and only if there exists a scalar c such that $\|x_k\| \leq c$ for all k. An infinite subset of a sequence $\{x_k\}$ is called a *subsequence* of $\{x_k\}$. Thus a subsequence can itself be viewed as a sequence, and can be represented as a set $\{x_k \mid k \in \mathcal{K}\}$, where $\mathcal{K}$ is an infinite subset of positive integers (the notation $\{x_k\}_\mathcal{K}$ will also be used).

A vector $x \in \Re^n$ is said to be a *limit point* of a sequence $\{x_k\}$ if there exists a subsequence of $\{x_k\}$ that converges to x.† The following is a classical result that will be used often.

Proposition 1.1.5: (Bolzano-Weierstrass Theorem) A bounded sequence in $\Re^n$ has at least one limit point.

$o(\cdot)$ Notation

For a positive integer p and a function $h : \Re^n \mapsto \Re^m$ we write

$$h(x) = o\big(\|x\|^p\big)$$

if

$$\lim_{k \to \infty} \frac{h(x_k)}{\|x_k\|^p} = 0,$$

for all sequences $\{x_k\}$ such that $x_k \to 0$ and $x_k \neq 0$ for all k.

Closed and Open Sets

We say that x is a *closure point of a subset* X of $\Re^n$ if there exists a sequence $\{x_k\} \subset X$ that converges to x. The *closure of* X, denoted cl(X), is the set of all closure points of X.

Definition 1.1.2: A subset X of $\Re^n$ is called *closed* if it is equal to its closure. It is called *open* if its complement, $\{x \mid x \notin X\}$, is closed. It is called *bounded* if there exists a scalar c such that $\|x\| \leq c$ for all $x \in X$. It is called *compact* if it is closed and bounded.

For any $\epsilon > 0$ and $x^* \in \Re^n$, consider the sets

$$\big\{x \mid \|x - x^*\| < \epsilon\big\}, \qquad \big\{x \mid \|x - x^*\| \leq \epsilon\big\}.$$

---

† Some authors prefer the alternative term "cluster point" of a sequence, and use the term "limit point of a set S" to indicate a point $\overline{x}$ such that $\overline{x} \notin S$ and there exists a sequence $\{x_k\} \subset S$ that converges to $\overline{x}$. With this terminology, $\overline{x}$ is a cluster point of a sequence $\{x_k \mid k = 1, 2, \dots\}$ if and only if $(\overline{x}, 0)$ is a limit point of the set $\big\{(x_k, 1/k) \mid k = 1, 2, \dots\big\}$. Our use of the term "limit point" of a sequence is quite popular in optimization and should not lead to any confusion.

The first set is open and is called an *open sphere* centered at x^*, while the second set is closed and is called a *closed sphere* centered at x^*. Sometimes the terms *open ball* and *closed ball* are used, respectively. A consequence of the definitions, is that a subset X of $\Re^n$ is open if and only if for every $x \in X$ there is an open sphere that is centered at x and is contained in X. A *neighborhood* of a vector x is an open set containing x.

Definition 1.1.3: We say that x is an *interior point* of a subset X of $\Re^n$ if there exists a neighborhood of x that is contained in X. The set of all interior points of X is called the *interior* of X, and is denoted by int(X). A vector $x \in \mathrm{cl}(X)$ which is not an interior point of X is said to be a *boundary point* of X. The set of all boundary points of X is called the *boundary* of X.

Proposition 1.1.6:

(a) The union of a finite collection of closed sets is closed.

(b) The intersection of any collection of closed sets is closed.

(c) The union of any collection of open sets is open.

(d) The intersection of a finite collection of open sets is open.

(e) A set is open if and only if all of its elements are interior points.

(f) Every subspace of $\Re^n$ is closed.

(g) A set X is compact if and only if every sequence of elements of X has a subsequence that converges to an element of X.

(h) If $\{X_k\}$ is a sequence of nonempty and compact sets such that $X_k \supset X_{k+1}$ for all k, then the intersection $\cap_{k=0}^{\infty} X_k$ is nonempty and compact.

The topological properties of sets in $\Re^n$, such as being open, closed, or compact, do not depend on the norm being used. This is a consequence of the following proposition, referred to as the *norm equivalence property in $\Re^n$*, which shows that if a sequence converges with respect to one norm, it converges with respect to all other norms.

Proposition 1.1.7: For any two norms $\| \cdot \|$ and $\| \cdot \|'$ on $\Re^n$, there exists a scalar c such that $\|x\| \leq c\|x\|'$ for all $x \in \Re^n$.

Using the preceding proposition, we obtain the following.

Proposition 1.1.8: If a subset of $\Re^n$ is open (respectively, closed, bounded, or compact) with respect to some norm, it is open (respectively, closed, bounded, or compact) with respect to all other norms.

Sequences of Sets

Let $\{X_k\}$ be a sequence of nonempty subsets of $\Re^n$. The *outer limit* of $\{X_k\}$, denoted $\limsup_{k\to\infty} X_k$, is the set of all $x \in \Re^n$ such that every neighborhood of x has a nonempty intersection with infinitely many of the sets X_k, $k = 1, 2, \ldots$. Equivalently, $\limsup_{k\to\infty} X_k$ is the set of all limit points of sequences $\{x_k\}$ such that $x_k \in X_k$ for all $k = 1, 2, \ldots$.

The *inner limit* of $\{X_k\}$, denoted $\liminf_{k\to\infty} X_k$, is the set of all $x \in \Re^n$ such that every neighborhood of x has a nonempty intersection with all except finitely many of the sets X_k, $k = 1, 2, \ldots$. Equivalently, $\liminf_{k\to\infty} X_k$ is the set of all limits of convergent sequences $\{x_k\}$ such that $x_k \in X_k$ for all $k = 1, 2, \ldots$.

The sequence $\{X_k\}$ is said to converge to a set X if

$$X = \liminf_{k\to\infty} X_k = \limsup_{k\to\infty} X_k.$$

In this case, X is called the *limit of* $\{X_k\}$, and is denoted by $\lim_{k\to\infty} X_k$.

The inner and outer limits are closed (possibly empty) sets. If each set X_k consists of a single point x_k, $\limsup_{k\to\infty} X_k$ is the set of limit points of $\{x_k\}$, while $\liminf_{k\to\infty} X_k$ is just the limit of $\{x_k\}$ if $\{x_k\}$ converges, and otherwise it is empty.

Continuity

Let $f : X \mapsto \Re^m$ be a function, where X is a subset of $\Re^n$, and let x be a vector in X. If there exists a vector $y \in \Re^m$ such that the sequence $\{f(x_k)\}$ converges to y for every sequence $\{x_k\} \subset X$ such that $\lim_{k\to\infty} x_k = x$, we write $\lim_{z\to x} f(z) = y$. If there exists a vector $y \in \Re^m$ such that the sequence $\{f(x_k)\}$ converges to y for every sequence $\{x_k\} \subset X$ such that $\lim_{k\to\infty} x_k = x$ and $x_k \leq x$ (respectively, $x_k \geq x$) for all k, we write $\lim_{z\uparrow x} f(z) = y$ [respectively, $\lim_{z\downarrow x} f(z)$].

Definition 1.1.4: Let X be a subset of $\Re^n$.

(a) A function $f : X \mapsto \Re^m$ is called *continuous* at a vector $x \in X$ if $\lim_{z \to x} f(z) = f(x)$.

(b) A function $f : X \mapsto \Re^m$ is called *right-continuous* (respectively, *left-continuous*) at a vector $x \in X$ if $\lim_{z \downarrow x} f(z) = f(x)$ [respectively, $\lim_{z \uparrow x} f(z) = f(x)$].

(c) A real-valued function $f : X \mapsto \Re$ is called *upper semicontinuous* (respectively, *lower semicontinuous*) at a vector $x \in X$ if $f(x) \geq \limsup_{k \to \infty} f(x_k)$ [respectively, $f(x) \leq \liminf_{k \to \infty} f(x_k)$] for every sequence $\{x_k\} \subset X$ that converges to x.

If $f : X \mapsto \Re^m$ is continuous at every vector in a subset of its domain X, we say that f *is continuous over that subset*. If $f : X \mapsto \Re^m$ is continuous at every vector in its domain X, we say that f *is continuous*. We use similar terminology for right-continuous, left-continuous, upper semicontinuous, and lower semicontinuous functions.

Proposition 1.1.9:

(a) Any vector norm on $\Re^n$ is a continuous function.

(b) Let $f : \Re^m \mapsto \Re^p$ and $g : \Re^n \mapsto \Re^m$ be continuous functions. The composition $f \cdot g : \Re^n \mapsto \Re^p$, defined by $(f \cdot g)(x) = f\big(g(x)\big)$, is a continuous function.

(c) Let $f : \Re^n \mapsto \Re^m$ be continuous, and let Y be an open (respectively, closed) subset of $\Re^m$. Then the inverse image of Y, $\big\{x \in \Re^n \mid f(x) \in Y\big\}$, is open (respectively, closed).

(d) Let $f : \Re^n \mapsto \Re^m$ be continuous, and let X be a compact subset of $\Re^n$. Then the image of X, $\big\{f(x) \mid x \in X\big\}$, is compact.

Matrix Norms

A norm $\| \cdot \|$ on the set of $n \times n$ matrices is a real-valued function that has the same properties as vector norms do when the matrix is viewed as a vector in $\Re^{n^2}$. The norm of an $n \times n$ matrix A is denoted by $\|A\|$.

An important class of matrix norms are *induced norms*, which are constructed as follows. Given any vector norm $\| \cdot \|$, the corresponding induced matrix norm, also denoted by $\| \cdot \|$, is defined by

$$\|A\| = \sup_{\|x\|=1} \|Ax\|.$$

It is easily verified that for any vector norm, the above equation defines a matrix norm.

Let $\|\cdot\|$ denote the Euclidean norm. Then by the Schwarz inequality (Prop. 1.1.2), we have

$$\|A\| = \sup_{\|x\|=1} \|Ax\| = \sup_{\|y\|=\|x\|=1} |y'Ax|.$$

By reversing the roles of x and y in the above relation and by using the equality $y'Ax = x'A'y$, it follows that $\|A\| = \|A'\|$.

1.1.3   Square Matrices

Definition 1.1.5: A square matrix A is called *singular* if its determinant is zero. Otherwise it is called *nonsingular* or *invertible*.

Proposition 1.1.10:

(a) Let A be an $n \times n$ matrix. The following are equivalent:

   (i) The matrix A is nonsingular.

   (ii) The matrix A' is nonsingular.

   (iii) For every nonzero $x \in \Re^n$, we have $Ax \neq 0$.

   (iv) For every $y \in \Re^n$, there is a unique $x \in \Re^n$ such that $Ax = y$.

   (v) There is an $n \times n$ matrix B such that $AB = I = BA$.

   (vi) The columns of A are linearly independent.

   (vii) The rows of A are linearly independent.

(b) Assuming that A is nonsingular, the matrix B of statement (v) (called the *inverse* of A and denoted by A^{-1}) is unique.

(c) For any two square invertible matrices A and B of the same dimensions, we have $(AB)^{-1} = B^{-1}A^{-1}$.

Definition 1.1.6: A symmetric $n \times n$ matrix A is called *positive definite* if $x'Ax > 0$ for all $x \in \Re^n$, $x \neq 0$. It is called *positive semidefinite* if $x'Ax \geq 0$ for all $x \in \Re^n$.

Throughout this book, the notion of positive definiteness applies exclusively to symmetric matrices. Thus *whenever we say that a matrix is positive (semi)definite, we implicitly assume that the matrix is symmetric*, although we usually add the term "symmetric" for clarity.

Proposition 1.1.11:

(a) A square matrix is symmetric and positive definite if and only if it is invertible and its inverse is symmetric and positive definite.

(b) The sum of two symmetric positive semidefinite matrices is positive semidefinite. If one of the two matrices is positive definite, the sum is positive definite.

(c) If A is a symmetric positive semidefinite $n \times n$ matrix and T is an $m \times n$ matrix, then the matrix TAT' is positive semidefinite. If A is positive definite and T is invertible, then TAT' is positive definite.

(d) If A is a symmetric positive definite $n \times n$ matrix, there exist positive scalars $\underline{\gamma}$ and $\overline{\gamma}$ such that

$$\underline{\gamma}\|x\|^2 \le x'Ax \le \overline{\gamma}\|x\|^2, \qquad \forall\, x \in \Re^n.$$

(e) If A is a symmetric positive definite $n \times n$ matrix, there exists a unique symmetric positive definite matrix that yields A when multiplied with itself. This matrix is called the *square root of A*. It is denoted by $A^{1/2}$, and its inverse is denoted by $A^{-1/2}$.

1.1.4   Derivatives

Let $f : \Re^n \mapsto \Re$ be some function, fix some $x \in \Re^n$, and consider the expression

$$\lim_{\alpha \to 0} \frac{f(x + \alpha e_i) - f(x)}{\alpha},$$

where e_i is the ith unit vector (all components are 0 except for the ith component which is 1). If the above limit exists, it is called the ith *partial derivative* of f at the vector x and it is denoted by $(\partial f / \partial x_i)(x)$ or $\partial f(x)/\partial x_i$ (x_i in this section will denote the ith component of the vector x). Assuming all of these partial derivatives exist, the *gradient* of f at x is defined as the column vector

$$\nabla f(x) = \begin{bmatrix} \frac{\partial f(x)}{\partial x_1} \\ \vdots \\ \frac{\partial f(x)}{\partial x_n} \end{bmatrix}.$$

For any $y \in \Re^n$, we define the one-sided *directional derivative* of f in the direction y, to be

$$f'(x; y) = \lim_{\alpha \downarrow 0} \frac{f(x + \alpha y) - f(x)}{\alpha},$$

provided that the limit exists.

If the directional derivative of f at a vector x exists in all directions y and $f'(x; y)$ is a linear function of y, we say that f is *differentiable* at x. This type of differentiability is also called *Gateaux differentiability*. It is seen that f is differentiable at x if and only if the gradient $\nabla f(x)$ exists and satisfies

$$\nabla f(x)'y = f'(x; y), \qquad \forall \, y \in \Re^n.$$

The function f is called *differentiable over a subset U of $\Re^n$* if it is differentiable at every $x \in U$. The function f is called *differentiable* (without qualification) if it is differentiable at all $x \in \Re^n$.

If f is differentiable over an open set U and $\nabla f(\cdot)$ is continuous at all $x \in U$, f is said to be *continuously differentiable over U*. It can then be shown that

$$\lim_{y \to 0} \frac{f(x + y) - f(x) - \nabla f(x)'y}{\|y\|} - 0, \qquad \forall \, x \in U, \qquad (1.1)$$

where $\| \cdot \|$ is an arbitrary vector norm. If f is continuously differentiable over $\Re^n$, then f is also called a *smooth* function. If f is not smooth, it is referred to as being *nonsmooth*.

The preceding equation can also be used as an alternative definition of differentiability. In particular, f is called *Frechet differentiable* at x if there exists a vector g satisfying Eq. (1.1) with $\nabla f(x)$ replaced by g. If such a vector g exists, it can be seen that all the partial derivatives $(\partial f/\partial x_i)(x)$ exist and that $g = \nabla f(x)$. Frechet differentiability implies (Gateaux) differentiability but not conversely (see for example Ortega and Rheinboldt [OrR70] for a detailed discussion). In this book, when dealing with a differentiable function f, we will always assume that f is continuously differentiable over some open set $[\nabla f(\cdot)$ is a continuous function over that set], in which case f is both Gateaux and Frechet differentiable, and the distinctions made above are of no consequence.

The definitions of differentiability of f at a vector x only involve the values of f in a neighborhood of x. Thus, these definitions can be used for functions f that are not defined on all of $\Re^n$, but are defined instead in a neighborhood of the vector at which the derivative is computed. In particular, for functions $f : X \mapsto \Re$, where X is a strict subset of $\Re^n$, we use the above definition of differentiability of f at a vector x, *provided x is an interior point of the domain X*. Similarly, we use the above definition of continuous differentiability of f over a subset U, *provided U is an open subset of the domain X*. Thus any mention of continuous differentiability of a function over a subset implicitly assumes that this subset is open.

Differentiation of Vector-Valued Functions

A vector-valued function $f : \Re^n \mapsto \Re^m$ is called differentiable (or smooth) if each component f_i of f is differentiable (or smooth, respectively). The *gradient matrix* of f, denoted $\nabla f(x)$, is the $n \times m$ matrix whose ith column is the gradient $\nabla f_i(x)$ of f_i. Thus,

$$\nabla f(x) = \Big[\nabla f_1(x) \cdots \nabla f_m(x) \Big].$$

The transpose of ∇f is called the *Jacobian* of f and is a matrix whose ijth entry is equal to the partial derivative $\partial f_i / \partial x_j$.

Now suppose that each one of the partial derivatives of a function $f : \Re^n \mapsto \Re$ is a smooth function of x. We use the notation $(\partial^2 f / \partial x_i \partial x_j)(x)$ to indicate the ith partial derivative of $\partial f / \partial x_j$ at a vector $x \in \Re^n$. The *Hessian* of f is the matrix whose ijth entry is equal to $(\partial^2 f / \partial x_i \partial x_j)(x)$, and is denoted by $\nabla^2 f(x)$. We have $(\partial^2 f / \partial x_i \partial x_j)(x) = (\partial^2 f / \partial x_j \partial x_i)(x)$ for every x, which implies that $\nabla^2 f(x)$ is symmetric.

If $f : \Re^{m+n} \mapsto \Re$ is a function of (x, y), where $x \in \Re^m$ and $y \in \Re^n$, and $x_1, \ldots, x_m$ and $y_1, \ldots, y_n$ denote the components of x and y, respectively, we write

$$\nabla_x f(x, y) = \begin{bmatrix} \frac{\partial f(x,y)}{\partial x_1} \\ \vdots \\ \frac{\partial f(x,y)}{\partial x_m} \end{bmatrix}, \qquad \nabla_y f(x, y) = \begin{bmatrix} \frac{\partial f(x,y)}{\partial y_1} \\ \vdots \\ \frac{\partial f(x,y)}{\partial y_n} \end{bmatrix}.$$

We denote by $\nabla_{xx}^2 f(x, y)$, $\nabla_{xy}^2 f(x, y)$, and $\nabla_{yy}^2 f(x, y)$ the matrices with components

$$\big[\nabla_{xx}^2 f(x, y)\big]_{ij} = \frac{\partial^2 f(x, y)}{\partial x_i \partial x_j}, \qquad \big[\nabla_{xy}^2 f(x, y)\big]_{ij} = \frac{\partial^2 f(x, y)}{\partial x_i \partial y_j},$$

$$\big[\nabla_{yy}^2 f(x, y)\big]_{ij} = \frac{\partial^2 f(x, y)}{\partial y_i \partial y_j}.$$

If $f : \Re^{m+n} \mapsto \Re^r$, and $f_1, f_2, \ldots, f_r$ are the component functions of f, we write

$$\nabla_x f(x, y) = \big[\nabla_x f_1(x, y) \cdots \nabla_x f_r(x, y) \big],$$

$$\nabla_y f(x, y) = \big[\nabla_y f_1(x, y) \cdots \nabla_y f_r(x, y) \big].$$

Let $f : \Re^k \mapsto \Re^m$ and $g : \Re^m \mapsto \Re^n$ be smooth functions, and let h be their composition, i.e.,

$$h(x) = g\big(f(x)\big).$$

Then, the *chain rule* for differentiation states that

$$\nabla h(x) = \nabla f(x) \nabla g(f(x)), \qquad \forall \, x \in \Re^k.$$

Some examples of useful relations that follow from the chain rule are:

$$\nabla(f(Ax)) = A'\nabla f(Ax), \qquad \nabla^2(f(Ax)) = A'\nabla^2 f(Ax)A,$$

where A is a matrix,

$$\nabla_x\left(f(h(x), y)\right) = \nabla h(x)\nabla_h f(h(x), y),$$

$$\nabla_x\left(f(h(x), g(x))\right) = \nabla h(x)\nabla_h f(h(x), g(x)) + \nabla g(x)\nabla_g f(h(x), g(x)).$$

Differentiation Theorems

We now state some theorems relating to differentiable functions that will be useful for our purposes.

Proposition 1.1.12: (Mean Value Theorem) Let $f : \Re^n \mapsto \Re$ be continuously differentiable over an open sphere S, and let x be a vector in S. Then for all y such that $x + y \in S$, there exists an $\alpha \in [0, 1]$ such that

$$f(x + y) = f(x) + \nabla f(x + \alpha y)'y.$$

Proposition 1.1.13: (Second Order Expansions) Let $f : \Re^n \mapsto \Re$ be twice continuously differentiable over an open sphere S, and let x be a vector in S. Then for all y such that $x + y \in S$:

(a) There exists an $\alpha \in [0, 1]$ such that

$$f(x + y) = f(x) + y'\nabla f(x) + \tfrac{1}{2}y'\nabla^2 f(x + \alpha y)y.$$

(b) We have

$$f(x + y) = f(x) + y'\nabla f(x) + \tfrac{1}{2}y'\nabla^2 f(x)y + o(\|y\|^2).$$

Proposition 1.1.14: (Implicit Function Theorem) Consider a function $f : \Re^{n+m} \mapsto \Re^m$ of $x \in \Re^n$ and $y \in \Re^m$ such that:

(1) $f(\overline{x}, \overline{y}) = 0$.

(2) f is continuous, and has a continuous and nonsingular gradient matrix $\nabla_y f(x, y)$ in an open set containing $(\overline{x}, \overline{y})$.

Then there exist open sets $S_{\overline{x}} \subset \Re^n$ and $S_{\overline{y}} \subset \Re^m$ containing $\overline{x}$ and $\overline{y}$, respectively, and a continuous function $\phi : S_{\overline{x}} \mapsto S_{\overline{y}}$ such that $\overline{y} = \phi(\overline{x})$ and $f(x, \phi(x)) = 0$ for all $x \in S_{\overline{x}}$. The function ϕ is unique in the sense that if $x \in S_{\overline{x}}$, $y \in S_{\overline{y}}$, and $f(x, y) = 0$, then $y = \phi(x)$. Furthermore, if for some integer $p > 0$, f is p times continuously differentiable the same is true for ϕ, and we have

$$\nabla \phi(x) = -\nabla_x f(x, \phi(x)) \left(\nabla_y f(x, \phi(x)) \right)^{-1}, \qquad \forall \, x \in S_{\overline{x}}.$$

As a final word of caution to the reader, let us mention that one can easily get confused with gradient notation and its use in various formulas, such as for example the order of multiplication of various gradients in the chain rule and the Implicit Function Theorem. Perhaps the safest guideline to minimize errors is to remember our conventions:

(a) A vector is viewed as a column vector (an $n \times 1$ matrix).

(b) The gradient ∇f of a scalar function $f : \Re^n \mapsto \Re$ is also viewed as a column vector.

(c) The gradient matrix ∇f of a vector function $f : \Re^n \mapsto \Re^m$ with components $f_1, \ldots, f_m$ is the $n \times m$ matrix whose columns are the vectors $\nabla f_1, \ldots, \nabla f_m$.

With these rules in mind, one can use "dimension matching" as an effective guide to writing correct formulas quickly.

1.2 CONVEX SETS AND FUNCTIONS

In this and the subsequent sections of this chapter, we introduce some of the basic notions relating to convex sets and functions. This material permeates all subsequent developments in this book, and will be used in the next chapter for the discussion of important issues in optimization.

We first define convex sets (see also Fig. 1.2.1).

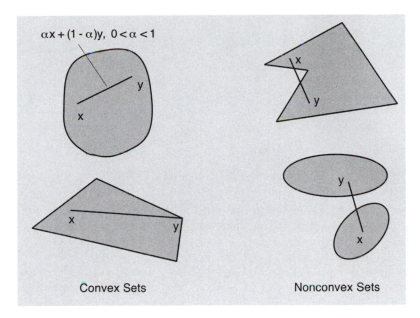

Figure 1.2.1. Illustration of the definition of a convex set. For convexity, linear interpolation between any two points of the set must yield points that lie within the set.

Definition 1.2.1: A subset C of $\Re^n$ is called *convex* if

$$\alpha x + (1 - \alpha)y \in C, \qquad \forall\ x, y \in C,\ \forall\ \alpha \in [0, 1].$$

Note that the empty set is by convention considered to be convex. Generally, when referring to a convex set, it will usually be apparent from the context whether this set can be empty, but we will often be specific in order to minimize ambiguities.

The following proposition lists some operations that preserve convexity of a set.

Proposition 1.2.1:

(a) The intersection $\cap_{i \in I} C_i$ of any collection $\{C_i \mid i \in I\}$ of convex sets is convex.

(b) The vector sum $C_1 + C_2$ of two convex sets C_1 and C_2 is convex.

(c) The set λC is convex for any convex set C and scalar λ. Furthermore, if C is a convex set and λ_1, λ_2 are positive scalars,

$$(\lambda_1 + \lambda_2)C = \lambda_1 C + \lambda_2 C.$$

(d) The closure and the interior of a convex set are convex.

(e) The image and the inverse image of a convex set under an affine function are convex.

Proof: The proof is straightforward using the definition of convexity. For example, to prove part (a), we take two points x and y from $\cap_{i \in I} C_i$, and we use the convexity of C_i to argue that the line segment connecting x and y belongs to all the sets C_i, and hence, to their intersection.

Similarly, to prove part (b), we take two points of $C_1 + C_2$, which we represent as $x_1 + x_2$ and $y_1 + y_2$, with $x_1, y_1 \in C_1$ and $x_2, y_2 \in C_2$. For any $\alpha \in [0, 1]$, we have

$$\alpha(x_1 + x_2) + (1 - \alpha)(y_1 + y_2) = \bigl(\alpha x_1 + (1 - \alpha)y_1\bigr) + \bigl(\alpha x_2 + (1 - \alpha)y_2\bigr).$$

By convexity of C_1 and C_2, the vectors in the two parentheses of the right-hand side above belong to C_1 and C_2, respectively, so that their sum belongs to $C_1 + C_2$. Hence $C_1 + C_2$ is convex. The proof of part (c) is left as Exercise 1.1. The proof of part (e) is similar to the proof of part (b).

To prove part (d), we take two points x and y from the closure of C, and sequences $\{x_k\} \subset C$ and $\{y_k\} \subset C$, such that $x_k \to x$ and $y_k \to y$. For any $\alpha \in [0, 1]$, the sequence $\{\alpha x_k + (1 - \alpha)y_k\}$, which belongs to C by the convexity of C, converges to $\alpha x + (1 - \alpha)y$. Hence $\alpha x + (1 - \alpha)y$ belongs to the closure of C, showing that the closure of C is convex. Similarly, we take two points x and y from the interior of C, and we consider open balls that are centered at x and y, and have sufficiently small radius r so that they are contained in C. For any $\alpha \in [0, 1]$, consider the open ball of radius r that is centered at $\alpha x + (1 - \alpha)y$. Any point in this ball, say $\alpha x + (1 - \alpha)y + z$, where $\|z\| < r$, belongs to X, because it can be expressed as the convex combination $\alpha(x + z) + (1 - \alpha)(y + z)$ of the vectors $x + z$ and $y + z$, which belong to X. Hence $\alpha x + (1 - \alpha)y$ belongs to the interior of C, showing that the interior of C is convex. **Q.E.D.**

A set C is said to be a *cone* if for all $x \in C$ and $\lambda > 0$, we have $\lambda x \in C$. A cone need not be convex and need not contain the origin, although the origin always lies in the closure of a nonempty cone (see Fig. 1.2.2). Several of the results of the preceding proposition have analogs for cones (see Exercise 1.2).

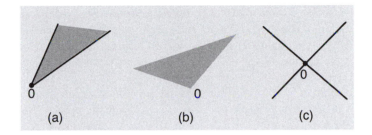

Figure 1.2.2. Illustration of convex and nonconvex cones. Cones (a) and (b) are convex, while cone (c), which consists of two lines passing through the origin, is not convex. Cone (b) does not contain the origin.

Convex Functions

The notion of a convex function is defined below and is illustrated in Fig. 1.2.3.

Definition 1.2.2: Let C be a convex subset of $\Re^n$. A function $f : C \mapsto \Re$ is called *convex* if

$$f\big(\alpha x + (1-\alpha)y\big) \leq \alpha f(x) + (1-\alpha)f(y), \qquad \forall\, x, y \in C,\ \forall\, \alpha \in [0,1]. \tag{1.2}$$

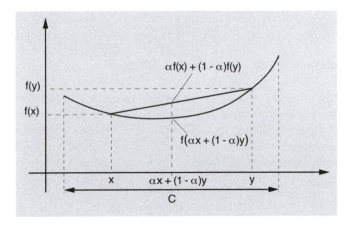

Figure 1.2.3. Illustration of the definition of a function $f : C \mapsto \Re$ that is convex. The linear interpolation $\alpha f(x) + (1-\alpha)f(y)$ overestimates the function value $f\big(\alpha x + (1-\alpha)y\big)$ for all $\alpha \in [0,1]$.

We introduce some more definitions that involve variations of the basic definition of convexity. A convex function $f : C \mapsto \Re$ is called *strictly convex* if the inequality (1.2) is strict for all $x, y \in C$ with $x \neq y$, and all $\alpha \in (0, 1)$. A function $f : C \mapsto \Re$, where C is a convex set, is called *concave* if $-f$ is convex.

Note that, according to our definition, convexity of the domain C is a prerequisite for calling a function $f : C \mapsto \Re$ "convex." Sometimes we will deal with functions $f : X \mapsto \Re$ that are defined over a (possibly nonconvex) domain X but are convex when restricted to a convex subset of their domain. The following definition formalizes this case.

Definition 1.2.3: Let C and X be subsets of $\Re^n$ such that C is nonempty and convex, and $C \subset X$. A function $f : X \mapsto \Re$ is called *convex over* C if Eq. (1.2) holds, i.e., when the domain of f is restricted to C, f becomes convex.

If $f : C \mapsto \Re$ is a function and γ is a scalar, the sets $\{x \in C \mid f(x) \leq \gamma\}$ and $\{x \in C \mid f(x) < \gamma\}$, are called *level sets* of f. If f is a convex function, then all its level sets are convex. To see this, note that if $x, y \in C$ are such that $f(x) \leq \gamma$ and $f(y) \leq \gamma$, then for any $\alpha \in [0, 1]$, we have $\alpha x + (1 - \alpha)y \in C$, by the convexity of C, and we have

$$f(\alpha x + (1 - \alpha)y) \leq \alpha f(x) + (1 - \alpha)f(y) \leq \gamma,$$

by the convexity of f. A similar proof can also be used to show that the level sets $\{x \in C \mid f(x) < \gamma\}$ are convex when f is convex. Note, however, that convexity of the level sets does not imply convexity of the function; for example, the scalar function $f(x) = \sqrt{|x|}$ has convex level sets but is not convex.

Extended Real-Valued Convex Functions

We generally prefer to deal with convex functions that are real-valued and are defined over the entire space $\Re^n$ (rather than over just a convex subset). However, in some situations, prominently arising in the context of optimization and duality, we will encounter operations on real-valued functions that produce extended real-valued functions. As an example, a function of the form

$$f(x) = \sup_{i \in I} f_i(x),$$

where I is an infinite index set, can take the value ∞ even if the functions f_i are real-valued.

Furthermore, we will encounter functions f that are convex over a convex subset C and cannot be extended to functions that are real-valued and convex over the entire space $\Re^n$ [e.g., the function $f : (0, \infty) \mapsto \Re$ defined by $f(x) = 1/x$]. In such situations, it may be convenient, instead of restricting the domain of f to the subset C where f takes real values, to extend the domain to all of $\Re^n$, but allow f to take infinite values.

We are thus motivated to introduce *extended real-valued* functions that can take the values of $-\infty$ and ∞ at some points. Such functions can be characterized using the notions of epigraph and effective domain, which we now introduce.

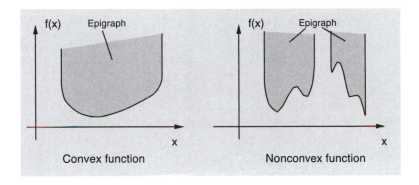

Figure 1.2.4. Illustration of the epigraphs of extended real-valued convex and nonconvex functions.

We define the *epigraph* of a function $f : X \mapsto [-\infty, \infty]$, where $X \subset \Re^n$, to be the subset of $\Re^{n+1}$ given by

$$\text{epi}(f) = \big\{(x, w) \mid x \in X,\ w \in \Re,\ f(x) \leq w\big\};$$

(see Fig. 1.2.4). We define the *effective domain* of f to be the set

$$\text{dom}(f) = \big\{x \in X \mid f(x) < \infty\big\}.$$

It can be seen that

$$\text{dom}(f) = \big\{x \mid \text{there exists } w \in \Re \text{ such that } (x, w) \in \text{epi}(f)\big\},$$

i.e., $\text{dom}(f)$ is obtained by a projection of $\text{epi}(f)$ on $\Re^n$ (the space of x). Note that if we restrict f to the set of x for which $f(x) < \infty$, its epigraph remains unaffected. Similarly, if we enlarge the domain of f by defining $f(x) = \infty$ for $x \notin X$, the epigraph remains unaffected.

It is often important to exclude the degenerate case where f is identically equal to ∞ [which is true if and only if $\text{epi}(f)$ is empty], and the case where the function takes the value $-\infty$ at some point [which is true if and only if $\text{epi}(f)$ contains a vertical line]. We will thus say that f is *proper* if $f(x) < \infty$ for at least one $x \in X$ and $f(x) > -\infty$ for all $x \in X$, and we will say that f *improper* if it is not proper. In words, a function is proper if and only if its epigraph is nonempty and does not contain a vertical line.

A difficulty in defining extended real-valued convex functions f that can take both values $-\infty$ and ∞ is that the term $\alpha f(x) + (1 - \alpha)f(y)$ arising in our earlier definition for the real-valued case may involve the forbidden sum $-\infty + \infty$ (this, of course, may happen only if f is improper, but improper functions may arise on occasion in proofs or other analyses, so we do not wish to exclude *a priori* such functions). The epigraph provides an effective way of dealing with this difficulty.

Definition 1.2.4: Let C be a convex subset of $\Re^n$. An extended real-valued function $f : C \mapsto [-\infty, \infty]$ is called *convex* if epi(f) is a convex subset of $\Re^{n+1}$.

It can be easily verified that, according to the above definition, convexity of f implies that its effective domain dom(f) and its level sets $\{x \in C \mid f(x) \le \gamma\}$ and $\{x \in C \mid f(x) < \gamma\}$ are convex sets for all scalars γ. Furthermore, if $f(x) < \infty$ for all x, or $f(x) > -\infty$ for all x, we have

$$f\big(\alpha x + (1-\alpha)y\big) \le \alpha f(x) + (1-\alpha)f(y), \qquad \forall\, x, y \in C, \ \forall\, \alpha \in [0, 1], \quad (1.3)$$

so the preceding definition is consistent with the earlier definition of convexity for real-valued functions.

A convex function $f : C \mapsto (-\infty, \infty]$ is called *strictly convex* if the inequality (1.3) is strict for all $x, y \in$ dom(f) with $x \ne y$, and all $\alpha \in (0, 1)$. A function $f : C \mapsto [-\infty, \infty]$, where C is a convex set, is called *concave* if the function $-f : C \mapsto [-\infty, \infty]$ is convex as per Definition 1.2.4.

The following definition deals with the case where an extended real-valued function becomes convex when restricted to a subset of its domain.

Definition 1.2.5: Let C and X be subsets of $\Re^n$ such that C is nonempty and convex, and $C \subset X$. An extended real-valued function $f : X \mapsto [-\infty, \infty]$ is called *convex over C* if f becomes convex when the domain of f is restricted to C, i.e., if the function $\tilde{f} : C \mapsto [-\infty, \infty]$, defined by $\tilde{f}(x) = f(x)$ for all $x \in C$, is convex.

Note that by replacing the domain of an extended real-valued proper convex function with its effective domain, we can convert it to a real-valued function. In this way, we can use results stated in terms of real-valued functions, and we can also avoid calculations with ∞. Thus, the entire subject of convex functions can be developed without resorting to extended real-valued functions. The reverse is also true, namely that extended real-valued functions can be adopted as the norm; for example, this approach is followed by Rockafellar [Roc70].

Generally, functions that are real-valued over the entire space $\Re^n$ are more convenient (and even essential) in numerical algorithms and also in optimization analyses where a calculus-oriented approach based on differentiability is adopted. This is typically the case in nonconvex optimization, where nonlinear equality and nonconvex inequality constraints are involved (see Chapter 5). On the other hand, extended real-valued functions offer notational advantages in convex optimization, and in fact may be more natural because some basic constructions around duality involve extended real-valued functions (see Chapters 6 and 7). Since we plan to deal with nonconvex as well as convex problems, and with duality theory as well as numerical methods, we will adopt a flexible approach, and use both real-valued and extended real-valued functions.

Lower Semicontinuity and Closedness of Convex Functions

An extended real-valued function $f : X \mapsto [-\infty, \infty]$ is called *lower semicontinuous* at a vector $x \in X$ if $f(x) \leq \liminf_{k \to \infty} f(x_k)$ for every sequence $\{x_k\} \subset X$ with $x_k \to x$. This is consistent with the corresponding definition for real-valued functions [cf. Definition 1.1.4(c)]. If f is lower semicontinuous at every x in a subset U of X, we say that f is *lower semicontinuous over U*. The following proposition relates lower semicontinuity of f with closedness of its epigraph and its level sets.

Proposition 1.2.2: For a function $f : \Re^n \mapsto [-\infty, \infty]$, the following are equivalent:

(i) The level set $\{x \mid f(x) \leq \gamma\}$ is closed for every scalar γ.

(ii) f is lower semicontinuous over $\Re^n$.

(iii) epi(f) is closed.

Proof: If $f(x) = \infty$ for all x, the result trivially holds. We thus assume that $f(x) < \infty$ for at least one $x \in \Re^n$, so that epi(f) is nonempty and there exist level sets of f that are nonempty.

We first show that (i) implies (ii). Assume that the level set $\{x \mid f(x) \leq \gamma\}$ is closed for every scalar γ. Suppose, to arrive at a contradiction, that $f(\overline{x}) > \liminf_{k \to \infty} f(x_k)$ for some $\overline{x}$ and sequence $\{x_k\}$ converging to $\overline{x}$, and let γ be a scalar such that

$$f(\overline{x}) > \gamma > \liminf_{k \to \infty} f(x_k).$$

Then, there exists a subsequence $\{x_k\}_\mathcal{K}$ such that $f(x_k) \leq \gamma$ for all $k \in \mathcal{K}$. Since the set $\{x \mid f(x) \leq \gamma\}$ is closed, $\overline{x}$ must belong to this set, so $f(\overline{x}) \leq \gamma$, a contradiction.

We next show that (ii) implies (iii). Assume that f is lower semicontinuous over $\Re^n$, and let $(\overline{x}, \overline{w})$ be the limit of a sequence $\{(x_k, w_k)\} \subset$ epi(f). Then we have $f(x_k) \leq w_k$, and by taking the limit as $k \to \infty$ and by using the lower semicontinuity of f at $\overline{x}$, we obtain $f(\overline{x}) \leq \liminf_{k\to\infty} f(x_k) \leq \overline{w}$. Hence, $(\overline{x}, \overline{w}) \in$ epi(f) and epi(f) is closed.

We finally show that (iii) implies (i). Assume that epi(f) is closed, and let $\{x_k\}$ be a sequence that converges to some $\overline{x}$ and belongs to the level set $\{x \mid f(x) \leq \gamma\}$ for some scalar γ. Then $(x_k, \gamma) \in$ epi(f) for all k and $(x_k, \gamma) \to (\overline{x}, \gamma)$, so since epi$(f)$ is closed, we have $(\overline{x}, \gamma) \in$ epi(f). Hence, $\overline{x}$ belongs to the level set $\{x \mid f(x) \leq \gamma\}$, implying that this set is closed. **Q.E.D.**

If the epigraph of a function $f : X \mapsto [-\infty, \infty]$ is a closed set, we say that f is a *closed* function. To understand the relation between closedness and lower semicontinuity, let us extend the domain of f to $\Re^n$ and consider the function $\tilde{f} : \Re^n \mapsto [-\infty, \infty]$ given by

$$\tilde{f}(x) = \begin{cases} f(x) & \text{if } x \in X, \\ \infty & \text{if } x \notin X. \end{cases}$$

Then, we see that f and $\tilde{f}$ have the same epigraph, and according to the preceding proposition, f *is closed if and only if* $\tilde{f}$ *is lower semicontinuous over* $\Re^n$.

Note, however, that if f is lower semicontinuous over dom(f), it is not necessarily closed; take for example f to be constant for x in some nonclosed set and ∞ otherwise. Furthermore, if f is closed, dom(f) need not be closed; for example, the function

$$f(x) = \begin{cases} \frac{1}{x} & \text{if } x > 0, \\ \infty & \text{otherwise,} \end{cases}$$

is closed but dom(f) is the open half-line of positive numbers. On the other hand, if dom(f) is closed and f is lower semicontinuous over dom(f), then f is closed because epi(f) is closed, as can be seen by reviewing the proof that (ii) implies (iii) in Prop. 1.2.2. We state this as a proposition.

Proposition 1.2.3: Let $f : X \mapsto [-\infty, \infty]$ be a function. If dom(f) is closed and f is lower semicontinuous over dom(f), then f is closed.

We finally note that an improper closed convex function is very peculiar: it cannot take a finite value at any point, so it has the form

$$f(x) = \begin{cases} -\infty & \text{if } x \in \text{dom}(f), \\ \infty & \text{if } x \notin \text{dom}(f). \end{cases}$$

To see this, consider an improper closed convex function $f : \Re^n \mapsto [-\infty, \infty]$, and assume that there exists an x such that $f(x)$ is finite. Let $\overline{x}$ be such that $f(\overline{x}) = -\infty$ (such a point must exist since f is improper and f is not identically equal to ∞). Because f is convex, it can be seen that every point of the form

$$x_k = \frac{k-1}{k}x + \frac{1}{k}\overline{x}, \qquad \forall\, k = 1, 2, \ldots$$

satisfies $f(x_k) = -\infty$, while we have $x_k \to x$. Since f is closed, this implies that $f(x) = -\infty$, which is a contradiction. In conclusion, a closed convex function that is improper cannot take a finite value anywhere.

Recognizing Convex Functions

We can verify the convexity of a given function in a number of ways. Several commonly encountered functions are convex. For example, affine functions and norms are convex; this is straightforward to verify using the definition of convexity. In particular, for any $x, y \in \Re^n$ and any $\alpha \in [0, 1]$, by using the triangle inequality, we have

$$\|\alpha x + (1-\alpha)y\| \le \|\alpha x\| + \|(1-\alpha)y\| = \alpha\|x\| + (1-\alpha)\|y\|,$$

so the norm function $\|\cdot\|$ is convex. The exercises provide further examples of useful convex functions.

Starting with some known convex functions, we can generate other convex functions by using some common algebraic operations that preserve convexity of a function. The following proposition provides some of the necessary machinery.

Proposition 1.2.4:

(a) Let $f_i : \Re^n \mapsto (-\infty, \infty]$, $i = 1, \ldots, m$, be given functions, let $\lambda_1, \ldots, \lambda_m$ be positive scalars, and consider the function $g : \Re^n \mapsto (-\infty, \infty]$ given by

$$g(x) = \lambda_1 f_1(x) + \cdots + \lambda_m f_m(x).$$

If $f_1, \ldots, f_m$ are convex, then g is also convex, while if $f_1, \ldots, f_m$ are closed, then g is also closed.

(b) Let $f : \Re^m \mapsto (-\infty, \infty]$ be a given function, let A be an $m \times n$ matrix, and consider the function $g : \Re^n \mapsto (-\infty, \infty]$ given by

$$g(x) = f(Ax).$$

If f is convex, then g is also convex, while if f is closed, then g is also closed.

(c) Let $f_i : \Re^n \mapsto (-\infty, \infty]$ be given functions for $i \in I$, where I is an arbitrary index set, and consider the function $g : \Re^n \mapsto (-\infty, \infty]$ given by

$$g(x) = \sup_{i \in I} f_i(x).$$

If f_i, $i \in I$, are convex, then g is also convex, while if f_i, $i \in I$, are closed, then g is also closed.

Proof: (a) Let $f_1, \ldots, f_m$ be convex. We use the definition of convexity to write for any $x, y \in \Re^n$ and $\alpha \in [0, 1]$,

$$g\big(\alpha x + (1 - \alpha)y\big) = \sum_{i=1}^m \lambda_i f_i\big(\alpha x + (1 - \alpha)y\big)$$

$$\leq \sum_{i=1}^m \lambda_i \big(\alpha f_i(x) + (1 - \alpha)f_i(y)\big)$$

$$= \alpha \sum_{i=1}^m \lambda_i f_i(x) + (1 - \alpha) \sum_{i=1}^m \lambda_i f_i(y)$$

$$= \alpha g(x) + (1 - \alpha)g(y).$$

Hence g is convex.

Let the functions $f_1, \ldots, f_m$ be closed. Then they are lower semicontinuous at every $x \in \Re^n$ (cf. Prop. 1.2.2), so for every sequence $\{x_k\}$ converging to x, we have $f_i(x) \leq \liminf_{k \to \infty} f_i(x_k)$ for all i. Hence

$$g(x) \leq \sum_{i=1}^m \lambda_i \liminf_{k \to \infty} f_i(x_k) \leq \liminf_{k \to \infty} \sum_{i=1}^m \lambda_i f_i(x_k) = \liminf_{k \to \infty} g(x_k),$$

where we have used the assumption $\lambda_i > 0$ and Prop. 1.1.4(d) (the sum of the lower limits of sequences is less than or equal to the lower limit of the sum sequence). Therefore, g is lower semicontinuous at all $x \in \Re^n$, so by Prop. 1.2.2, it is closed.

(b) This is straightforward, along the lines of the proof of part (a).

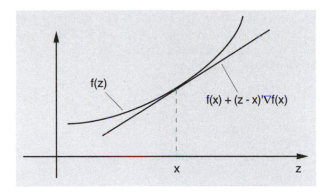

Figure 1.2.5. Characterization of convexity in terms of first derivatives. The condition $f(z) \geq f(x) + (z-x)'\nabla f(x)$ states that a linear approximation, based on the gradient, underestimates a convex function.

(c) A pair (x, w) belongs to epi(g) if and only if $g(x) \leq w$, which is true if and only if $f_i(x) \leq w$ for all $i \in I$, or equivalently $(x, w) \in \cap_{i \in I} \text{epi}(f_i)$. Therefore,

$$\text{epi}(g) = \cap_{i \in I} \text{epi}(f_i).$$

If the f_i are convex, the epigraphs epi(f_i) are convex, so epi(g) is convex, and g is convex. If the f_i are closed, then, by definition, the epigraphs epi(f_i) are closed, so epi(g) is closed, and g is closed.   **Q.E.D.**

For once or twice differentiable functions, there are some additional useful criteria for verifying convexity, as we now proceed to discuss.

Characterizations of Differentiable Convex Functions

For differentiable functions, a useful alternative characterization of convexity is given in the following proposition and is illustrated in Fig. 1.2.5.

Proposition 1.2.5: Let C be a convex subset of $\Re^n$ and let $f : \Re^n \mapsto \Re$ be differentiable over $\Re^n$.

(a) f is convex over C if and only if

$$f(z) \geq f(x) + (z-x)'\nabla f(x), \qquad \forall \; x, z \in C. \qquad (1.4)$$

(b) f is strictly convex over C if and only if the above inequality is strict whenever $x \neq z$.

Proof: The ideas of the proof are geometrically illustrated in Fig. 1.2.6. We prove (a) and (b) simultaneously. Assume that the inequality (1.4) holds. Choose any $x, y \in C$ and $\alpha \in [0, 1]$, and let $z = \alpha x + (1-\alpha)y$. Using the inequality (1.4) twice, we obtain

$$f(x) \geq f(z) + (x - z)'\nabla f(z),$$

$$f(y) \geq f(z) + (y - z)'\nabla f(z).$$

We multiply the first inequality by α, the second by $(1-\alpha)$, and add them to obtain

$$\alpha f(x) + (1-\alpha)f(y) \geq f(z) + \big(\alpha x + (1-\alpha)y - z\big)'\nabla f(z) = f(z),$$

which proves that f is convex. If the inequality (1.4) is strict as stated in part (b), then if we take $x \neq y$ and $\alpha \in (0,1)$ above, the three preceding inequalities become strict, thus showing the strict convexity of f.

Conversely, assume that f is convex, let x and z be any vectors in C with $x \neq z$, and for $\alpha \in (0, 1)$, consider the function

$$g(\alpha) = \frac{f\big(x + \alpha(z - x)\big) - f(x)}{\alpha}, \qquad \alpha \in (0, 1].$$

We will show that $g(\alpha)$ is monotonically increasing with α, and is strictly monotonically increasing if f is strictly convex. This will imply that

$$(z - x)'\nabla f(x) = \lim_{\alpha \downarrow 0} g(\alpha) \leq g(1) = f(z) - f(x),$$

with strict inequality if g is strictly monotonically increasing, thereby showing that the desired inequality (1.4) holds, and holds strictly if f is strictly convex. Indeed, consider any α_1, α_2, with $0 < \alpha_1 < \alpha_2 < 1$, and let

$$\overline{\alpha} = \frac{\alpha_1}{\alpha_2}, \qquad \overline{z} = x + \alpha_2(z - x). \tag{1.5}$$

We have

$$f\big(x + \overline{\alpha}(\overline{z} - x)\big) \leq \overline{\alpha}f(\overline{z}) + (1 - \overline{\alpha})f(x),$$

or

$$\frac{f\big(x + \overline{\alpha}(\overline{z} - x)\big) - f(x)}{\overline{\alpha}} \leq f(\overline{z}) - f(x), \tag{1.6}$$

and the above inequalities are strict if f is strictly convex. Substituting the definitions (1.5) in Eq. (1.6), we obtain after a straightforward calculation

$$\frac{f\big(x + \alpha_1(z - x)\big) - f(x)}{\alpha_1} \leq \frac{f\big(x + \alpha_2(z - x)\big) - f(x)}{\alpha_2},$$

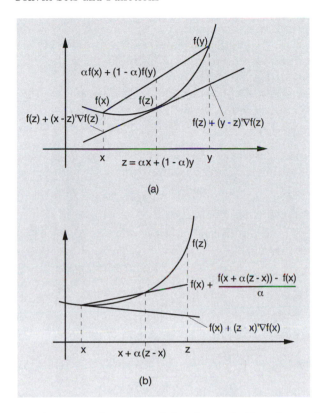

Figure 1.2.6. Geometric illustration of the ideas underlying the proof of Prop. 1.2.5. In figure (a), we linearly approximate f at $z = \alpha x + (1-\alpha)y$. The inequality (1.4) implies that

$$f(x) \geq f(z) + (x - z)'\nabla f(z),$$

$$f(y) \geq f(z) + (y - z)'\nabla f(z).$$

As can be seen from the figure, it follows that $\alpha f(x) + (1-\alpha)f(y)$ lies above $f(z)$, so f is convex.

In figure (b), we assume that f is convex, and from the figure's geometry, we note that

$$f(x) + \frac{f\big(x + \alpha(z - x)\big) - f(x)}{\alpha}$$

lies below $f(z)$, is monotonically nonincreasing as $\alpha \downarrow 0$, and converges to $f(x) + (z - x)'\nabla f(x)$. It follows that $f(z) \geq f(x) + (z - x)'\nabla f(x)$.

or

$$g(\alpha_1) \leq g(\alpha_2),$$

with strict inequality if f is strictly convex. Hence g is monotonically increasing with α, and strictly so if f is strictly convex. **Q.E.D.**

Note a simple consequence of Prop. 1.2.5(a): if $f : \Re^n \mapsto \Re$ is a convex function and $\nabla f(x^*) = 0$, then x^* minimizes f over $\Re^n$. This is a classical sufficient condition for unconstrained optimality, originally formulated (in one dimension) by Fermat in 1637.

For twice differentiable convex functions, there is another characterization of convexity as shown by the following proposition.

Proposition 1.2.6: Let C be a convex subset of $\Re^n$ and let $f : \Re^n \mapsto \Re$ be twice continuously differentiable over $\Re^n$.

(a) If $\nabla^2 f(x)$ is positive semidefinite for all $x \in C$, then f is convex over C.

(b) If $\nabla^2 f(x)$ is positive definite for all $x \in C$, then f is strictly convex over C.

(c) If C is open and f is convex over C, then $\nabla^2 f(x)$ is positive semidefinite for all $x \in C$.

Proof: (a) By Prop. 1.1.13(b), for all $x, y \in C$ we have

$$f(y) = f(x) + (y - x)'\nabla f(x) + \tfrac{1}{2}(y - x)'\nabla^2 f\big(x + \alpha(y - x)\big)(y - x)$$

for some $\alpha \in [0, 1]$. Therefore, using the positive semidefiniteness of $\nabla^2 f$, we obtain
$$f(y) \geq f(x) + (y - x)'\nabla f(x), \qquad \forall \; x, y \in C.$$

From Prop. 1.2.5(a), we conclude that f is convex.

(b) Similar to the proof of part (a), we have $f(y) > f(x) + (y - x)'\nabla f(x)$ for all $x, y \in C$ with $x \neq y$, and the result follows from Prop. 1.2.5(b).

(c) Assume, to obtain a contradiction, that there exist some $x \in C$ and some $z \in \Re^n$ such that $z'\nabla^2 f(x)z < 0$. Since C is open and $\nabla^2 f$ is continuous, we can choose z to have small enough norm so that $x + z \in C$ and $z'\nabla^2 f(x + \alpha z)z < 0$ for every $\alpha \in [0, 1]$. Then, using again Prop. 1.1.13(b), we obtain $f(x + z) < f(x) + z'\nabla f(x)$, which, in view of Prop. 1.2.5(a), contradicts the convexity of f over C. **Q.E.D.**

As an example, consider the quadratic function

$$f(x) = x'Qx + a'x,$$

where Q is a symmetric $n \times n$ matrix and b is a vector in $\Re^n$. Since $\nabla^2 f(x) = 2Q$, it follows by using Prop. 1.2.6, that f is convex if and only if Q is positive semidefinite, and it is strictly convex if and only if Q is positive definite.

If f is convex over a strict subset $C \subset \Re^n$, it is not necessarily true that $\nabla^2 f(x)$ is positive semidefinite at any point of C [take for example $n = 2$, $C = \{(x_1, 0) \mid x_1 \in \Re\}$, and $f(x) = x_1^2 - x_2^2$]. The relation of convexity and twice differentiability is further considered in Exercises 1.8 and 1.9. In particular, it can be shown that the conclusion of Prop. 1.2.6(c) also holds if C has nonempty interior instead of being open.

1.3 CONVEX AND AFFINE HULLS

Let X be a nonempty subset of $\Re^n$. A *convex combination* of elements of X is a vector of the form $\sum_{i=1}^m \alpha_i x_i$, where m is a positive integer, $x_1, \ldots, x_m$ belong to X, and $\alpha_1, \ldots, \alpha_m$ are scalars such that

$$\alpha_i \geq 0, \quad i = 1, \ldots, m, \qquad \sum_{i=1}^m \alpha_i = 1.$$

Note that if X is convex, then a convex combination belongs to X (see the construction of Fig. 1.3.1).

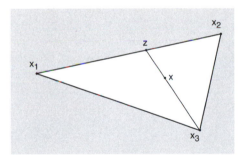

Figure 1.3.1. Illustration of the construction of a convex combination of m vectors by forming a sequence of $m - 1$ convex combinations of pairs of vectors. For example, we have

$$x = \alpha_1 x_1 + \alpha_2 x_2 + \alpha_3 x_3 = (\alpha_1 + \alpha_2)\left(\frac{\alpha_1}{\alpha_1 + \alpha_2} x_1 + \frac{\alpha_2}{\alpha_1 + \alpha_2} x_2\right) + \alpha_3 x_3,$$

so the convex combination $\alpha_1 x_1 + \alpha_2 x_2 + \alpha_3 x_3$ can be obtained by forming the convex combination

$$z = \frac{\alpha_1}{\alpha_1 + \alpha_2} x_1 + \frac{\alpha_2}{\alpha_1 + \alpha_2} x_2,$$

and then by forming the convex combination

$$x = (\alpha_1 + \alpha_2)z + \alpha_3 x_3$$

as shown in the figure. The construction shows among other things that a convex combination of a collection of vectors from a convex set belongs to the convex set.

For any function $f : \Re^n \mapsto \Re$ that is convex over X, we have

$$f\left(\sum_{i=1}^{m} \alpha_i x_i\right) \leq \sum_{i=1}^{m} \alpha_i f(x_i). \tag{1.7}$$

This follows by using repeatedly the definition of convexity together with the construction of Fig. 1.3.1. The preceding relation is a special case of a relation known as *Jensen's inequality*, and can be used to prove a number of interesting relations in applied mathematics and probability theory.

The *convex hull* of a set X, denoted conv(X), is the intersection of all convex sets containing X, and is a convex set by Prop. 1.2.1(a). It is straightforward to verify that the set of all convex combinations of elements of X is convex, and is equal to conv(X) (Exercise 1.14). In particular, if X consists of a finite number of vectors $x_1, \ldots, x_m$, its convex hull is

$$\text{conv}\big(\{x_1, \ldots, x_m\}\big) = \left\{\sum_{i=1}^{m} \alpha_i x_i \;\middle|\; \alpha_i \geq 0,\, i = 1, \ldots, m,\, \sum_{i=1}^{m} \alpha_i = 1\right\}.$$

We recall that an affine set M in $\Re^n$ is a set of the form $x + S$, where x is some vector and S is a subspace uniquely determined by M and called the *subspace parallel to* M. If X is a subset of $\Re^n$, the *affine hull* of X, denoted aff(X), is the intersection of all affine sets containing X. Note that aff(X) is itself an affine set and that it contains conv(X). It can be seen that

$$\text{aff}(X) = \text{aff}\big(\text{conv}(X)\big) = \text{aff}\big(\text{cl}(X)\big),$$

(see Exercise 1.18). Furthermore, in the case where $0 \in X$, aff(X) is the subspace generated by X. For a convex set C, the *dimension* of C is defined to be the dimension of aff(C).

Given a nonempty subset X of $\Re^n$, a *nonnegative combination* of elements of X is a vector of the form $\sum_{i=1}^{m} \alpha_i x_i$, where m is a positive integer, $x_1, \ldots, x_m$ belong to X, and $\alpha_1, \ldots, \alpha_m$ are nonnegative scalars. If the scalars α_i are all positive, $\sum_{i=1}^{m} \alpha_i x_i$ is said to be a *positive combination*. The *cone generated by* X, denoted cone(X), is the set of all nonnegative combinations of elements of X. It is easily seen that cone(X) is a convex cone containing the origin, although it need not be closed even if X is compact, as shown in Fig. 1.3.2 [it can be proved that cone(X) is closed in special cases, such as when X consists of a finite number of elements – this is one of the central results of polyhedral convexity, which will be shown in Section 3.2].

The following is a fundamental characterization of convex hulls (see Fig. 1.3.3).

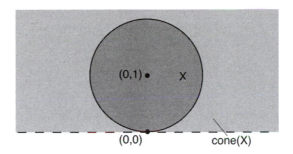

Figure 1.3.2. An example in $\Re^2$ where X is convex and compact, but $\text{cone}(X)$ is not closed. Here

$$X = \big\{(x_1, x_2) \mid x_1^2 + (x_2 - 1)^2 \leq 1\big\},$$

$$\text{cone}(X) = \big\{(x_1, x_2) \mid x_2 > 0\big\} \cup \big\{(0,0)\big\}.$$

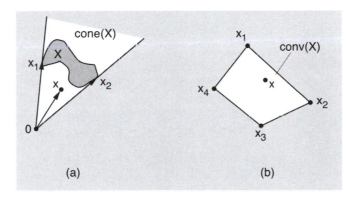

Figure 1.3.3. Illustration of Caratheodory's Theorem. In (a), X is a nonconvex set in $\Re^2$, and a point $x \in \text{cone}(X)$ is represented as a positive combination of the two linearly independent vectors $x_1, x_2 \in X$. In (b), X consists of four points x_1, x_2, x_3, x_4 in $\Re^2$, and the point $x \in \text{conv}(X)$ shown in the figure can be represented as a convex combination of the three vectors x_1, x_2, x_3. Note that the vectors $x_2 - x_1, x_3 - x_1$ are linearly independent. Note also that x can alternatively be represented as a convex combination of the vectors x_1, x_2, x_4, so the representation is not unique.

Proposition 1.3.1: (Caratheodory's Theorem) Let X be a non-empty subset of $\Re^n$.

  (a) Every x in $\text{cone}(X)$ can be represented as a positive combination of vectors $x_1, \ldots, x_m$ from X that are linearly independent.

(b) Every x in $\text{conv}(X)$ can be represented as a convex combination of vectors $x_1, \ldots, x_m$ from X such that $x_2 - x_1, \ldots, x_m - x_1$ are linearly independent.

Proof: (a) Let x be a nonzero vector in $\text{cone}(X)$, and let m be the smallest integer such that x has the form $\sum_{i=1}^{m} \alpha_i x_i$, where $\alpha_i > 0$ and $x_i \in X$ for all $i = 1, \ldots, m$. If the vectors x_i were linearly dependent, there would exist scalars $\lambda_1, \ldots, \lambda_m$, with $\sum_{i=1}^{m} \lambda_i x_i = 0$ and at least one λ_i is positive. Consider the linear combination $\sum_{i=1}^{m} (\alpha_i - \overline{\gamma}\lambda_i) x_i$, where $\overline{\gamma}$ is the largest γ such that $\alpha_i - \gamma\lambda_i \geq 0$ for all i. This combination provides a representation of x as a positive combination of fewer than m vectors of X – a contradiction. Therefore, $x_1, \ldots, x_m$, are linearly independent.

(b) The proof will be obtained by applying part (a) to the subset of $\Re^{n+1}$ given by

$$Y = \big\{(x, 1) \mid x \in X\big\}.$$

If $x \in \text{conv}(X)$, then for some positive integer I and some positive scalars γ_i, $i = 1, \ldots, I$, with $1 = \sum_{i=1}^{I} \gamma_i$, we have $x = \sum_{i=1}^{I} \gamma_i x_i$, so that $(x, 1) \in \text{cone}(Y)$. By part (a), we have $(x, 1) = \sum_{i=1}^{m} \alpha_i (x_i, 1)$ for some positive scalars $\alpha_1, \ldots, \alpha_m$ and some linearly independent vectors $(x_1, 1), \ldots, (x_m, 1)$, i.e.,

$$x = \sum_{i=1}^{m} \alpha_i x_i, \qquad 1 = \sum_{i=1}^{m} \alpha_i.$$

Assume, to arrive at a contradiction, that $x_2 - x_1, \ldots, x_m - x_1$ are linearly dependent, so that there exist $\lambda_2, \ldots, \lambda_m$, not all 0, with

$$\sum_{i=2}^{m} \lambda_i (x_i - x_1) = 0.$$

Equivalently, defining $\lambda_1 = -(\lambda_2 + \cdots + \lambda_m)$, we have

$$\sum_{i=1}^{m} \lambda_i (x_i, 1) = 0,$$

which contradicts the linear independence of $(x_1, 1), \ldots, (x_m, 1)$. **Q.E.D.**

Note that in view of the linear independence assertions in Caratheodory's Theorem, a vector in $\text{cone}(X)$ [or $\text{conv}(X)$] may be represented by no more than n (or $n + 1$, respectively) vectors of X. Note also that the proof of the theorem suggests an algorithm to obtain a representation of a vector $x \in \text{cone}(X)$ in terms of linearly independent vectors. The typical step in this algorithm is the proof's construction, which starts with a representation involving linearly dependent vectors, and yields another representation involving fewer vectors.

Caratheodory's Theorem can be used to prove several other important results. An example is the following proposition.

Proposition 1.3.2: The convex hull of a compact set is compact.

Proof: Let X be a compact subset of $\Re^n$. To show that $\text{conv}(X)$ is compact, we will take a sequence in $\text{conv}(X)$ and show that it has a convergent subsequence whose limit is in $\text{conv}(X)$. Indeed, by Caratheodory's Theorem, a sequence in $\text{conv}(X)$ can be expressed as $\left\{ \sum_{i=1}^{n+1} \alpha_i^k x_i^k \right\}$, where for all k and i, $\alpha_i^k \geq 0$, $x_i^k \in X$, and $\sum_{i=1}^{n+1} \alpha_i^k = 1$. Since the sequence

$$\left\{ (\alpha_1^k, \ldots, \alpha_{n+1}^k, x_1^k, \ldots, x_{n+1}^k) \right\}$$

is bounded, it has a limit point $\left\{ (\alpha_1, \ldots, \alpha_{n+1}, x_1, \ldots, x_{n+1}) \right\}$, which must satisfy $\sum_{i=1}^{n+1} \alpha_i = 1$, and $\alpha_i \geq 0$, $x_i \in X$ for all i. Thus, the vector $\sum_{i=1}^{n+1} \alpha_i x_i$, which belongs to $\text{conv}(X)$, is a limit point of the sequence $\left\{ \sum_{i=1}^{n+1} \alpha_i^k x_i^k \right\}$, showing that $\text{conv}(X)$ is compact.   **Q.E.D.**

Note that it is not generally true that the convex hull of a closed set is closed. As an example, for the closed subset of $\Re^2$

$$X = \left\{ (0,0) \right\} \cup \left\{ (x_1, x_2) \mid x_1 x_2 \geq 1, \, x_1 \geq 0, \, x_2 \geq 0 \right\},$$

the convex hull is

$$\text{conv}(X) = \left\{ (0,0) \right\} \cup \left\{ (x_1, x_2) \mid x_1 > 0, \, x_2 > 0 \right\},$$

which is not closed.

1.4   RELATIVE INTERIOR, CLOSURE, AND CONTINUITY

We now consider some generic topological properties of convex sets and functions. Let C be a nonempty convex subset of $\Re^n$. The closure of C is also a nonempty convex set (Prop. 1.2.1). While the interior of C may be empty, it turns out that convexity implies the existence of interior points relative to the affine hull of C. This is an important property, which we now formalize.

Let C be a nonempty convex set. We say that x is a *relative interior point* of C, if $x \in C$ and there exists an open sphere S centered at x such that $S \cap \text{aff}(C) \subset C$, i.e., x is an interior point of C relative to $\text{aff}(C)$. The set of all relative interior points of C is called the *relative interior of C*, and is denoted by $\text{ri}(C)$. The set C is said to be *relatively open* if $\text{ri}(C) = C$. A vector in the closure of C which is not a relative interior point of C is said to be a *relative boundary point* of C. The set of all relative boundary points of C is called the *relative boundary* of C.

For example, if C is a line segment connecting two distinct points in the plane, then $\text{ri}(C)$ consists of all points of C except for the two end points. The relative boundary of C consists of the two end points.

The following proposition gives some basic facts about relative interior points.

Proposition 1.4.1: Let C be a nonempty convex set.

(a) (*Line Segment Principle*) If $x \in \text{ri}(C)$ and $\overline{x} \in \text{cl}(C)$, then all points on the line segment connecting x and $\overline{x}$, except possibly $\overline{x}$, belong to $\text{ri}(C)$.

(b) (*Nonemptiness of Relative Interior*) $\text{ri}(C)$ is a nonempty convex set, and has the same affine hull as C. In fact, if m is the dimension of $\text{aff}(C)$ and $m > 0$, there exist vectors $x_0, x_1, \ldots, x_m \in \text{ri}(C)$ such that $x_1 - x_0, \ldots, x_m - x_0$ span the subspace parallel to $\text{aff}(C)$.

(c) $x \in \text{ri}(C)$ if and only if every line segment in C having x as one endpoint can be prolonged beyond x without leaving C [i.e., for every $\overline{x} \in C$, there exists a $\gamma > 1$ such that $x + (\gamma - 1)(x - \overline{x}) \in C$].

Proof: (a) For the case where $\overline{x} \in C$, the proof is given in Fig. 1.4.1. Consider the case where $\overline{x} \notin C$. To show that for any $\alpha \in (0, 1]$ we have $x_\alpha = \alpha x + (1 - \alpha)\overline{x} \in \text{ri}(C)$, consider a sequence $\{x_k\} \subset C$ that converges to $\overline{x}$, and let $x_{k,\alpha} = \alpha x + (1 - \alpha)x_k$. Then as in Fig. 1.4.1, we see that $\{z \mid \|z - x_{k,\alpha}\| < \alpha\epsilon\} \cap \text{aff}(C) \subset C$ for all k, where ϵ is such that the open sphere $S = \{z \mid \|z - x\| < \epsilon\}$ satisfies $S \cap \text{aff}(C) \subset C$. Since $x_{k,\alpha} \to x_\alpha$, for large enough k, we have

$$\{z \mid \|z - x_\alpha\| < \alpha\epsilon/2\} \subset \{z \mid \|z - x_{k,\alpha}\| < \alpha\epsilon\}.$$

It follows that $\{z \mid \|z - x_\alpha\| < \alpha\epsilon/2\} \cap \text{aff}(C) \subset C$, which shows that $x_\alpha \in \text{ri}(C)$.

(b) Convexity of $\text{ri}(C)$ follows from the Line Segment Principle of part (a). By using a translation argument if necessary, we assume without loss of generality that $0 \in C$. Then, the affine hull of C is a subspace whose dimension will be denoted by m. If $m = 0$, then C and $\text{aff}(C)$ consist of a single point, which is a unique relative interior point. If $m > 0$, we

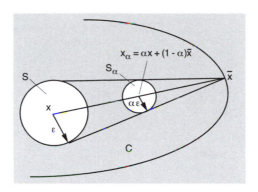

Figure 1.4.1. Proof of the Line Segment Principle for the case where $\overline{x} \in C$. Since $x \in \text{ri}(C)$, there exists an open sphere $S = \{z \mid \|z - x\| < \epsilon\}$ such that $S \cap \text{aff}(C) \subset C$. For all $\alpha \in (0, 1]$, let $x_\alpha = \alpha x + (1 - \alpha)\overline{x}$ and let $S_\alpha = \{z \mid \|z - x_\alpha\| < \alpha\epsilon\}$. It can be seen that each point of $S_\alpha \cap \text{aff}(C)$ is a convex combination of $\overline{x}$ and some point of $S \cap \text{aff}(C)$. Therefore, by the convexity of C, $S_\alpha \cap \text{aff}(C) \subset C$, implying that $x_\alpha \in \text{ri}(C)$.

can find m linearly independent vectors $z_1, \ldots, z_m$ in C that span $\text{aff}(C)$; otherwise there would exist $r < m$ linearly independent vectors in C whose span contains C, contradicting the fact that the dimension of $\text{aff}(C)$ is m. Thus $z_1, \ldots, z_m$ form a basis for $\text{aff}(C)$.

Consider the set

$$X = \left\{ x \mid x = \sum_{i=1}^{m} \alpha_i z_i, \ \sum_{i=1}^{m} \alpha_i < 1, \ \alpha_i > 0, \ i = 1, \ldots, m \right\}$$

(see Fig. 1.4.2). We claim that this set is open relative to $\text{aff}(C)$, i.e., for every vector $\overline{x} \in X$, there exists an open ball B centered at $\overline{x}$ such that $\overline{x} \in B$ and $B \cap \text{aff}(C) \subset X$. To see this, fix $\overline{x} \in X$ and let x be another vector in $\text{aff}(C)$. We have $\overline{x} = Z\overline{\alpha}$ and $x = Z\alpha$, where Z is the $n \times m$ matrix whose columns are the vectors $z_1, \ldots, z_m$, and $\overline{\alpha}$ and α are suitable m-dimensional vectors, which are unique since $z_1, \ldots, z_m$ form a basis for $\text{aff}(C)$. Since Z has linearly independent columns, the matrix $Z'Z$ is symmetric and positive definite, so by Prop. 1.1.11(d), we have for some positive scalar γ, which is independent of x and $\overline{x}$,

$$\|x - \overline{x}\|^2 = (\alpha - \overline{\alpha})'Z'Z(\alpha - \overline{\alpha}) \geq \gamma\|\alpha - \overline{\alpha}\|^2. \tag{1.8}$$

Since $\overline{x} \in X$, the corresponding vector $\overline{\alpha}$ lies in the open set

$$A = \left\{ (\alpha_1, \ldots, \alpha_m) \mid \sum_{i=1}^{m} \alpha_i < 1, \ \alpha_i > 0, \ i = 1, \ldots, m \right\}.$$

From Eq. (1.8), we see that if x lies in a suitably small ball centered at $\overline{x}$, the corresponding vector α lies in A, implying that $x \in X$. Hence

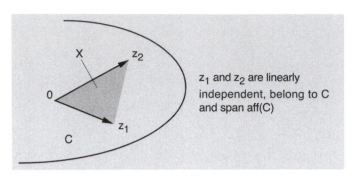

Figure 1.4.2. Construction of the relatively open set X in the proof of nonemptiness of the relative interior of a convex set C that contains the origin, assuming that $m > 0$. We choose m linearly independent vectors $z_1, \ldots, z_m \in C$, where m is the dimension of $\text{aff}(C)$, and we let

$$X = \left\{ \sum_{i=1}^{m} \alpha_i z_i \,\Bigm|\, \sum_{i=1}^{m} \alpha_i < 1, \, \alpha_i > 0, \, i = 1, \ldots, m \right\}.$$

X contains the intersection of $\text{aff}(C)$ and an open ball centered at $\bar{x}$, so X is open relative to $\text{aff}(C)$. It follows that all points of X are relative interior points of C, so that $\text{ri}(C)$ is nonempty. Also, since by construction, $\text{aff}(X) = \text{aff}(C)$ and $X \subset \text{ri}(C)$, we see that $\text{ri}(C)$ and C have the same affine hull.

To show the last assertion of part (b), consider vectors

$$x_0 = \alpha \sum_{i=1}^{m} z_i, \qquad x_i = x_0 + \alpha z_i, \quad i = 1, \ldots, m,$$

where α is a positive scalar such that $\alpha(m+1) < 1$. The vectors $x_0, \ldots, x_m$ belong to X, and since $X \subset \text{ri}(C)$, they also belong to $\text{ri}(C)$. Furthermore, because $x_i - x_0 = \alpha z_i$ for all i and the vectors $z_1, \ldots, z_m$ span $\text{aff}(C)$, the vectors $x_1 - x_0, \ldots, x_m - x_0$ also span $\text{aff}(C)$.

(c) If $x \in \text{ri}(C)$, the given condition clearly holds, using the definition of relative interior point. Conversely, let x satisfy the given condition. We will show that $x \in \text{ri}(C)$. By part (b), there exists a vector $\bar{x} \in \text{ri}(C)$. We may assume that $\bar{x} \neq x$, since otherwise we are done. By the given condition, since $\bar{x}$ is in C, there is a $\gamma > 1$ such that $y = x + (\gamma - 1)(x - \bar{x}) \in C$. Then we have $x = (1 - \alpha)\bar{x} + \alpha y$, where $\alpha = 1/\gamma \in (0, 1)$, so by the Line Segment Principle, we obtain $x \in \text{ri}(C)$. **Q.E.D.**

We will see in the following chapters that the notion of relative interior is pervasive in convex optimization and duality theory. As an example, we provide an important characterization of the set of optimal solutions in the case where the cost function is concave.

Proposition 1.4.2: Let X be a nonempty convex subset of $\Re^n$, let $f : X \mapsto \Re$ be a concave function, and let X^* be the set of vectors where f attains a minimum over X, i.e.,

$$X^* = \left\{ x \in X \mid f(x^*) = \inf_{x \in X} f(x) \right\}.$$

If X^* contains a relative interior point of X, then f must be constant over X, i.e., $X^* = X$.

Proof: Let x^* belong to $X^* \cap \mathrm{ri}(X)$, and let x be any vector in X. By Prop. 1.4.1(c), there exists a $\gamma > 1$ such that the vector

$$\hat{x} = x^* + (\gamma - 1)(x^* - x)$$

belongs to X, implying that

$$x^* = \frac{1}{\gamma}\hat{x} + \frac{\gamma - 1}{\gamma}x.$$

By the concavity of the function f, we have

$$f(x^*) \geq \frac{1}{\gamma}f(\hat{x}) + \frac{\gamma - 1}{\gamma}f(x),$$

and since $f(\hat{x}) \geq f(x^*)$ and $f(x) \geq f(x^*)$, we obtain

$$f(x^*) \geq \frac{1}{\gamma}f(\hat{x}) + \frac{\gamma - 1}{\gamma}f(x) \geq f(x^*).$$

Hence $f(x) = f(x^*)$.   **Q.E.D.**

One consequence of the preceding proposition is that a linear cost function $f(x) = c'x$, with $c \neq 0$, cannot attain a minimum at some interior point of a constraint set, since such a function cannot be constant over an open sphere.

Operations with Relative Interiors and Closures

To deal with set operations such as intersection, vector sum, linear transformation in the analysis of convex optimization problems, we need tools for calculating the corresponding relative interiors and closures. These tools are provided in the next three propositions.

Proposition 1.4.3: Let C be a nonempty convex set.

(a) We have $\mathrm{cl}(C) = \mathrm{cl}(\mathrm{ri}(C))$.

(b) We have $\mathrm{ri}(C) = \mathrm{ri}(\mathrm{cl}(C))$.

(c) Let $\overline{C}$ be another nonempty convex set. Then the following three conditions are equivalent:

    (i) C and $\overline{C}$ have the same relative interior.

    (ii) C and $\overline{C}$ have the same closure.

    (iii) $\mathrm{ri}(C) \subset \overline{C} \subset \mathrm{cl}(C)$.

Proof: (a) Since $\mathrm{ri}(C) \subset C$, we have $\mathrm{cl}(\mathrm{ri}(C)) \subset \mathrm{cl}(C)$. Conversely, let $\overline{x} \in \mathrm{cl}(C)$. We will show that $\overline{x} \in \mathrm{cl}(\mathrm{ri}(C))$. Let x be any point in $\mathrm{ri}(C)$ [there exists such a point by Prop. 1.4.1(b)], and assume that $\overline{x} \neq x$ (otherwise we are done). By the Line Segment Principle [Prop. 1.4.1(a)], we have $\alpha x + (1 - \alpha)\overline{x} \in \mathrm{ri}(C)$ for all $\alpha \in (0, 1]$. Thus, $\overline{x}$ is the limit of the sequence $\{(1/k)x + (1 - 1/k)\overline{x} \mid k \geq 1\}$ that lies in $\mathrm{ri}(C)$, so $\overline{x} \in \mathrm{cl}(\mathrm{ri}(C))$.

(b) The inclusion $\mathrm{ri}(C) \subset \mathrm{ri}(\mathrm{cl}(C))$ follows from the definition of a relative interior point and the fact $\mathrm{aff}(C) = \mathrm{aff}(\mathrm{cl}(C))$ (see Exercise 1.18). To prove the reverse inclusion, let $z \in \mathrm{ri}(\mathrm{cl}(C))$. We will show that $z \in \mathrm{ri}(C)$. By Prop. 1.4.1(b), there exists an $x \in \mathrm{ri}(C)$. We may assume that $x \neq z$ (otherwise we are done). We choose $\gamma > 1$, with γ sufficiently close to 1 so that the vector $y = z + (\gamma - 1)(z - x)$ belongs to $\mathrm{cl}(C)$ [cf. Prop. 1.4.1(c)]. Then we have $z = (1 - \alpha)x + \alpha y$ where $\alpha = 1/\gamma \in (0, 1)$, so by the Line Segment Principle [Prop. 1.4.1(a)], we obtain $z \in \mathrm{ri}(C)$.

(c) If $\mathrm{ri}(C) = \mathrm{ri}(\overline{C})$, part (a) implies that $\mathrm{cl}(C) = \mathrm{cl}(\overline{C})$. Similarly, if $\mathrm{cl}(C) = \mathrm{cl}(\overline{C})$, part (b) implies that $\mathrm{ri}(C) = \mathrm{ri}(\overline{C})$. Thus, (i) and (ii) are equivalent. Also, (i), (ii), and the relation $\mathrm{ri}(\overline{C}) \subset \overline{C} \subset \mathrm{cl}(\overline{C})$ imply condition (iii). Finally, let condition (iii) hold. Then by taking closures, we have $\mathrm{cl}(\mathrm{ri}(C)) \subset \mathrm{cl}(\overline{C}) \subset \mathrm{cl}(C)$, and by using part (a), we obtain $\mathrm{cl}(C) \subset \mathrm{cl}(\overline{C}) \subset \mathrm{cl}(C)$. Hence $\mathrm{cl}(\overline{C}) = \mathrm{cl}(C)$, i.e., (ii) holds. **Q.E.D.**

Proposition 1.4.4: Let C be a nonempty convex subset of $\Re^n$ and let A be an $m \times n$ matrix.

(a) We have $A \cdot \mathrm{ri}(C) = \mathrm{ri}(A \cdot C)$.

(b) We have $A \cdot \mathrm{cl}(C) \subset \mathrm{cl}(A \cdot C)$. Furthermore, if C is bounded, then $A \cdot \mathrm{cl}(C) = \mathrm{cl}(A \cdot C)$.

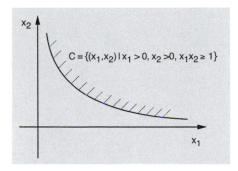

Figure 1.4.3. An example of a closed convex set C whose image $A \cdot C$ under a linear transformation A is not closed. Here

$$C = \big\{ (x_1, x_2) \mid x_1 > 0,\ x_2 > 0,\ x_1 x_2 \geq 1 \big\}$$

and A acts as projection on the horizontal axis, i.e., $A \cdot (x_1, x_2) = (x_1, 0)$. Then $A \cdot C$ is the (nonclosed) halfline $\big\{ (x_1, x_2) \mid x_1 > 0,\ x_2 = 0 \big\}$.

Proof: (a) For any set X, we have $A \cdot \mathrm{cl}(X) \subset \mathrm{cl}(A \cdot X)$, since if a sequence $\{x_k\} \subset X$ converges to some $x \in \mathrm{cl}(X)$ then the sequence $\{Ax_k\}$, which belongs to $A \cdot X$, converges to Ax, implying that $Ax \in \mathrm{cl}(A \cdot X)$. We use this fact and Prop. 1.4.3(a) to write

$$A \cdot \mathrm{ri}(C) \subset A \cdot C \subset A \cdot \mathrm{cl}(C) = A \cdot \mathrm{cl}\big(\mathrm{ri}(C)\big) \subset \mathrm{cl}\big(A \cdot \mathrm{ri}(C)\big).$$

Thus the convex set $A \cdot C$ lies between the convex set $A \cdot \mathrm{ri}(C)$ and the closure of that set, implying that the relative interiors of the sets $A \cdot C$ and $A \cdot \mathrm{ri}(C)$ are equal [Prop. 1.4.3(c)]. Hence $\mathrm{ri}(A \cdot C) \subset A \cdot \mathrm{ri}(C)$. To show the reverse inclusion, we take any $z \in A \cdot \mathrm{ri}(C)$ and we show that $z \in \mathrm{ri}(A \cdot C)$. Let x be any vector in $A \cdot C$, and let $\overline{z} \in \mathrm{ri}(C)$ and $\overline{x} \in C$ be such that $A\overline{z} = z$ and $A\overline{x} = x$. By Prop. 1.4.1(c), there exists a $\gamma > 1$ such that the vector $\overline{y} = \overline{z} + (\gamma - 1)(\overline{z} - \overline{x})$ belongs to C. Thus we have $A\overline{y} \in A \cdot C$ and $A\overline{y} = z + (\gamma - 1)(z - x)$, so by Prop. 1.4.1(c), it follows that $z \in \mathrm{ri}(A \cdot C)$.

(b) By the argument given in part (a), we have $A \cdot \mathrm{cl}(C) \subset \mathrm{cl}(A \cdot C)$. To show the converse, assuming that C is bounded, choose any $z \in \mathrm{cl}(A \cdot C)$. Then, there exists a sequence $\{x_k\} \subset C$ such that $Ax_k \to z$. Since C is bounded, $\{x_k\}$ has a subsequence that converges to some $x \in \mathrm{cl}(C)$, and we must have $Ax = z$. It follows that $z \in A \cdot \mathrm{cl}(C)$. **Q.E.D.**

Note that if C is closed and convex but unbounded, the set $A \cdot C$ need not be closed [cf. part (b) of the above proposition]. An example is given in Fig. 1.4.3.

Proposition 1.4.5: Let C_1 and C_2 be nonempty convex sets.

(a) We have

$$\text{ri}(C_1) \cap \text{ri}(C_2) \subset \text{ri}(C_1 \cap C_2), \qquad \text{cl}(C_1 \cap C_2) \subset \text{cl}(C_1) \cap \text{cl}(C_2).$$

Furthermore, if the sets $\text{ri}(C_1)$ and $\text{ri}(C_2)$ have a nonempty intersection, then

$$\text{ri}(C_1 \cap C_2) = \text{ri}(C_1) \cap \text{ri}(C_2), \qquad \text{cl}(C_1 \cap C_2) = \text{cl}(C_1) \cap \text{cl}(C_2).$$

(b) We have

$$\text{ri}(C_1 + C_2) = \text{ri}(C_1) + \text{ri}(C_2), \qquad \text{cl}(C_1) + \text{cl}(C_2) \subset \text{cl}(C_1 + C_2).$$

Furthermore, if at least one of the sets C_1 and C_2 is bounded, then

$$\text{cl}(C_1) + \text{cl}(C_2) = \text{cl}(C_1 + C_2).$$

Proof: (a) Take any $x \in \text{ri}(C_1) \cap \text{ri}(C_2)$ and any $y \in C_1 \cap C_2$. By Prop. 1.4.1(c), it can be seen that the line segment connecting x and y can be prolonged beyond x by a small amount without leaving C_1 and also by another small amount without leaving C_2. Thus, by the same proposition, it follows that $x \in \text{ri}(C_1 \cap C_2)$, so that $\text{ri}(C_1) \cap \text{ri}(C_2) \subset \text{ri}(C_1 \cap C_2)$. Also, since the set $C_1 \cap C_2$ is contained in the closed set $\text{cl}(C_1) \cap \text{cl}(C_2)$, we have

$$\text{cl}(C_1 \cap C_2) \subset \text{cl}(C_1) \cap \text{cl}(C_2).$$

Assume now that $\text{ri}(C_1) \cap \text{ri}(C_2)$ is nonempty. Let $y \in \text{cl}(C_1) \cap \text{cl}(C_2)$, and let $x \in \text{ri}(C_1) \cap \text{ri}(C_2)$. By the Line Segment Principle [Prop. 1.4.1(a)], the vector $\alpha x + (1 - \alpha)y$ belongs to $\text{ri}(C_1) \cap \text{ri}(C_2)$ for all $\alpha \in (0, 1]$. Hence, y is the limit of a sequence $\alpha_k x + (1 - \alpha_k)y \subset \text{ri}(C_1) \cap \text{ri}(C_2)$ with $\alpha_k \to 0$, implying that $y \in \text{cl}\big(\text{ri}(C_1) \cap \text{ri}(C_2)\big)$. Thus,

$$\text{cl}(C_1) \cap \text{cl}(C_2) \subset \text{cl}\big(\text{ri}(C_1) \cap \text{ri}(C_2)\big) \subset \text{cl}(C_1 \cap C_2).$$

We showed earlier that $\text{cl}(C_1 \cap C_2) \subset \text{cl}(C_1) \cap \text{cl}(C_2)$, so equality holds throughout in the preceding relation, and therefore $\text{cl}(C_1 \cap C_2) = \text{cl}(C_1) \cap \text{cl}(C_2)$. Furthermore, the sets $\text{ri}(C_1) \cap \text{ri}(C_2)$ and $C_1 \cap C_2$ have the same closure. Therefore, by Prop. 1.4.3(c), they have the same relative interior, implying that

$$\text{ri}(C_1 \cap C_2) \subset \text{ri}(C_1) \cap \text{ri}(C_2).$$

We showed earlier the reverse inclusion, so we obtain $\mathrm{ri}(C_1 \cap C_2) = \mathrm{ri}(C_1) \cap \mathrm{ri}(C_2)$.

(b) Consider the linear transformation $A : \Re^{2n} \mapsto \Re^n$ given by $A(x_1, x_2) = x_1 + x_2$ for all $x_1, x_2 \in \Re^n$. The relative interior of the Cartesian product $C_1 \times C_2$ (viewed as a subset of $\Re^{2n}$) is $\mathrm{ri}(C_1) \times \mathrm{ri}(C_2)$ (see Exercise 1.37). Since $A(C_1 \times C_2) = C_1 + C_2$, from Prop. 1.4.4(a), we obtain $\mathrm{ri}(C_1 + C_2) = \mathrm{ri}(C_1) + \mathrm{ri}(C_2)$.

Similarly, the closure of the Cartesian product $C_1 \times C_2$ is $\mathrm{cl}(C_1) \times \mathrm{cl}(C_2)$ (see Exercise 1.37). From Prop. 1.4.4(b), we have $A \cdot \mathrm{cl}(C_1 \times C_2) \subset \mathrm{cl}(A \cdot (C_1 \times C_2))$, or equivalently, $\mathrm{cl}(C_1) + \mathrm{cl}(C_2) \subset \mathrm{cl}(C_1 + C_2)$.

Finally, we show the reverse inclusion, assuming that C_1 is bounded. Indeed, if $x \in \mathrm{cl}(C_1 + C_2)$, there exist sequences $\{x_k^1\} \subset C_1$ and $\{x_k^2\} \subset C_2$ such that $x_k^1 + x_k^2 \to x$. Since $\{x_k^1\}$ is bounded, it follows that $\{x_k^2\}$ is also bounded. Thus, $\{(x_k^1, x_k^2)\}$ has a subsequence that converges to a vector (x^1, x^2), and we have $x^1 + x^2 = x$. Since $x^1 \in \mathrm{cl}(C_1)$ and $x^2 \in \mathrm{cl}(C_2)$, it follows that $x \in \mathrm{cl}(C_1) + \mathrm{cl}(C_2)$. Hence $\mathrm{cl}(C_1 + C_2) \subset \mathrm{cl}(C_1) + \mathrm{cl}(C_2)$. **Q.E.D.**

The requirement that $\mathrm{ri}(C_1) \cap \mathrm{ri}(C_2) \neq \emptyset$ is essential in part (a) of the preceding proposition. As an example, consider the following subsets of $\Re$:

$$C_1 = \{x \mid x \geq 0\}, \qquad C_2 = \{x \mid x \leq 0\}.$$

Then we have $\mathrm{ri}(C_1 \cap C_2) = \{0\} \neq \emptyset = \mathrm{ri}(C_1) \cap \mathrm{ri}(C_2)$. Also, consider the following subsets of $\Re$:

$$C_1 = \{x \mid x > 0\}, \qquad C_2 = \{x \mid x < 0\}.$$

Then we have $\mathrm{cl}(C_1 \cap C_2) = \emptyset \neq \{0\} = \mathrm{cl}(C_1) \cap \mathrm{cl}(C_2)$.

The requirement that at least one of the sets C_1 and C_2 be bounded is essential in part (b) of the preceding proposition. This is illustrated by the example of Fig. 1.4.4.

We note that the results regarding the closure of the image of a closed set under a linear transformation [cf. Prop. 1.4.4(b)] and the related result regarding the closure of the vector sum of two closed sets [cf. Prop. 1.4.5(b)] will be refined in the next section by using the machinery of recession cones, which will be developed in that section.

Continuity of Convex Functions

We close this section with a basic result on the continuity properties of convex functions.

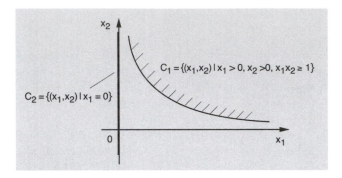

Figure 1.4.4. An example where the sum of two closed convex sets C_1 and C_2 is not closed. Here

$$C_1 = \big\{(x_1, x_2) \mid x_1 > 0, \, x_2 > 0, \, x_1 x_2 \geq 1\big\}, \qquad C_2 = \big\{(x_1, x_2) \mid x_1 = 0\big\},$$

and $C_1 + C_2$ is the open halfspace $\big\{(x_1, x_2) \mid x_1 > 0\big\}$.

Proposition 1.4.6: If $f : \Re^n \mapsto \Re$ is convex, then it is continuous. More generally, if $f : \Re^n \mapsto (-\infty, \infty]$ is a proper convex function, then f, restricted to $\mathrm{dom}(f)$, is continuous over the relative interior of $\mathrm{dom}(f)$.

Proof: Restricting attention to the affine hull of $\mathrm{dom}(f)$ and using a transformation argument if necessary, we assume without loss of generality, that the origin is an interior point of $\mathrm{dom}(f)$ and that the unit cube $X = \{x \mid \|x\|_\infty \leq 1\}$ is contained in $\mathrm{dom}(f)$. It will suffice to show that f is continuous at 0, i.e., that for any sequence $\{x_k\} \subset \Re^n$ that converges to 0, we have $f(x_k) \to f(0)$.

Let e_i, $i = 1, \ldots, 2^n$, be the corners of X, i.e., each e_i is a vector whose entries are either 1 or -1. It can be seen that any $x \in X$ can be expressed in the form $x = \sum_{i=1}^{2^n} \alpha_i e_i$, where each α_i is a nonnegative scalar and $\sum_{i=1}^{2^n} \alpha_i = 1$. Let $A = \max_i f(e_i)$. From Jensen's inequality [Eq. (1.7)], it follows that $f(x) \leq A$ for every $x \in X$.

For the purpose of proving continuity at 0, we can assume that $x_k \in X$ and $x_k \neq 0$ for all k. Consider the sequences $\{y_k\}$ and $\{z_k\}$ given by

$$y_k = \frac{x_k}{\|x_k\|_\infty}, \qquad z_k = -\frac{x_k}{\|x_k\|_\infty};$$

(cf. Fig. 1.4.5). Using the definition of a convex function for the line segment that connects y_k, x_k, and 0, we have

$$f(x_k) \leq \big(1 - \|x_k\|_\infty\big) f(0) + \|x_k\|_\infty f(y_k).$$

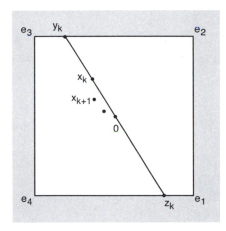

Figure 1.4.5. Construction for proving continuity of a convex function (cf. Prop. 1.4.6).

Since $\|x_k\|_\infty \to 0$ and $f(y_k) \le A$ for all k, by taking the limit as $k \to \infty$, we obtain

$$\limsup_{k \to \infty} f(x_k) \le f(0).$$

Using the definition of a convex function for the line segment that connects x_k, 0, and z_k, we have

$$f(0) \le \frac{\|x_k\|_\infty}{\|x_k\|_\infty + 1} f(z_k) + \frac{1}{\|x_k\|_\infty + 1} f(x_k)$$

and letting $k \to \infty$, we obtain

$$f(0) \le \liminf_{k \to \infty} f(x_k).$$

Thus, $\lim_{k \to \infty} f(x_k) = f(0)$ and f is continuous at zero.   **Q.E.D.**

A straightforward consequence of the continuity of a real-valued function f that is convex over $\Re^n$ is that its epigraph as well as the level sets $\{x \mid f(x) \le \gamma\}$ for all scalars γ are closed and convex (cf. Prop. 1.2.2). Thus, a real-valued convex function is closed.

1.5 RECESSION CONES

Some of the preceding results [Props. 1.3.2, 1.4.4(b)] have illustrated how closedness and compactness of convex sets are affected by various operations such as linear transformations. In this section we take a closer look at this issue. In the process, we develop some important convexity topics that are broadly useful in optimization.

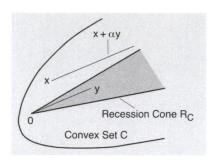

Figure 1.5.1. Illustration of the recession cone R_C of a convex set C. A direction of recession y has the property that $x + \alpha y \in C$ for all $x \in C$ and $\alpha \geq 0$.

We will first introduce the principal notions of this section: directions of recession and lineality space. We will then show how these concepts play an important role in various conditions that guarantee:

(a) The nonemptiness of the intersection of a sequence of closed convex sets.

(b) The closedness of the image of a closed convex set under a linear transformation.

It turns out that these issues lie at the heart of important questions relating to the existence of solutions of convex optimization problems, to minimax theory, and to duality theory, as we will see in subsequent chapters.

Given a nonempty convex set C, we say that a vector y is a *direction of recession* of C if $x + \alpha y \in C$ for all $x \in C$ and $\alpha \geq 0$. Thus, y is a direction of recession of C if starting at any x in C and going indefinitely along y, we never cross the relative boundary of C to points outside C. The set of all directions of recession is a cone containing the origin. It is called the *recession cone* of C and it is denoted by R_C (see Fig. 1.5.1). The following proposition gives some properties of recession cones.

Proposition 1.5.1: (Recession Cone Theorem) Let C be a nonempty closed convex set.

(a) The recession cone R_C is a closed convex cone.

(b) A vector y belongs to R_C if and only if there exists a vector $x \in C$ such that $x + \alpha y \in C$ for all $\alpha \geq 0$.

(c) R_C contains a nonzero direction if and only if C is unbounded.

(d) The recession cones of C and ri(C) are equal.

(e) If D is another closed convex set such that $C \cap D \neq \emptyset$, we have

$$R_{C \cap D} = R_C \cap R_D.$$

> More generally, for any collection of closed convex sets C_i, $i \in I$,
> where I is an arbitrary index set and $\cap_{i \in I} C_i$ is nonempty, we
> have
> $$R_{\cap_{i \in I} C_i} = \cap_{i \in I} R_{C_i}.$$

Proof: (a) If y_1, y_2 belong to R_C and λ_1, λ_2 are positive scalars such that $\lambda_1 + \lambda_2 = 1$, we have for any $x \in C$ and $\alpha \geq 0$

$$x + \alpha(\lambda_1 y_1 + \lambda_2 y_2) = \lambda_1 (x + \alpha y_1) + \lambda_2 (x + \alpha y_2) \in C,$$

where the last inclusion holds because C is convex, and $x + \alpha y_1$ and $x + \alpha y_2$ belong to C by the definition of R_C. Hence $\lambda_1 y_1 + \lambda_2 y_2 \in R_C$, implying that R_C is convex.

Let y be in the closure of R_C, and let $\{y_k\} \subset R_C$ be a sequence converging to y. For any $x \in C$ and $\alpha \geq 0$ we have $x + \alpha y_k \in C$ for all k, and because C is closed, we have $x + \alpha y \in C$. This implies that $y \in R_C$ and that R_C is closed.

(b) If $y \in R_C$, every vector $x \in C$ has the required property by the definition of R_C. Conversely, let y be such that there exists a vector $x \in C$ with $x + \alpha y \in C$ for all $\alpha \geq 0$. With no loss of generality, we assume that $y \neq 0$. We fix $\overline{x} \in C$ and $\alpha > 0$, and we show that $\overline{x} + \alpha y \in C$. It is sufficient to show that $\overline{x} + y \in C$, i.e., to assume that $\alpha = 1$, since the general case where $\alpha > 0$ can be reduced to the case where $\alpha = 1$ by replacing y with y/α.

Let

$$z_k = x + ky, \qquad k = 1, 2, \ldots$$

and note that $z_k \in C$ for all k, since $x \in C$ and $y \in R_C$. If $\overline{x} = z_k$ for some k, then $\overline{x} + y = x + (k+1)y$, which belongs to C and we are done. We thus assume that $\overline{x} \neq z_k$ for all k, and we define

$$y_k = \frac{z_k - \overline{x}}{\|z_k - \overline{x}\|} \|y\|, \qquad k = 1, 2, \ldots$$

so that $\overline{x} + y_k$ lies on the line that starts at $\overline{x}$ and passes through z_k (see the construction of Fig. 1.5.2).

We have

$$\frac{y_k}{\|y\|} = \frac{\|z_k - x\|}{\|z_k - \overline{x}\|} \cdot \frac{z_k - x}{\|z_k - x\|} + \frac{x - \overline{x}}{\|z_k - \overline{x}\|} = \frac{\|z_k - x\|}{\|z_k - \overline{x}\|} \cdot \frac{y}{\|y\|} + \frac{x - \overline{x}}{\|z_k - \overline{x}\|}.$$

Because $\{z_k\}$ is an unbounded sequence,

$$\frac{\|z_k - x\|}{\|z_k - \overline{x}\|} \to 1, \qquad \frac{x - \overline{x}}{\|z_k - \overline{x}\|} \to 0,$$

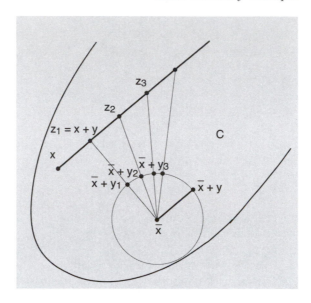

Figure 1.5.2. Construction for the proof of Prop. 1.5.1(b).

so by combining the preceding relations, we have $y_k \to y$. The vector $\overline{x} + y_k$ lies between $\overline{x}$ and z_k in the line segment connecting $\overline{x}$ and z_k for all k such that $\|z_k - \overline{x}\| \geq \|y\|$, so by the convexity of C, we have $\overline{x} + y_k \in C$ for all sufficiently large k. Since $\overline{x} + y_k \to \overline{x} + y$ and C is closed, it follows that $\overline{x} + y$ must belong to C.

(c) Assuming that C is unbounded, we will show that R_C contains a nonzero direction (the reverse implication is clear). Choose any $\overline{x} \in C$ and any unbounded sequence $\{z_k\} \subset C$. Consider the sequence $\{y_k\}$, where

$$y_k = \frac{z_k - \overline{x}}{\|z_k - \overline{x}\|},$$

and let y be a limit point of $\{y_k\}$ (compare with the construction of Fig. 1.5.2). For any fixed $\alpha \geq 0$, the vector $\overline{x} + \alpha y_k$ lies between $\overline{x}$ and z_k in the line segment connecting $\overline{x}$ and z_k for all k such that $\|z_k - \overline{x}\| \geq \alpha$. Hence by the convexity of C, we have $\overline{x} + \alpha y_k \in C$ for all sufficiently large k. Since $\overline{x} + \alpha y$ is a limit point of $\{\overline{x} + \alpha y_k\}$, and C is closed, we have $\overline{x} + \alpha y \in C$. Hence, using also part (b), it follows that the nonzero vector y is a direction of recession.

(d) If $y \in R_{\mathrm{ri}(C)}$, then for a fixed $x \in \mathrm{ri}(C)$ and all $\alpha \geq 0$, we have $x + \alpha y \in \mathrm{ri}(C) \subset C$. Hence, by part (b), we have $y \in R_C$. Conversely, if $y \in R_C$, for any $x \in \mathrm{ri}(C)$, we have $x + \alpha y \in C$ for all $\alpha \geq 0$. It follows from the Line Segment Principle [cf. Prop. 1.4.1(a)] that $x + \alpha y \in \mathrm{ri}(C)$ for all $\alpha \geq 0$, so that y belongs to $R_{\mathrm{ri}(C)}$.

(e) By the definition of direction of recession, $y \in R_{C \cap D}$ implies that $x + \alpha y \in C \cap D$ for all $x \in C \cap D$ and all $\alpha \geq 0$. By part (b), this in turn implies that $y \in R_C$ and $y \in R_D$, so that $R_{C \cap D} \subset R_C \cap R_D$. Conversely, by the definition of direction of recession, if $y \in R_C \cap R_D$ and $x \in C \cap D$, we have $x + \alpha y \in C \cap D$ for all $\alpha > 0$, so $y \in R_{C \cap D}$. Thus, $R_C \cap R_D \subset R_{C \cap D}$. The preceding argument can also be simply adapted to show that $R_{\cap_{i \in I} C_i} = \cap_{i \in I} R_{C_i}$.   **Q.E.D.**

It is essential to assume that the set C is closed in the above proposition. For an example where part (a) fails without this assumption, consider the set

$$C = \big\{ (x_1, x_2) \mid 0 < x_1,\, 0 < x_2 \big\} \cup \big\{ (0,0) \big\}.$$

Its recession cone is equal to C, which is not closed. For an example where parts (b)-(e) fail, consider the unbounded convex set

$$C = \big\{ (x_1, x_2) \mid 0 \leq x_1 < 1,\, 0 \leq x_2 \big\} \cup \big\{ (1,0) \big\}.$$

By using the definition of direction of recession, it can be verified that C has no nonzero directions of recession, so parts (b) and (c) of the proposition fail. It can also be verified that $(0,1)$ is a direction of recession of $\mathrm{ri}(C)$, so part (d) also fails. Finally, by letting

$$D = \big\{ (x_1, x_2) \mid -1 \leq x_1 \leq 0,\, 0 \leq x_2 \big\},$$

it can be seen that part (e) fails as well.

Note that part (e) of the preceding proposition implies that if C and D are nonempty closed and convex sets such that $C \subset D$, then $R_C \subset R_D$. This can be seen by using part (e) to write $R_C = R_{C \cap D} = R_C \cap R_D$, from which we obtain $R_C \subset R_D$. It is essential that the sets C and D be closed in order for this property to hold.

Note also that part (c) of the above proposition yields a characterization of compact and convex sets, namely that a closed convex set C is bounded if and only if $R_C = \{0\}$. The following is a useful generalization.

Proposition 1.5.2: Let C be a nonempty closed convex subset of $\Re^n$, let W be a nonempty convex compact subset of $\Re^m$, and let A be an $m \times n$ matrix. Consider the set

$$V = \{ x \in C \mid Ax \in W \},$$

and assume that it is nonempty. Then, V is closed and convex, and its recession cone is $R_C \cap N(A)$, where $N(A)$ is the nullspace of A. Furthermore, V is compact if and only if

$$R_C \cap N(A) = \{0\}.$$

Proof: We note that $V = C \cap \overline{V}$, where $\overline{V}$ is the set

$$\overline{V} = \{x \in \Re^n \mid Ax \in W\},$$

which is closed and convex since it is the inverse image of the closed convex set W under the continuous linear transformation A. Hence, V is closed and convex.

The recession cone of $\overline{V}$ is $N(A)$ [clearly $N(A) \subset R_{\overline{V}}$; if $y \notin N(A)$ but $y \in R_{\overline{V}}$, then for all $x \in \overline{V}$, we must have

$$Ax + \alpha Ay \in W, \qquad \forall\, \alpha > 0,$$

which contradicts the boundedness of W since $Ay \neq 0$]. Hence, since $V = C \cap \overline{V}$, V is nonempty, and the sets C and $\overline{V}$ are closed and convex, by Prop. 1.5.1(e), the recession cone of V is $R_C \cap N(A)$. Since V is closed and convex, by Prop. 1.5.1(c), it follows that V is compact if and only if $R_C \cap N(A) = \{0\}$.   **Q.E.D.**

Lineality Space

A subset of the recession cone of a convex set C that plays an important role in a number of interesting contexts is its *lineality space*, denoted by L_C. It is defined as the set of directions of recession y whose opposite, $-y$, are also directions of recession:

$$L_C = R_C \cap (-R_C).$$

Thus, if $y \in L_C$, then for every $x \in C$, the line $\{x + \alpha y \mid \alpha \in \Re\}$ is contained in C.

The lineality space inherits several of the properties of the recession cone that we have shown in the preceding two propositions. We collect these properties in the following proposition.

Proposition 1.5.3: Let C be a nonempty closed convex subset of $\Re^n$.

  (a) The lineality space of C is a subspace of $\Re^n$.

  (b) The lineality spaces of C and $\mathrm{ri}(C)$ are equal.

  (c) If D is another closed convex set such that $C \cap D \neq \emptyset$, we have

$$L_{C \cap D} = L_C \cap L_D.$$

More generally, for any collection of closed convex sets C_i, $i \in I$, where I is an arbitrary index set and $\cap_{i \in I} C_i$ is nonempty, we have

$$L_{\cap_{i \in I} C_i} = \cap_{i \in I} L_{C_i}.$$

(d) Let W be a convex and compact subset of $\Re^m$, and let A be an $m \times n$ matrix. If the set

$$V = \{x \in C \mid Ax \in W\}$$

is nonempty, it is closed and convex, and its lineality space is $L_C \cap N(A)$, where $N(A)$ is the nullspace of A.

Proof: (a) Let y_1 and y_2 belong to L_C, and let α_1 and α_2 be nonzero scalars. We will show that $\alpha_1 y_1 + \alpha_2 y_2$ belongs to L_C. Indeed, we have

$$\alpha_1 y_1 + \alpha_2 y_2 = |\alpha_1|\big(\mathrm{sgn}(\alpha_1)y_1\big) + |\alpha_2|\big(\mathrm{sgn}(\alpha_2)y_2\big) \tag{1.9}$$
$$= \big(|\alpha_1| + |\alpha_2|\big)\big(\alpha \bar{y}_1 + (1-\alpha)y_2\big),$$

where

$$\alpha = \frac{|\alpha_1|}{|\alpha_1| + |\alpha_2|}, \qquad \bar{y}_1 = \mathrm{sgn}(\alpha_1)y_1, \quad \bar{y}_2 = \mathrm{sgn}(\alpha_2)y_2,$$

and for a nonzero scalar s, we use the notation $\mathrm{sgn}(s) = 1$ or $\mathrm{sgn}(s) = -1$ depending on whether s is positive or negative, respectively. We now note that L_C is a convex cone, being the intersection of the convex cones R_C and $-R_C$. Hence, since $\bar{y}_1$ and $\bar{y}_2$ belong to L_C, any positive multiple of a convex combination of $\bar{y}_1$ and $\bar{y}_2$ belongs to L_C. It follows from Eq. (1.9) that $\alpha_1 y_1 + \alpha_2 y_2 \in L_C$.

(b) We have

$$L_{\mathrm{ri}(C)} = R_{\mathrm{ri}(C)} \cap \big(-R_{\mathrm{ri}(C)}\big) = R_C \cap (-R_C) = L_C,$$

where the second equality follows from Prop. 1.5.1(d).

(c) We have
$$L_{\cap_{i \in I} C_i} = \big(R_{\cap_{i \in I} C_i}\big) \cap \big(-R_{\cap_{i \in I} C_i}\big)$$
$$= (\cap_{i \in I} R_{C_i}) \cap (-\cap_{i \in I} R_{C_i})$$
$$= \cap_{i \in I}\big(R_{C_i} \cap (-R_{C_i})\big)$$
$$= \cap_{i \in I} L_{C_i},$$

where the second equality follows from Prop. 1.5.1(e).

(d) We have

$$L_V = R_V \cap (-R_V)$$
$$= \big(R_C \cap N(A)\big) \cap \big((-R_C) \cap N(A)\big)$$
$$= \big(R_C \cap (-R_C)\big) \cap N(A)$$
$$= L_C \cap N(A),$$

where the second equality follows from Prop. 1.5.2. **Q.E.D.**

Let us also prove a useful result that allows the decomposition of a convex set along a subspace of its lineality space (possibly the entire lineality space) and its orthogonal complement (see Fig. 1.5.3).

Proposition 1.5.4: (Decomposition of a Convex Set) Let C be a nonempty convex subset of $\Re^n$. Then, for every subspace S that is contained in the lineality space L_C, we have

$$C = S + (C \cap S^\perp).$$

Proof: We can decompose $\Re^n$ as the sum of the subspace S and its orthogonal complement $S^\perp$. Let $x \in C$, so that $x = y + z$ for some $y \in S$ and $z \in S^\perp$. Because $-y \in S$ and $S \subset L_C$, the vector $-y$ is a direction of recession of C, so the vector $x - y$, which is equal to z, belongs to C. Thus, $z \in C \cap S^\perp$, and we have $x = y + z$ with $y \in S$ and $z \in C \cap S^\perp$. This shows that $C \subset S + (C \cap S^\perp)$.

Conversely, if $x \in S + (C \cap S^\perp)$, then $x = y + z$ with $y \in S$ and $z \in C \cap S^\perp$. Thus, we have $z \in C$. Furthermore, because $S \subset L_C$, the vector y is a direction of recession of C, implying that $y + z \in C$. Hence $x \in C$, showing that $S + (C \cap S^\perp) \subset C$. **Q.E.D.**

1.5.1   Nonemptiness of Intersections of Closed Sets

The notions of recession cone and lineality space can be used to generalize some of the fundamental properties of compact sets to closed convex sets. One such property is that the intersection of a nested sequence of nonempty compact sets is nonempty and compact [cf. Prop. 1.1.6(h)]. Another property is that the image of a compact set under a linear transformation is compact [cf. Prop. 1.1.9(d)]. These properties fail for general closed sets, but it turns out that they hold under some assumptions involving convexity and directions of recession.

In what follows in this section, we will generalize the properties of compact sets just mentioned to closed convex sets. In subsequent chapters,

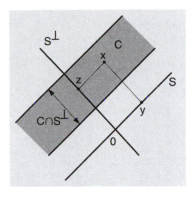

Figure 1.5.3. Illustration of the decomposition of a convex set C as

$$C = S + (C \cap S^\perp),$$

where S is a subspace contained in the lineality space L_C. A vector $x \in C$ is expressed as $x = y + z$ with $y \in S$ and $z \in C \cap S^\perp$, as shown.

we will translate these properties into important results relating to the existence of solutions of convex optimization problems, and to fundamental issues in minimax theory and duality theory. For a glimpse into this connection, note that the set of minimizing points of a function is equal to the intersection of its nonempty level sets, so the question of existence of a solution to an optimization problem reduces to a question of nonemptiness of a set intersection.

We consider a nested sequence $\{C_k\}$ of nonempty closed convex sets, and in the subsequent propositions, we will derive several alternative conditions under which the intersection $\cap_{k=0}^\infty C_k$ is nonempty. These conditions involve a variety of assumptions about the recession cones, the lineality spaces, and the structure of the sets C_k.

The following result makes no assumptions about the structure of the sets C_k, other than closedness and convexity.

Proposition 1.5.5: Let $\{C_k\}$ be a sequence of nonempty closed convex subsets of $\Re^n$ such that $C_{k+1} \subset C_k$ for all k. Let R_k and L_k be the recession cone and the lineality space of C_k, respectively, and let

$$R = \cap_{k=0}^\infty R_k, \qquad L = \cap_{k=0}^\infty L_k.$$

Assume that

$$R = L.$$

Then the intersection $\cap_{k=0}^\infty C_k$ is nonempty and has the form

$$\cap_{k=0}^\infty C_k = L + \tilde{C},$$

where $\tilde{C}$ is some nonempty and compact set.

Proof: Since the sets C_k are nested, the lineality spaces L_k are also nested [cf. Prop. 1.5.1(e)]. Since each L_k is a subspace, it follows that for all k sufficiently large, we have $L_k = L$. Thus, we may assume without loss of generality that

$$L_k = L, \qquad \forall \, k.$$

We next show by contradiction that for all sufficiently large k, we have $R_k \cap L^\perp = \{0\}$. Indeed, suppose that this is not so. Then since the R_k are nested [cf. Prop. 1.5.1(e)], for each k there exists some $y_k \in R_k \cap L^\perp$ such that $\|y_k\| = 1$. Hence the set $\{y \mid \|y\| = 1\} \cap R_k \cap L^\perp$ is nonempty, and since it is also compact, the intersection $\{y \mid \|y\| = 1\} \cap (\cap_{k=0}^\infty R_k) \cap L^\perp$ is nonempty. This intersection is equal to $\{y \mid \|y\| = 1\} \cap L \cap L^\perp$, since, by hypothesis, we have $\cap_{k=0}^\infty R_k = R = L$. But this is a contradiction since $L \cap L^\perp = \{0\}$. We may thus assume without loss of generality that

$$R_k \cap L^\perp = \{0\}, \qquad \forall \, k.$$

By the Recession Cone Theorem [Prop. 1.5.1(e)], for each k, the recession cone of $C_k \cap L^\perp$ is given by

$$R_{C_k \cap L^\perp} = R_k \cap R_{L^\perp},$$

and since $R_{L^\perp} = L^\perp$ and $R_k \cap L^\perp = \{0\}$, it follows that

$$R_{C_k \cap L^\perp} = \{0\}, \qquad \forall \, k.$$

Hence, by the Recession Cone Theorem [Prop. 1.5.1(c)], the sets $C_k \cap L^\perp$ are compact, as well as nested, so that their intersection

$$\tilde{C} = \cap_{k=0}^\infty (C_k \cap L^\perp) \tag{1.10}$$

is nonempty and compact, which implies in particular that the intersection $\cap_{k=0}^\infty C_k$ is nonempty. Furthermore, since L is the lineality space of all the sets C_k, it is also the lineality space of $\cap_{k=0}^\infty C_k$ [cf. Prop. 1.5.3(c)]. Therefore, by using the decomposition property of Prop. 1.5.4, we have

$$\cap_{k=0}^\infty C_k = L + \left(\cap_{k=0}^\infty C_k\right) \cap L^\perp,$$

implying, by Eq. (1.10), that $\cap_{k=0}^\infty C_k = L + \tilde{C}$, as required. **Q.E.D.**

Note that in the special case where $\cap_{k=0}^\infty R_k = \{0\}$, the preceding proposition shows that the intersection $\cap_{k=0}^\infty C_k$ is nonempty and compact. In fact, the proof of the proposition shows that the set C_k is compact for all sufficiently large k.

In the following two propositions, we consider the intersection of sets that are defined, at least in part, in terms of linear and/or quadratic inequalities.

Proposition 1.5.6: Let $\{C_k\}$ be a sequence of closed convex subsets of $\Re^n$, and let X be a subset of $\Re^n$ specified by linear inequality constraints, i.e.,

$$X = \{x \mid a_j'x \le b_j, \, j = 1, \ldots, r\}, \qquad (1.11)$$

where a_j are vectors in $\Re^n$ and b_j are scalars. Assume that:

(1) $C_{k+1} \subset C_k$ for all k.

(2) The intersection $X \cap C_k$ is nonempty for all k.

(3) We have

$$R_X \cap R \subset L,$$

where R_X is the recession cone of X, and

$$R = \cap_{k=0}^\infty R_k, \qquad L = \cap_{k=0}^\infty L_k,$$

with R_k and L_k denoting the recession cone and the lineality space of C_k, respectively.

Then the intersection $X \cap \left(\cap_{k=0}^\infty C_k\right)$ is nonempty.

Proof: We use induction on the dimension of the set X. Suppose that the dimension of X is 0. Then, X consists of a single point. By assumption (2), this point belongs to $X \cap C_k$ for all k, and hence belongs to the intersection $X \cap \left(\cap_{k=0}^\infty C_k\right)$.

Assume that, for some $l < n$, the intersection $\overline{X} \cap \left(\cap_{k=0}^\infty C_k\right)$ is nonempty for every set $\overline{X}$ of dimension less than or equal to l that is specified by linear inequality constraints, and is such that $\overline{X} \cap C_k$ is nonempty for all k and $R_{\overline{X}} \cap R \subset L$. Let X be of the form (1.11), be such that $X \cap C_k$ is nonempty for all k, satisfy $R_X \cap R \subset L$, and have dimension $l + 1$. We will show that the intersection $X \cap \left(\cap_{k=0}^\infty C_k\right)$ is nonempty.

If $L_X \cap L = R_X \cap R$, then by Prop. 1.5.5 applied to the sets $X \cap C_k$, we have that $X \cap \left(\cap_{k=0}^\infty C_k\right)$ is nonempty, and we are done. We may thus assume that $L_X \cap L \ne R_X \cap R$.

Since we always have $L_X \cap L \subset R_X \cap R$, from the assumption $R_X \cap R \subset L$ it follows that there exists a nonzero direction $\overline{y} \in R_X \cap R$ such that $\overline{y} \notin L_X$, i.e.,

$$\overline{y} \in R_X, \qquad -\overline{y} \notin R_X, \qquad \overline{y} \in L.$$

Using Prop. 1.5.1(e), it is seen that the recession cone of X is

$$R_X = \{y \mid a_j'y \le 0, \, j = 1, \ldots, r\},$$

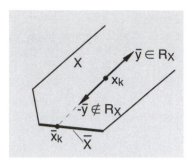

Figure 1.5.4. Construction used in the induction proof of Prop. 1.5.6.

so the fact $\overline{y} \in R_X$ implies that

$$a_j'\overline{y} \le 0, \qquad \forall\, j = 1, \dots, r,$$

while the fact $-\overline{y} \notin R_X$ implies that the index set

$$J = \{j \mid a_j'\overline{y} < 0\}$$

is nonempty.

By assumption (2), we may select a sequence $\{x_k\}$ such that

$$x_k \in X \cap C_k, \qquad \forall\, k.$$

We then have

$$a_j'x_k \le b_j, \qquad \forall\, j = 1, \dots, r, \quad \forall\, k.$$

We may assume that

$$a_j'x_k < b_j, \qquad \forall\, j \in J, \quad \forall\, k;$$

otherwise we can replace x_k with $x_k + \overline{y}$, which belongs to $X \cap C_k$ (since $\overline{y} \in R_X$ and $\overline{y} \in L$).

Suppose that for each k, we start at x_k and move along $-\overline{y}$ as far as possible without leaving the set X, up to the point where we encounter the vector

$$\overline{x}_k = x_k - \beta_k \overline{y},$$

where β_k is the positive scalar given by

$$\beta_k = \min_{j \in J} \frac{a_j'x_k - b_j}{a_j'\overline{y}}$$

(see Fig. 1.5.4). Since $a_j'\overline{y} = 0$ for all $j \notin J$, we have $a_j'\overline{x}_k = a_j'x_k$ for all $j \notin J$, so the number of linear inequalities of X that are satisfied by $\overline{x}_k$ as equalities is strictly larger than the number of those satisfied by x_k. Thus, there exists $j_0 \in J$ such that $a_{j_0}'\overline{x}_k = b_{j_0}$ for all k in an infinite index set $\mathcal{K} \subset \{0, 1, \dots\}$. By reordering the linear inequalities if necessary, we can assume that $j_0 = 1$, i.e.,

$$a_1'\overline{x}_k = b_1, \qquad a_1'x_k < b_1, \qquad \forall\, k \in \mathcal{K}.$$

To apply the induction hypothesis, consider the set

$$\overline{X} = \{x \mid a_1'x = b_1, \; a_j'x \leq b_j, \; j = 2,\ldots,r\},$$

and note that $\{\overline{x}_k\}_\mathcal{K} \subset \overline{X}$. Since $\overline{x}_k = x_k - \beta_k \overline{y}$ with $x_k \in C_k$ and $\overline{y} \in L$, we have $\overline{x}_k \in C_k$ for all k, implying that $\overline{x}_k \in \overline{X} \cap C_k$ for all $k \in \mathcal{K}$. Thus, $\overline{X} \cap C_k \neq \emptyset$ for all k. Because the sets C_k are nested, so are the sets $\overline{X} \cap C_k$. Furthermore, the recession cone of $\overline{X}$ is

$$R_{\overline{X}} = \{y \mid a_1'y = 0, \; a_j'y \leq 0, \; j = 2,\ldots,r\},$$

which is contained in R_X, so that

$$R_{\overline{X}} \cap R \subset R_X \cap R \subset L.$$

Finally, to show that the dimension of $\overline{X}$ is smaller than the dimension of X, note that the set $\{x \mid a_1'x = b_1\}$ contains $\overline{X}$, so that a_1 is orthogonal to the subspace $S_{\overline{X}}$ that is parallel to $\text{aff}(\overline{X})$. Since $a_1'\overline{y} < 0$, it follows that $\overline{y} \notin S_{\overline{X}}$. On the other hand, $\overline{y}$ belongs to S_X, the subspace that is parallel to $\text{aff}(X)$, since for all k, we have $x_k \in X$ and $x_k - \beta_k \overline{y} \in X$.

Based on the preceding, we can use the induction hypothesis to assert that the intersection $\overline{X} \cap \left(\cap_{k=0}^{\infty} C_k\right)$ is nonempty. Since $\overline{X} \subset X$, it follows that $X \cap \left(\cap_{k=0}^{\infty} C_k\right)$ is nonempty.   **Q.E.D.**

Figure 1.5.5 illustrates the need for the assumptions of the preceding proposition.

Proposition 1.5.7: Let $\{C_k\}$ be a sequence of subsets of $\Re^n$ given by

$$C_k = \{x \mid x'Qx + a'x + b \leq w_k\},$$

where Q is a symmetric positive semidefinite $n \times n$ matrix, a is a vector in $\Re^n$, b is a scalar, and $\{w_k\}$ is a nonincreasing scalar sequence that converges to 0. Let also X be a subset of $\Re^n$ of the form

$$X = \{x \mid x'Q_jx + a_j'x + b_j \leq 0, \; j = 1,\ldots,r\}, \qquad (1.12)$$

where Q_j are symmetric positive semidefinite $n \times n$ matrices, a_j are vectors in $\Re^n$, and b_j are scalars. Assume further that $X \cap C_k$ is nonempty for all k. Then, the intersection $X \cap \left(\cap_{k=0}^{\infty} C_k\right)$ is nonempty.

Proof: We note that X and all the sets C_k are closed, and that by the positive semidefiniteness of Q and Q_j, the set X and all the sets C_k are

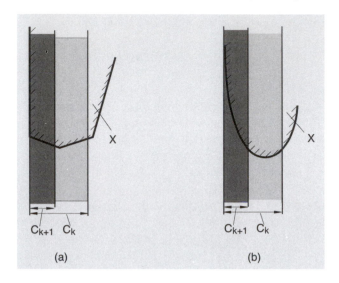

Figure 1.5.5. Illustration of the issues regarding the nonemptiness of the intersection $X \cap \left(\cap_{k=0}^{\infty} C_k\right)$ in Prop. 1.5.6, under the assumption $R_X \cap R \subset L$. Here the intersection $\cap_{k=0}^{\infty} C_k$ is equal to the left vertical line. In the figure on the left, X is specified by linear constraints and the intersection $X \cap \left(\cap_{k=0}^{\infty} C_k\right)$ is nonempty. In the figure on the right, X is specified by a nonlinear constraint, and the intersection $X \cap \left(\cap_{k=0}^{\infty} C_k\right)$ is empty.

convex [cf. Prop. 1.2.6(a)]. Furthermore, all the sets C_k have the same recession cone R and the same lineality space L, which are given by

$$R = \{y \mid Qy = 0, \ a'y \leq 0\}, \qquad L = \{y \mid Qy = 0, \ a'y = 0\}.$$

[To see this, note that $y \in R$ if and only if $x + \alpha y \in R$ for all $x \in C_k$ and $\alpha > 0$, or equivalently, $(x + \alpha y)'Q(x + \alpha y) + a'(x + \alpha y) + b \leq w_k$, i.e.,

$$x'Qx + a'x + \alpha(2x'Qy + a'y) + \alpha^2 y'Qy + b \leq w_k, \qquad \forall \ \alpha \geq 0, \ \forall \ x \in C_k.$$

Since Q is positive semidefinite, this relation implies that $y'Qy = 0$ and that y is in the nullspace of Q, so that we must also have $a'y \leq 0$. Conversely, if $y'Qy = 0$ and $a'y \leq 0$, the above relation holds and $y \in R$.] Similarly, the recession cone of X is

$$R_X = \{y \mid Q_j y = 0, \ a_j' y \leq 0, \ j = 1, \ldots, r\}.$$

We will prove the result by induction on the number r of quadratic functions that define X. For $r = 0$, we have $X \cap C_k = C_k$ for all k, and by our assumption that $X \cap C_k \neq \emptyset$, we have $C_k \neq \emptyset$ for all k. Furthermore, since $\{w_k\}$ is nonincreasing, the sets C_k are nested. If $R = L$, then by Prop.

1.5.5, it follows that $\cap_{k=0}^{\infty} C_k$ is nonempty and we are done, so assume that $R \neq L$. Since we always have $L \subset R$, there exists a nonzero vector $\overline{y}$ such that $\overline{y} \in R$ while $-\overline{y} \notin R$, i.e.,

$$Q\overline{y} = 0, \qquad c'\overline{y} < 0.$$

Consider a point of the form $x + \alpha\overline{y}$ for some $x \in \Re^n$ and $\alpha > 0$. We have

$$(x + \alpha\overline{y})'Q(x + \alpha\overline{y}) + a'(x + \alpha\overline{y}) = x'Qx + a'x + \alpha c'\overline{y}.$$

Since $c'\overline{y} < 0$, we can choose $\alpha > 0$ so that

$$x'Qx + a'x + \alpha a'\overline{y} + b < 0,$$

and because $\{w_k\}$ is nonincreasing, it follows that $x + \alpha\overline{y} \in \cap_{k=0}^{\infty} C_k$.

Assume that $\overline{X} \cap \left(\cap_{k=0}^{\infty} C_k\right)$ is nonempty whenever $\overline{X}$ is given by Eq. (1.12) in terms of at most r convex quadratic functions and is such that $\overline{X} \cap C_k$ is nonempty for all k. Suppose that the set X is given by Eq. (1.12) in terms of $r+1$ convex quadratic functions. If $R_X \cap R = L_X \cap L$, by using Prop. 1.5.5, we see that $X \cap \left(\cap_{k=0}^{\infty} C_k\right) \neq \varnothing$ and we are done, so assume that there exists a direction $\overline{y}$ such that $\overline{y} \in R_X \cap R$ but $-\overline{y} \notin R_X \cap R$. For any $x \in X$, and any $\overline{y} \in R_X \cap R$ and $\alpha > 0$, we have

$$x + \alpha\overline{y} \in X,$$

$$(x + \alpha\overline{y})'Q(x + \alpha\overline{y}) + a'(x + \alpha\overline{y}) = x'Qx + a'x + \alpha a'\overline{y}.$$

If $-\overline{y} \notin R$, i.e., $c'\overline{y} < 0$, it follows that for some sufficiently large α,

$$x'Qx + a'x + \alpha a'\overline{y} + b \leq 0,$$

implying that $x + \alpha\overline{y} \in \cap_{k=0}^{\infty} C_k$. Since $x + \alpha\overline{y} \in X$ for all α, it follows that $x + \alpha\overline{y} \in X \cap \left(\cap_{k=0}^{\infty} C_k\right)$. Thus, if $\overline{y} \in R_X \cap R$ but $-\overline{y} \notin R$, then $X \cap \left(\cap_{k=0}^{\infty} C_k\right)$ is nonempty.

Assume now that $\overline{y} \in R_X \cap R$ and $-\overline{y} \notin R_X$. Then $Q_j\overline{y} = 0$ and $a_j'\overline{y} \leq 0$ for all j, while $a_j'\overline{y} < 0$ for at least one j. For convenience, let us reorder the inequalities of X so that

$$Q_j\overline{y} = 0, \qquad a_j'\overline{y} = 0, \qquad \forall\, j = 1, \ldots, \overline{r}, \tag{1.13}$$

$$Q_j\overline{y} = 0, \qquad a_j'\overline{y} < 0, \qquad \forall\, j = \overline{r}+1, \ldots, r+1, \tag{1.14}$$

where $\overline{r}$ is an integer with $0 \leq \overline{r} < r+1$.

Consider now the set

$$\overline{X} = \left\{ x \mid x'Q_j x + a_j'x + b_j \leq 0,\ j = 1, \ldots, \overline{r} \right\},$$

where $\overline{X} = \Re^n$ if $\overline{r} = 0$. Since $X \subset \overline{X}$ and $X \cap C_k \neq \emptyset$ for all k, the set $\overline{X} \cap C_k$ is nonempty for all k. Thus, by the induction hypothesis, it follows that the intersection

$$\overline{X} \cap \left(\cap_{k=0}^{\infty} C_k \right)$$

is nonempty. Let $\tilde{x}$ be a point in this set. Since $\overline{x} \in \cap_{k=0}^{\infty} C_k$ and $\overline{y} \in R$, it follows that for all $\alpha \geq 0$,

$$\overline{x} + \alpha \overline{y} \in \cap_{k=0}^{\infty} C_k.$$

Furthermore, since $\overline{x} \in \overline{X}$, by Eq. (1.13), we have that for all $\alpha \geq 0$,

$$(\overline{x} + \alpha \overline{y})' Q_j (\overline{x} + \alpha \overline{y}) + a_j' (\overline{x} + \alpha \overline{y}) + b_j \leq 0, \qquad \forall \, j = 1, \ldots, \overline{r}.$$

Finally, in view of Eq. (1.14), we can choose a sufficiently large $\overline{\alpha} > 0$ so that

$$(\overline{x} + \overline{\alpha} \, \overline{y})' Q_j (\overline{x} + \overline{\alpha} \, \overline{y}) + a_j' (\overline{x} + \overline{\alpha} \, \overline{y}) + b_j \leq 0, \qquad \forall \, j = \overline{r}+1, \ldots, r+1.$$

The preceding three relations imply that $\overline{x} + \overline{\alpha} \, \overline{y} \in X \cap \left(\cap_{k=0}^{\infty} C_k \right)$, showing that $X \cap \left(\cap_{k=0}^{\infty} C_k \right)$ is nonempty.   **Q.E.D.**

Note that it is essential to require that $\{w_k\}$ is convergent in Prop. 1.5.7. As an example, consider the subsets of $\Re^2$ given by

$$X = \left\{ (x_1, x_2) \mid x_1^2 \leq x_2 \right\}, \qquad C_k = \left\{ (x_1, x_2) \mid x_1 \leq -k \right\}, \quad k = 0, 1, \ldots$$

Then all the assumptions of Prop. 1.5.7 are satisfied, except that the right-hand side, $-k$, of the quadratic inequality that defines C_k does not converge to a scalar. It can be seen that the intersection $X \cap \left(\cap_{k=0}^{\infty} C_k \right)$ is empty, since $\cap_{k=0}^{\infty} C_k$ is empty.

1.5.2   Closedness Under Linear Transformations

The conditions just obtained regarding the nonemptiness of the intersection of a sequence of closed convex sets can be translated to conditions guaranteeing the closedness of the image, AC, of a closed convex set C under a linear transformation A. This is the subject of the following proposition.

Proposition 1.5.8: Let C be a nonempty closed convex subset of $\Re^n$, and let A be an $m \times n$ matrix with nullspace denoted by $N(A)$.

(a) If $R_C \cap N(A) \subset L_C$, then the set AC is closed.

(b) Let X be a nonempty subset of $\Re^n$ specified by linear inequality constraints, i.e.,

$$X = \{x \mid a_j'x \leq b_j, \, j = 1, \ldots, r\},$$

where a_j are vectors in $\Re^n$ and b_j are scalars. If

$$R_X \cap R_C \cap N(A) \subset L_C,$$

then the set $A(X \cap C)$ is closed.

(c) Let C be specified by convex quadratic inequalities, i.e.,

$$C = \{x \mid x'Q_j x + a_j'x + b_j \leq 0, \, j = 1, \ldots, r\},$$

where Q_j are symmetric positive semidefinite $n \times n$ matrices, a_j are vectors in $\Re^n$, and b_j are scalars. Then the set AC is closed.

Proof: (a) Let $\{y_k\}$ be a sequence of points in AC converging to some $\overline{y} \in \Re^n$. We will prove that AC is closed by showing that $\overline{y} \in AC$.

We introduce the sets

$$W_k = \{z \mid \|z - \overline{y}\| \leq \|y_k - \overline{y}\|\},$$

and

$$C_k = \{x \in C \mid Ax \in W_k\}$$

(see Fig. 1.5.6). To show that $\overline{y} \in AC$, it is sufficient to prove that the intersection $\cap_{k=0}^\infty C_k$ is nonempty, since every $\overline{x} \in \cap_{k=0}^\infty C_k$ satisfies $\overline{x} \in C$ and $A\overline{x} = \overline{y}$ (because $y_k \to \overline{y}$). To show that $\cap_{k=0}^\infty C_k$ is nonempty, we will use Prop. 1.5.5.

Each set C_k is nonempty (since $y_k \in AC$ and $y_k \in W_k$), and it is convex and closed by Prop. 1.5.2. By taking an appropriate subsequence if necessary, we may assume that the sets C_k are nested. It can be seen that all C_k have the same recession cone, denoted by R, and the same lineality space, denoted by L, which by Props. 1.5.2 and 1.5.3(d), are given by

$$R = R_C \cap N(A), \qquad L = L_C \cap N(A). \tag{1.15}$$

Since $R_C \cap N(A) \subset L_C$, we have $R_C \cap N(A) \subset L_C \cap N(A)$, and in view of the relation $L_C \cap N(A) \subset R_C \cap N(A)$, which always holds, it follows that

$$R_C \cap N(A) = L_C \cap N(A).$$

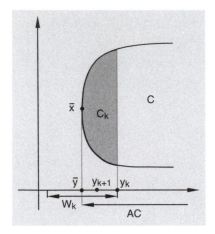

Figure 1.5.6. Construction used in the proof of Prop. 1.5.8(a). Here A is the projection on the horizontal axis of points in the plane.

This relation and Eq. (1.15) imply that $R = L$, so that by Prop. 1.5.5, the intersection $\cap_{k=0}^{\infty} C_k$ is nonempty.

(b) We use a similar argument to the one used for part (a), except that we assume that $\{y_k\} \subset A(X \cap C)$, and we use Prop. 1.5.6 to show that $X \cap \left(\cap_{k=0}^{\infty} C_k \right)$ is nonempty.

Let W_k and C_k be defined as in part (a). By our choice of $\{y_k\}$, the sets C_k are nested, so that assumption (1) of Prop. 1.5.6 is satisfied. Since $y_k \in A(X \cap C)$ and $y_k \in W_k$, it follows that $X \cap C_k$ is nonempty for all k. Thus assumption (2) of Prop. 1.5.6 is also satisfied.

Since $R_X \cap R_C \cap N(A) \subset L_C$, we also have

$$R_X \cap R_C \cap N(A) \subset L_C \cap N(A).$$

In view of this relation and Eq. (1.15), it follows that $R_X \cap R \subset L$, thus implying that assumption (3) of Prop. 1.5.6 is satisfied. Therefore, by applying Prop. 1.5.6 to the sets $X \cap C_k$, we see that the intersection $X \cap \left(\cap_{k=0}^{\infty} C_k \right)$ is nonempty. Every point in this intersection is such that $x \in X$ and $x \in C$ with $Ax = \overline{y}$, showing that $\overline{y} \in A(X \cap C)$.

(c) Similar to part (a), we let $\{y_k\}$ be a sequence in AC converging to some $\overline{y} \in \Re^n$. We will show that $\overline{y} \in AC$. We let

$$C_k = \left\{ x \mid \|Ax - \overline{y}\|^2 \leq \|y_k - \overline{y}\|^2 \right\},$$

or equivalently

$$C_k = \left\{ x \mid x'A'Ax - 2(A'\overline{y})'x + \|\overline{y}\|^2 \leq \|y_k - \overline{y}\|^2 \right\}.$$

Thus, C_k has the form given in Prop. 1.5.7, with

$$Q = A'A, \qquad a = -2A'\overline{y}, \qquad b = \|\overline{y}\|^2, \qquad w_k = \|y_k - \overline{y}\|^2,$$

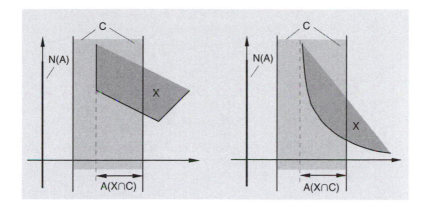

Figure 1.5.7. Illustration of the need to assume that the set X is specified by linear inequalities in Prop. 1.5.8(b). In both examples shown, the matrix A is the projection onto the horizontal axis, and its nullspace is the vertical axis. The condition $R_X \cap R_C \cap N(A) \subset L_C$ is satisfied. However, in the example on the right, X is not specified by linear inequalities, and the set $A(X \cap C)$ is not closed.

and $w_k \to 0$. By applying Prop. 1.5.7, with $X = C$, we see that the intersection $X \cap \left(\cap_{k=0}^{\infty} C_k \right)$ is nonempty. For any x in this intersection, we have $x \in C$ and $Ax = \overline{y}$ (since $y_k \to \overline{y}$), showing that $y \in AC$, and implying that AC is closed.   **Q.E.D.**

Figure 1.5.7 illustrates the need for the assumptions of part (b) of the preceding proposition. Part (a) of the proposition implies that if

$$R_C \cap N(A) = \{0\},$$

i.e., there is no nonzero direction of recession of C that lies in the nullspace of A, then AC is closed. This fact can be applied to obtain conditions that guarantee the closedness of the vector sum of closed convex sets. The idea is that the vector sum of a finite number of sets can be viewed as the image of their Cartesian product under a special type of linear transformation, as can be seen from the proof of the following proposition.

Proposition 1.5.9: Let $C_1, \ldots, C_m$ be nonempty closed convex subsets of $\Re^n$ such that the equality $y_1 + \cdots + y_m = 0$ for some vectors $y_i \in R_{C_i}$ implies that $y_i = 0$ for all $i = 1, \ldots, m$. Then the vector sum $C_1 + \cdots + C_m$ is a closed set.

Proof: Let C be the Cartesian product $C_1 \times \cdots \times C_m$ viewed as a subset of $\Re^{mn}$, and let A be the linear transformation that maps a vector $(x_1, \ldots, x_m) \in \Re^{mn}$ into $x_1 + \cdots + x_m$. Then it can be verified that C is closed and convex [see Exercise 1.37(a)]. We have

$$R_C = R_{C_1} \times \cdots \times R_{C_m}$$

[see Exercise 1.37(c)] and

$$N(A) = \big\{ (y_1, \ldots, y_m) \mid y_1 + \cdots + y_m = 0, \ y_i \in \Re^n \big\},$$

so under the given condition, we obtain $R_C \cap N(A) = \{0\}$. Since $AC = C_1 + \cdots + C_m$, the result follows from Prop. 1.5.8(a).   **Q.E.D.**

When specialized to just two sets, the above proposition implies that if C_1 and $-C_2$ are closed convex sets, then $C_1 - C_2$ is closed if there is no common nonzero direction of recession of C_1 and C_2, i.e.

$$R_{C_1} \cap R_{C_2} = \{0\}.$$

This result can be generalized by using Prop. 1.5.8(a). In particular, by using the argument of Prop. 1.5.9, we can show that if C_1 and C_2 are closed convex sets, the set $C_1 - C_2$ is closed if

$$R_{C_1} \cap R_{C_2} = L_{C_1} \cap L_{C_2}$$

(see Exercise 1.43).

Some other conditions asserting the closedness of vector sums can be derived from Prop. 1.5.8. For example, by applying Prop. 1.5.8(b), we can show that if X is specified by linear inequality constraints, and C is a closed convex set, then $X + C$ is closed if every direction of recession of X whose opposite is a direction of recession of C lies also in the lineality space of C. Furthermore, if the sets $C_1, \ldots, C_m$ are specified by convex quadratic inequalities as in Prop. 1.5.8(c), then, similar to Prop. 1.5.9, we can show that the vector sum $C_1 + \cdots + C_m$ is closed (see Exercise 1.44).

As an illustration of the need for the assumptions of Prop. 1.5.9, consider the example of Fig. 1.4.4, where C_1 and C_2 are closed sets, but $C_1 + C_2$ is not closed. In this example, the set C_1 has a nonzero direction of recession, which is the opposite of a direction of recession of C_2.

1.6 NOTES, SOURCES, AND EXERCISES

Among early classical works on convexity, we mention Caratheodory [Car11], Minkowski [Min11], and Steinitz [Ste13], [Ste14], [Ste16]. In particular, Caratheodory gave the theorem on convex hulls that carries his name, while Steinitz developed the theory of relative interiors and recession cones. Minkowski is credited with initiating the theory of hyperplane separation of convex sets and the theory of support functions (a precursor to conjugate convex functions). Furthermore, Minkowski and Farkas (whose work, published in Hungarian, spans a 30-year period starting around 1894), are credited with laying the foundations of polyhedral convexity.

The work of Fenchel was instrumental in launching the modern era of convex analysis, when the subject came to a sharp focus thanks to its rich applications in optimization and game theory. In his 1951 lecture notes [Fen51], Fenchel laid the foundations of convex duality theory, and together with related works by von Neumann [Neu28], [Neu37] on saddle points and game theory, and Kuhn and Tucker on nonlinear programming [KuT51], inspired much subsequent work on convexity and its connections with optimization. Furthermore, Fenchel introduced several of the topics that are fundamental in our exposition, such as the theory of subdifferentiability and the theory of conjugate convex functions.

There are several books that relate to both convex analysis and optimization. The book by Rockafellar [Roc70], widely viewed as the classic convex analysis text, contains a detailed development of convexity and convex optimization (it does not cross over into nonconvex optimization). The book by Rockafellar and Wets [RoW98] is an extensive treatment of "variational analysis," a broad spectrum of topics that integrate classical analysis, convexity, and optimization of both convex and nonconvex (possibly nonsmooth) functions. The normal cone, introduced by Mordukhovich [Mor76] and discussed in Chapter 4, and the work of Clarke on nonsmooth analysis [Cla83] play a central role in this subject.

Among other books with detailed accounts of convexity and optimization, Stoer and Witzgall [StW70] discuss similar topics as Rockafellar [Roc70] but less comprehensively. Ekeland and Temam [EkT76] develop the subject in infinite dimensional spaces. Hiriart-Urruty and Lemarechal [HiL93] emphasize algorithms for dual and nondifferentiable optimization. Rockafellar [Roc84] focuses on convexity and duality in network optimization, and an important generalization, called monotropic programming. Bertsekas [Ber98] also gives a detailed coverage of this material, which owes much to the early work of Minty [Min60] on network optimization. Bonnans and Shapiro [BoS00] emphasize sensitivity analysis and discuss infinite dimensional problems as well. Borwein and Lewis [BoL00] develop many of the concepts in Rockafellar and Wets [RoW98], but more succinctly. Schrijver [Sch86] provides an extensive account of polyhedral convexity with applications to integer programming and combinatorial optimization, and gives many historical references. Ben-Tal and Nemirovski [BeN01] focus on conic and semidefinite programming. Auslender and Teboulle [AuT03] emphasize the question of existence of solutions for convex as well as nonconvex optimization problems, and associated issues in duality theory and variational inequalities. Finally, let us note a few books that focus primarily on the geometry and other properties of convex sets, but have limited connection with duality, game theory, and optimization: Bonnesen and

Fenchel [BoF34], Eggleston [Egg58], Grunbaum [Gru67], Klee [Kle63], and Valentine [Val64].

The development of this chapter mostly follows well-established lines. The only exception are the conditions guaranteeing the nonemptiness of a closed set intersection and the closedness of the image of a closed convex set under a linear transformation (Props. 1.5.5-1.5.8). The associated line of analysis, together with its use in minimax theory and duality theory in subsequent chapters (Sections 2.3, 2.6, and 6.5), have not received much attention, and are largely new (Nedić and Bertsekas [BeN02]). Our Props. 1.5.5 and 1.5.6 are new in the level of generality given here. Our Prop. 1.5.7 is due to Luo (see Luo and Zhang [LuZ99]). A generalization of this result to nonquadratic functions is given in Exercise 2.7 of Chapter 2 (see the material on bidirectionally flat functions). Our Prop. 1.5.6 may be derived from a special form of Helly's Theorem (Th. 21.5 in Rockafellar [Roc70], which deals with the intersection of a possibly uncountable family of sets; see also Rockafellar [Roc65]). Our induction proof of Prop. 1.5.6 is more elementary, and relies on our assumption that the family of sets is countable, which is sufficient for the analyses of this book. We note that the history and range of applications of Helly's Theorem are discussed, among others, by Danzer, Grunbaum, and Klee [DGK63], and Valentine [Val63], [Val64]. The use of recession cones in proving refined versions of Helly's Theorem and closedness of images of sets under a linear transformation was first studied by Fenchel [Fen51].

EXERCISES

1.1

Let C be a nonempty subset of $\Re^n$, and let λ_1 and λ_2 be positive scalars. Show that if C is convex, then $(\lambda_1 + \lambda_2)C = \lambda_1 C + \lambda_2 C$ [cf. Prop. 1.2.1(c)]. Show by example that this need not be true when C is not convex.

1.2 (Properties of Cones)

Show that:

  (a) The intersection $\cap_{i \in I} C_i$ of a collection $\{C_i \mid i \in I\}$ of cones is a cone.

  (b) The Cartesian product $C_1 \times C_2$ of two cones C_1 and C_2 is a cone.

  (c) The vector sum $C_1 + C_2$ of two cones C_1 and C_2 is a cone.

  (d) The closure of a cone is a cone.

(e) The image and the inverse image of a cone under a linear transformation is a cone.

1.3 (Lower Semicontinuity under Composition)

(a) Let $f : \Re^n \mapsto \Re^m$ be a continuous function and $g : \Re^m \mapsto \Re$ be a lower semicontinuous function. Show that the function h defined by $h(x) = g\big(f(x)\big)$ is lower semicontinuous.

(b) Let $f : \Re^n \mapsto \Re$ be a lower semicontinuous function, and $g : \Re \mapsto \Re$ be a lower semicontinuous and monotonically nondecreasing function. Show that the function h defined by $h(x) = g\big(f(x)\big)$ is lower semicontinuous. Give an example showing that the monotonic nondecrease assumption is essential.

1.4 (Convexity under Composition)

Let C be a nonempty convex subset of $\Re^n$

(a) Let $f : C \mapsto \Re$ be a convex function, and $g : \Re \mapsto \Re$ be a function that is convex and monotonically nondecreasing over a convex set that contains the set of values that f can take, $\big\{ f(x) \mid x \in C \big\}$. Show that the function h defined by $h(x) = g\big(f(x)\big)$ is convex over C. In addition, if g is monotonically increasing and f is strictly convex, then h is strictly convex.

(b) Let $f = (f_1, \ldots, f_m)$, where each $f_i : C \mapsto \Re$ is a convex function, and let $g : \Re^m \mapsto \Re$ be a function that is convex and monotonically nondecreasing over a convex set that contains the set $\big\{ f(x) \mid x \in C \big\}$, in the sense that for all $u, \bar{u}$ in this set such that $u \leq \bar{u}$, we have $g(u) \leq g(\bar{u})$. Show that the function h defined by $h(x) = g\big(f(x)\big)$ is convex over $C \times \cdots \times C$.

1.5 (Examples of Convex Functions)

Show that the following functions from $\Re^n$ to $(-\infty, \infty]$ are convex:

(a)
$$f_1(x_1, \ldots, x_n) = \begin{cases} -(x_1 x_2 \cdots x_n)^{\frac{1}{n}} & \text{if } x_1 > 0, \ldots, x_n > 0, \\ \infty & \text{otherwise.} \end{cases}$$

(b) $f_2(x) = \ln\big(e^{x_1} + \cdots + e^{x_n}\big)$.

(c) $f_3(x) = \|x\|^p$ with $p \geq 1$.

(d) $f_4(x) = \frac{1}{f(x)}$, where f is concave and $f(x)$ is a positive number for all x.

(e) $f_5(x) = \alpha f(x) + \beta$, where $f : \Re^n \mapsto \Re$ is a convex function, and α and β are scalars such that $\alpha \geq 0$.

(f) $f_6(x) = e^{\beta x' A x}$, where A is a positive semidefinite symmetric $n \times n$ matrix and β is a positive scalar.

(g) $f_7(x) = f(Ax + b)$, where $f : \Re^m \mapsto \Re$ is a convex function, A is an $m \times n$ matrix, and b is a vector in $\Re^m$.

1.6 (Ascent/Descent Behavior of a Convex Function)

Let $f : \Re \mapsto \Re$ be a convex function.

(a) *(Monotropic Property)* Use the definition of convexity to show that f is "turning upwards" in the sense that if x_1, x_2, x_3 are three scalars such that $x_1 < x_2 < x_3$, then

$$\frac{f(x_2) - f(x_1)}{x_2 - x_1} \leq \frac{f(x_3) - f(x_2)}{x_3 - x_2}.$$

(b) Use part (a) to show that there are four possibilities as x increases to ∞: (1) $f(x)$ decreases monotonically to $-\infty$, (2) $f(x)$ decreases monotonically to a finite value, (3) $f(x)$ reaches some value and stays at that value, (4) $f(x)$ increases monotonically to ∞ when $x \geq \bar{x}$ for some $\bar{x} \in \Re$.

1.7 (Characterization of Differentiable Convex Functions)

Let $f : \Re^n \mapsto \Re$ be a differentiable function. Show that f is convex over a nonempty convex set C if and only if

$$\big(\nabla f(x) - \nabla f(y)\big)'(x - y) \geq 0, \qquad \forall\, x, y \in C.$$

Note: The condition above says that the function f, restricted to the line segment connecting x and y, has monotonically nondecreasing gradient.

1.8 (Characterization of Twice Continuously Differentiable Convex Functions)

Let C be a nonempty convex subset of $\Re^n$ and let $f : \Re^n \mapsto \Re$ be twice continuously differentiable over $\Re^n$. Let S be the subspace that is parallel to the affine hull of C. Show that f is convex over C if and only if $y'\nabla^2 f(x)y \geq 0$ for all $x \in C$ and $y \in S$. [In particular, when C has nonempty interior, f is convex over C if and only if $\nabla^2 f(x)$ is positive semidefinite for all $x \in C$.]

1.9 (Strong Convexity)

Let $f : \Re^n \mapsto \Re$ be a differentiable function. We say that f *is strongly convex with coefficient* α if

$$\big(\nabla f(x) - \nabla f(y)\big)'(x - y) \geq \alpha\|x - y\|^2, \qquad \forall\, x, y \in \Re^n, \qquad (1.16)$$

where α is some positive scalar.

(a) Show that if f is strongly convex with coefficient α, then f is strictly convex.

(b) Assume that f is twice continuously differentiable. Show that strong convexity of f with coefficient α is equivalent to the positive semidefiniteness of $\nabla^2 f(x) - \alpha I$ for every $x \in \Re^n$, where I is the identity matrix.

1.10 (Posynomials)

A *posynomial* is a function of positive scalar variables $y_1, \ldots, y_n$ of the form

$$g(y_1, \ldots, y_n) = \sum_{i=1}^{m} \beta_i y_1^{a_{i1}} \cdots y_n^{a_{in}},$$

where a_{ij} and β_i are scalars, such that $\beta_i > 0$ for all i. Show the following:

(a) A posynomial need not be convex.

(b) By a logarithmic change of variables, where we set

$$f(x) = \ln\big(g(y_1, \ldots, y_n)\big), \qquad b_i = \ln \beta_i, \ \forall \ i, \qquad x_j = \ln y_j, \ \forall \ j,$$

we obtain a convex function

$$f(x) = \ln \exp(Ax + b), \qquad \forall \ x \in \Re^n,$$

where $\exp(z) = e^{z_1} + \cdots + e^{z_m}$ for all $z \in \Re^m$, A is an $m \times n$ matrix with entries a_{ij}, and $b \in \Re^m$ is a vector with components b_i.

(c) Every function $g : \Re^n \mapsto \Re$ of the form

$$g(y) = g_1(y)^{\gamma_1} \cdots g_r(y)^{\gamma_r},$$

where g_k is a posynomial and $\gamma_k > 0$ for all k, can be transformed by a logarithmic change of variables into a convex function f given by

$$f(x) = \sum_{k=1}^{r} \gamma_k \ln \exp(A_k x + b_k),$$

with the matrix A_k and the vector b_k being associated with the posynomial g_k for each k.

1.11 (Arithmetic-Geometric Mean Inequality)

Show that if $\alpha_1, \ldots, \alpha_n$ are positive scalars with $\sum_{i=1}^{n} \alpha_i = 1$, then for every set of positive scalars $x_1, \ldots, x_n$, we have

$$x_1^{\alpha_1} x_2^{\alpha_2} \cdots x_n^{\alpha_n} \leq \alpha_1 x_1 + \alpha_2 x_2 + \cdots + \alpha_n x_n,$$

with equality if and only if $x_1 = x_2 = \cdots = x_n$. *Hint:* Show that $-\ln x$ is a strictly convex function on $(0, \infty)$.

1.12 (Young and Holder Inequalities)

Use the result of Exercise 1.11 to verify Young's inequality

$$xy \leq \frac{x^p}{p} + \frac{y^q}{q}, \qquad \forall \, x \geq 0, \, \forall \, y \geq 0,$$

where $p > 0$, $q > 0$, and

$$1/p + 1/q = 1.$$

Then, use Young's inequality to verify Holder's inequality

$$\sum_{i=1}^{n} |x_i y_i| \leq \left(\sum_{i=1}^{n} |x_i|^p \right)^{1/p} \left(\sum_{i=1}^{n} |y_i|^q \right)^{1/q}.$$

1.13

Let C be a nonempty convex set in $\Re^{n+1}$, and let $f : \Re^n \mapsto [-\infty, \infty]$ be the function defined by

$$f(x) = \inf\{w \mid (x, w) \in C\}, \qquad x \in \Re^n.$$

Show that f is convex.

1.14

Show that the convex hull of a nonempty set coincides with the set of all convex combinations of its elements.

1.15

Let C be a nonempty convex subset of $\Re^n$. Show that

$$\text{cone}(C) = \cup_{x \in C}\{\gamma x \mid \gamma \geq 0\}.$$

1.16 (Convex Cones)

Show that:

(a) For any collection of vectors $\{a_i \mid i \in I\}$, the set $C = \{x \mid a_i'x \leq 0, \, i \in I\}$ is a closed convex cone.

(b) A cone C is convex if and only if $C + C \subset C$.

(c) For any two convex cones C_1 and C_2 containing the origin, we have

$$C_1 + C_2 = \text{conv}(C_1 \cup C_2),$$

$$C_1 \cap C_2 = \bigcup_{\alpha \in [0,1]} \left(\alpha C_1 \cap (1 - \alpha) C_2 \right).$$

1.17

Let $\{C_i \mid i \in I\}$ be an arbitrary collection of convex sets in $\Re^n$, and let C be the convex hull of the union of the collection. Show that

$$C = \bigcup_{\overline{I} \subset I,\ \overline{I}:\ \text{finite set}} \left\{ \sum_{i \in \overline{I}} \alpha_i C_i \ \Big| \ \sum_{i \in \overline{I}} \alpha_i = 1,\ \alpha_i \geq 0,\ \forall\, i \in \overline{I} \right\},$$

i.e., the convex hull of the union of the C_i is equal to the set of all convex combinations of vectors from the C_i.

1.18 (Convex Hulls, Affine Hulls, and Generated Cones)

Let X be a nonempty set. Show that:

(a) X, $\mathrm{conv}(X)$, and $\mathrm{cl}(X)$ have the same affine hull.

(b) $\mathrm{cone}(X) = \mathrm{cone}\big(\mathrm{conv}(X)\big)$.

(c) $\mathrm{aff}\big(\mathrm{conv}(X)\big) \subset \mathrm{aff}\big(\mathrm{cone}(X)\big)$. Give an example where the inclusion is strict, i.e., $\mathrm{aff}\big(\mathrm{conv}(X)\big)$ is a strict subset of $\mathrm{aff}\big(\mathrm{cone}(X)\big)$.

(d) If the origin belongs to $\mathrm{conv}(X)$, then $\mathrm{aff}\big(\mathrm{conv}(X)\big) = \mathrm{aff}\big(\mathrm{cone}(X)\big)$.

1.19

Let $\{f_i \mid i \in I\}$ be an arbitrary collection of proper convex functions $f_i : \Re^n \mapsto (-\infty, \infty]$. Define

$$f(x) = \inf \big\{ w \mid (x, w) \in \mathrm{conv}\big(\cup_{i \in I} \mathrm{epi}(f_i)\big) \big\}, \qquad x \in \Re^n.$$

Show that $f(x)$ is given by

$$f(x) = \inf \left\{ \sum_{i \in \overline{I}} \alpha_i f_i(x_i) \ \Big| \ \sum_{i \in \overline{I}} \alpha_i x_i = x,\ x_i \in \Re^n,\ \sum_{i \in \overline{I}} \alpha_i = 1,\ \alpha_i \geq 0,\ \forall\, i \in \overline{I}, \right.$$

$$\left. \overline{I} \subset I,\ \overline{I}:\ \text{finite} \right\}.$$

1.20 (Convexification of Nonconvex Functions)

Let X be a nonempty subset of $\Re^n$ and let $f : X \mapsto \Re$ be a function that is bounded below over X. Define the function $F : \mathrm{conv}(X) \mapsto \Re$ by

$$F(x) = \inf \big\{ w \mid (x, w) \in \mathrm{conv}\big(\mathrm{epi}(f)\big) \big\}.$$

Show that:

(a) F is convex over $\text{conv}(X)$ and it is given by

$$F(x) = \inf\left\{\sum_i \alpha_i f(x_i) \,\Big|\, \sum_i \alpha_i x_i = x,\ x_i \in X,\ \sum_i \alpha_i = 1,\ \alpha_i \geq 0,\ \forall\, i\right\},$$

where the infimum is taken over all representations of x as a convex combination of elements of X (i.e., with finitely many nonzero coefficients α_i).

(b)
$$\inf_{x\in\text{conv}(X)} F(x) = \inf_{x\in X} f(x).$$

(c) Every $x^* \in X$ that attains the minimum of f over X, i.e., $f(x^*) = \inf_{x\in X} f(x)$, also attains the minimum of F over $\text{conv}(X)$.

1.21 (Minimization of Linear Functions)

Show that minimization of a linear function over a set is equivalent to minimization over its convex hull. In particular, if $X \subset \Re^n$ and $c \in \Re^n$, then

$$\inf_{x\in\text{conv}(X)} c'x = \inf_{x\in X} c'x.$$

Furthermore, the infimum in the left-hand side above is attained if and only if the infimum in the right-hand side is attained.

1.22 (Extension of Caratheodory's Theorem)

Let X_1 and X_2 be nonempty subsets of $\Re^n$, and let $X = \text{conv}(X_1) + \text{cone}(X_2)$. Show that every vector x in X can be represented in the form

$$x = \sum_{i=1}^{k} \alpha_i x_i + \sum_{i=k+1}^{m} \alpha_i y_i,$$

where m is a positive integer with $m \leq n+1$, the vectors $x_1, \ldots, x_k$ belong to X_1, the vectors $y_{k+1}, \ldots, y_m$ belong to X_2, and the scalars $\alpha_1, \ldots, \alpha_m$ are nonnegative with $\alpha_1 + \cdots + \alpha_k = 1$. Furthermore, the vectors $x_2 - x_1, \ldots, x_k - x_1, y_{k+1}, \ldots, y_m$ are linearly independent.

1.23

Let X be a nonempty bounded subset of $\Re^n$. Show that

$$\text{cl}\big(\text{conv}(X)\big) = \text{conv}\big(\text{cl}(X)\big).$$

In particular, if X is compact, then $\text{conv}(X)$ is compact (cf. Prop. 1.3.2).

1.24 (Radon's Theorem)

Let $x_1, \ldots, x_m$ be vectors in $\Re^n$, where $m \geq n+2$. Show that there exists a partition of the index set $\{1, \ldots, m\}$ into two disjoint sets I and J such that

$$\text{conv}\big(\{x_i \mid i \in I\}\big) \cap \text{conv}\big(\{x_j \mid j \in J\}\big) \neq \emptyset.$$

Hint: The system of $n+1$ equations in the m unknowns $\lambda_1, \ldots, \lambda_m$,

$$\sum_{i=1}^{m} \lambda_i x_i = 0, \qquad \sum_{i=1}^{m} \lambda_i = 0,$$

has a nonzero solution λ^*. Let $I = \{i \mid \lambda_i^* \geq 0\}$ and $J = \{j \mid \lambda_j^* < 0\}$.

1.25 (Helly's Theorem [Hel21])

Consider a finite collection of convex subsets of $\Re^n$, and assume that the intersection of every subcollection of $n+1$ (or fewer) sets has nonempty intersection. Show that the entire collection has nonempty intersection. *Hint*: Use induction. Assume that the conclusion holds for every collection of M sets, where $M \geq n+1$, and show that the conclusion holds for every collection of $M+1$ sets. In particular, let $C_1, \ldots, C_{M+1}$ be a collection of $M+1$ convex sets, and consider the collection of $M+1$ sets $B_1, \ldots, B_{M+1}$, where

$$B_j = \cap_{\substack{i=1,\ldots,M+1 \\ i \neq j}} C_i, \qquad j = 1, \ldots, M+1.$$

Note that, by the induction hypothesis, each set B_j is the intersection of a collection of M sets that have the property that every subcollection of $n+1$ (or fewer) sets has nonempty intersection. Hence each set B_j is nonempty. Let x_j be a vector in B_j. Apply Radon's Theorem (Exercise 1.24) to the vectors $x_1, \ldots, x_{M+1}$. Show that any vector in the intersection of the corresponding convex hulls belongs to the intersection of $C_1, \ldots, C_{M+1}$.

1.26

Consider the problem of minimizing over $\Re^n$ the function

$$\max\big\{f_1(x), \ldots, f_M(x)\big\},$$

where $f_i : \Re^n \mapsto (-\infty, \infty]$, $i = 1, \ldots, M$, are convex functions, and assume that the optimal value, denoted f^*, is finite. Show that there exists a subset I of $\{1, \ldots, M\}$, containing no more than $n+1$ indices, such that

$$\inf_{x \in \Re^n} \Big\{ \max_{i \in I} f_i(x) \Big\} = f^*.$$

Hint: Consider the convex sets $X_i = \big\{x \mid f_i(x) < f^*\big\}$, argue by contradiction, and apply Helly's Theorem (Exercise 1.25).

1.27

Let C be a nonempty convex subset of $\Re^n$, and let $f : \Re^n \mapsto (-\infty, \infty]$ be a convex function such that $f(x)$ is finite for all $x \in C$. Show that if for some scalar γ, we have $f(x) \geq \gamma$ for all $x \in C$, then we also have $f(x) \geq \gamma$ for all $x \in \mathrm{cl}(C)$.

1.28

Let C be a nonempty convex set, and let S be the subspace that is parallel to the affine hull of C. Show that

$$\mathrm{ri}(C) = \mathrm{int}(C + S^{\perp}) \cap C.$$

1.29

Let $x_0, \ldots, x_m$ be vectors in $\Re^n$ such that $x_1 - x_0, \ldots, x_m - x_0$ are linearly independent. The convex hull of $x_0, \ldots, x_m$ is called an *m-dimensional simplex*, and $x_0, \ldots, x_m$ are called the *vertices* of the simplex.

(a) Show that the dimension of a convex set is the maximum of the dimensions of all the simplices contained in the set.

(b) Use part (a) to show that a nonempty convex set has a nonempty relative interior.

1.30

Let C_1 and C_2 be two nonempty convex sets such that $C_1 \subset C_2$.

(a) Give an example showing that $\mathrm{ri}(C_1)$ need not be a subset of $\mathrm{ri}(C_2)$.

(b) Assuming that the sets C_1 and C_2 have the same affine hull, show that $\mathrm{ri}(C_1) \subset \mathrm{ri}(C_2)$.

(c) Assuming that the sets $\mathrm{ri}(C_1)$ and $\mathrm{ri}(C_2)$ have nonempty intersection, show that $\mathrm{ri}(C_1) \subset \mathrm{ri}(C_2)$.

(d) Assuming that the sets C_1 and $\mathrm{ri}(C_2)$ have nonempty intersection, show that the set $\mathrm{ri}(C_1) \cap \mathrm{ri}(C_2)$ is nonempty.

1.31

Let C be a nonempty convex set.

(a) Show the following refinement of Prop. 1.4.1(c): $x \in \mathrm{ri}(C)$ if and only if for every $\bar{x} \in \mathrm{aff}(C)$, there exists a $\gamma > 1$ such that $x + (\gamma - 1)(x - \bar{x}) \in C$.

(b) Assuming that the origin lies in $\mathrm{ri}(C)$, show that $\mathrm{cone}(C)$ coincides with $\mathrm{aff}(C)$.

(c) Show the following extension of part (b) to a nonconvex set: If X is a nonempty set such that the origin lies in the relative interior of $\mathrm{conv}(X)$, then $\mathrm{cone}(X)$ coincides with $\mathrm{aff}(X)$.

1.32

Let C be a nonempty set.

  (a) If C is convex and compact, and the origin is not in the relative boundary of C, then $\mathrm{cone}(C)$ is closed.

  (b) Give examples showing that the assertion of part (a) fails if C is unbounded or the origin is in the relative boundary of C.

  (c) If C is compact and the origin is not in the relative boundary of $\mathrm{conv}(C)$, then $\mathrm{cone}(C)$ is closed. *Hint*: Use part (a) and Exercise 1.18(b).

1.33

  (a) Let C be a nonempty convex cone. Show that $\mathrm{ri}(C)$ is also a convex cone.

  (b) Let $C = \mathrm{cone}\big(\{x_1,\ldots,x_m\}\big)$. Show that

$$\mathrm{ri}(C) = \left\{ \sum_{i=1}^{m} \alpha_i x_i \;\middle|\; \alpha_i > 0, \; i = 1,\ldots,m \right\}.$$

1.34

Let A be an $m \times n$ matrix and let C be a nonempty convex set in $\Re^m$. Assuming that $A^{-1} \cdot \mathrm{ri}(C)$ is nonempty, show that

$$\mathrm{ri}(A^{-1} \cdot C) = A^{-1} \cdot \mathrm{ri}(C), \qquad \mathrm{cl}(A^{-1} \cdot C) = A^{-1} \cdot \mathrm{cl}(C).$$

(Compare these relations with those of Prop. 1.4.4.)

1.35 (Closure of a Convex Function)

Consider a proper convex function $f : \Re^n \mapsto (-\infty, \infty]$ and the function whose epigraph is the closure of the epigraph of f. This function is called the *closure of f* and is denoted by $\mathrm{cl}\, f$. Show that:

  (a) $\mathrm{cl}\, f$ is the greatest lower semicontinuous function majorized by f, i.e., if $g : \Re^n \mapsto [-\infty, \infty]$ is lower semicontinuous and satisfies $g(x) \le f(x)$ for all $x \in \Re^n$, then $g(x) \le (\mathrm{cl} f)(x)$ for all $x \in \Re^n$.

  (b) $\mathrm{cl}\, f$ is a closed proper convex function and

$$(\mathrm{cl}\, f)(x) = f(x), \qquad \forall\, x \in \mathrm{ri}\big(\mathrm{dom}(f)\big).$$

(c) If $x \in \text{ri}\big(\text{dom}(f)\big)$ and $y \in \text{dom}(\text{cl} f)$, we have

$$(\text{cl} f)(y) = \lim_{\alpha \downarrow 0} f\big(y + \alpha(x - y)\big).$$

(d) Assume that $f = f_1 + \cdots + f_m$, where $f_i : \Re^n \mapsto (-\infty, \infty]$, $i = 1, \ldots, m$, are proper convex functions such that $\cap_{i=1}^m \text{ri}\big(\text{dom}(f_i)\big) \neq \emptyset$. Show that

$$(\text{cl} f)(x) = (\text{cl} f_1)(x) + \cdots + (\text{cl} f_m)(x), \qquad \forall\, x \in \Re^n.$$

1.36

Let C be a convex set and let M be an affine set such that the intersection $C \cap M$ is nonempty and bounded. Show that for every affine set $\overline{M}$ that is parallel to M, the intersection $C \cap \overline{M}$ is bounded.

1.37 (Properties of Cartesian Products)

Given nonempty sets $X_i \subset \Re^{n_i}$, $i = 1, \ldots, m$, let $X = X_1 \times \cdots \times X_m$ be their Cartesian product. Show that:

(a) The convex hull (closure, affine hull) of X is equal to the Cartesian product of the convex hulls (closures, affine hulls, respectively) of the X_i.

(b) If all the sets $X_1, \ldots, X_m$ contain the origin, then

$$\text{cone}(X) = \text{cone}(X_1) \times \cdots \times \text{cone}(X_m).$$

Furthermore, the result fails if one of the sets does not contain the origin.

(c) If all the sets $X_1, \ldots, X_m$ are convex, then the relative interior (recession cone) of X is equal to the Cartesian product of the relative interiors (recession cones, respectively) of the X_i.

1.38 (Recession Cones of Nonclosed Sets)

Let C be a nonempty convex set.

(a) Show that
$$R_C \subset R_{\text{cl}(C)}, \qquad \text{cl}(R_C) \subset R_{\text{cl}(C)}.$$

Give an example where the inclusion $\text{cl}(R_C) \subset R_{\text{cl}(C)}$ is strict.

(b) Let $\overline{C}$ be a closed convex set such that $C \subset \overline{C}$. Show that $R_C \subset R_{\overline{C}}$. Give an example showing that the inclusion can fail if $\overline{C}$ is not closed.

1.39 (Recession Cones of Relative Interiors)

Let C be a nonempty convex set.

(a) Show that $R_{\text{ri}(C)} = R_{\text{cl}(C)}$.

(b) Show that a vector y belongs to $R_{\text{ri}(C)}$ if and only if there exists a vector $x \in \text{ri}(C)$ such that $x + \alpha y \in \text{ri}(C)$ for every $\alpha \geq 0$.

(c) Let $\overline{C}$ be a convex set such that $\overline{C} = \text{ri}(\overline{C})$ and $C \subset \overline{C}$. Show that $R_C \subset R_{\overline{C}}$. Give an example showing that the inclusion can fail if $\overline{C} \neq \text{ri}(\overline{C})$.

1.40

This exercise is a refinement of Prop. 1.5.6. Let $\{X_k\}$ and $\{C_k\}$ be sequences of closed convex subsets of $\Re^n$, such that the intersection

$$X = \cap_{k=0}^{\infty} X_k$$

is specified by linear inequality constraints as in Prop. 1.5.6. Assume that:

(1) $X_{k+1} \subset X_k$ and $C_{k+1} \subset C_k$ for all k.

(2) $X_k \cap C_k$ is nonempty for all k.

(3) We have

$$R_X = L_X, \qquad R_X \cap R_C \subset L_C,$$

where

$$R_X = \cap_{k=0}^{\infty} R_{X_k}, \qquad L_X = \cap_{k=0}^{\infty} L_{X_k},$$
$$R_C = \cap_{k=0}^{\infty} R_{C_k}, \qquad L_C = \cap_{k=0}^{\infty} L_{C_k}.$$

Then the intersection $\cap_{k=0}^{\infty}(X_k \cap C_k)$ is nonempty. *Hint:* Consider the sets $\overline{C}_k = X_k \cap C_k$ and the intersection $X \cap (\cap_{k=0}^{\infty} \overline{C}_k)$. Apply Prop. 1.5.6.

1.41

Let C be a nonempty convex subset of $\Re^n$ and let A be an $m \times n$ matrix. Show that if $R_{\text{cl}(C)} \cap N(A) = \{0\}$, then

$$\text{cl}(A \cdot C) = A \cdot \text{cl}(C), \qquad A \cdot R_{\text{cl}(C)} = R_{A \cdot \text{cl}(C)}.$$

Give an example showing that $A \cdot R_{\text{cl}(C)}$ and $R_{A \cdot \text{cl}(C)}$ may differ when $R_{\text{cl}(C)} \cap N(A) \neq \{0\}$.

1.42

Let C be a nonempty convex subset of $\Re^n$. Show the following refinement of Prop. 1.5.8(a) and Exercise 1.41: if A is an $m \times n$ matrix and $R_{\text{cl}(C)} \cap N(A)$ is a subspace of the lineality space of $\text{cl}(C)$, then

$$\text{cl}(A \cdot C) = A \cdot \text{cl}(C), \qquad A \cdot R_{\text{cl}(C)} = R_{A \cdot \text{cl}(C)}.$$

1.43 (Recession Cones of Vector Sums)

This exercise is a refinement of Prop. 1.5.9.

(a) Let $C_1, \ldots, C_m$ be nonempty closed convex subsets of $\Re^n$ such that the equality $y_1 + \cdots + y_m = 0$ with $y_i \in R_{C_i}$ implies that each y_i belongs to the lineality space of C_i. Then, the vector sum $C_1 + \cdots + C_m$ is a closed set and
$$R_{C_1+\cdots+C_m} = R_{C_1} + \cdots + R_{C_m}.$$

(b) Show the following extension of part (a) to nonclosed sets: Let $C_1, \ldots, C_m$ be nonempty convex subsets of $\Re^n$ such that the equality $y_1 + \cdots + y_m = 0$ with $y_i \in R_{\mathrm{cl}(C_i)}$ implies that each y_i belongs to the lineality space of $\mathrm{cl}(C_i)$. Then, we have

$$\mathrm{cl}(C_1 + \cdots + C_m) = \mathrm{cl}(C_1) + \cdots + \mathrm{cl}(C_m),$$

$$R_{\mathrm{cl}(C_1+\cdots+C_m)} = R_{\mathrm{cl}(C_1)} + \cdots + R_{\mathrm{cl}(C_m)}.$$

1.44

Let $C_1, \ldots, C_m$ be nonempty subsets of $\Re^n$ that are specified by convex quadratic inequalities, i.e., for all $i = 1, \ldots, n$,

$$C_i = \left\{ x \mid x'Q_{ij}x + a'_{ij}x + b_{ij} \leq 0, \ j = 1, \ldots, r_i \right\},$$

where Q_{ij} are symmetric positive semidefinite $n \times n$ matrices, a_{ij} are vectors in $\Re^n$, and b_{ij} are scalars. Show that the vector sum $C_1 + \cdots + C_m$ is a closed set.

1.45 (Set Intersection and Helly's Theorem)

Show that the conclusions of Props. 1.5.5 and 1.5.6 hold if the assumption that the sets C_k are nonempty and nested is replaced by the weaker assumption that any subcollection of $n + 1$ (or fewer) sets from the sequence $\{C_k\}$ has nonempty intersection. *Hint*: Consider the sets $\overline{C}_k$ given by

$$\overline{C}_k = \cap_{i=1}^k C_i, \qquad \forall \, k = 1, 2, \ldots,$$

and use Helly's Theorem (Exercise 1.25) to show that they are nonempty.

2

Convexity and Optimization

Contents

2.1. Global and Local Minima . . . . p. 84
2.2. The Projection Theorem . . . . p. 88
2.3. Directions of Recession and Existence of Optimal Solutions p. 92
    2.3.1. Existence of Solutions of Convex Programs . . . . . p. 94
    2.3.2. Unbounded Optimal Solution Sets . . . . p. 97
    2.3.3. Partial Minimization of Convex Functions . . . . p. 101
2.4. Hyperplanes . . . . p. 107
2.5. An Elementary Form of Duality . . . . p. 117
    2.5.1. Nonvertical Hyperplanes . . . . p. 117
    2.5.2. Min Common/Max Crossing Duality . . . . . . p. 120
2.6. Saddle Point and Minimax Theory . . . . p. 128
    2.6.1. Min Common/Max Crossing Framework for Minimax p. 133
    2.6.2. Minimax Theorems . . . . p. 139
    2.6.3. Saddle Point Theorems . . . . p. 143
2.7. Notes, Sources, and Exercises . . . . p. 151

In this chapter we discuss applications of convexity to some basic optimization issues, such as the existence of optimal solutions. Furthermore, we develop the theory of hyperplanes and we highlight its special role in duality theory. In particular, we introduce an elementary form of duality, and we use it for the analysis of minimax problems and the existence of saddle points.

2.1 GLOBAL AND LOCAL MINIMA

Consider a nonempty subset X of $\Re^n$ and a function $f : \Re^n \mapsto (-\infty, \infty]$. We focus on the problem of minimizing $f(x)$ over $x \in X$. In this context, we say that any vector $x \in X$ is a *feasible solution* of the problem (we also use the terms *feasible vector* or *feasible point*). If there is at least one feasible solution, we say that the problem is *feasible*; otherwise we say that the problem is *infeasible*. We say that a vector $x^* \in X$ is a *minimum of f over X* if $f(x^*) = \inf_{x \in X} f(x)$. We also call x^* a *minimizing point* or *minimizer* or *global minimum over X*. Alternatively, we say that f *attains a minimum over X at x^**, and we indicate this by writing

$$x^* \in \arg \min_{x \in X} f(x).$$

If x^* is known to be the unique minimizer of f over X, by slight abuse of notation, we also write

$$x^* = \arg \min_{x \in X} f(x).$$

We use similar terminology for maxima, i.e., a vector $x^* \in X$ such that $f(x^*) = \sup_{x \in X} f(x)$ is said to be a *maximum of f over X*, and we indicate this by writing

$$x^* \in \arg \max_{x \in X} f(x).$$

If $X = \Re^n$ or if the domain of f is the set X (instead of $\Re^n$), we also call x^* a (global) minimum or (global) maximum of f (without the qualifier "over X").

A basic question in optimization problems is whether an optimal solution exists. It can be seen that the set of minima of f over X is equal to the intersection of X and the nonempty level sets of f. Using this fact, it can be proved that the set of minima is nonempty if the sets of the form

$$\{x \in X \mid f(x) \leq \gamma\},$$

where γ is a scalar, are closed and at least one of them is nonempty and compact. This is the essence of the classical theorem of Weierstrass, which states that a continuous function attains a minimum over a compact set.

We will provide a more general version of this theorem, and to this end, we introduce some terminology. We say that a function $f : \Re^n \mapsto (-\infty, \infty]$ is *coercive over a set* $X \subset \Re^n$ if for every sequence $\{x_k\} \subset X$ such that $\|x_k\| \to \infty$, we have $\lim_{k\to\infty} f(x_k) = \infty$. In the case where $X = \Re^n$, we just say that f is *coercive*. Note that as a consequence of the definition, all the nonempty level sets of a coercive function f are bounded.

Proposition 2.1.1: (Weierstrass' Theorem) Consider a closed proper function $f : \Re^n \mapsto (-\infty, \infty]$, and assume that one of the following three conditions holds:

(1) $\mathrm{dom}(f)$ is bounded.

(2) There exists a scalar γ such that the level set

$$\{x \mid f(x) \leq \gamma\}$$

is nonempty and bounded.

(3) f is coercive.

Then the set of minima of f over $\Re^n$ is nonempty and compact.

Proof: Assume that condition (1) holds. Note that since f is proper, $\mathrm{dom}(f)$ is nonempty. Consider a sequence $\{x_k\} \subset \mathrm{dom}(f)$ such that

$$\lim_{k\to\infty} f(x_k) = \inf_{x\in\Re^n} f(x).$$

Since $\mathrm{dom}(f)$ is bounded, this sequence has at least one limit point x^* [Prop. 1.1.5(a)]. Since f is closed, it is lower semicontinuous at x^* [cf. Prop. 1.2.2(b)], so that $f(x^*) \leq \lim_{k\to\infty} f(x_k) = \inf_{x\in\Re^n} f(x)$, and x^* is a minimum of f. Thus, X^*, the set of minima of f over $\Re^n$, is nonempty. Since $X^* \subset \mathrm{dom}(f)$ and $\mathrm{dom}(f)$ is bounded, X^* is bounded. Also X^* is the intersection of all the level sets $\{x \mid f(x) \leq \gamma\}$ where $\gamma > \inf_{x\in\Re^n} f(x)$. These level sets are closed since f is closed [cf. Prop. 1.2.2(c)], so X^* is closed, and hence compact.

Assume that condition (2) holds. Replace f by the function

$$\tilde{f}(x) = \begin{cases} f(x) & \text{if } f(x) \leq \gamma, \\ \infty & \text{otherwise.} \end{cases}$$

The domain of $\tilde{f}$ is the set $\{x \mid f(x) \leq \gamma\}$, which is bounded by assumption and closed by the closedness of f. Hence, using also the closedness of f, it follows that the function $\tilde{f}$ is closed (cf. Prop. 1.2.3). Furthermore, the set of minima of $\tilde{f}$ is the same as the one of f. Thus the result shown under condition (1) applies.

Assume that condition (3) holds. Since f is proper, it has some nonempty level sets. Since f is coercive, its nonempty level sets are bounded, so condition (2) is satisfied. **Q.E.D.**

The most common application of the above proposition is when we want to minimize a real-valued function $f : \Re^n \mapsto \Re$ over a nonempty set X. Then, by applying the proposition to the extended real-valued function

$$\tilde{f}(x) = \begin{cases} f(x) & \text{if } x \in X, \\ \infty & \text{otherwise,} \end{cases}$$

we see that the set of minima of f over X is nonempty and compact if X is closed, f is lower semicontinuous over X (implying, by Prop. 1.2.3, that $\tilde{f}$ is closed), and one of the following conditions holds:

(1) X is bounded.

(2) Some level set $\{x \in X \mid f(x) \leq \gamma\}$ is nonempty and bounded.

(3) $\tilde{f}$ is coercive, i.e., for every sequence $\{x_k\} \subset X$ such that $\|x_k\| \to \infty$, we have $\lim_{k\to\infty} f(x_k) = \infty$.

Note that with appropriate adjustments, the preceding analysis applies to the existence of maxima of f over X. For example, if a real-valued function f is upper semicontinuous at all points of a compact set X, then the set of maxima of f over X is nonempty and compact.

Local Minima

Often in optimization problems we have to deal with a weaker form of minimum, one that is optimum only when compared with points that are "nearby." In particular, given a subset X of $\Re^n$ and a function $f : \Re^n \mapsto (-\infty, \infty]$, we say that a vector $x^* \in X$ is a *local minimum of f over X* if there exists some $\epsilon > 0$ such that

$$f(x^*) \leq f(x), \qquad \forall\, x \in X \text{ with } \|x - x^*\| \leq \epsilon.$$

If $X = \Re^n$ or the domain of f is the set X (instead of $\Re^n$), we also call x^* a local minimum of f (without the qualifier "over X"). A local minimum x^* is said to be *strict* if there is no other local minimum within a neighborhood of x^*. Local maxima are defined similarly.

In practical applications we are typically interested in global minima, yet we have to contend with local minima because of the inability of many optimality conditions and algorithms to distinguish between the two types of minima. This can be a major practical difficulty, but an important implication of convexity of f and X is that all local minima are also global, as shown in the following proposition and in Fig. 2.1.1.

Proposition 2.1.2: If X is a convex subset of $\Re^n$ and $f : \Re^n \mapsto (-\infty, \infty]$ is a proper convex function, then a local minimum of f over X is also a global minimum of f over X. If in addition f is strictly convex, then there exists at most one global minimum of f over X.

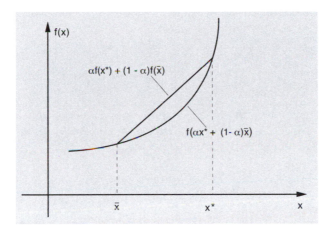

Figure 2.1.1. Illustration of why local minima of convex functions are also global (cf. Prop. 2.1.2). If x^* is a local minimum of f over X that is not global, there must exist an $\overline{x} \in X$ such that $f(\overline{x}) < f(x^*)$. By convexity, for all $\alpha \in (0,1)$,

$$f\big(\alpha x^* + (1 - \alpha)\overline{x}\big) \le \alpha f(x^*) + (1 - \alpha)f(\overline{x}) < f(x^*),$$

which contradicts the local minimality of x^*.

Proof: Let f be convex, and assume to arrive at a contradiction, that x^* is a local minimum of f over X that is not global (see Fig. 2.1.1). Then, there must exist an $\overline{x} \in X$ such that $f(\overline{x}) < f(x^*)$. By convexity, for all $\alpha \in (0,1)$,

$$f\big(\alpha x^* + (1 - \alpha)\overline{x}\big) \le \alpha f(x^*) + (1 - \alpha)f(\overline{x}) < f(x^*).$$

Thus, f has strictly lower value than $f(x^*)$ at every point on the line segment connecting x^* with $\overline{x}$, except at x^*. Since X is convex, the line segment belongs to X, thereby contradicting the local minimality of x^*.

Let f be strictly convex, and assume to arrive at a contradiction, that two distinct global minima of f over X, x and y, exist. Then the midpoint $(x+y)/2$ must belong to X, since X is convex. Furthermore, the value of f must be smaller at the midpoint than at x and y by the strict convexity of f. Since x and y are global minima, we obtain a contradiction. **Q.E.D.**

2.2 THE PROJECTION THEOREM

In this section we develop a basic result of analysis and optimization. It
deals with the problem of finding a vector in a given nonempty closed
convex set C, which is at minimum Euclidean distance from another given
vector x. It turns out that such a vector, called the *projection of x on
C*, exists and is unique. This result is very useful in theoretical analyses,
and has wide-ranging applications in a variety of fields, such as function
approximation, least-squares estimation, etc.

Proposition 2.2.1: (Projection Theorem) Let C be a nonempty
closed convex subset of $\Re^n$.

(a) For every $x \in \Re^n$, there exists a unique vector that minimizes
$\|z - x\|$ over all $z \in C$. This vector, called the *projection of x on
C*, is denoted by $P_C(x)$:

$$P_C(x) = \arg \min_{z \in C} \|z - x\|.$$

(b) For every $x \in \Re^n$, a vector $z \in C$ is equal to $P_C(x)$ if and only if

$$(y - z)'(x - z) \leq 0, \qquad \forall\, y \in C.$$

In the case where C is an affine set, the above condition is equivalent to

$$(x - z) \in S^\perp,$$

where S is the subspace that is parallel to C.

(c) The function $P_C : \Re^n \mapsto C$ is continuous and nonexpansive, i.e.,

$$\big\|P_C(y) - P_C(x)\big\| \leq \|y - x\|, \qquad \forall\, x, y \in \Re^n.$$

(d) The distance function $d : \Re^n \mapsto \Re$, defined by

$$d(x, C) = \min_{z \in C} \|z - x\|,$$

is convex.

Proof: (a) Fix x and let w be some element of C. Minimizing $\|x - z\|$
over all $z \in C$ is equivalent to minimizing the continuous function

$$g(z) = \tfrac{1}{2}\|z - x\|^2$$

over the set of all $z \in C$ such that $\|x-z\| \leq \|x-w\|$, which is a compact set. Therefore, by Weierstrass' Theorem (Prop. 2.1.1), there exists a minimizing vector.

To prove uniqueness, note that $g(\cdot)$ is a strictly convex function because its Hessian matrix is the identity matrix, which is positive definite [Prop. 1.2.6(b)], so its minimum is attained at a unique vector (Prop. 2.1.2).

(b) For all y and z in C, we have

$$\|y-x\|^2 = \|y-z\|^2 + \|z-x\|^2 - 2(y-z)'(x-z) \geq \|z-x\|^2 - 2(y-z)'(x-z).$$

Therefore, if z is such that $(y - z)'(x - z) \leq 0$ for all $y \in C$, we have $\|y - x\|^2 \geq \|z - x\|^2$ for all $y \in C$, implying that $z = P_C(x)$.

Conversely, let $z = P_C(x)$, consider any $y \in C$, and for $\alpha > 0$, define $y_\alpha = \alpha y + (1 - \alpha)z$. We have

$$
\begin{aligned}
\|x - y_\alpha\|^2 &= \left\| \alpha(x - y) + (1 - \alpha)(x - z) \right\|^2 \\
&= \alpha^2 \|x - y\|^2 + (1 - \alpha)^2 \|x - z\|^2 + 2\alpha(1 - \alpha)(x - y)'(x - z).
\end{aligned}
$$

Viewing $\|x - y_\alpha\|^2$ as a function of α, we have

$$\frac{\partial}{\partial \alpha}\{\|x - y_\alpha\|^2\}\Big|_{\alpha=0} = -2\|x - z\|^2 + 2(x - y)'(x - z) = -2(y - z)'(x - z).$$

Since $\alpha = 0$ minimizes $\|x - y_\alpha\|^2$ over $\alpha \in [0, 1]$, we have

$$\frac{\partial}{\partial \alpha}\{\|x - y_\alpha\|^2\}\Big|_{\alpha=0} \geq 0,$$

or equivalently $(y - z)'(x - z) \leq 0$.

Let $z = P_C(x)$. If C is affine and is parallel to the subspace S, we have that $y \in C$ if and only if $y - z \in S$. Hence, the condition $(y - z)'(x - z) \leq 0$ for all $y \in C$ is equivalent to $w'(x - z) \leq 0$ for all $w \in S$, or $(x - z) \perp S$.

(c) Let x and y be elements of $\Re^n$. From part (b), we have

$$\left(w - P_C(x)\right)'\left(x - P_C(x)\right) \leq 0, \qquad \forall\, w \in C.$$

Since $P_C(y) \in C$, we obtain

$$\left(P_C(y) - P_C(x)\right)'\left(x - P_C(x)\right) \leq 0.$$

Similarly,

$$\left(P_C(x) - P_C(y)\right)'\left(y - P_C(y)\right) \leq 0.$$

Adding these two inequalities, we obtain

$$\left(P_C(y) - P_C(x)\right)'\left(x - P_C(x) - y + P_C(y)\right) \leq 0.$$

By rearranging and by using the Schwarz inequality, we have

$$\left\| P_C(y) - P_C(x) \right\|^2 \leq \left(P_C(y) - P_C(x) \right)'(y - x) \leq \left\| P_C(y) - P_C(x) \right\| \cdot \| y - x \|,$$

showing that $P_C(\cdot)$ is nonexpansive and therefore also continuous.

(d) Assume, to arrive at a contradiction, that there exist $x_1, x_2 \in \Re^n$ and an $\alpha \in (0,1)$ such that

$$d\big(\alpha x_1 + (1 - \alpha)x_2, C\big) > \alpha d(x_1, C) + (1 - \alpha)d(x_2, C).$$

Then, there must exist $z_1, z_2 \in C$ such that

$$d\big(\alpha x_1 + (1 - \alpha)x_2, C\big) > \alpha \|z_1 - x_1\| + (1 - \alpha)\|z_2 - x_2\|,$$

which implies that

$$\left\| \alpha z_1 + (1 - \alpha)z_2 - \alpha x_1 - (1 - \alpha)x_2 \right\| > \alpha \|x_1 - z_1\| + (1 - \alpha)\|x_2 - z_2\|.$$

This contradicts the triangle inequality. **Q.E.D.**

Figure 2.2.1 illustrates the necessary and sufficient condition of part (b) of the Projection Theorem.

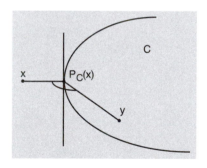

Figure 2.2.1. Illustration of the condition satisfied by the projection $P_C(x)$. For each vector $y \in C$, the vectors $x - P_C(x)$ and $y - P_C(x)$ form an angle greater than or equal to $\pi/2$ or, equivalently,

$$\big(y - P_C(x)\big)'\big(x - P_C(x)\big) \leq 0.$$

Problems of minimization of a strictly convex quadratic function over a nonempty closed convex set can be viewed as problems of projection. The reason is that a quadratic function $\frac{1}{2}x'Qx + c'x$, where Q is a positive definite symmetric matrix, can be written as $\frac{1}{2}\|y - \overline{y}\|^2 - \frac{1}{2}\|\overline{y}\|^2$, where $\overline{y} = -Q^{-1/2}c$, by using the transformation $y = Q^{1/2}x$ [here $Q^{1/2}$ denotes the unique square root of Q, the unique positive definite matrix that yields Q when multiplied with itself; cf. Prop. 1.1.11(e)]. Thus, minimizing $\frac{1}{2}x'Qx + c'x$ over a closed convex set X is equivalent to finding the projection of $\overline{y}$ on the set $Q^{1/2}X$.

Problems involving minimization of a quadratic function subject to linear inequality constraints:

$$\text{minimize} \quad \tfrac{1}{2}x'Qx + c'x$$
$$\text{subject to} \quad a_j'x \leq b_j, \qquad j = 1, \ldots, r,$$

are called *quadratic programming problems*, and will be discussed in some detail in the next section. When the quadratic cost function is strictly convex, these problems can be viewed as problems of projection, and thus have a unique solution. In practice, this solution usually cannot be obtained analytically. An exception is when the constraints are equalities, as discussed in the following example.

Example 2.2.1: (Equality-Constrained Quadratic Programming)

Consider the quadratic programming problem

$$\text{minimize} \quad \tfrac{1}{2}\|x\|^2 + c'x$$
$$\text{subject to} \quad Ax = 0, \tag{2.1}$$

where c is a given vector in $\Re^n$ and A is an $m \times n$ matrix of rank m.

By adding the constant term $\tfrac{1}{2}\|c\|^2$ to the cost function, we can equivalently write this problem as

$$\text{minimize} \quad \tfrac{1}{2}\|c + x\|^2$$
$$\text{subject to} \quad Ax = 0,$$

which is the problem of projecting the vector $-c$ on the subspace $X = \{x \mid Ax = 0\}$. By Prop. 2.2.1(b), a vector x^* such that $Ax^* = 0$ is the unique projection if and only if

$$(c + x^*)'x = 0, \qquad \forall \ x \text{ with } Ax = 0.$$

It can be seen that the vector

$$x^* = -\big(I - A'(AA')^{-1}A\big)c \tag{2.2}$$

satisfies this condition and is thus the unique solution of the quadratic programming problem (2.1). (The matrix AA' is invertible because A has rank m.)

Consider now the more general quadratic program

$$\text{minimize} \quad \tfrac{1}{2}(x - \bar{x})'Q(x - \bar{x}) + c'(x - \bar{x})$$
$$\text{subject to} \quad Ax = b, \tag{2.3}$$

where c and A are as before, Q is a symmetric positive definite matrix, b is a given vector in $\Re^m$, and $\bar{x}$ is a given vector in $\Re^n$, which is feasible, that is, it satisfies $A\bar{x} = b$. By introducing the transformation $y = Q^{1/2}(x - \bar{x})$, we can write this problem as

$$\text{minimize} \quad \tfrac{1}{2}\|y\|^2 + \big(Q^{-1/2}c\big)' y$$
$$\text{subject to} \quad AQ^{-1/2}y = 0.$$

Using Eq. (2.2) we see that the solution of this problem is

$$y^* = - \left(I - Q^{-1/2} A' \left(AQ^{-1} A' \right)^{-1} AQ^{-1/2} \right) Q^{-1/2} c$$

and by passing to the x-coordinate system through the transformation $x^* - \bar{x} = Q^{-1/2} y^*$, we obtain the optimal solution

$$x^* = \bar{x} - Q^{-1}(c - A'\lambda), \tag{2.4}$$

where the vector λ is given by

$$\lambda = \left(AQ^{-1} A' \right)^{-1} AQ^{-1} c. \tag{2.5}$$

The quadratic program (2.3) contains as a special case the program

$$\begin{aligned} \text{minimize} \quad & \tfrac{1}{2} x'Qx + c'x \\ \text{subject to} \quad & Ax = b. \end{aligned} \tag{2.6}$$

This special case is obtained when $\bar{x}$ is given by

$$\bar{x} = Q^{-1} A' (AQ^{-1} A')^{-1} b. \tag{2.7}$$

Indeed $\bar{x}$ as given above satisfies $A\bar{x} = b$ as required, and for all x with $Ax = b$, we have

$$x'Q\bar{x} = x'A'(AQ^{-1}A')^{-1}b = b'(AQ^{-1}A')^{-1}b,$$

which implies that for all x with $Ax = b$,

$$\tfrac{1}{2}(x-\bar{x})'Q(x-\bar{x})+c'(x-\bar{x}) = \tfrac{1}{2}x'Qx+c'x+\left(\tfrac{1}{2}\bar{x}'Q\bar{x}-c'\bar{x}-b'(AQ^{-1}A')^{-1}b\right).$$

The last term in parentheses on the right-hand side above is constant, thus establishing that the programs (2.3) and (2.6) have the same optimal solution when $\bar{x}$ is given by Eq. (2.7). By combining Eqs. (2.4) and (2.7), we obtain the optimal solution of program (2.6):

$$x^* = -Q^{-1} \left(c - A'\lambda - A'(AQ^{-1}A')^{-1}b \right),$$

where λ is given by Eq. (2.5).

2.3 DIRECTIONS OF RECESSION AND EXISTENCE OF OPTIMAL SOLUTIONS

The question of existence of optimal solutions of convex optimization problems can be conveniently analyzed by using the theory of recession cone developed in Section 1.5. A key fact is that a convex function can be described in terms of its epigraph, which is a convex set. The recession cone of the epigraph can be used to obtain the directions along which the function decreases monotonically. This is the idea underlying the following proposition.

Proposition 2.3.1: Let $f : \Re^n \mapsto (-\infty, \infty]$ be a closed proper convex function and consider the level sets $V_\gamma = \{x \mid f(x) \le \gamma\}$, where γ is a scalar. Then:

(a) All the nonempty level sets V_γ have the same recession cone, given by

$$R_{V_\gamma} = \{y \mid (y, 0) \in R_{\mathrm{epi}(f)}\},$$

where $R_{\mathrm{epi}(f)}$ is the recession cone of the epigraph of f.

(b) If one nonempty level set V_γ is compact, then all of the level sets V_γ are compact.

Proof: (a) Fix a γ such that V_γ is nonempty. Let y be a vector in the recession cone R_{V_γ}. Then we have $f(x + \alpha y) \le \gamma$ for all $x \in R_{V_\gamma}$ and $\alpha \ge 0$, so that $(x + \alpha y, \gamma) \in \mathrm{epi}(f)$ for all $\alpha \ge 0$. It follows, using also the Recession Cone Theorem [Prop. 1.5.1(b), which applies since $\mathrm{epi}(f)$ is closed in view of the closedness of f], that $(y, 0) \in R_{\mathrm{epi}(f)}$, and

$$R_{V_\gamma} \subset \{y \mid (y, 0) \in R_{\mathrm{epi}(f)}\}.$$

Conversely, let $(0, y) \in R_{\mathrm{epi}(f)}$, and choose a vector $(x, \gamma) \in \mathrm{epi}(f)$ (there exists such a vector since V_γ is nonempty). We have $(x + \alpha y, \gamma) \in \mathrm{epi}(f)$ for all $\alpha \ge 0$, so that $f(x + \alpha y) \le \gamma$ and $(x + \alpha y) \in V_\gamma$ for all $\alpha \ge 0$. By the Recession Cone Theorem [Prop. 1.5.1(b), which applies since V_γ is closed in view of the closedness of f], it follows that $y \in R_{V_\gamma}$. Thus,

$$R_{V_\gamma} \supset \{y \mid (y, 0) \in R_{\mathrm{epi}(f)}\}.$$

(b) From Prop. 1.5.1(c), a nonempty level set V_γ is compact if and only if the recession cone R_{V_γ} contains no nonzero direction. By part (a), all nonempty level sets V_γ have the same recession cone, so if one of them is compact, all of them are compact. **Q.E.D.**

For a closed proper convex function $f : \Re^n \mapsto (-\infty, \infty]$, the (common) recession cone of the nonempty level sets

$$V_\gamma = \{x \mid f(x) \le \gamma\}, \qquad \gamma \in \Re,$$

is called the *recession cone of* f, and is denoted by R_f (see Fig. 2.3.1). Each $y \in R_f$ is called a *direction of recession of* f. Note that the requirement

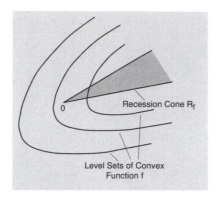

Figure 2.3.1. Illustration of the recession cone R_f of a proper closed convex function f. It is the (common) recession cone of the nonempty level sets of f.

that f be proper, implies that $f(x) < \infty$ for at least one $x \in \Re^n$, and guarantees that there exist some nonempty level sets, so that the preceding definition makes sense. Note also that the requirement that f be closed is essential for the level sets V_γ to have a common recession cone, as per Prop. 2.3.1(a). The reader may verify this by using as an example the convex but not closed function $f : \Re^2 \mapsto (-\infty, \infty]$ given by

$$f(x_1, x_2) = \begin{cases} -x_1 & \text{if } x_1 > 0, \ x_2 \geq 0, \\ x_2 & \text{if } x_1 = 0, \ x_2 \geq 0, \\ \infty & \text{if } x_1 < 0 \text{ or } x_2 < 0. \end{cases}$$

Here, for $\gamma < 0$, we have $V_\gamma = \{(x_1, x_2) \mid x_1 \geq -\gamma, \ x_2 \geq 0\}$, so that $(0, 1) \in R_{V_\gamma}$, but $V_0 = \{(x_1, x_2) \mid x_1 > 0, \ x_2 \geq 0\} \cup \{(0, 0)\}$, so that $(0, 1) \notin R_{V_0}$.

2.3.1 Existence of Solutions of Convex Programs

We will now discuss the use of directions of recession in asserting the existence of solutions of convex optimization problems. The most intuitive way to look at directions of recession of a convex function f is from a descent viewpoint: if we start at any $x \in \Re^n$ and move indefinitely along a direction of recession y, we must stay within each level set that contains x, or equivalently we must encounter exclusively points z with $f(z) \leq f(x)$. In words, *a direction of recession of f is a direction of continuous nonascent for f.* Conversely, if we start at some $x \in \Re^n$ and while moving along a direction y, we encounter a point z with $f(z) > f(x)$, then y cannot be a direction of recession. By the convexity of the level sets of f, once we cross the relative boundary of a level set, we never cross it back again, and with a little thought, it can be seen that *a direction that is not a direction of recession of f is a direction of eventual continuous ascent of f* [see Figs. 2.3.2(e),(f)].

In view of the preceding observations, it is not surprising that directions of recession play a prominent role in characterizing the existence of

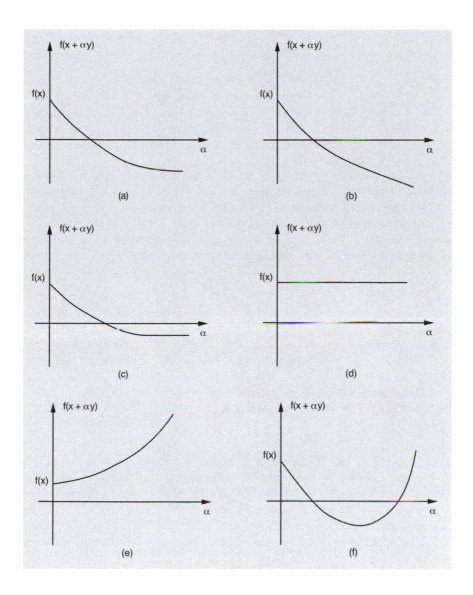

Figure 2.3.2. Ascent/descent behavior of a closed proper convex function starting at some $x \in \Re^n$ and moving along a direction y (see also Exercise 1.6). If y is a direction of recession of f, there are two possibilities: either f decreases monotonically to a finite value or $-\infty$ [figures (a) and (b), respectively], or f reaches a value that is less or equal to $f(x)$ and stays at that value [figures (c) and (d)]. If y is *not* a direction of recession of f, then eventually f increases monotonically to ∞ [figures (e) and (f)], i.e., for some $\overline{\alpha} \geq 0$ and all $\alpha_1, \alpha_2 \geq \overline{\alpha}$ with $\alpha_1 < \alpha_2$, we have

$$f(x + \alpha_1 y) < f(x + \alpha_2 y).$$

solutions of convex optimization problems. The following proposition deals with the case where the optimal solution set is bounded.

Proposition 2.3.2: Let X be a closed convex subset of $\Re^n$, and let $f : \Re^n \mapsto (-\infty, \infty]$ be a closed convex function such that $X \cap \mathrm{dom}(f) \neq \emptyset$. The set of minimizing points of f over X is nonempty and compact if and only if X and f have no common nonzero direction of recession.

Proof: Let X^* denote the set of minimizing points of f over X. We first prove the result for the special case where $X = \Re^n$. We then use the result for this special case to prove the result for the general case where $X \neq \Re^n$.

Assume that $X = \Re^n$. We must then show that X^* is nonempty and compact if and only if f has no nonzero directions of recession. Indeed if X^* is nonempty and compact, we have

$$X^* = \left\{ x \;\middle|\; f(x) \le \inf_{x \in \Re^n} f(x) \right\},$$

so the recession cones of X^* and f coincide, and consist of just the zero vector because X^* is compact [cf. Prop. 1.5.1(c)]. Conversely, if f has no nonzero direction of recession, all its nonempty level sets $\{x \mid f(x) \le \gamma\}$ are compact, so X^* is nonempty and compact by Weierstrass' Theorem (Prop. 2.1.1).

Consider now the more general case where $X \neq \Re^n$. We introduce the function $\tilde{f} : \Re^n \mapsto (-\infty, \infty]$ defined by

$$\tilde{f}(x) = \begin{cases} f(x) & \text{if } x \in X, \\ \infty & \text{otherwise.} \end{cases}$$

This function is closed and convex, and it is proper since its domain is $X \cap \mathrm{dom}(f)$, which is nonempty by assumption. Furthermore, the set of minimizing points of $\tilde{f}$ over $\Re^n$ is X^*. By the result already shown for the case where $X = \Re^n$, X^* is nonempty and compact if and only if $\tilde{f}$ has no nonzero direction of recession. The latter statement is equivalent to X and f having no common nonzero direction of recession, since for any scalar γ, we have

$$\{x \mid \tilde{f}(x) \le \gamma\} = X \cap \{x \mid f(x) \le \gamma\},$$

so the recession cones of $\tilde{f}$, X, and f are related by $R_{\tilde{f}} = R_X \cap R_f$ [cf. Props. 1.5.1(e) and 2.3.1(a), which apply since X is closed and convex, and f and $\tilde{f}$ are closed proper and convex]. **Q.E.D.**

If the closed convex set X and the closed proper convex function f of the above proposition have a common direction of recession, then either the optimal solution set is empty [take for example, $X = (-\infty, 0]$ and $f(x) = e^x$] or else it is nonempty and unbounded [take for example, $X = (-\infty, 0]$ and $f(x) = \max\{0, x\}$].

2.3.2   Unbounded Optimal Solution Sets

In our existence analysis so far (Weierstrass' Theorem and Prop. 2.3.2), we have aimed to derive conditions under which the set of optimal solutions is not only nonempty, but also compact. We now proceed to obtain conditions that guarantee the attainment of the minimum over a possibly unbounded set.

We consider the minimization of a closed proper convex function $f : \Re^n \mapsto (-\infty, \infty]$ over a closed convex set X. The key idea of our approach is to reduce the question of existence of an optimal solution to a question about the nonemptiness of a set intersection. In particular, let

$$f^* = \inf_{x \in X} f(x),$$

let $\{\gamma_k\}$ be a scalar sequence such that $\gamma_k \downarrow f^*$, and consider the level sets

$$V_k = \big\{x \in \Re^n \mid f(x) \le \gamma_k\big\}.$$

Then the set of minimizers of f over X is equal to $\cap_{k=1}^{\infty}(X \cap V_k)$. The question of nonemptiness of this set can be addressed using the theory developed in Section 1.5. This theory involves assumptions about the recession cones and the lineality spaces of X and the level sets V_k. We introduce the corresponding terminology, and then we prove an existence result under three different sets of conditions that correspond to the three cases discussed in Section 1.5 (cf. Props. 1.5.5-1.5.7).

We recall from Section 1.5 that the lineality space of X, denoted by L_X, is the subspace of all directions y such that both $y \in R_X$ and $-y \in R_X$. The lineality space of the recession cone R_f of the function f, denoted by L_f, is the set of all $y \in \Re^n$ such that both y and $-y$ are directions of recession of f, i.e.,

$$L_f = R_f \cap (-R_f).$$

Equivalently, $y \in L_f$ if and only if both y and $-y$ are directions of recession of each of the nonempty level sets $\{x \mid f(x) \le \gamma\}$ (cf. Prop. 2.3.1). In view of the convexity of f, we see that $y \in L_f$ if and only if

$$f(x + \alpha y) = f(x), \qquad \forall\, x \in \mathrm{dom}(f),\ \forall\, \alpha \in \Re.$$

Consequently, any $y \in L_f$ is called a *direction in which f is constant*, and L_f is also called the *constancy space of f*.

For example, if f is a linear function given by

$$f(x) = c'x,$$

where $c \in \Re^n$, then its recession cone and constancy space are

$$R_f = \{y \mid c'y \le 0\}, \qquad L_f = \{y \mid c'y = 0\}.$$

If f is a quadratic function given by

$$f(x) = x'Qx + c'x + b,$$

where Q is a symmetric positive semidefinite $n \times n$ matrix, c is a vector in $\Re^n$, and b is a scalar, then its recession cone and constancy space are

$$R_f = \{y \mid Qy = 0, \ c'y \leq 0\}, \qquad L_f = \{y \mid Qy = 0, \ c'y = 0\}.$$

The following result provides various conditions for the set of minima of f over X to be nonempty (but possibly unbounded). The proof is a straightforward application of Props. 1.5.5-1.5.7, which provide conditions for nonemptiness of the intersection of closed sets. Some refinements of this proposition, are given in Exercises 2.6-2.8. In particular, Exercise 2.6 provides a substantial extension of condition (2) of the proposition.

Proposition 2.3.4: Let X be a closed convex subset of $\Re^n$, let $f : \Re^n \mapsto (-\infty, \infty]$ be a closed convex function such that $X \cap \mathrm{dom}(f) \neq \emptyset$, and denote $f^* = \inf_{x \in X} f(x)$. The set of minimizing points of f over X is nonempty under any one of the following conditions:

(1) $R_X \cap R_f = L_X \cap L_f$.

(2) $R_X \cap R_f \subset L_f$, and X is specified by linear inequality constraints, i.e.,

$$X = \{x \mid a_j'x \leq b_j, \ j = 1, \ldots, r\},$$

where a_j are vectors in $\Re^n$ and b_j are scalars.

(3) $f^* > -\infty$, the function f is of the form

$$f(x) = x'Qx + c'x,$$

and X is of the form

$$X = \{x \mid x'Q_jx + a_j'x + b_j \leq 0, \ j = 1, \ldots, r\},$$

where Q and Q_j are symmetric positive semidefinite $n \times n$ matrices, c and a_j are vectors in $\Re^n$, and b_j are scalars.

Under condition (1), the set of minimizing points is of the form

$$\tilde{X} + (L_X \cap L_f),$$

where $\tilde{X}$ is some nonempty and compact set.

Proof: We choose a scalar sequence $\{\gamma_k\}$ such that $\gamma_k \downarrow f^*$, and we consider the (nonempty) level sets

$$V_k = \{x \in \Re^n \mid f(x) \leq \gamma_k\}.$$

Under each of the conditions (1)-(3), we show that the set of minimizers of f over X, which is

$$X^* = \cap_{k=1}^{\infty}(X \cap V_k),$$

is nonempty.

Let condition (1) hold. The sets $X \cap V_k$ are nonempty, closed, convex, and nested. Furthermore, they have the same recession cone, $R_X \cap R_f$, and the same lineality space $L_X \cap L_f$, while by assumption, $R_X \cap R_f = L_X \cap L_f$. It follows by Prop. 1.5.5 that X^* is nonempty and in fact has the form

$$X^* = \tilde{X} + (L_X \cap L_f),$$

where $\tilde{X}$ is some nonempty compact set.

Let condition (2) hold. The sets V_k are nested and the intersection $X \cap V_k$ is nonempty for all k. Furthermore, the sets V_k have the same recession cone, R_f, and the same lineality space, L_f, while by assumption, $R_X \cap R_f \subset L_f$ and X is specified by linear inequality constraints. By Prop. 1.5.6, it follows that X^* is nonempty.

Let condition (3) hold. The sets V_k have the form

$$V_k = \{x \in \Re^n \mid x'Qx + c'x \leq \gamma_k\},$$

where γ_k converges to the scalar f^*. Furthermore, X is specified by convex quadratic inequalities, and the intersection $X \cap V_k$ is nonempty for all k. By Prop. 1.5.7, it follows that X^* is nonempty.   **Q.E.D.**

Figure 2.3.3(b) provides a counterexample showing that if X is specified by nonlinear constraints, the condition

$$R_X \cap R_f \subset L_f$$

is not sufficient to guarantee the existence of optimal solutions or even the finiteness of f^*.

As can be seen from the example of Fig. 2.3.3(b), it is possible for a convex function to be bounded below and to attain a minimum (in fact be constant) along any direction of recession of a convex constraint set, but not to attain a minimum over the entire set. In the special cases of linear and quadratic programming problems, boundedness from below of the cost function over the constraint set guarantees the existence of an optimal solution, as shown by Prop. 2.3.4 under condition (3). The following proposition gives an alternative proof of this fact, and also provides a necessary and sufficient condition for attainment of the minimum.

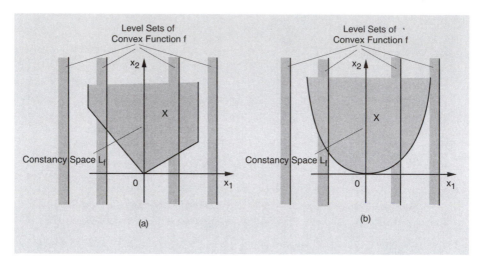

Figure 2.3.3. Illustration of the issues regarding existence of an optimal solution assuming $R_X \cap R_f \subset L_f$, i.e., that every common direction of recession of X and f is a direction in which f is constant [cf. Prop. 2.3.4 under condition (2)].

In both problems illustrated in (a) and (b), the cost function is

$$f(x_1, x_2) = e^{x_1}.$$

In the problem of (a), the constraints are linear, while in the problem of (b), X is specified by a quadratic inequality:

$$X = \left\{ (x_1, x_2) \mid x_1^2 \leq x_2 \right\}.$$

In both cases, we have

$$R_X = \left\{ (y_1, y_2) \mid y_1 = 0, \ y_2 \geq 0 \right\},$$

$$R_f = \left\{ (y_1, y_2) \mid y_1 \leq 0, \ y_2 \in \Re \right\}, \qquad L_f = \left\{ (y_1, y_2) \mid y_1 = 0, \ y_2 \in \Re \right\},$$

so that $R_X \cap R_f \subset L_f$.

In the problem of (a), it can be seen that an optimal solution exists. In the problem of (b), however, we have $f(x_1, x_2) > 0$ for all (x_1, x_2), while for $x_1 = -\sqrt{x_2}$ where $x_2 \geq 0$, we have $(x_1, x_2) \in X$ with

$$\lim_{x_2 \to \infty} f\left(-\sqrt{x_2}, x_2\right) = \lim_{x_2 \to \infty} e^{-\sqrt{x_2}} = 0,$$

implying that $f^* = 0$. Thus f cannot attain the minimum value f^* over X. Note that f attains a minimum over the intersection of any line with X.

If in the problem of (b) the cost function were instead $f(x_1, x_2) = x_1$, we would still have $R_X \cap R_f \subset L_f$ and f would still attain a minimum over the intersection of any line with X, but it can be seen that $f^* = -\infty$. If the constraint set were instead $X = \left\{ (x_1, x_2) \mid |x_1| \leq x_2 \right\}$, which can be specified by linear inequalities, we would still have $f^* = -\infty$, but then the condition $R_X \cap R_f \subset L_f$ would be violated.

Proposition 2.3.5: (Existence of Solutions of Quadratic Programs) Let $f : \Re^n \mapsto \Re$ be a quadratic function of the form

$$f(x) = x'Qx + c'x,$$

where Q is a symmetric positive semidefinite $n \times n$ matrix and c is a vector in $\Re^n$. Let also X be a nonempty set of the form

$$X = \{x \mid Ax \leq b\},$$

where A is an $m \times n$ matrix and b is a vector in $\Re^m$. The following are equivalent:

(i) f attains a minimum over X.

(ii) $f^* > -\infty$.

(iii) For all y such that $Ay \leq 0$ and $y \in N(Q)$, we have $c'y \geq 0$.

Proof: Clearly (i) implies (ii). We next show that (ii) implies (iii). For all $x \in X$, $y \in N(Q)$ with $Ay \leq 0$, and $\alpha \geq 0$, we have $x + \alpha y \in X$ and

$$f(x + \alpha y) = (x + \alpha y)'Q(x + \alpha y) + c'(x + \alpha y) = f(x) + \alpha c'y.$$

If $c'y < 0$, then $\lim_{\alpha \to \infty} f(x + \alpha y) = -\infty$, so that $f^* = -\infty$, which contradicts (ii). Hence, (ii) implies that $c'y \geq 0$ for all $y \in N(Q)$ with $Ay \leq 0$.

We finally show that (iii) implies (i). Indeed, we have $R_X = \{y \mid Ay \leq 0\}$. Furthermore, if $y \in R_f$, we must have $y \in N(Q)$ (by the positive semidefiniteness of Q), and $c'y \leq 0$ [since for $y \in N(Q)$, f becomes linear along the direction y, i.e., $f(x + \alpha y) = f(x) + \alpha c'y$ for all $x \in \Re^n$ and $\alpha \in \Re$]. Thus $R_f = N(Q) \cap \{y \mid c'y \leq 0\}$, so that

$$R_X \cap R_f = \{y \mid Ay \leq 0\} \cap N(Q) \cap \{y \mid c'y \leq 0\}.$$

In view of condition (iii), it follows that if $y \in R_X \cap R_f$, then we have $c'y = 0$, so that f is constant in the direction y. Thus, $R_X \cap R_f \subset L_f$, and by Prop. 2.3.4 [under condition (2) of that proposition], f attains a minimum over X.   **Q.E.D.**

2.3.3   Partial Minimization of Convex Functions

In our development of minimax and duality theory, we will often encounter functions obtained by minimizing other functions partially, i.e., with respect to some of their variables. In particular, starting with a function

$F : \Re^{n+m} \mapsto [-\infty, \infty]$ of two vectors $x \in \Re^n$ and $z \in \Re^m$, we may consider the function $f : \Re^n \mapsto [-\infty, \infty]$ given by

$$f(x) = \inf_{z \in \Re^m} F(x, z).$$

It is then useful to be able to deduce properties of f, such as convexity and closedness, from corresponding properties of F.

We will show in the following proposition that convexity of F implies convexity of f. Furthermore, we will show that closedness of F together with conditions that generally imply the attainment of the infimum over z (cf. Props. 2.3.2 and 2.3.4) also imply closedness of f. We will see that this similarity can be traced to a unifying principle: both questions of existence of optimal solutions and of preservation of closedness under partial minimization can be reduced to questions of nonemptiness of intersections of closed sets, which can be addressed using the theory of Section 1.5.

Proposition 2.3.6: Let $F : \Re^{n+m} \mapsto [-\infty, \infty]$ be a convex function. Then the function f given by

$$f(x) = \inf_{z \in \Re^m} F(x, z), \qquad x \in \Re^n,$$

is convex.

Proof: If $f(x) = \infty$ for all $x \in \Re^n$, then the epigraph of f is empty and hence convex, so f is convex. Assume that $\text{epi}(f) \neq \emptyset$, and let $(\overline{x}, \overline{w})$ and $(\tilde{x}, \tilde{w})$ be two points in $\text{epi}(f)$. Then $f(\overline{x}) < \infty$, $f(\tilde{x}) < \infty$, and there exist sequences $\{\overline{z}_k\}$ and $\{\tilde{z}_k\}$ such that

$$F(\overline{x}, \overline{z}_k) \to f(\overline{x}), \qquad F(\tilde{x}, \tilde{z}_k) \to f(\tilde{x}).$$

Using the definition of f and the convexity of F, we have for all $\alpha \in [0, 1]$ and k,

$$f\big(\alpha\overline{x} + (1 - \alpha)\tilde{x}\big) \leq F\big(\alpha\overline{x} + (1 - \alpha)\tilde{x}, \alpha\overline{z}_k + (1 - \alpha)\tilde{z}_k\big)$$
$$\leq \alpha F(\overline{x}, \overline{z}_k) + (1 - \alpha)F(\tilde{x}, \tilde{z}_k).$$

By taking the limit as $k \to \infty$, we obtain

$$f\big(\alpha\overline{x} + (1 - \alpha)\tilde{x}\big) \leq \alpha f(\overline{x}) + (1 - \alpha)f(\tilde{x}) \leq \alpha\overline{w} + (1 - \alpha)\tilde{w}.$$

It follows that the point $\alpha(\overline{x}, \overline{w}) + (1 - \alpha)(\tilde{x}, \tilde{w})$ belongs to $\text{epi}(f)$. Thus $\text{epi}(f)$ is convex, implying that f is convex. **Q.E.D.**

It is not necessarily true that closedness of F implies the closedness of f [this is contrary to the case where F is *maximized* over z; cf. Prop. 1.2.4(c)]. The reason is that the level sets $\{x \mid f(x) \le \gamma\}$ are obtained by projection of level sets of F on $\Re^n$: for any $\gamma \in \Re$ and sequence $\{\gamma_k\}$ with $\gamma_k \downarrow \gamma$,

$$\{x \mid f(x) \le \gamma\} = \cap_{k=1}^{\infty}\{x \mid \text{there exists } (x, z) \text{ with } F(x, z) \le \gamma_k\}, \quad (2.8)$$

and the set in the right-hand side is the projection on the space of x of the level set $\{(x, z) \mid F(x, z) \le \gamma_k\}$. We know from our discussion in Section 1.5 that the projection operation (and more generally a linear transformation) on an unbounded closed set need not preserve closedness – some conditions involving directions of recession are required for this. Thus, f is not guaranteed to be closed, even when F is closed. However, the results of Section 1.5 (Prop. 1.5.8) can be used to address this situation, as in the following four propositions, which correspond to different sets of assumptions that guarantee that the projections of the level sets of F in Eq. (2.8) are closed.

Proposition 2.3.7: Let $F : \Re^{n+m} \mapsto (-\infty, \infty]$ be a closed proper convex function, and consider the function f given by

$$f(x) = \inf_{z \in \Re^m} F(x, z), \qquad x \in \Re^n.$$

Assume that there exists a vector $\overline{x} \in \Re^n$ and a scalar $\overline{\gamma}$ such that the set

$$\{z \mid F(\overline{x}, z) \le \overline{\gamma}\}$$

is nonempty and compact. Then f is convex, closed, and proper. Furthermore, for each $x \in \text{dom}(f)$, the set of points that attain the infimum of $F(x, \cdot)$ over $\Re^m$ is nonempty and compact.

Proof: We first note that by Prop. 2.3.6, f is convex. Consider any nonempty level set

$$V_\gamma = \{(x, z) \mid F(x, z) \le \gamma\}.$$

Any direction of recession of V_γ of the form $(0, y)$ must also be a direction of recession of the level set

$$V_{\overline{\gamma}} = \{(x, z) \mid F(x, z) \le \overline{\gamma}\},$$

[cf. Prop. 2.3.1(a)], and hence must satisfy

$$F(\overline{x}, z + \alpha y) \le \overline{\gamma}, \qquad \forall \, \alpha \ge 0.$$

Since, by assumption, the set $\{z \mid F(\overline{x}, z) \leq \overline{\gamma}\}$ is compact, it follows that $y = 0$. Thus there is no nonzero direction of recession of V_γ of the form $(0, y)$.

We now note that the nullspace of the projection operation onto the space of x [i.e., the mapping $(x, z) \mapsto x$] is the set of vectors of the form $(0, y)$, so this nullspace does not contain any nonzero direction of recession of V_γ. We can thus use Prop. 1.5.8(a) to assert that the projection of any level set V_γ onto the space of x is closed. From Eq. (2.8), we see that the level sets $\{x \mid f(x) \leq \gamma\}$ of f are closed and that f is a closed function (cf. Prop. 1.2.2).

Finally, we have shown that for every $x \in \text{dom}(f)$, the function $F(x, \cdot)$ has no nonzero direction of recession, so it follows from Prop. 2.3.2 that the set of points attaining the infimum of $F(x, \cdot)$ is nonempty and compact. Thus, $f(x)$ is finite for all $x \in \text{dom}(f)$. Furthermore, since $f(\overline{x}) \leq \overline{\gamma}$, it follows that $\overline{x}$ belongs to $\text{dom}(f)$ and $f(\overline{x})$ is finite. Hence, f is proper [an improper closed convex function takes an infinite value (∞ or $-\infty$) at every point]. **Q.E.D.**

Proposition 2.3.8: Let $F : \Re^{n+m} \mapsto (-\infty, \infty]$ be a closed proper convex function, and consider the function f given by

$$f(x) = \inf_{z \in \Re^m} F(x, z), \qquad x \in \Re^n.$$

Assume that there exists a vector $\overline{x} \in \Re^n$ and a scalar $\overline{\gamma}$ such that the set

$$\{z \mid F(\overline{x}, z) \leq \overline{\gamma}\}$$

is nonempty and its recession cone is equal to its lineality space. Then f is convex, closed, and proper. Furthermore, for each $x \in \text{dom}(f)$, the set of points that attain the infimum of $F(x, \cdot)$ over $\Re^m$ is nonempty.

Proof: The proof is similar to the one of the preceding proposition. We first note that by Prop. 2.3.6, f is convex. Consider any nonempty level set

$$V_\gamma = \{(x, z) \mid F(x, z) \leq \gamma\}.$$

Any direction of recession of V_γ of the form $(0, y)$ must also be a direction of recession of the level set

$$V_{\overline{\gamma}} = \{(x, z) \mid F(x, z) \leq \overline{\gamma}\},$$

[cf. Prop. 2.3.1(a)], and hence must satisfy

$$F(\overline{x}, z + \alpha y) \leq \overline{\gamma}, \qquad \forall \, \alpha \geq 0.$$

Hence $(0, y)$ is a direction of recession of $F(\overline{x}, \cdot)$, so by assumption, it is a direction along which $F(\overline{x}, \cdot)$ is constant. It follows that $(0, y)$ belongs to the lineality space of $V_{\overline{\gamma}}$, and hence also to the lineality space of V_{γ}.

In conclusion, we have shown that any direction of recession of V_{γ} of the form $(0, y)$ is also in the lineality space of V_{γ}. We can thus use Prop. 1.5.8(a) to assert that the projection of any level set V_{γ} onto the space of x is closed, which by Eq. (2.8), implies that f is closed. Finally, we also use Prop. 2.3.4 [condition (1)] to assert that the set of points attaining the infimum of $F(x, \cdot)$ is nonempty for all $x \in \text{dom}(f)$. In particular, this implies that $f(\overline{x})$ is finite, which in view of the closedness of f, implies that f is proper.   **Q.E.D.**

The following two propositions apply to the case where, for each x, a real-valued function $\overline{F}(x, z)$ is partially minimized over all z such that $(x, z) \in C$, where C is either specified by linear inequalities,

$$C = \big\{ (x, z) \mid a'_j(x, z) \le b_j, \ j = 1, \ldots, r \big\}, \tag{2.9}$$

or is specified by convex quadratic inequalities,

$$C = \big\{ (x, z) \mid (x, z)' Q_j (x, z) + a'_j(x, z) + b_j \le 0, \ j = 1, \ldots, r \big\}, \tag{2.10}$$

where a_j are vectors in $\Re^{n+m}$, b_j are scalars, and Q_j are symmetric positive semidefinite $(n + m) \times (n + m)$ matrices.

Proposition 2.3.9: Let $F : \Re^{n+m} \mapsto (-\infty, \infty]$ be a function of the form

$$F(x, z) = \begin{cases} \overline{F}(x, z) & \text{if } (x, z) \in C, \\ \infty & \text{otherwise,} \end{cases}$$

where $\overline{F} : \Re^{n+m} \mapsto (-\infty, \infty]$ is a closed proper convex function on $\Re^{n+m}$ and C is a subset of $\Re^{n+m}$ that is specified by linear inequalities as in Eq. (2.9). Consider the function f given by

$$f(x) = \inf_{z \in \Re^m} F(x, z), \qquad x \in \Re^n.$$

Assume that there exists a vector $\overline{x} \in \text{dom}(f)$ such that every direction of recession of the function $F(\overline{x}, \cdot)$ is a direction along which the function $\overline{F}(\overline{x}, \cdot)$ is constant. Then f is convex, closed, and proper. Furthermore, for each $x \in \text{dom}(f)$, the set of points that attain the infimum of $F(x, \cdot)$ over $\Re^m$ is nonempty.

Proof: The proof is similar to the proofs of the preceding two propositions. By Prop. 2.3.6, it follows that f is convex. To show that f is closed, we first note that every level set of F is given by

$$V_\gamma = C \cap \big\{ (x, z) \mid \overline{F}(x, z) \le \gamma \big\}.$$

We then show that every direction of recession of a nonempty level set V_γ that has the form $(0, y)$ (i.e., belongs to the nullspace of the projection onto the space of x) is in the lineality space of the level set

$$\big\{ (x, z) \mid \overline{F}(x, z) \le \gamma \big\}.$$

We then use Prop. 1.5.8(b) to assert that the projection of any level set V_γ onto the space of x is closed, which by Eq. (2.8), implies that f is closed. We also use Prop. 2.3.4 [condition (2)] to assert that the set of points attaining the infimum of $F(x, \cdot)$ is nonempty for all $x \in \mathrm{dom}(f)$, which implies in particular that $f(\overline{x})$ is finite and that f is proper.   **Q.E.D.**

Proposition 2.3.10: Let $F : \Re^{n+m} \mapsto (-\infty, \infty]$ be a function of the form

$$F(x, z) = \begin{cases} \overline{F}(x, z) & \text{if } (x, z) \in C, \\ \infty & \text{otherwise,} \end{cases}$$

where $\overline{F}$ is a quadratic convex function on $\Re^{n+m}$ and C is a subset of $\Re^{n+m}$ that is specified by convex quadratic inequalities as in Eq. (2.10). Consider the function f given by

$$f(x) = \inf_{z \in \Re^m} F(x, z), \qquad x \in \Re^n.$$

Assume that there exists a vector $\overline{x} \in \Re^n$ such that $-\infty < f(\overline{x}) < \infty$. Then f is convex, closed, and proper. Furthermore, for each $x \in \mathrm{dom}(f)$, the set of points that attain the infimum of $F(x, \cdot)$ over $\Re^m$ is nonempty.

Proof: Similar to the proofs of the preceding three propositions, we use Prop. 2.3.6 to assert that f is convex, and we use Prop. 1.5.8(c) to assert that the projection of any level set V_γ onto the space of x is closed. This implies that f is closed, and it is also proper because, by assumption, there exists an $\overline{x}$ such that $f(\overline{x})$ is finite. We also use Prop. 2.3.4 [condition (3)] to assert that the set of points attaining the infimum of $F(x, \cdot)$ is nonempty for all $x \in \mathrm{dom}(f)$.   **Q.E.D.**

An interesting generalization of the preceding proposition is given in Exercise 2.7. In particular, it is shown that f is closed if C and $\overline{F}$ are defined in terms of convex functions from a class called *bidirectionally flat*, which contains the class of convex quadratic functions as a special case.

For another view of the question of closedness of a partially minimized function

$$f(x) = \inf_{z \in \Re^m} F(x, z),$$

note that we can prove the following relation:

$$P\big(\text{epi}(F)\big) \subset \text{epi}(f) \subset \text{cl}\big(P\big(\text{epi}(F)\big)\big), \tag{2.11}$$

where $P(\cdot)$ denotes projection on the space of (x, w), i.e., $P(x, z, w) = (x, w)$. [The left-hand side of this relation follows from the definition

$$\text{epi}(f) = \left\{ (x, w) \mid \inf_{z \in \Re^m} F(x, z) \leq w \right\}.$$

To show the right-hand side, note that for any $(x, w) \in \text{epi}(f)$ and every k, there exists a z_k such that $(x, z_k, w + 1/k) \in \text{epi}(F)$, so that $(x, w + 1/k) \in P\big(\text{epi}(F)\big)$ and $(x, w) \in \text{cl}\big(P\big(\text{epi}(F)\big)\big)$.] Equation (2.11), provides some intuition into Prop. 2.3.6 (convexity of F implies convexity of f). The equation also implies that if $P\big(\text{epi}(F)\big)$ is a closed set, then f is closed, thereby providing an alternative line of proof of the preceding four propositions based on the analysis of Section 1.5 (cf. Prop. 1.5.8).

2.4 HYPERPLANES

Some of the most important principles in convexity and optimization, including duality, revolve around the use of hyperplanes, i.e., $(n - 1)$-dimensional affine sets. A hyperplane has the property that it divides the space into two halfspaces. We will see shortly that a closed convex set has an important characterization: it is equal to the intersection of all the halfspaces that contain it. Thus, any closed convex set can be described in dual fashion:

(a) As the closure of the union of all line segments connecting the points of the set.

(b) As the intersection of all closed halfspaces containing the set.

This fundamental principle carries over to a closed convex function, once the function is specified in terms of its epigraph. In this section, we develop the principle just outlined, and in the following section, we apply it to some geometrical constructions that are central in duality theory.

A *hyperplane* in $\Re^n$ is a set of the form $\{x \mid a'x = b\}$, where a is nonzero vector in $\Re^n$ and b is a scalar. If $\overline{x}$ is any vector in a hyperplane $H = \{x \mid a'x = b\}$, then we must have $a'\overline{x} = b$, so the hyperplane can be equivalently described as

$$H = \{x \mid a'x = a'\overline{x}\},$$

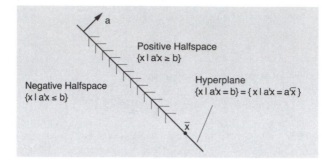

Figure 2.4.1. Illustration of the hyperplane $H = \{x \mid a'x = b\}$. If $\overline{x}$ is any vector in the hyperplane, then the hyperplane can be equivalently described as

$$H = \{x \mid a'x = a'\overline{x}\} = \overline{x} + \{x \mid a'x = 0\}.$$

The hyperplane divides the space into two halfspaces as illustrated.

or

$$H = \overline{x} + \{x \mid a'x = 0\}.$$

Thus, H is an affine set that is parallel to the subspace $\{x \mid a'x = 0\}$. This subspace is orthogonal to the vector a, and consequently, a is called the *normal* vector of H; see Fig. 2.4.1.

The sets

$$\{x \mid a'x \geq b\}, \qquad \{x \mid a'x \leq b\},$$

are called the *closed halfspaces* associated with the hyperplane (also referred to as the *positive and negative halfspaces*, respectively). The sets

$$\{x \mid a'x > b\}, \qquad \{x \mid a'x < b\},$$

are called the *open halfspaces* associated with the hyperplane.

We say that two sets C_1 and C_2 are *separated by a hyperplane* $H = \{x \mid a'x = b\}$ if each lies in a different closed halfspace associated with H, i.e., if either

$$a'x_1 \leq b \leq a'x_2, \qquad \forall\ x_1 \in C_1,\ \forall\ x_2 \in C_2,$$

or

$$a'x_2 \leq b \leq a'x_1, \qquad \forall\ x_1 \in C_1,\ \forall\ x_2 \in C_2.$$

We then also say that the hyperplane H *separates* C_1 and C_2, or that H is a *separating hyperplane* of C_1 and C_2. We use several different variants of this terminology. For example, the statement that two sets C_1 and C_2 *can be separated by a hyperplane* or that *there exists a hyperplane separating C_1 and C_2*, means that there exists a vector $a \neq 0$ such that

$$\sup_{x \in C_1} a'x \leq \inf_{x \in C_2}\ a'x, \qquad \forall\ x_1 \in C_1,\ \forall\ x_2 \in C_2;$$

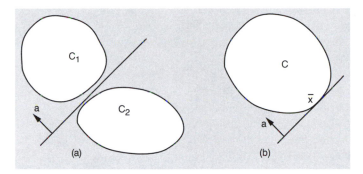

Figure 2.4.2. (a) Illustration of a hyperplane separating two sets C_1 and C_2. (b) Illustration of a hyperplane supporting a set C at a point $\overline{x}$ that belongs to the closure of C.

(see Fig. 2.4.2).

If a vector $\overline{x}$ belongs to the closure of a set C, a hyperplane that separates C and the singleton set $\{\overline{x}\}$ is said to be *supporting C at $\overline{x}$.* Thus the statement that *there exists a supporting hyperplane of C at $\overline{x}$* means that there exists a vector $a \neq 0$ such that

$$a'\overline{x} \leq a'x, \qquad \forall\, x \in C,$$

or equivalently, since $\overline{x}$ is a closure point of C,

$$a'\overline{x} = \inf_{x \in C} a'x.$$

As illustrated in Fig. 2.4.2, a supporting hyperplane is a hyperplane that "just touches" the set C.

We will prove several results regarding the existence of hyperplanes that separate two convex sets. Some of these results assert the existence of separating hyperplanes with special properties that will prove useful in various specialized contexts to be described later. The following proposition deals with the basic case where one of the two sets consists of a single vector. The proof is based on the Projection Theorem and is illustrated in Fig. 2.4.3.

Proposition 2.4.1: (Supporting Hyperplane Theorem) Let C be a nonempty convex subset of $\Re^n$ and let $\overline{x}$ be a vector in $\Re^n$. If $\overline{x}$ is not an interior point of C, there exists a hyperplane that passes through $\overline{x}$ and contains C in one of its closed halfspaces, i.e., there exists a vector $a \neq 0$ such that

$$a'\overline{x} \leq a'x, \qquad \forall\, x \in C. \tag{2.12}$$

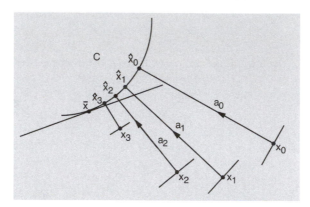

Figure 2.4.3. Illustration of the proof of the Supporting Hyperplane Theorem for the case where the vector $\bar{x}$ belongs to $\mathrm{cl}(C)$, the closure of C. We choose a sequence $\{x_k\}$ of vectors that do not belong to $\mathrm{cl}(C)$, with $x_k \to \bar{x}$, and we project x_k on $\mathrm{cl}(C)$. We then consider, for each k, the hyperplane that is orthogonal to the line segment connecting x_k and its projection $\hat{x}_k$, and passes through x_k. These hyperplanes "converge" to a hyperplane that supports C at $\bar{x}$.

Proof: Consider $\mathrm{cl}(C)$, the closure of C, which is a convex set by Prop. 1.2.1(d). Let $\{x_k\}$ be a sequence of vectors such that $x_k \to \bar{x}$ and $x_k \notin \mathrm{cl}(C)$ for all k; such a sequence exists because $\bar{x}$ does not belong to the interior of C. If $\hat{x}_k$ is the projection of x_k on $\mathrm{cl}(C)$, we have by part (b) of the Projection Theorem (Prop. 2.2.1)

$$(\hat{x}_k - x_k)'(x - \hat{x}_k) \geq 0, \qquad \forall\ x \in \mathrm{cl}(C).$$

Hence we obtain for all $x \in \mathrm{cl}(C)$ and all k,

$$(\hat{x}_k - x_k)'x \geq (\hat{x}_k - x_k)'\hat{x}_k = (\hat{x}_k - x_k)'(\hat{x}_k - x_k) + (\hat{x}_k - x_k)'x_k \geq (\hat{x}_k - x_k)'x_k.$$

We can write this inequality as

$$a_k'x \geq a_k'x_k, \qquad\qquad \forall\ x \in \mathrm{cl}(C),\ \forall\ k, \qquad\qquad (2.13)$$

where

$$a_k = \frac{\hat{x}_k - x_k}{\|\hat{x}_k - x_k\|}.$$

We have $\|a_k\| = 1$ for all k, and hence the sequence $\{a_k\}$ has a subsequence that converges to a nonzero limit a. By considering Eq. (2.13) for all a_k belonging to this subsequence and by taking the limit as $k \to \infty$, we obtain Eq. (2.12). **Q.E.D.**

Note that if $\bar{x}$ is a closure point of C, then the hyperplane of the preceding proposition supports C at $\bar{x}$. Note also that if C has empty interior, then any vector $\bar{x}$ can be separated from C as in the proposition.

Proposition 2.4.2: (Separating Hyperplane Theorem) Let C_1 and C_2 be two nonempty convex subsets of $\Re^n$. If C_1 and C_2 are disjoint, there exists a hyperplane that separates them, i.e., there exists a vector $a \neq 0$ such that

$$a'x_1 \leq a'x_2, \qquad \forall \, x_1 \in C_1, \, \forall \, x_2 \in C_2. \qquad (2.14)$$

Proof: Consider the convex set

$$C = C_2 - C_1 = \{x \mid x = x_2 - x_1, \, x_1 \in C_1, \, x_2 \in C_2\}.$$

Since C_1 and C_2 are disjoint, the origin does not belong to C, so by the Supporting Hyperplane Theorem (Prop. 2.4.1), there exists a vector $a \neq 0$ such that

$$0 \leq a'x, \qquad \forall \, x \in C,$$

which is equivalent to Eq. (2.14).   **Q.E.D.**

We next consider a stronger form of separation of two sets C_1 and C_2 in $\Re^n$. We say that a hyperplane $\{x \mid a'x = b\}$ *strictly separates C_1 and C_2* if it separates C_1 and C_2 while containing neither a point of C_1 nor a point of C_2, i.e.,

$$a'x_1 < b < a'x_2, \qquad \forall \, x_1 \in C_1, \, \forall \, x_2 \in C_2.$$

Clearly, C_1 and C_2 must be disjoint in order that they can be strictly separated. However, this is not sufficient to guarantee strict separation (see Fig. 2.4.4). The following proposition provides conditions that guarantee the existence of a strictly separating hyperplane.

Proposition 2.4.3: (Strict Separation Theorem) Let C_1 and C_2 be two disjoint nonempty convex sets. There exists a hyperplane that strictly separates C_1 and C_2 under any one of the following five conditions:

(1) $C_2 - C_1$ is closed.

(2) C_1 is closed and C_2 is compact.

(3) C_1 and C_2 are closed, and

$$R_{C_1} \cap R_{C_2} = L_{C_1} \cap L_{C_2},$$

where R_{C_i} and L_{C_i} denote the recession cone and the lineality space of C_i, $i = 1, 2$.

(4) C_1 is closed, and C_2 is specified by a finite number of linear inequality constraints:

$$C_2 = \{x \mid a_j'x \leq b_j, \ j = 1, \ldots, r\},$$

where a_j are some vectors and b_j are some scalars. Furthermore, every common direction of recession of C_1 and C_2 belongs to the lineality space of C_1.

(5) C_1 and C_2 are specified by a finite number of convex quadratic inequality constraints:

$$C_i = \{x \mid x'Q_{ij}x + a_{ij}'x \leq b_{ij}, \ j = 1, \ldots, r_i\}, \qquad i = 1, 2,$$

where Q_{ij} are symmetric positive semidefinite matrices, a_{ij} are some vectors, and b_{ij} are some scalars.

Proof: We will show the result under condition (1). The result will then follow under conditions (2)-(5), because these conditions imply condition (1) (see Prop. 1.5.9 and the discussion following that proposition).

Assume that $C_2 - C_1$ is closed, and consider the vector of minimum norm (projection of the origin) in $C_2 - C_1$. This vector is of the form $\overline{x}_2 - \overline{x}_1$, where $\overline{x}_1 \in C_1$ and $\overline{x}_2 \in C_2$. Let

$$a = \frac{\overline{x}_2 - \overline{x}_1}{2}, \qquad \overline{x} = \frac{\overline{x}_1 + \overline{x}_2}{2}, \qquad b = a'\overline{x}.$$

Then, $a \neq 0$, since $\overline{x}_1 \in C_1$, $\overline{x}_2 \in C_2$, and C_1 and C_2 are disjoint. We will show that the hyperplane

$$\{x \mid a'x = b\}$$

strictly separates C_1 and C_2, i.e., that

$$a'x_1 < b < a'x_2, \qquad \forall \ x_1 \in C_1, \ \forall \ x_2 \in C_2. \tag{2.15}$$

To this end we note that $\overline{x}_1$ is the projection of $\overline{x}$ on $\mathrm{cl}(C_1)$, while $\overline{x}_2$ is the projection of $\overline{x}$ on $\mathrm{cl}(C_2)$ [see Fig. 2.4.4(b)]. Thus, we have

$$(\overline{x} - \overline{x}_1)'(x_1 - \overline{x}_1) \leq 0, \qquad \forall \ x_1 \in C_1,$$

or equivalently, since $\overline{x} - \overline{x}_1 = a$,

$$a'x_1 \leq a'\overline{x}_1 = a'\overline{x} + a'(\overline{x}_1 - \overline{x}) = b - \|a\|^2 < b, \qquad \forall \ x_1 \in C_1.$$

Thus, the left-hand side of Eq. (2.15) is proved. The right-hand side is proved similarly. **Q.E.D.**

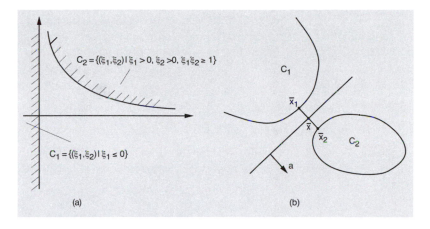

Figure 2.4.4. (a) An example of two disjoint convex sets that cannot be strictly separated. (b) Illustration of the construction of a strictly separating hyperplane of two disjoint closed convex sets C_1 and C_2.

Note that as a corollary of the preceding proposition, a closed set C can be strictly separated from a vector $\overline{x} \notin C$, i.e., from the singleton set $\{\overline{x}\}$. We will use this fact to provide a fundamental characterization of closed convex sets.

Proposition 2.4.4: The closure of the convex hull of a set C is the intersection of the closed halfspaces that contain C. In particular, a closed and convex set is the intersection of the closed halfspaces that contain it.

Proof: Assume first that C is closed and convex. Then, C is contained in the intersection of the closed halfspaces that contain C, so we focus on proving the reverse inclusion. Let $x \notin C$. By applying the Strict Separation Theorem [Prop. 2.4.3(b)] to the sets C and $\{x\}$, we see that there exists a closed halfspace containing C but not containing x. Hence, if $x \notin C$, then x cannot belong to the intersection of the closed halfspaces containing C, proving that C contains that intersection. Thus the result is proved for the case where C is closed and convex.

Consider the case of a general set C. Each closed halfspace H that contains C must also contain $\mathrm{conv}(C)$ (since H is convex), as well as $\mathrm{cl}\big(\mathrm{conv}(C)\big)$ (since H is closed). Hence, the intersection of all closed halfspaces containing C and the intersection of all closed halfspaces containing $\mathrm{cl}\big(\mathrm{conv}(C)\big)$ coincide. From what has been proved for the case of a closed convex set, the latter intersection is equal to $\mathrm{cl}\big(\mathrm{conv}(C)\big)$.   **Q.E.D.**

The dual characterization of a convex set given in the above proposition is fundamental for our optimization-related purposes. In the following section, we use this characterization to investigate two simple geometrical problems, dual to each other, which capture the essence of two important optimization topics:

(a) Saddle point and minimax theory (see Sections 2.6 and 3.5).

(b) Lagrangian duality theory (see Chapter 6).

Before doing so, we discuss one additional form of hyperplane separation, called *proper*, which turns out to be useful in some important optimization contexts.

Proper Hyperplane Separation

Let C_1 and C_2 be two subsets of $\Re^n$. We say that a hyperplane *properly separates C_1 and C_2* if it separates C_1 and C_2 and does not fully contain both C_1 and C_2. Thus there exists a hyperplane that properly separates C_1 and C_2 if and only if there is a vector a such that

$$\sup_{x_1 \in C_1} a'x_1 \leq \inf_{x_2 \in C_2} a'x_2, \qquad \inf_{x_1 \in C_1} a'x_1 < \sup_{x_2 \in C_2} a'x_2;$$

(see Fig. 2.4.5).

Note that a convex set in $\Re^n$ that has nonempty interior (and hence has dimension n) cannot be fully contained in a hyperplane (which has dimension $n-1$). Thus, in view of the Separating Hyperplane Theorem (Prop. 2.4.2), two disjoint convex sets one of which has nonempty interior can be properly separated. Similarly and more generally, two disjoint convex sets such that the affine hull of their union has dimension n can be properly separated. Figure 2.4.5(b) provides an example of two convex sets that cannot be properly separated.

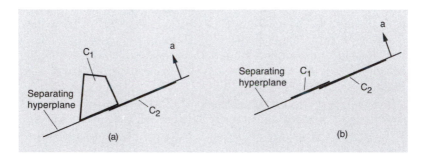

Figure 2.4.5. (a) Illustration of a properly separating hyperplane. (b) Illustration of two convex sets that cannot be properly separated.

The existence of a hyperplane that properly separates two convex sets is intimately tied to conditions involving the relative interiors of the sets. To understand why, note that, given a nonempty convex set C and a hyperplane H that contains C in one of its closed halfspaces, we have

$$C \subset H \quad \text{if and only if} \quad \text{ri}(C) \cap H \neq \emptyset. \quad (2.16)$$

To see this, let us represent H as $H = \{x \mid a'x = b\}$ with $a'x \geq b$ for all $x \in C$. Then for a vector $\bar{x} \in \text{ri}(C)$, we have $\bar{x} \in H$ if and only if $a'\bar{x} = b$, i.e., $a'x$ attains its minimum over C at $\bar{x}$. By Prop. 1.4.2, this is so if and only if $a'x = b$ for all $x \in C$, i.e., $C \subset H$.

If C is a subset of $\Re^n$ and $\bar{x}$ is a vector in $\Re^n$, we say that a hyperplane *properly separates* C *and* $\bar{x}$ if it properly separates C and the singleton set $\{\bar{x}\}$. Thus there exists a hyperplane that properly separates C and $\bar{x}$ if and only if there exists a vector a such that

$$a'\bar{x} \leq \inf_{x \in C} a'x, \qquad a'\bar{x} < \sup_{x \in C} a'x.$$

Based on Eq. (2.16), it can be seen that given a convex set C and a vector $\bar{x}$ in $\Re^n$, and a hyperplane H that separates them, we have

H is properly separating   if and only if $\bar{x} \notin H$ or $\text{ri}(C) \cap H = \emptyset$.

In particular, if
$$\bar{x} \in \text{ri}(C),$$

there cannot exist a hyperplane that properly separates C and $\bar{x}$.

The following propositions formalize the preceding discussion and provide relative interior assumptions that guarantee the existence of properly separating hyperplanes.

Proposition 2.4.5: (Proper Separation Theorem) Let C be a nonempty convex subset of $\Re^n$ and let $\bar{x}$ be a vector in $\Re^n$. There exists a hyperplane that properly separates C and $\bar{x}$ if and only if

$$\bar{x} \notin \text{ri}(C).$$

Proof: If there exists a hyperplane that properly separates C and $\bar{x}$, then $\bar{x} \notin \text{ri}(C)$, by the discussion that precedes the proposition.

Conversely, assume that $\bar{x}$ is not a relative interior point of C. To show the existence of a properly separating hyperplane, we consider two cases (see Fig. 2.4.6):

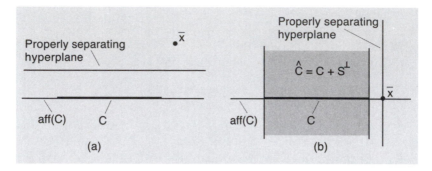

Figure 2.4.6. Illustration of the construction of a hyperplane that properly separates a convex set C and a point $\bar{x}$ that is not in the relative interior of C (cf. the proof of Prop. 2.4.5). In case (a), where $\bar{x} \notin \text{aff}(C)$, the hyperplane is constructed as shown. In case (b), where $\bar{x} \in \text{aff}(C)$, we consider the subspace S that is parallel to $\text{aff}(C)$, we set $\hat{C} = C + S^{\perp}$, and we use the Supporting Hyperplane Theorem (Prop. 2.4.1).

(a) $\bar{x} \notin \text{aff}(C)$. In this case, since $\text{aff}(C)$ is closed and convex, by the Strict Separation Theorem [Prop. 2.4.3 under condition (2)] there exists a hyperplane that separates $\{\bar{x}\}$ and $\text{aff}(C)$ strictly, and hence also properly separates C and $\bar{x}$.

(b) $\bar{x} \in \text{aff}(C)$. In this case, let S be the subspace that is parallel to $\text{aff}(C)$, and consider the set $\hat{C} = C + S^{\perp}$. It can be seen that the interior of $\hat{C}$ is $\text{ri}(C) + S^{\perp}$, so that $\bar{x}$ is not an interior point of $\hat{C}$ [otherwise there must exist a vector $x \in \text{ri}(C)$ such that $x - \bar{x} \in S^{\perp}$, which, since $x \in \text{aff}(C)$, $\bar{x} \in \text{aff}(C)$, and $x - \bar{x} \in S$, implies that $x = 0$, thereby contradicting the hypothesis $\bar{x} \notin \text{ri}(C)$]. By the Supporting Hyperplane Theorem (Prop. 2.4.1), it follows that there exists a vector $a \neq 0$ such that $a'x \geq a'\bar{x}$ for all $x \in \hat{C}$. Since $\hat{C}$ has nonempty interior, $a'x$ cannot be constant over $\hat{C}$, and

$$a'\bar{x} < \sup_{x \in \hat{C}} a'x = \sup_{x \in C,\, z \in S^{\perp}} a'(x+z) = \sup_{x \in C} a'x + \sup_{z \in S^{\perp}} a'z. \quad (2.17)$$

If we had $a'\bar{z} \neq 0$ for some $\bar{z} \in S^{\perp}$, we would also have $\inf_{\alpha \in \Re} a'(x + \alpha\bar{z}) = -\infty$, which contradicts the fact $a'(x + z) \geq a'\bar{x}$ for all $x \in C$ and $z \in S^{\perp}$. It follows that

$$a'z = 0, \qquad \forall\, z \in S^{\perp},$$

which when combined with Eq. (2.17), yields

$$a'\bar{x} < \sup_{x \in C} a'x.$$

Thus the hyperplane $\{x \mid a'x = a'\bar{x}\}$ properly separates C and $\bar{x}$.
Q.E.D.

Proposition 2.4.6: (Proper Separation of Two Convex Sets)
Let C_1 and C_2 be two nonempty convex subsets of $\Re^n$. There exists a hyperplane that properly separates C_1 and C_2 if and only if

$$\mathrm{ri}(C_1) \cap \mathrm{ri}(C_2) = \varnothing.$$

Proof: Consider the convex set

$$C = C_2 - C_1 = \{x \mid x = x_2 - x_1,\ x_1 \in C_1,\ x_2 \in C_2\}.$$

By Prop. 1.4.5(b), we have

$$\mathrm{ri}(C) = \big\{x \mid x = x_2 - x_1,\ x_1 \in \mathrm{ri}(C_1),\ x_2 \in \mathrm{ri}(C_2)\big\},$$

so the assumption $\mathrm{ri}(C_1) \cap \mathrm{ri}(C_2) = \varnothing$ is equivalent to $0 \notin \mathrm{ri}(C)$. By using Prop. 2.4.5, it follows that there exists a hyperplane properly separating C and the origin, or

$$0 \le \inf_{x_1 \in C_1, x_2 \in C_2} a'(x_2 - x_1), \qquad 0 < \sup_{x_1 \in C_1, x_2 \in C_2} a'(x_2 - x_1),$$

if and only if $\mathrm{ri}(C_1) \cap \mathrm{ri}(C_2) = \varnothing$. This is equivalent to the desired assertion. **Q.E.D.**

2.5 AN ELEMENTARY FORM OF DUALITY

In this section we introduce a simple type of duality, which forms the foundation for our development of minimax theory and constrained convex optimization. This duality is based on geometric constructions involving special forms of convex sets and hyperplanes in $\Re^{n+1}$, which we now introduce.

2.5.1 Nonvertical Hyperplanes

In the context of optimization, supporting hyperplanes are often used in conjunction with the epigraph of a function $f : \Re^n \mapsto (-\infty, \infty]$. Since the epigraph is a subset of $\Re^{n+1}$, we associate a corresponding hyperplane in $\Re^{n+1}$ with a nonzero vector of the form (μ, β), where $\mu \in \Re^n$ and $\beta \in \Re$. We say that the hyperplane is *horizontal* if $\mu = 0$ and we say that it is *vertical* if $\beta = 0$.

Note that if a hyperplane with normal (μ, β) is nonvertical (i.e., $\beta \neq 0$), then it crosses the $(n+1)$st axis (the axis associated with w) at a unique point. In particular, if $(\overline{u}, \overline{w})$ is any vector on the hyperplane, the crossing point has the form $(0, \xi)$, where

$$\xi = \frac{\mu'}{\beta}\overline{u} + \overline{w},$$

since from the hyperplane equation, we have $(0, \xi)'(\mu, \beta) = (\overline{u}, \overline{w})'(\mu, \beta)$. If the hyperplane is vertical, it either contains the entire $(n+1)$st axis, or else it does not cross it at all; see Fig. 2.5.1.

Vertical lines in $\Re^{n+1}$ are sets of the form $\{(\overline{u}, w) \mid w \in \Re\}$, where $\overline{u}$ is a fixed vector in $\Re^n$. It can be seen that vertical hyperplanes, as well as the corresponding closed halfspaces, consist of the union of the vertical lines that pass through their points. If $f : \Re^n \mapsto (-\infty, \infty]$ is a proper function, then $\mathrm{epi}(f)$ cannot contain a vertical line, and it appears plausible that $\mathrm{epi}(f)$ is contained in a closed halfspace corresponding to some nonvertical hyperplane. We prove this fact in greater generality in the following proposition, which will also be useful as a first step in the subsequent development.

Proposition 2.5.1: Let C be a nonempty convex subset of $\Re^{n+1}$ that contains no vertical lines. Let the vectors in $\Re^{n+1}$ be denoted by (u, w), where $u \in \Re^n$ and $w \in \Re$. Then:

(a) C is contained in a closed halfspace corresponding to a nonvertical hyperplane, i.e., there exist a vector $\mu \in \Re^n$, a scalar $\beta \neq 0$, and a scalar γ such that

$$\mu'u + \beta w \geq \gamma, \qquad \forall \ (u, w) \in C.$$

(b) If $(\overline{u}, \overline{w})$ does not belong to $\mathrm{cl}(C)$, there exists a nonvertical hyperplane strictly separating $(\overline{u}, \overline{w})$ and C.

Proof: (a) Assume, to arrive at a contradiction, that every hyperplane containing C in one of its closed halfspaces is vertical. Then every hyperplane containing $\mathrm{cl}(C)$ in one of its closed halfspaces must also be vertical. By Prop. 2.4.4, $\mathrm{cl}(C)$ is the intersection of all closed halfspaces that contain it, so we have

$$\mathrm{cl}(C) = \cap_{i \in I}\{(u, w) \mid \mu'_i u \geq \gamma_i\}$$

for a collection of nonzero vectors μ_i, $i \in I$, and scalars γ_i, $i \in I$. Then, for every $(\overline{u}, \overline{w}) \in \mathrm{cl}(C)$, the vertical line $\{(\overline{u}, w) \mid w \in \Re\}$ also belongs to $\mathrm{cl}(C)$. Since the recession cones of $\mathrm{cl}(C)$ and $\mathrm{ri}(C)$ coincide [cf. Prop.

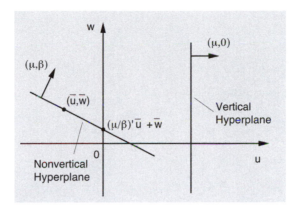

Figure 2.5.1. Illustration of vertical and nonvertical hyperplanes in $\Re^{n+1}$. A hyperplane with normal (μ, β) is nonvertical if $\beta \neq 0$, or, equivalently, if it intersects the $(n+1)$st axis at the unique point $\xi = (\mu/\beta)'\overline{u} + \overline{w}$, where $(\overline{u}, \overline{w})$ is any vector on the hyperplane.

1.5.1(d)], for every $(\overline{u}, \overline{w}) \in \mathrm{ri}(C)$, the vertical line $\{(\overline{u}, w) \mid w \in \Re\}$ belongs to $\mathrm{ri}(C)$ and hence to C. This contradicts the assumption that C does not contain a vertical line.

(b) If $(\overline{u}, \overline{w}) \notin \mathrm{cl}(C)$, then there exists a hyperplane strictly separating $(\overline{u}, \overline{w})$ and $\mathrm{cl}(C)$ [cf. Prop. 2.4.3(b)]. If this hyperplane is nonvertical, since $C \subset \mathrm{cl}(C)$, we are done, so assume otherwise. Then, we have a nonzero vector $\overline{\mu}$ and a scalar $\overline{\gamma}$ such that

$$\overline{\mu}'u > \overline{\gamma} > \overline{\mu}'\overline{u}, \qquad \forall\ (u, w) \in \mathrm{cl}(C). \tag{2.18}$$

The idea now is to combine this vertical hyperplane with a suitably constructed nonvertical hyperplane in order to construct a nonvertical hyperplane that strictly separates $(\overline{u}, \overline{w})$ from $\mathrm{cl}(C)$ (see Fig. 2.5.2).

Since, by assumption, C does not contain a vertical line, $\mathrm{ri}(C)$ also does not contain a vertical line. Since the recession cones of $\mathrm{cl}(C)$ and $\mathrm{ri}(C)$ coincide [cf. Prop. 1.5.1(d)], it follows that $\mathrm{cl}(C)$ does not contain a vertical line. Hence, by part (a), there exists a nonvertical hyperplane containing $\mathrm{cl}(C)$ in one of its closed halfspaces, so that for some (μ, β) and γ, with $\beta \neq 0$, we have

$$\mu'u + \beta w \geq \gamma, \qquad \forall\ (u, w) \in \mathrm{cl}(C).$$

By multiplying this relation with an $\epsilon > 0$ and combining it with Eq. (2.18), we obtain

$$(\overline{\mu} + \epsilon\mu)'u + \epsilon\beta w > \overline{\gamma} + \epsilon\gamma, \qquad \forall\ (u, w) \in \mathrm{cl}(C),\ \forall\ \epsilon > 0.$$

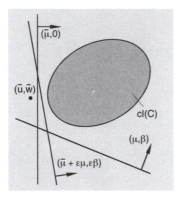

Figure 2.5.2. Construction of a strictly separating nonvertical hyperplane in the proof of Prop. 2.5.1(b).

Since $\overline{\gamma} > \overline{\mu}'\,\overline{u}$, there is a small enough ϵ such that

$$\overline{\gamma} + \epsilon\gamma > (\overline{\mu} + \epsilon\mu)'\overline{u} + \epsilon\beta\overline{w}.$$

From the above two relations, we obtain

$$(\overline{\mu} + \epsilon\mu)'u + \epsilon\beta w > (\overline{\mu} + \epsilon\mu)'\overline{u} + \epsilon\beta\overline{w}, \qquad \forall\, (u,w) \in \mathrm{cl}(C),$$

implying that there is a nonvertical hyperplane with normal $(\overline{\mu} + \epsilon\mu, \epsilon\beta)$ that strictly separates $(\overline{u},\overline{w})$ and $\mathrm{cl}(C)$. Since $C \subset \mathrm{cl}(C)$, this hyperplane also strictly separates $(\overline{u},\overline{w})$ and C. **Q.E.D.**

2.5.2 Min Common/Max Crossing Duality

Hyperplanes allow an insightful visualization of duality concepts. This is particularly so in a construction involving two simple optimization problems, which we now discuss. These problems will form the basis for minimax and saddle point theorems that will be developed in the next section, and will also be central in the development of Lagrangian duality (see Chapter 6). They will also serve a variety of other analytical purposes.

Let M be a nonempty subset of $\Re^{n+1}$ and consider the following two problems.

(a) *Min Common Point Problem*: Among all vectors that are common to M and the $(n + 1)$st axis, we want to find one whose $(n + 1)$st component is minimum.

(b) *Max Crossing Point Problem*: Consider nonvertical hyperplanes that contain M in their corresponding "upper" closed halfspace, i.e., the closed halfspace that contains the vertical halfline $\{(0, w) \mid w \geq 0\}$ in its recession cone (see Fig. 2.5.3). We want to find the maximum crossing point of the $(n + 1)$st axis with such a hyperplane.

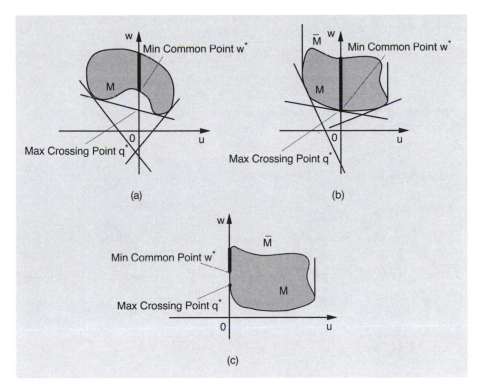

Figure 2.5.3. Illustration of the optimal values of the min common and max crossing problems. In (a), the two optimal values are not equal. In (b), when M is "extended upwards" along the $(n+1)$st axis, it yields the set

$$\overline{M} = \left\{ (u, w) \mid \text{there exists } \overline{w} \text{ with } \overline{w} \le w \text{ and } (u, \overline{w}) \in M \right\},$$

which is convex and admits a nonvertical supporting hyperplane passing through $(0, w^*)$. As a result, the two optimal values are equal. In (c), the set $\overline{M}$ is convex but not closed, and there are points $(0, \overline{w})$ on the vertical axis with $\overline{w} < w^*$ that lie in the closure of $\overline{M}$. Here q^* is equal to the minimum such value of $\overline{w}$, and we have $q^* < w^*$.

Figure 2.5.3 suggests that the optimal value of the max crossing problem is no larger than the optimal value of the min common problem, and that under favorable circumstances the two optimal values are equal. We will now formalize the analysis of the two problems and provide conditions that guarantee equality of their optimal values.

The min common problem is

$$\text{minimize} \quad w$$

$$\text{subject to} \quad (0, w) \in M,$$

and its optimal value is denoted by w^*, i.e.,

$$w^* = \inf_{(0,w) \in M} w.$$

Given a nonvertical hyperplane in $\Re^{n+1}$, multiplication of its normal vector (μ, β) by a nonzero scalar produces a vector that is also normal to the same hyperplane. Hence, the set of nonvertical hyperplanes, where $\beta \neq 0$, can be equivalently described as the set of all hyperplanes with normals of the form $(\mu, 1)$. A hyperplane of this type crosses the $(n+1)$st axis at some vector $(0, \xi)$ and is of the form

$$H_{\mu, \xi} = \big\{ (u, w) \mid w + \mu'u = \xi \big\}.$$

In order for M to be contained in the closed halfspace that corresponds to the hyperplane $H_{\mu, \xi}$ and contains the vertical halfline $\big\{ (0, w) \mid w \geq 0 \big\}$ in its recession cone, we must have

$$\xi \leq w + \mu'u, \qquad \forall \, (u, w) \in M.$$

The maximum crossing level ξ over all hyperplanes $H_{\mu, \xi}$ with the same normal $(\mu, 1)$ is given by

$$q(\mu) = \inf_{(u,w) \in M} \{w + \mu'u\}; \tag{2.19}$$

(see Fig. 2.5.4). The problem of maximizing the crossing level over all nonvertical hyperplanes is to maximize over all $\mu \in \Re^n$ the maximum crossing level corresponding to μ, i.e.,

$$\begin{aligned} &\text{maximize} \quad q(\mu) \\ &\text{subject to} \quad \mu \in \Re^n. \end{aligned} \tag{2.20}$$

We denote by q^* the corresponding optimal value,

$$q^* = \sup_{\mu \in \Re^n} q(\mu).$$

The following proposition gives a basic property of q.

Proposition 2.5.2: The cost function q of the max crossing problem is concave and upper semicontinuous over $\Re^n$.

Proof: By definition [cf. Eq. (2.19)], q is the infimum of a collection of affine functions, and the result follows from Prop. 1.2.4(c). **Q.E.D.**

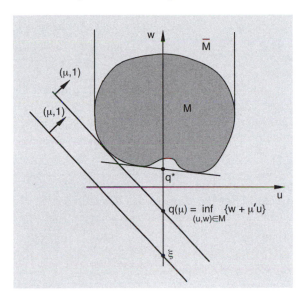

Figure 2.5.4. Mathematical specification of the max crossing problem. For each $\mu \in \Re^n$, we consider $q(\mu)$, the highest crossing level over hyperplanes, which have normal $(\mu, 1)$ and are such that M is contained in their positive halfspace [the one that contains the vertical halfline $\{(0, w) \mid w \geq 0\}$ in its recession cone]. The max crossing point q^* is the supremum over $\mu \in \Re^n$ of the crossing levels $q(\mu)$.

We are now ready to establish a generic relation between q^* and w^*, which is intuitively apparent from Fig. 2.5.3.

Proposition 2.5.3: (Weak Duality Theorem) For the min common and max crossing problems we have

$$q^* \leq w^*. \tag{2.21}$$

Proof: For every $(u, w) \in M$ and every $\mu \in \Re^n$, we have

$$q(\mu) = \inf_{(u,w)\in M} \{w + \mu'u\} \leq \inf_{(0,w)\in M} w = w^*,$$

so by taking the supremum of the left-hand side over $\mu \in \Re^n$, we obtain $q^* \leq w^*$. **Q.E.D.**

We now turn to establishing conditions under which we have $q^* = w^*$, in which case we say that *strong duality holds* or that *there is no duality gap*. To avoid degenerate cases, we will generally exclude the case $w^* = \infty$, when the min common problem is infeasible, i.e., $\{w \mid (0, w) \in M\} = \emptyset$.

An important point, around which much of our analysis revolves, is that when w^* is a scalar, the vector $(0, w^*)$ is a closure point of the set M, so *if we assume that M is convex and closed, and admits a nonvertical supporting hyperplane at $(0, w^*)$, then we have $q^* = w^*$ and the optimal values q^* and w^* are attained*. Between the "unfavorable" case where $q^* < w^*$, and the "most favorable" case where $q^* = w^*$ while the optimal values q^* and w^* are attained, there are several intermediate cases. The following proposition provides a necessary and sufficient condition for $q^* = w^*$, but does not address the attainment of the optimal values.

Proposition 2.5.4: (Min Common/Max Crossing Theorem I)
Consider the min common and max crossing problems, and assume the following:

(1) $w^* < \infty$.

(2) The set

$$\overline{M} = \big\{ (u, w) \mid \text{there exists } \overline{w} \text{ with } \overline{w} \le w \text{ and } (u, \overline{w}) \in M \big\}$$

is convex.

Then, we have $q^* = w^*$ if and only if for every sequence $\{(u_k, w_k)\} \subset M$ with $u_k \to 0$, there holds $w^* \le \liminf_{k\to\infty} w_k$.

Proof: If $w^* = -\infty$, by the Weak Duality Theorem (Prop. 2.5.3), we also have $q^* = -\infty$ and $q(\mu) = -\infty$ for all $\mu \in \Re^n$, so the conclusion trivially follows. We thus focus on the case where w^* is a real number. Assume that for every sequence $\{(u_k, w_k)\} \subset M$ with $u_k \to 0$, there holds $w^* \le \liminf_{k\to\infty} w_k$. We first note that $(0, w^*)$ is a closure point of $\overline{M}$, since by the definition of w^*, there exists a sequence $\{(0, w_k)\}$ that belongs to M, and hence also to $\overline{M}$, and is such that $w_k \to w^*$.

We next show by contradiction that $\overline{M}$ does not contain any vertical lines. If this were not so, by convexity of $\overline{M}$, the direction $(0, -1)$ would be a direction of recession of $\mathrm{cl}(\overline{M})$ (although not necessarily a direction of recession of $\overline{M}$), and hence also a direction of recession of $\mathrm{ri}(\overline{M})$ [cf. Prop. 1.5.1(d)]. Because $(0, w^*)$ is a closure point of $\overline{M}$, it is also a closure point of $\mathrm{ri}(\overline{M})$ [cf. Prop. 1.4.3(a)], and therefore, there exists a sequence $\{(u_k, w_k)\} \subset \mathrm{ri}(\overline{M})$ converging to $(0, w^*)$. Since $(0, -1)$ is a direction of recession of $\mathrm{ri}(\overline{M})$, the sequence $\{(u_k, w_k - 1)\}$ belongs to $\mathrm{ri}(\overline{M})$ and consequently, $\{(u_k, w_k - 1)\} \subset \overline{M}$. In view of the definition of $\overline{M}$, there is a sequence $\{(u_k, \overline{w}_k)\} \subset M$ with $\overline{w}_k \le w_k - 1$ for all k, so that $\liminf_{k\to\infty} \overline{w}_k \le w^* - 1$. This contradicts the assumption $w^* \le \liminf_{k\to\infty} w_k$, since $u_k \to 0$.

We now prove that the vector $(0, w^* - \epsilon)$ does not belong to $\mathrm{cl}(\overline{M})$ for any $\epsilon > 0$. To arrive at a contradiction, suppose that $(0, w^* - \epsilon)$ is a closure point of $\overline{M}$ for some $\epsilon > 0$, so that there exists a sequence $\{(u_k, w_k)\} \subset \overline{M}$ converging to $(0, w^* - \epsilon)$. In view of the definition of $\overline{M}$, this implies the existence of another sequence $\{(u_k, \overline{w}_k)\} \subset M$ with $u_k \to 0$ and $\overline{w}_k \leq w_k$ for all k, and we have that $\liminf_{k\to\infty} \overline{w}_k \leq w^* - \epsilon$, which contradicts the assumption $w^* \leq \liminf_{k\to\infty} w_k$.

Since, as shown above, $\overline{M}$ does not contain any vertical lines and the vector $(0, w^* - \epsilon)$ does not belong to $\mathrm{cl}(\overline{M})$ for any $\epsilon > 0$, by Prop. 2.5.1(b), it follows that there exists a nonvertical hyperplane strictly separating $(0, w^* - \epsilon)$ and $\overline{M}$. This hyperplane crosses the $(n+1)$st axis at a unique vector $(0, \xi)$, which must lie between $(0, w^* - \epsilon)$ and $(0, w^*)$, i.e., $w^* - \epsilon \leq \xi \leq w^*$. Furthermore, ξ cannot exceed the optimal value q^* of the max crossing problem, which, together with weak duality ($q^* \leq w^*$), implies that $w^* - \epsilon \leq q^* \leq w^*$. Since ϵ can be arbitrarily small, it follows that $q^* = w^*$.

Conversely, assume that $q^* = w^*$. Let $\{(u_k, w_k)\}$ be any sequence in M, which is such that $u_k \to 0$. Then,

$$q(\mu) = \inf_{(u,w)\in M} \{w + \mu'u\} \leq w_k + \mu'u_k, \qquad \forall\, k, \quad \forall\, \mu \in \Re^n.$$

Taking the limit as $k \to \infty$, we obtain

$$q(\mu) \leq \liminf_{k\to\infty} w_k, \qquad \forall\, \mu \in \Re^n,$$

implying that

$$w^* = q^* = \sup_{\mu \in \Re^n} q(\mu) \leq \liminf_{k\to\infty} w_k.$$

Q.E.D.

We now provide another version of the Min Common/Max Crossing Theorem, which in addition to the equality $q^* = w^*$, guarantees the attainment of the maximum crossing point by a nonvertical hyperplane under an additional relative interior assumption [see condition (3) of the proposition].

Proposition 2.5.5: (Min Common/Max Crossing Theorem II)
Consider the min common and max crossing problems, and assume the following:

(1) $-\infty < w^*$.

(2) The set

$$\overline{M} = \big\{ (u, w) \mid \text{there exists } \overline{w} \text{ with } \overline{w} \le w \text{ and } (u, \overline{w}) \in M \big\}$$

is convex.

(3) The set

$$D = \big\{ u \mid \text{there exists } w \in \Re \text{ with } (u, w) \in \overline{M} \big\}$$

contains the origin in its relative interior.

Then $q^* = w^*$ and the optimal solution set of the max crossing problem, $Q^* = \{ \mu \mid q(\mu) = q^* \}$, has the form

$$Q^* = \big(\text{aff}(D) \big)^\perp + \tilde{Q},$$

where $\tilde{Q}$ is a nonempty, convex, and compact set, and $\big(\text{aff}(D) \big)^\perp$ is the orthogonal complement of $\text{aff}(D)$ [which is a subspace by assumption (3)]. Furthermore, Q^* is nonempty and compact if and only if D contains the origin in its interior.

Proof: We first show that $q^* = w^*$ and that Q^* is nonempty. We note that condition (3) implies that $w^* < \infty$, so in view of condition (1), w^* is a real number. Since w^* is the optimal min common value and the line $\big\{ (0, w) \mid w \in \Re \big\}$ is contained in the affine hull of $\overline{M}$, it follows that $(0, w^*)$ is not a relative interior point of $\overline{M}$. Therefore, by the Proper Separation Theorem (cf. Prop. 2.4.5), there exists a hyperplane that passes through $(0, w^*)$, contains $\overline{M}$ in one of its closed halfspaces, but does not fully contain $\overline{M}$, i.e., there exists a vector (μ, β) such that

$$\beta w^* \le \mu' u + \beta w, \qquad \forall \, (u, w) \in \overline{M}, \tag{2.22}$$

$$\beta w^* < \sup_{(u,w) \in \overline{M}} \{ \mu' u + \beta w \}. \tag{2.23}$$

Since for any $(\overline{u}, \overline{w}) \in M$, the set $\overline{M}$ contains the halfline $\big\{ (\overline{u}, w) \mid \overline{w} \le w \big\}$, it follows from Eq. (2.22) that $\beta \ge 0$. If $\beta = 0$, then from Eq. (2.22), we have

$$0 \le \mu' u, \qquad \forall \, u \in D.$$

Thus, the linear function $\mu' u$ attains its minimum over the set D at 0, which is a relative interior point of D by condition (3). Since D is convex, being the projection on the space of u of the set $\overline{M}$, which is convex by assumption (2), it follows by Prop. 1.4.2 that $\mu'u$ is constant over D, i.e.,

$$\mu'u = 0, \qquad \forall\, u \in D.$$

This, however, contradicts Eq. (2.23). Therefore, we must have $\beta > 0$, and by appropriate normalization if necessary, we can assume that $\beta = 1$. From Eq. (2.22), we then obtain

$$w^* \le \inf_{(u,w)\in\overline{M}} \{\mu'u + w\} \le \inf_{(u,w)\in M} \{\mu'u + w\} = q(\mu) \le q^*.$$

Since the inequality $q^* \le w^*$ holds always [cf. Eq. (2.21)], equality holds throughout in the above relation, and we must have $q(\mu) = q^* = w^*$. Thus Q^* is nonempty, and since $Q^* = \{\mu \mid q(\mu) \ge q^*\}$ and q is concave and upper semicontinuous (cf. Prop. 2.5.2), it follows that Q^* is also convex and closed.

We next show that $Q^* = \big(\mathrm{aff}(D)\big)^{\perp} + \tilde{Q}$. We first prove that the recession cone R_{Q^*} and the lineality space L_{Q^*} of Q^* are both equal to $\big(\mathrm{aff}(D)\big)^{\perp}$. The proof of this is based on the generic relation $L_{Q^*} \subset R_{Q^*}$ and the following two relations

$$\big(\mathrm{aff}(D)\big)^{\perp} \subset L_{Q^*}, \qquad R_{Q^*} \subset \big(\mathrm{aff}(D)\big)^{\perp},$$

which we show next.

Let y be a vector in $\big(\mathrm{aff}(D)\big)^{\perp}$, so that $y'u = 0$ for all $u \in D$. For any vector $\mu \in Q^*$ and any scalar α, we then have

$$q(\mu + \alpha y) = \inf_{(u,w)\in\overline{M}} \{(\mu + \alpha y)'u + w\} = \inf_{(u,w)\in\overline{M}} \{\mu'u + w\} = q(\mu),$$

implying that $\mu + \alpha y$ is in Q^*. Hence $y \in L_{Q^*}$, and it follows that $\big(\mathrm{aff}(D)\big)^{\perp} \subset L_{Q^*}$.

Let y be a vector in R_{Q^*}, so that for any $\mu \in Q^*$ and $\alpha \ge 0$,

$$q(\mu + \alpha y) = \inf_{(u,w)\in\overline{M}} \{(\mu + \alpha y)'u + w\} = q^*.$$

Since $0 \in \mathrm{ri}(D)$, for any $u \in \mathrm{aff}(D)$, there exists a positive scalar γ such that the vectors γu and $-\gamma u$ are in D. By the definition of D, there exist scalars w^+ and w^- such that the pairs $(\gamma u, w^+)$ and $(-\gamma u, w^-)$ are in $\overline{M}$. Using the preceding equation, it follows that for any $\mu \in Q^*$, we have

$$(\mu + \alpha y)'(\gamma u) + w^+ \ge q^*, \qquad \forall\, \alpha \ge 0,$$

$$(\mu + \alpha y)'(-\gamma u) + w^- \ge q^*, \qquad \forall\, \alpha \ge 0.$$

If $y'u \neq 0$, then for sufficiently large $\alpha \geq 0$, one of the preceding two relations will be violated. Thus we must have $y'u = 0$, showing that $y \in \left(\text{aff}(D)\right)^{\perp}$ and implying that

$$R_{Q^*} \subset \left(\text{aff}(D)\right)^{\perp}.$$

This relation, together with the generic relation $L_{Q^*} \subset R_{Q^*}$ and the relation $\left(\text{aff}(D)\right)^{\perp} \subset L_{Q^*}$ shown earlier, shows that

$$\left(\text{aff}(D)\right)^{\perp} \subset L_{Q^*} \subset R_{Q^*} \subset \left(\text{aff}(D)\right)^{\perp}.$$

Therefore

$$L_{Q^*} = R_{Q^*} = \left(\text{aff}(D)\right)^{\perp}.$$

Based on the decomposition result of Prop. 1.5.5, we have

$$Q^* = L_{Q^*} + (Q^* \cap L_{Q^*}^{\perp}).$$

Since $L_{Q^*} = \left(\text{aff}(D)\right)^{\perp}$, we obtain

$$Q^* = \left(\text{aff}(D)\right)^{\perp} + \tilde{Q},$$

where $\tilde{Q} = Q^* \cap \text{aff}(D)$. Furthermore, by the Recession Cone Theorem [Prop. 1.5.1(e)], we have

$$R_{\tilde{Q}} = R_{Q^*} \cap R_{\text{aff}(D)}.$$

Since $R_{Q^*} = \left(\text{aff}(D)\right)^{\perp}$, as shown earlier, and $R_{\text{aff}(D)} = \text{aff}(D)$, the recession cone $R_{\tilde{Q}}$ consists of the zero vector only, implying by the Recession Cone Theorem [Prop. 1.5.1(c)] that the set $\tilde{Q}$ is compact.

Finally, to show the last statement in the proposition, we note that 0 is an interior point of D if and only if $\text{aff}(D) = \Re^n$, which in turn is equivalent to Q^* being equal to the compact set $\tilde{Q}$.   **Q.E.D.**

A third version of the min common/max crossing theorem will be given in the next chapter (Section 3.5), following the development of polyhedral convexity concepts.

2.6 SADDLE POINT AND MINIMAX THEORY

In this section we consider a function $\phi : X \times Z \mapsto \Re$, where X and Z are nonempty subsets of $\Re^n$ and $\Re^m$, respectively. We wish to either

$$\text{minimize}\quad \sup_{z \in Z} \phi(x, z)$$

$$\text{subject to}\ \ x \in X$$

or

$$\text{maximize} \quad \inf_{x \in X} \phi(x, z)$$

$$\text{subject to} \quad z \in Z.$$

These problems are encountered in at least four major optimization contexts:

(1) *Worst-case design*, whereby we view z as a parameter and we wish to minimize over x a cost function, assuming the worst possible value of z. A special case of this is the *discrete minimax problem*, where we want to minimize over $x \in X$

$$\max\{f_1(x), \ldots, f_m(x)\},$$

where f_i are some given functions. Here, Z is the finite set $\{1, \ldots, m\}$. We will revisit this problem in Chapter 4, where we will discuss the differentiability properties of the function $\max_{z \in Z} \phi(x, z)$.

(2) *Exact penalty functions*, which can be used for example to convert the constrained optimization problem

$$\begin{aligned} &\text{minimize} \quad f(x) \\ &\text{subject to} \quad x \in X, \qquad g_j(x) \le 0, \quad j = 1, \ldots, r, \end{aligned} \tag{2.24}$$

to the (less constrained) minimax problem

$$\begin{aligned} &\text{minimize} \quad f(x) + c\max\{0, g_1(x), \ldots, g_r(x)\} \\ &\text{subject to} \quad x \in X, \end{aligned}$$

where c is a large positive penalty parameter. This conversion is useful for both analytical and computational purposes, and will be discussed in Chapters 5 and 7.

(3) *Duality theory*, where using problem (2.24) as an example, we introduce a vector $\mu = (\mu_1, \ldots, \mu_r) \in \Re^r$, and the function

$$L(x, \mu) = f(x) + \sum_{j=1}^{r} \mu_j g_j(x),$$

referred to as the *Lagrangian function*. We then consider the problem

$$\begin{aligned} &\text{maximize} \quad \inf_{x \in X} L(x, \mu) \\ &\text{subject to} \quad \mu \ge 0, \end{aligned} \tag{2.25}$$

referred to as the *dual problem*. Note that the original problem (2.24) (referred to as the *primal problem*) can also be written as

$$\begin{aligned} &\text{minimize} \quad \sup_{\mu \ge 0} L(x, \mu) \\ &\text{subject to} \quad x \in X, \end{aligned}$$

[if x violates any of the constraints $g_j(x) \leq 0$, we have $\sup_{\mu \geq 0} L(x, \mu) = \infty$, and if it does not, we have $\sup_{\mu \geq 0} L(x, \mu) = f(x)$]. Thus the primal and the dual problems (2.24) and (2.25) can be viewed in terms of minimax problems.

(4) *Zero sum games*, where there are two players: the first may choose one out of n moves and the second may choose one out of m moves. If moves i and j are selected by the first and the second player, respectively, the first player gives a specified amount a_{ij} to the second. The objective of the first player is to minimize the amount given to the other player, and the objective of the second player is to maximize this amount. The players use mixed strategies, whereby the first player selects a probability distribution $x = (x_1, \ldots, x_n)$ over his n possible moves and the second player selects a probability distribution $z = (z_1, \ldots, z_m)$ over his m possible moves. Since the probability of selecting i and j is $x_i z_j$, the expected amount to be paid by the first player to the second is $\sum_{i,j} a_{ij} x_i z_j$ or $x'Az$, where A is the $n \times m$ matrix with elements a_{ij}. If each player adopts a worst case viewpoint, whereby he optimizes his choice against the worst possible selection by the other player, the first player must minimize $\max_z x'Az$ and the second player must maximize $\min_x x'Az$. The main result, a special case of the results we will prove shortly, is that these two optimal values are equal, implying that there is an amount that can be meaningfully viewed as the value of the game for its participants.

We will now derive conditions guaranteeing that

$$\sup_{z \in Z} \inf_{x \in X} \phi(x, z) = \inf_{x \in X} \sup_{z \in Z} \phi(x, z), \tag{2.26}$$

and that the infimum and the supremum above are attained. This is significant in the zero sum game context discussed above, and it is also a major issue in duality theory because it connects the primal and the dual problems [cf. Eqs. (2.24) and (2.25)] through their optimal values and optimal solutions. In particular, when we discuss duality in Chapter 6, we will see that a major question is whether there is no duality gap, i.e., whether the optimal primal and dual values are equal. This is so if and only if

$$\sup_{\mu \geq 0} \inf_{x \in X} L(x, \mu) = \inf_{x \in X} \sup_{\mu \geq 0} L(x, \mu). \tag{2.27}$$

We will prove in this section a classical result, the Saddle Point Theorem, which guarantees the equality (2.26), as well as the attainment of the infimum and the supremum, assuming convexity/concavity assumptions on ϕ, and (essentially) compactness assumptions on X and Z. Unfortunately, the Saddle Point Theorem is only partially adequate for the development of duality theory, because compactness of Z and, to some extent, compactness of X turn out to be restrictive assumptions [for example Z corresponds to the set $\{\mu \mid \mu \geq 0\}$ in Eq. (2.27), which is not compact]. We will prove two versions of another major result, the Minimax Theorem, which is more relevant to duality theory and gives conditions guaranteeing the minimax equality (2.26), although it need not guarantee the attainment of the infimum and the supremum. We will also give additional theorems of the minimax type in Chapter 6, when we discuss duality and we make a closer connection between saddle points and Lagrange multipliers.

A first observation regarding the potential validity of the minimax equality (2.26) is that we always have the inequality

$$\sup_{z \in Z} \inf_{x \in X} \phi(x, z) \leq \inf_{x \in X} \sup_{z \in Z} \phi(x, z), \tag{2.28}$$

[for every $\overline{z} \in Z$, write

$$\inf_{x \in X} \phi(x, \overline{z}) \leq \inf_{x \in X} \sup_{z \in Z} \phi(x, z)$$

and take the supremum over $\overline{z} \in Z$ of the left-hand side]. We refer to this relation as the *minimax inequality*. It is sufficient to show the reverse inequality in order for the minimax equality (2.26) to hold. However, special conditions are required for the reverse inequality to be true.

The following definition formalizes pairs of vectors that attain the infimum and the supremum in the minimax equality (2.26).

Definition 2.6.1: A pair of vectors $x^* \in X$ and $z^* \in Z$ is called a *saddle point* of ϕ if

$$\phi(x^*, z) \leq \phi(x^*, z^*) \leq \phi(x, z^*), \qquad \forall \, x \in X, \, \forall \, z \in Z.$$

Note that (x^*, z^*) is a saddle point if and only if $x^* \in X$, $z^* \in Z$, and

$$\sup_{z \in Z} \phi(x^*, z) = \phi(x^*, z^*) = \inf_{x \in X} \phi(x, z^*). \tag{2.29}$$

This equality implies that

$$\inf_{x \in X} \sup_{z \in Z} \phi(x, z) \leq \sup_{z \in Z} \phi(x^*, z) = \phi(x^*, z^*) = \inf_{x \in X} \phi(x, z^*) \leq \sup_{z \in Z} \inf_{x \in X} \phi(x, z),$$
$$\tag{2.30}$$

which combined with the minimax inequality (2.28), proves that the minimax equality holds. Thus if a saddle point exists, the minimax equality holds. More generally, we have the following characterization of a saddle point.

> **Proposition 2.6.1:** A pair (x^*, z^*) is a saddle point of ϕ if and only if the minimax equality (2.26) holds, and x^* is an optimal solution of the problem
>
> $$\text{minimize} \quad \sup_{z \in Z} \phi(x, z)$$
>
> $$\text{subject to} \quad x \in X, \tag{2.31}$$
>
> while z^* is an optimal solution of the problem
>
> $$\text{maximize} \quad \inf_{x \in X} \phi(x, z)$$
>
> $$\text{subject to} \quad z \in Z. \tag{2.32}$$

Proof: Suppose that x^* is an optimal solution of problem (2.31) and z^* is an optimal solution of problem (2.32). Then we have

$$\sup_{z \in Z} \inf_{x \in X} \phi(x, z) = \inf_{x \in X} \phi(x, z^*) \leq \phi(x^*, z^*) \leq \sup_{z \in Z} \phi(x^*, z) = \inf_{x \in X} \sup_{z \in Z} \phi(x, z).$$

If the minimax equality [cf. Eq. (2.26)] holds, then equality holds throughout above, so that

$$\sup_{z \in Z} \phi(x^*, z) = \phi(x^*, z^*) = \inf_{x \in X} \phi(x, z^*),$$

i.e., (x^*, z^*) is a saddle point of ϕ [cf. Eq. (2.29)].

Conversely, if (x^*, z^*) is a saddle point, then Eq. (2.30) holds, and together with the minimax inequality (2.28), implies that the minimax equality (2.26) holds. Therefore, equality holds throughout in Eq. (2.30), which implies that x^* and z^* are optimal solutions of problems (2.31) and (2.32), respectively. **Q.E.D.**

Note a simple consequence of the above proposition: the set of saddle points, *when nonempty*, is the Cartesian product $X^* \times Z^*$, where X^* and Z^* are the sets of optimal solutions of problems (2.31) and (2.32), respectively. In other words x^* and z^* can be *independently* chosen from the sets X^* and Z^*, respectively, to form a saddle point. Note also that if the minimax equality (2.26) does not hold, there is no saddle point, even if the sets X^* and Z^* are nonempty.

One can visualize saddle points in terms of the sets of minimizing points over X for fixed $z \in Z$ and maximizing points over Z for fixed $x \in X$:

$$\hat{X}(z) = \left\{ \hat{x} \mid \hat{x} \text{ minimizes } \phi(x, z) \text{ over } X \right\},$$

$$\hat{Z}(x) = \left\{ \hat{z} \mid \hat{z} \text{ maximizes } \phi(x, z) \text{ over } Z \right\}.$$

The definition implies that the pair (x^*, z^*) is a saddle point if and only if it is a "point of intersection" of $\hat{X}(\cdot)$ and $\hat{Z}(\cdot)$ in the sense that

$$x^* \in \hat{X}(z^*), \qquad z^* \in \hat{Z}(x^*);$$

see Fig. 2.6.1.

2.6.1 Min Common/Max Crossing Framework for Minimax

We will now provide conditions implying that the minimax equality (2.26) holds. In our analysis, a critical role is played by the min common/max crossing framework of Section 2.5 and by the function $p : \Re^m \mapsto [-\infty, \infty]$ given by

$$p(u) = \inf_{x \in X} \sup_{z \in Z} \{\phi(x, z) - u'z\}, \qquad u \in \Re^m. \tag{2.33}$$

This function characterizes how the "infsup" of the function ϕ changes when the linear perturbation term $u'z$ is subtracted from ϕ. It turns out that if p changes in a "regular" manner to be specified shortly, the minimax equality (2.26) is guaranteed. The significance of the function p will be elaborated on and clarified further in connection with convex programming and duality in Chapters 6 and 7.

In the subsequent applications of the min common/max crossing framework, the set M will be taken to be the epigraph of p,

$$M = \mathrm{epi}(p),$$

so that the min common value w^* will be equal to $p(0)$, which by the definition of p, is also equal to the "infsup" value

$$w^* = p(0) = \inf_{x \in X} \sup_{z \in Z} \phi(x, z). \tag{2.34}$$

Under some convexity assumptions with respect to x (see the subsequent Lemma 2.6.1), we will show that p is convex, so that M is convex, which satisfies a major assumption for the application of the min common/max crossing theorems of the preceding section (with M equal to an epigraph of a function, the sets M and $\overline{M}$ appearing in the min common/max crossing theorems coincide).

The corresponding max crossing problem is [cf. Eqs. (2.20) and (2.19)]

$$\text{maximize} \quad q(\mu)$$
$$\text{subject to} \quad \mu \in \Re^n,$$

where

$$q(\mu) = \inf_{(u,w) \in \mathrm{epi}(p)} \{w + \mu'u\} = \inf_{\{(u,w) | p(u) \leq w\}} \{w + \mu'u\} = \inf_{u \in \Re^m} \{p(u) + \mu'u\}.$$

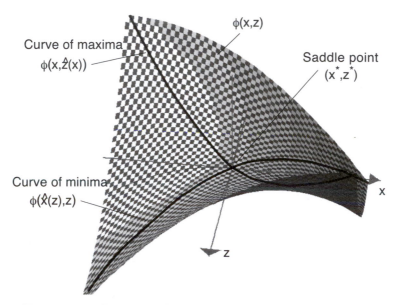

Figure 2.6.1. Illustration of a saddle point of a function $\phi(x, z)$ over $x \in X$ and $z \in Z$; the function plotted here is $\phi(x, z) = \frac{1}{2}(x^2 + 2xz - z^2)$. Let

$$\hat{x}(z) \in \arg\min_{x \in X} \phi(x, z), \qquad \hat{z}(x) \in \arg\max_{z \in Z} \phi(x, z).$$

In the case illustrated, $\hat{x}(z)$ and $\hat{z}(x)$ consist of unique minimizing and maximizing points, respectively, so $\hat{x}(z)$ and $\hat{z}(x)$ are viewed as (single-valued) functions; otherwise $\hat{x}(z)$ and $\hat{z}(x)$ should be viewed as set-valued mappings. We consider the corresponding curves $\phi\big(\hat{x}(z), z\big)$ and $\phi\big(x, \hat{z}(x)\big)$. By definition, a pair (x^*, z^*) is a saddle point if and only if

$$\max_{z \in Z} \phi(x^*, z) = \phi(x^*, z^*) = \min_{x \in X} \phi(x, z^*),$$

or equivalently, if (x^*, z^*) lies on both curves $[x^* = \hat{x}(z^*)$ and $z^* = \hat{z}(x^*)]$. At such a pair, we also have

$$\max_{z \in Z} \phi\big(\hat{x}(z), z\big) = \max_{z \in Z} \min_{x \in X} \phi(x, z) = \phi(x^*, z^*) = \min_{x \in X} \max_{z \in Z} \phi(x, z) = \min_{x \in X} \phi\big(x, \hat{z}(x)\big),$$

so that

$$\phi\big(\hat{x}(z), z\big) \le \phi(x^*, z^*) \le \phi\big(x, \hat{z}(x)\big), \qquad \forall\, x \in X, z \in Z$$

(see Prop. 2.6.1). Visually, the curve of maxima $\phi\big(x, \hat{z}(x)\big)$ must lie "above" the curve of minima $\phi\big(\hat{x}(z), z\big)$ completely (i.e., for all $x \in X$ and $z \in Z$), but the two curves should meet at (x^*, z^*).

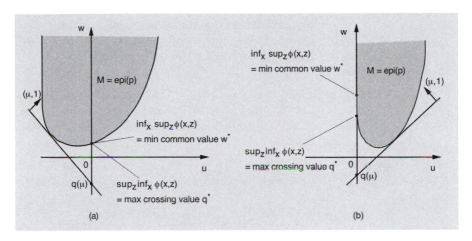

Figure 2.6.2. Min common/max crossing framework for minimax theory. The set M is taken to be the epigraph of the function

$$p(u) = \inf_{x \in X} \sup_{z \in Z} \{\phi(x, z) - u'z\}.$$

The "Infsup" value of ϕ is equal to the min common value w^*. Under suitable assumptions, the "supinf" values of ϕ will turn out to be equal to the max crossing value q^*. Figures (a) and (b) illustrate cases where p is convex, and the minimax equality (2.26) holds and does not hold, respectively.

By using this relation and the definition of p, we obtain

$$q(\mu) = \inf_{u \in \Re^m} \inf_{x \in X} \sup_{z \in Z} \{\phi(x, z) + u'(\mu - z)\}.$$

For every $\mu \in Z$, by setting $z = \mu$ in the right-hand side above, we obtain

$$\inf_{x \in X} \phi(x, \mu) \le q(\mu), \qquad \forall\, \mu \in Z.$$

Thus, using also Eq. (2.34) and the weak duality relation $q^* \le w^*$ (Prop. 2.5.3), we have

$$\sup_{z \in Z} \inf_{x \in X} \phi(x, z) \le \sup_{\mu \in \Re^m} q(\mu) = q^* \le w^* = p(0) = \inf_{x \in X} \sup_{z \in Z} \phi(x, z). \quad (2.35)$$

This inequality indicates a generic connection of the minimax and the min common/max crossing frameworks. In particular, if the minimax equality

$$\sup_{z \in Z} \inf_{x \in X} \phi(x, z) = \inf_{x \in X} \sup_{z \in Z} \phi(x, z)$$

holds, then $q^* = w^*$, i.e., that the optimal values of the min common and max crossing problems are equal.

An even stronger connection between the minimax and the min common/max crossing frameworks holds under some convexity and semicontinuity assumptions, as shown in the following two lemmas. Loosely phrased, these lemmas assert that:

(a) Convexity with respect to x [convexity of X and $\phi(\cdot, z)$ for all $z \in Z$] guarantees that epi(p) is a convex set, thereby allowing the use of the two Min Common/Max Crossing Theorems of the preceding section (Props. 2.5.4 and 2.5.5).

(b) Concavity and semicontinuity with respect to z [convexity of Z, and concavity and upper semicontinuity of $\phi(x, \cdot)$ for all $x \in X$] guarantee that

$$q(\mu) = \inf_{x \in X} \phi(x, \mu), \qquad q^* = \sup_{z \in Z} \inf_{x \in X} \phi(x, z),$$

as indicated in Fig. 2.6.2. Thus, under these conditions, the minimax equality (2.26) is equivalent to the equality $q^* = w^*$ in the corresponding min common/max crossing framework.

Thus, if ϕ is convex with respect to x, and concave and upper semicontinuous with respect to z, as specified in the following two lemmas, the min common/max crossing framework applies in its most powerful form and provides the answers to the most critical questions within the minimax framework.

Lemma 2.6.1: Let X be a nonempty convex subset of $\Re^n$, let Z be a nonempty subset of $\Re^m$, and let $\phi : X \times Z \mapsto \Re$ be a function. Assume that for each $z \in Z$, the function $\phi(\cdot, z) : X \mapsto \Re$ is convex. Then the function p of Eq. (2.33) is convex.

Proof: Since taking pointwise supremum preserves convexity [cf. Prop. 1.2.4(c)], our convexity assumption implies that the function $F : \Re^{m+n} \to (-\infty, \infty]$ given by

$$F(x, u) = \begin{cases} \sup_{z \in Z} \{\phi(x, z) - u'z\} & \text{if } x \in X, \\ \infty & \text{if } x \notin X, \end{cases}$$

is convex. Since

$$p(u) = \inf_{x \in \Re^n} F(x, u),$$

and partial minimization preserves convexity (cf. Prop. 2.3.6), the convexity of p follows from the convexity of F.   **Q.E.D.**

The following lemma shows that under some convexity and semicontinuity assumptions, p defines not only the "infsup" of the function ϕ [cf.

Eq. (2.34)], but through its epigraph, it also defines the "supinf" (cf. Fig. 2.6.2).

Lemma 2.6.2: Let X be a nonempty subset of $\Re^n$, let Z be a nonempty convex subset of $\Re^m$, and let $\phi : X \times Z \mapsto \Re$ be a function. Assume that for each $x \in X$, the function $-\phi(x, \cdot) : Z \mapsto \Re$ is closed and convex. Then the function $q : \Re^m \mapsto [-\infty, \infty]$ given by

$$q(\mu) = \inf_{(u,w) \in \mathrm{epi}(p)} \{w + u'\mu\}, \qquad \mu \in \Re^m,$$

where p is given by Eq. (2.33), satisfies

$$q(\mu) = \begin{cases} \inf_{x \in X} \phi(x, \mu) & \text{if } \mu \in Z, \\ -\infty & \text{if } \mu \notin Z. \end{cases} \tag{2.36}$$

Furthermore, we have $q^* = w^*$ if and only if the minimax equality (2.26) holds.

Proof: For every $\mu \in \Re^m$, we have

$$q(\mu) = \inf_{(u,w) \in \mathrm{epi}(p)} \{w + \mu'u\} = \inf_{\{(u,w) \mid p(u) \le w\}} \{w + \mu'u\} = \inf_{u \in \Re^m} \{p(u) + \mu'u\}.$$

By using this relation and the definition of p, we obtain for every $\mu \in \Re^m$,

$$\begin{aligned} q(\mu) &= \inf_{u \in \Re^m} \{p(u) + u'\mu\} \\ &= \inf_{u \in \Re^m} \inf_{x \in X} \sup_{z \in Z} \{\phi(x, z) + u'(\mu - z)\} \\ &= \inf_{x \in X} \inf_{u \in \Re^m} \sup_{z \in Z} \{\phi(x, z) + u'(\mu - z)\}. \end{aligned} \tag{2.37}$$

For $\mu \in Z$, we have

$$\sup_{z \in Z} \{\phi(x, z) + u'(\mu - z)\} \ge \phi(x, \mu), \qquad \forall \, x \in X, \quad \forall \, u \in \Re^m,$$

implying that

$$q(\mu) \ge \inf_{x \in X} \phi(x, \mu), \qquad \forall \, \mu \in Z.$$

Thus, to prove Eq. (2.36), we must show that

$$q(\mu) \le \inf_{x \in X} \phi(x, \mu), \qquad \forall \, \mu \in Z, \tag{2.38}$$

and

$$q(\mu) = -\infty, \qquad \forall \, \mu \notin Z.$$

For all $x \in X$ and $z \in Z$, denote

$$r_x(z) = -\phi(x, z),$$

so that the function $r_x : Z \mapsto \Re$ is closed and convex by assumption. We will consider separately the two cases where $\mu \in Z$ and $\mu \notin Z$. We first assume that $\mu \in Z$. We fix an arbitrary $x \in X$, and we note that by assumption, $\mathrm{epi}(r_x)$ is a closed convex set. Since $\mu \in Z$, the point $(\mu, r_x(\mu))$ belongs to $\mathrm{epi}(r_x)$. For some $\epsilon > 0$, we consider the point $(\mu, r_x(\mu) - \epsilon)$, which does not belong to $\mathrm{epi}(r_x)$. By the definition of r_x, $r_x(z)$ is finite for all $z \in Z$, Z is nonempty, and $\mathrm{epi}(r_x)$ is closed, so that $\mathrm{epi}(r_x)$ does not contain any vertical lines. Therefore, by Prop. 2.5.1(b), there exists a nonvertical hyperplane that strictly separates the point $(\mu, r_x(\mu) - \epsilon)$ from $\mathrm{epi}(r_x)$, i.e., a vector $(\overline{u}, \zeta)$ with $\zeta \neq 0$, and a scalar c such that

$$\overline{u}'\mu + \zeta\big(r_x(\mu) - \epsilon\big) < c < \overline{u}'z + \zeta w, \qquad \forall \, (z, w) \in \mathrm{epi}(r_x).$$

Since w can be made arbitrarily large, we have $\zeta > 0$, and without loss of generality, we can take $\zeta = 1$. In particular for $w = r_x(z)$, with $z \in Z$, we have

$$\overline{u}'\mu + \big(r_x(\mu) - \epsilon\big) < \overline{u}'z + r_x(z), \qquad \forall \, z \in Z,$$

or equivalently,

$$\phi(x, z) + \overline{u}'(\mu - z) < \phi(x, \mu) + \epsilon, \qquad \forall \, z \in Z.$$

Letting $\epsilon \downarrow 0$, we obtain for all $x \in X$

$$\inf_{u \in \Re^m} \sup_{z \in Z} \{\phi(x, z) + u'(\mu - z)\} \leq \sup_{z \in Z}\{\phi(x, z) + \overline{u}'(\mu - z)\} \leq \phi(x, \mu).$$

By taking the infimum over $x \in X$ in the above relation, and by using Eq. (2.37), we see that Eq. (2.38) holds.

We now assume that $\mu \notin Z$. We consider a sequence $\{w_k\}$ with $w_k \to \infty$ and we fix an arbitrary $x \in X$. Since $\mu \notin Z$, the sequence $\{(\mu, w_k)\}$ does not belong to $\mathrm{epi}(r_x)$. Therefore, similar to the argument above, there exists a sequence of nonvertical hyperplanes with normals $(u_k, 1)$ such that

$$w_k + u_k'\mu < -\phi(x, z) + u_k'z, \qquad \forall \, z \in Z, \quad \forall \, k,$$

implying that

$$\phi(x, z) + u_k'(\mu - z) < -w_k, \qquad \forall \, z \in Z, \quad \forall \, k.$$

Thus, we have

$$\inf_{u \in \Re^m} \sup_{z \in Z}\{\phi(x, z) + u'(\mu - z)\} \leq \sup_{z \in Z}\{\phi(x, z) + u_k'(\mu - z)\} \leq -w_k, \qquad \forall \, k,$$

and by taking the limit in the preceding inequality as $k \to \infty$, we obtain

$$\inf_{u \in \Re^m} \sup_{z \in Z} \{\phi(x,z) + u'(\mu - z)\} = -\infty, \qquad \forall\, x \in X.$$

Using Eq. (2.37), we see that $q(\mu) = -\infty$. Thus, $q(\mu)$ has the form given in Eq. (2.36). The equality $q^* = w^*$ and the minimax equality are equivalent in view of the discussion following Eq. (2.35).   **Q.E.D.**

The assumption that $-\phi(x,\cdot)$ is closed and convex in Lemma 2.6.2 is essential for Eq. (2.36) to hold. This can be seen by considering the special case where ϕ is independent of x, and by noting that q is concave and upper semicontinuous (cf. Prop. 2.5.2).

2.6.2   Minimax Theorems

We now use the preceding two lemmas and the Min Common/Max Crossing Theorem I (cf. Prop. 2.5.4) to prove the following proposition.

Proposition 2.6.2: (Minimax Theorem I) Let X and Z be nonempty convex subsets of $\Re^n$ and $\Re^m$, respectively, and let $\phi : X \times Z \mapsto \Re$ be a function. Assume that for each $z \in Z$, the function $\phi(\cdot,z) : X \mapsto \Re$ is convex, and for each $x \in X$, the function $-\phi(x,\cdot) : Z \mapsto \Re$ is closed and convex. Assume further that

$$\inf_{x \in X} \sup_{z \in Z} \phi(x,z) < \infty.$$

Then, the minimax equality holds, i.e.,

$$\sup_{z \in Z} \inf_{x \in X} \phi(x,z) = \inf_{x \in X} \sup_{z \in Z} \phi(x,z),$$

if and only if the function p of Eq. (2.33) is lower semicontinuous at $u = 0$, i.e., $p(0) \le \liminf_{k \to \infty} p(u_k)$ for all sequences $\{u_k\}$ with $u_k \to 0$.

Proof: The proof consists of showing that with an appropriate selection of the set M, the assumptions of the proposition are essentially equivalent to the corresponding assumptions of the Min Common/Max Crossing Theorem I.

We choose the set M (as well as the set $\overline{M}$) in the Min Common/Max Crossing Theorem I to be the epigraph of p,

$$M = \overline{M} = \big\{(u,w) \mid u \in \Re^m,\ p(u) \le w\big\},$$

which is convex in view of the assumed convexity of $\phi(\cdot, z)$ and Lemma 2.6.1. Thus, assumption (2) of the Min Common/Max Crossing Theorem I is satisfied.

From the definition of p, we have

$$w^* = p(0) = \inf_{x \in X} \sup_{z \in Z} \phi(x, z).$$

It follows that the assumption

$$\inf_{x \in X} \sup_{z \in Z} \phi(x, z) < \infty$$

is equivalent to the assumption $w^* < \infty$ of the Min Common/Max Crossing Theorem I.

Finally, the condition

$$p(0) \leq \liminf_{k \to \infty} p(u_k)$$

for all sequences $\{u_k\}$ with $u_k \to 0$ is equivalent to the condition of the Min Common/Max Crossing Theorem I that for every sequence $\{(u_k, w_k)\} \subset M$ with $u_k \to 0$, there holds $w^* \leq \liminf_{k \to \infty} w_k$. Thus, by the conclusion of that theorem, the condition $p(0) \leq \liminf_{k \to \infty} p(u_k)$ holds if and only if $q^* = w^*$, which in turn holds if and only if the minimax equality holds [cf. Lemma 2.6.2, which applies because of the assumed closedness and convexity of $-\phi(x, \cdot)$]. **Q.E.D.**

The proof of the preceding proposition can be easily modified to use the second Min Common/Max Crossing Theorem of the preceding section [cf. Prop. 2.5.5 and Eq. (2.36)]. What is needed is an assumption that $p(0)$ is finite and that 0 lies in the relative interior of the effective domain of p. We then obtain the following result, which also asserts that the supremum in the minimax equality is attained [this follows from the corresponding attainment assertion of Prop. 2.5.5 and Eq. (2.36)].

Proposition 2.6.3: (Minimax Theorem II) Let X and Z be nonempty convex subsets of $\Re^n$ and $\Re^m$, respectively, and let $\phi : X \times Z \mapsto \Re$ be a function. Assume that for each $z \in Z$, the function $\phi(\cdot, z) : X \mapsto \Re$ is convex, and for each $x \in X$, the function $-\phi(x, \cdot) : Z \mapsto \Re$ is closed and convex. Assume further that

$$-\infty < \inf_{x \in X} \sup_{z \in Z} \phi(x, z),$$

and that 0 lies in the relative interior of the effective domain of the function p of Eq. (2.33). Then, the minimax equality holds, i.e.,

$$\sup_{z \in Z} \inf_{x \in X} \phi(x, z) = \inf_{x \in X} \sup_{z \in Z} \phi(x, z),$$

and the supremum over Z in the left-hand side is finite and is attained. Furthermore, the set of $z \in Z$ attaining this supremum is compact if and only if 0 lies in the interior of the effective domain of p.

The preceding two Minimax Theorems indicate that, aside from convexity and semicontinuity assumptions, the properties of the function p around $u = 0$ are critical to guarantee the minimax equality. Here is an illustrative example:

Example 2.6.1:

Let

$$X = \left\{ x \in \Re^2 \mid x \geq 0 \right\}, \qquad Z = \{ z \in \Re \mid z \geq 0 \},$$

and let

$$\phi(x, z) = e^{-\sqrt{x_1 x_2}} + zx_1,$$

which can be shown to satisfy the convexity and closedness assumptions of Prop. 2.6.2. [For every $z \in Z$, the Hessian matrix of the function $\phi(\cdot, z)$ is positive definite within $\text{int}(X)$, so it can be seen that $\phi(\cdot, z)$ is convex over $\text{int}(X)$. Since $\phi(\cdot, z)$ is continuous, it is also convex over X.] For all $z \geq 0$, we have

$$\inf_{x \geq 0} \left\{ e^{-\sqrt{x_1 x_2}} + zx_1 \right\} = 0,$$

since the expression in braces is nonnegative for $x \geq 0$ and can approach zero by taking $x_1 \to 0$ and $x_1 x_2 \to \infty$. Hence,

$$\sup_{z \geq 0} \inf_{x \geq 0} \phi(x, z) = 0.$$

We also have for all $x \geq 0$,

$$\sup_{z \geq 0} \left\{ e^{-\sqrt{x_1 x_2}} + zx_1 \right\} = \begin{cases} 1 & \text{if } x_1 = 0, \\ \infty & \text{if } x_1 > 0. \end{cases}$$

Hence,

$$\inf_{x \geq 0} \sup_{z \geq 0} \phi(x, z) = 1,$$

so

$$\inf_{x \geq 0} \sup_{z \geq 0} \phi(x, z) > \sup_{z \geq 0} \inf_{x \geq 0} \phi(x, z).$$

Here, the function p is given by

$$p(u) = \inf_{x \geq 0} \sup_{z \geq 0} \left\{ e^{-\sqrt{x_1 x_2}} + z(x_1 - u) \right\} = \begin{cases} \infty & \text{if } u < 0, \\ 1 & \text{if } u = 0, \\ 0 & \text{if } u > 0, \end{cases}$$

(cf. Fig. 2.6.3). Thus, it can be seen that even though $p(0)$ is finite, p is not lower semicontinuous at 0. As a result the assumptions of Props. 2.6.2 are violated, and the minimax equality does not hold.

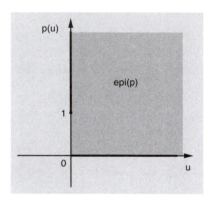

Figure 2.6.3. The function p for Example 2.6.1:

$$p(u) = \begin{cases} \infty & \text{if } u < 0, \\ 1 & \text{if } u = 0, \\ 0 & \text{if } u > 0. \end{cases}$$

Here p is not lower semicontinuous at 0, and the minimax equality does not hold.

The following example illustrates how the minimax equality may hold, while the supremum over $z \in Z$ is not attained because the relative interior assumption of Prop. 2.6.3 is not satisfied.

Example 2.6.2:

Let

$$X = \Re, \qquad Z = \{z \in \Re \mid z \geq 0\},$$

and let

$$\phi(x, z) = x + zx^2,$$

which satisfy the convexity and closedness assumptions of Prop. 2.6.3. For all $z \geq 0$, we have

$$\inf_{x \in \Re} \left\{ x + zx^2 \right\} = \begin{cases} -1/(4z) & \text{if } z > 0, \\ -\infty & \text{if } z = 0. \end{cases}$$

Hence,

$$\sup_{z \geq 0} \inf_{x \in \Re} \phi(x, z) = 0.$$

We also have for all $x \in \Re$,

$$\sup_{z \geq 0} \left\{ x + zx^2 \right\} = \begin{cases} 0 & \text{if } x = 0, \\ \infty & \text{if } x \neq 0. \end{cases}$$

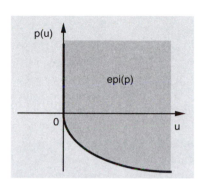

p(u)

epi(p)

0   u

Figure 2.6.4. The function p for Example 2.6.2:

$$p(u) = \begin{cases} -\sqrt{u} & \text{if } u \geq 0, \\ \infty & \text{if } u < 0. \end{cases}$$

Here p is lower semicontinuous at 0 and the minimax equality holds. However, 0 is not a relative interior point of $\text{dom}(p)$ and the supremum of $\inf_{x \in \Re} \phi(x, z)$, which is

$$\inf_{x \in \Re} \phi(x, z) = \begin{cases} -1/(4z) & \text{if } z > 0, \\ -\infty & \text{if } z = 0, \end{cases}$$

is not attained.

Hence,

$$\inf_{x \in \Re} \sup_{z \geq 0} \phi(x, z) = 0,$$

so $\inf_{x \in \Re} \sup_{z \geq 0} \phi(x, z) = \sup_{z \geq 0} \inf_{x \in \Re} \phi(x, z)$, i.e., the minimax equality holds. However, the problem

$$\text{maximize} \quad \inf_{x \in \Re} \phi(x, z)$$

$$\text{subject to } z \in Z$$

does not have an optimal solution. Here we have

$$\sup_{z \geq 0} \{x + zx^2 - uz\} = \begin{cases} x & \text{if } x^2 \leq u, \\ \infty & \text{if } x^2 > u, \end{cases}$$

and

$$p(u) = \inf_{x \in \Re} \sup_{z \geq 0} \{x + zx^2 - uz\} = \begin{cases} -\sqrt{u} & \text{if } u \geq 0, \\ \infty & \text{if } u < 0, \end{cases}$$

(cf. Fig. 2.6.4). It can be seen that 0 is not a relative interior point of the effective domain of p, thus violating the relative interior assumption of Prop. 2.6.3.

2.6.3   Saddle Point Theorems

The first minimax theorem (Prop. 2.6.2) provides a general method to ascertain the validity of the minimax equality

$$\sup_{z \in Z} \inf_{x \in X} \phi(x, z) = \inf_{x \in X} \sup_{z \in Z} \phi(x, z), \tag{2.39}$$

namely showing that p is closed and convex. By noting that p is obtained by the partial minimization

$$p(u) = \inf_{x \in \Re^n} F(x, u), \tag{2.40}$$

where

$$F(x, u) = \begin{cases} \sup_{z \in Z}\{\phi(x, z) - u'z\} & \text{if } x \in X, \\ \infty & \text{if } x \notin X, \end{cases} \qquad (2.41)$$

we see that the closedness of p can be guaranteed by using the theory of Section 2.3, i.e., the conditions of Props. 2.3.7-2.3.10. These conditions simultaneously guarantee that the infimum in Eq. (2.40) is attained, and in particular (for $u = 0$) that the infimum over X in the minimax equality of Eq. (2.39) is attained. Furthermore, when the roles of minimization and maximization are reversed, these conditions can be used to guarantee that the supremum over Z in the minimax equality of Eq. (2.39) is attained, thereby providing conditions for the existence of a saddle point.

To formulate the corresponding results, for each $z \in Z$, we introduce the function $t_z : \Re^n \mapsto (-\infty, \infty]$ defined by

$$t_z(x) = \begin{cases} \phi(x, z) & \text{if } x \in X, \\ \infty & \text{if } x \notin X, \end{cases}$$

and for each $x \in X$, we introduce the function $r_x : \Re^m \mapsto (-\infty, \infty]$ defined by

$$r_x(z) = \begin{cases} -\phi(x, z) & \text{if } z \in Z, \\ \infty & \text{if } z \notin Z. \end{cases}$$

We also consider the functions $t : \Re^n \mapsto (-\infty, \infty]$ and $r : \Re^m \mapsto (-\infty, \infty]$ given by

$$t(x) = \sup_{z \in Z} t_z(x), \qquad x \in \Re^n,$$

and

$$r(z) = \sup_{x \in X} r_x(z), \qquad z \in \Re^m.$$

We will use the following assumption.

Assumption 2.6.1: (Closedness and Convexity) For each $z \in Z$, the function t_z is closed and convex, and for each $x \in X$, the function r_x is closed and convex.

Note that if the preceding assumption holds, the function t is closed and convex [cf. Prop. 1.2.4(c)]. Furthermore, we have $t(x) > -\infty$ for all x, so for t to be proper, we must have $t(x) < \infty$ for some $x \in X$, or equivalently

$$\inf_{x \in X} \sup_{z \in Z} \phi(x, z) < \infty.$$

Similarly the function r is closed and convex, and it is proper if

$$-\infty < \sup_{z \in Z} \inf_{x \in X} \phi(x, z).$$

We note that if the minimax equality holds, the set of saddle points is the set $X^* \times Z^*$, where X^* and Z^* are the sets of minimizers of the functions t and r, respectively (cf. Prop. 2.6.1). Thus, conditions that guarantee that a saddle point exists, must also guarantee that the set of minimizers of t is nonempty. Examples of such conditions are that t is proper with compact level sets, and other conditions for existence of minima from Section 2.3.1. It turns out that these conditions not only guarantee that the set of minimizers of t is nonempty, but they also guarantee that the function p is lower semicontinuous at 0, and that the minimax equality holds. This is the subject of the next four propositions, which correspond to the four conditions, given in Section 2.3.1, for attainment of minima of convex functions (see Props. 2.3.2 and 2.3.4).

Proposition 2.6.4: Let Assumption 2.6.1 (Closedness and Convexity) hold, and assume that

$$\inf_{x \in X} \sup_{z \in Z} \phi(x, z) < \infty,$$

and that the level sets $\big\{ x \mid t(x) \leq \gamma \big\}$, $\gamma \in \Re$, of the function t are compact. Then the minimax equality

$$\sup_{z \in Z} \inf_{x \in X} \phi(x, z) = \inf_{x \in X} \sup_{z \in Z} \phi(x, z)$$

holds and the infimum over X in the right-hand side above is attained at a set of points that is nonempty and compact.

Proof: The function p is defined by partial minimization of the function F of Eq. (2.41), i.e.,

$$p(u) = \inf_{x \in \Re^n} F(x, u)$$

[cf. Eq. (2.40)]. We note that F is proper, since t is proper, which follows from the assumption $\inf_{x \in X} \sup_{z \in Z} \phi(x, z) < \infty$. Furthermore, we have

$$t(x) = F(x, 0),$$

so the compactness assumption on the level sets of t can be translated to the compactness assumption of Prop. 2.3.7 (with 0 playing the role of the vector $\overline{x}$ in Prop. 2.3.7). It follows from the result of the proposition that p is closed and proper. By the Minimax Theorem I (Prop. 2.6.2), using also the closedness of p and the assumption $\inf_{x \in X} \sup_{z \in Z} \phi(x, z) < \infty$, it follows that the minimax equality holds. Finally, the infimum over X in the right-hand side of the minimax equality is attained at the set of minimizing points of the function t, which is nonempty and compact since t is proper and has compact level sets. **Q.E.D.**

Proposition 2.6.5: Let Assumption 2.6.1 (Closedness and Convexity) hold, and assume that

$$\inf_{x \in X} \sup_{z \in Z} \phi(x, z) < \infty,$$

and that the recession cone and the constancy space of the function t are equal. Then the minimax equality

$$\sup_{z \in Z} \inf_{x \in X} \phi(x, z) = \inf_{x \in X} \sup_{z \in Z} \phi(x, z)$$

holds, and the infimum over X in the right-hand side above is attained.

Proof: The proof is similar to the one of Prop. 2.6.4. We use Prop. 2.3.8 in place of Prop. 2.3.7. **Q.E.D.**

The next two propositions deal with the case where ϕ, X, and Z are such that the partial minimization of F over x is subject to inequality constraints involving linear or convex quadratic functions.

Proposition 2.6.6: Let Assumption 2.6.1 (Closedness and Convexity) hold, and assume that

$$\inf_{x \in X} \sup_{z \in Z} \phi(x, z) < \infty.$$

Assume further that the function

$$F(x, u) = \begin{cases} \sup_{z \in Z} \{\phi(x, z) - u'z\} & \text{if } x \in X, \\ \infty & \text{if } x \notin X, \end{cases}$$

[cf. Eq. (2.41)] has the form

$$F(x, u) = \begin{cases} \overline{F}(x, u) & \text{if } (x, u) \in C, \\ \infty & \text{if } (x, u) \notin C, \end{cases}$$

where $\overline{F}$ is a closed proper convex function on $\Re^{m+n}$ and C is specified by linear inequalities, i.e.,

---

$$C = \{(x, u) \mid Ax + Bu \le b\},$$

where A and B are matrices, and b is a vector. Assume further that every common direction of recession of C and $\overline{F}$ is a direction along which $\overline{F}$ is constant. Then the minimax equality

$$\sup_{z \in Z} \inf_{x \in X} \phi(x, z) = \inf_{x \in X} \sup_{z \in Z} \phi(x, z)$$

holds, and the infimum over X in the right-hand side above is attained.

Proof: The proof is similar to the one of Prop. 2.6.4. We use Prop. 2.3.9 in place of Prop. 2.3.7. **Q.E.D.**

Proposition 2.6.7: Let Assumption 2.6.1 (Closedness and Convexity) hold, and assume that

$$-\infty < \inf_{x \in X} \sup_{z \in Z} \phi(x, z) < \infty.$$

Assume further that the function ϕ is quadratic of the form

$$\phi(x, z) = x'Qx + c'x + z'Mx - z'Rz - d'z,$$

where Q and R are symmetric matrices, M is a matrix, and c and d are vectors. Assume further that $Z = \Re^m$, Q is positive semidefinite, R is positive definite, and X is of the form

$$X = \{x \mid x'Q_j x + a_j'x + b_j \le 0, \ j = 1, \dots, r\},$$

where Q_j are symmetric positive semidefinite $n \times n$ matrices, c and a_j are vectors in $\Re^n$, and b_j are scalars. Then the minimax equality

$$\sup_{z \in Z} \inf_{x \in X} \phi(x, z) = \inf_{x \in X} \sup_{z \in Z} \phi(x, z)$$

holds, and the infimum over X in the right-hand side above is attained.

Proof: Because $Z = \Re^m$ and R is positive definite, for each (x, u), the supremum in the expression

$$\sup_{z \in \Re^m} \{\phi(x, z) - u'z\}, \qquad (2.42)$$

is attained at some point $z(x, u)$ that depends linearly on (x, u). In particular, by setting to 0 the gradient with respect to x of $\phi(x, z) - u'z$, we obtain

$$z(x, u) = \tfrac{1}{2} R^{-1}(c + Mx - d - u).$$

When this expression for $z(x, u)$ is substituted in place of z in Eq. (2.42), we obtain a quadratic function of (x, u), which is convex (since it is the pointwise supremum of convex functions). Thus the function F of Eq. (2.41) has the form

$$F(x, u) = \begin{cases} \overline{F}(x, u) & \text{if } x \in X, \\ \infty & \text{if } x \notin X, \end{cases}$$

where $\overline{F}$ is a convex quadratic function in (x, u). Using this fact, the proof is similar to the one of Prop. 2.6.4. We use Prop. 2.3.10 in place of Prop. 2.3.7. Note that the assumption $-\infty < \inf_{x \in X} \sup_{z \in Z} \phi(x, z) < \infty$ is equivalent to $-\infty < p(0) < \infty$, so the vector $u = 0$ can play the role of $\overline{x}$ in Prop. 2.3.10. **Q.E.D.**

By reversing the roles of minimization and maximization, we can use the preceding propositions to guarantee that the supremum over Z in the minimax equality is attained, thereby providing conditions for the existence of a saddle point.

Proposition 2.6.8: Let Assumption 2.6.1 (Closedness and Convexity) hold. Assume also that either

$$\inf_{x \in X} \sup_{z \in Z} \phi(x, z) < \infty,$$

or

$$-\infty < \sup_{z \in Z} \inf_{x \in X} \phi(x, z).$$

(a) If the level sets $\{x \mid t(x) \le \gamma\}$ and $\{x \mid r(x) \le \gamma\}$, $\gamma \in \Re$, of the functions t and r are compact, the set of saddle points of ϕ is nonempty and compact.

(b) If the recession cones of the functions t and r are equal to the constancy spaces of t and r, respectively, the set of saddle points of ϕ is nonempty.

Proof: (a) If

$$\inf_{x \in X} \sup_{z \in Z} \phi(x, z) < \infty,$$

we apply Prop. 2.6.4 to show that the minimax equality holds, and that the infimum over X is finite and is attained at a nonempty and compact set. Therefore, we have

$$-\infty < \sup_{z \in Z} \inf_{x \in X} \phi(x, z) = \inf_{x \in X} \sup_{z \in Z} \phi(x, z) < \infty.$$

We can thus reverse the roles of x and z and the sign of ϕ, and apply Prop. 2.6.4 again to show that the supremum over Z is attained at a nonempty and compact set.

If

$$\inf_{x \in X} \sup_{z \in Z} \phi(x, z) = \infty,$$

then we must have

$$-\infty < \sup_{z \in Z} \inf_{x \in X} \phi(x, z).$$

Therefore

$$\inf_{z \in Z} \sup_{x \in X} \left(-\phi(x, z)\right) < \infty.$$

We then reverse the roles of x and z, and apply the preceding argument in conjunction with Prop. 2.6.4.

(b) The proof is similar to the one of part (a), except that we use Prop. 2.6.5 instead of Prop. 2.6.4.   **Q.E.D.**

The existence of a saddle point of ϕ can also be shown under assumptions other than those of Prop. 2.6.8 by combining in various ways the assumptions of Props. 2.6.4-2.6.7, as suggested by the preceding proof.

Note that, under the Closedness and Convexity Assumption, the nonempty level sets $\{x \mid t(x) \leq \gamma\}$ and $\{z \mid r(z) \leq \gamma\}$ are compact [cf. Prop. 2.6.8(a)] if and only if the set of common directions of recession of all the functions t_z, $z \in Z$, consists of the zero vector only, and the set of common directions of recession of all the functions r_x, $x \in X$, consists of the zero vector only. To see this note that, in view of the definition

$$t(x) = \sup_{z \in Z} t_z(x),$$

a nonempty level set $\{x \mid t(x) \leq \gamma\}$, is equal to the intersection of the level sets $\{x \mid t_z(x) \leq \gamma\}$, as z ranges over Z. Since the latter level sets are closed (in view of the closedness of t_z), the recession cone of their intersection, is equal to the intersection of their recession cones [Prop. 1.5.1(e)], which consists of just the zero vector. This implies that the nonempty level sets of the function t are compact if and only if there is no nonzero direction of recession common to all the functions t_z, $z \in Z$. (Note, however, that even if this directions of recession assumption holds, there is no guarantee that t has some nonempty level sets, i.e., that t is proper.)

The compactness of the level sets of the function t can be guaranteed by simpler sufficient conditions. In particular, under the Closedness and Convexity Assumption, the level sets $\{x \mid t(x) \leq \gamma\}$ are compact under any one of the following two conditions:

(1) The set X is compact [since the level sets $\{x \mid t(x) \leq \gamma\}$ are closed, by the closedness of t, and are contained in X].

(2) There exists a vector $\bar{z} \in Z$ and a scalar $\bar{\gamma}$ such that the set

$$\{x \in X \mid \phi(x, \bar{z}) \leq \bar{\gamma}\}$$

is nonempty and compact [since then all the level sets $\{x \in X \mid \phi(x, \bar{z}) \leq \gamma\}$ are compact, and a nonempty level set $\{x \mid t(x) \leq \gamma\}$ is contained in the level set $\{x \in X \mid \phi(x, \bar{z}) \leq \gamma\}$].

Furthermore, any one of above two conditions also guarantees that the function r is proper; for example under condition (2), the supremum and infimum over $x \in X$ in the relations

$$r(\bar{z}) = \sup_{x \in X} \left(-\phi(x, \bar{z}) \right) = - \inf_{x \in X} \phi(x, \bar{z})$$

is attained by Weierstrass' Theorem, so that $r(\bar{z}) < \infty$.

By a symmetric argument, we also see that, under the Closedness and Convexity Assumption, the level sets $\{x \mid t(x) \leq \gamma\}$ are compact under any one of the following two conditions:

(1) The set Z is compact.

(2) There exists a vector $\bar{x} \in X$ and a scalar $\bar{\gamma}$ such that the set

$$\{z \in Z \mid \phi(\bar{x}, z) \geq \bar{\gamma}\}$$

is nonempty and compact.

Furthermore, any one of above two conditions also guarantees that the function t is proper.

Thus, by combining the preceding discussion and Prop. 2.6.8(a), we obtain the following result, an extension of the classical theorem of von Neumann [Neu28], which provides sufficient conditions for the existence of a saddle point.

Proposition 2.6.9: (Saddle Point Theorem) Let Assumption 2.6.1 (Closedness and Convexity) hold. The set of saddle points of ϕ is nonempty and compact under any one of the following conditions:

(1) X and Z are compact.

(2) Z is compact and there exists a vector $\bar{z} \in Z$ and a scalar γ such that the level set

$$\big\{ x \in X \mid \phi(x, \bar{z}) \leq \gamma \big\}$$

is nonempty and compact.

(3) X is compact and there exists a vector $\bar{x} \in X$ and a scalar γ such that the level set

$$\big\{ z \in Z \mid \phi(\bar{x}, z) \geq \gamma \big\}$$

is nonempty and compact.

(4) There exist vectors $\bar{x} \in X$ and $\bar{z} \in Z$, and a scalar γ such that the level sets

$$\big\{ x \in X \mid \phi(x, \bar{z}) \leq \gamma \big\}, \qquad \big\{ z \in Z \mid \phi(\bar{x}, z) \geq \gamma \big\},$$

are nonempty and compact.

Proof: From the discussion preceding the proposition, it is seen that, under the Closedness and Convexity Assumption, the level sets of t and r are compact. Furthermore, the functions t and r are proper, which is equivalent to $\inf_{x \in X} \sup_{z \in Z} \phi(x, z) < \infty$, and $-\infty < \sup_{z \in Z} \inf_{x \in X} \phi(x, z)$, respectively. The result follows from Prop. 2.6.8(a). **Q.E.D.**

2.7 NOTES, SOURCES, AND EXERCISES

Our definition and analysis of the recession cone of a function has been restricted to closed convex functions. This is the most interesting case because the analysis of the existence of optimal solutions of convex optimization problems typically assumes that the functions involved are closed. The recession cone of a convex function that is not closed can be defined, but the corresponding analysis is more complicated. The reason is that the Recession Cone Theorem (Prop. 2.7.1), which applies to closed convex sets, cannot be invoked in the proof of Prop. 2.3.1, and the subsequent definition of the recession cone of a function.

The result on the attainment of the minimum of a closed convex function over sets specified by linear inequalities (Prop. 2.3.4 for the case of linear constraints) is due to Rockafellar ([Roc70], Th. 27.3). His existence proof is based on a refined version of Helly's Theorem, which applies to an infinite collection of closed convex sets that are specified in part by linear inequalities. Our proof is based instead on the more elementary set intersection result of Prop. 1.5.6. An extension involving directions along which the cost function is flat is given in Exercise 2.6. The extensions to quasiconvex functions of the result on existence of optimal solutions for linear constraints, and the other existence results of Section 2.3.1 [Exercise 2.8(b)], are new.

The result on existence of optimal solutions for the case of quadratic cost and quadratic constraints (cf. Prop. 2.3.4) is due to Terlaky [Ter85], who gave a different proof than the one given here. Another proof of the same result is given by Luo and Zhang [LuZ99], who also consider extensions to some nonconvex problems with quadratic constraints. More general versions of parts of Prop. 2.3.4 and some related results are given by Auslender [Aus96], [Aus97]. The generalization involving bidirectionally flat functions (Exercise 2.7) is based on the work of Auslender [Aus00], and Tseng and the authors [TBN02]. Proposition 2.3.5 was shown by Frank and Wolfe [FrW56] in greater generality, whereby f is a quadratic but not necessarily convex function. Their proof uses the machinery of polyhedral convexity (see Exercise 3.15 in Chapter 3).

The closedness of the function $f(x) = \inf_z F(x, z)$, obtained by partial minimization of a function F, has been studied by several authors, but under assumptions that are too restrictive for our intended use in minimax and duality theory. In particular, our analysis of Props. 2.3.7-2.3.10, and its use in guaranteeing that the minimax equality holds (Section 2.6), are new to our knowledge.

Caratheodory [Car11] and Minkowski [Min11] are credited with the first investigations of separating and supporting hyperplanes. Klee [Kle68] and Rockafellar [Roc70] describe and analyze several different notions of separating hyperplanes. The proper separation theorem (Prop. 2.4.5) is due to Fenchel [Fen51].

The min common/max crossing duality framework was formalized in Bertsekas, Nedić, and Ozdaglar [BNO02]. There are many works that treat duality, including the textbooks by Rockafellar [Roc70], Stoer and Witzgall [StW70], Ekeland and Temam [EkT76], Rockafellar [Roc84], Hiriart-Urruty and Lemarechal [HiL93], Rockafellar and Wets [RoW98], Bertsekas [Ber99a], Bonnans and Shapiro [BoS00], Borwein and Lewis [BoL00], and Auslender and Teboulle [AuT03]. The constructions involved in the min common and max crossing problems are implicit in these duality analyses and have been used for visualization purposes (see e.g., [Ber99a], Section 5.1). However, the two problems have not been used as an analytical framework for constrained optimization duality, saddle point theory, or other contexts, except implicitly through the machinery of conjugate convex functions, and the complicated theory of convex bifunctions and conjugate saddle functions (given by Rockafellar in [Roc70]). By using the min common/max crossing duality framework as the foundation of our analysis, we are able to postpone the introduction of conjugate convex functions until they are needed for the limited purposes of Fenchel duality (Chapter 7), and to avoid altogether dealing with conjugate saddle functions. This allows a more intuitive and accessible exposition.

Saddle point analysis originated with von Neumann [Neu28], who established the existence of a saddle point in finite zero sum games (see the discussion in Section 2.6). Kakutani [Kak41] extended this result to infinite games, where X and Z are arbitrary convex and compact sets. These results were sharpened by Moreau [Mor64] and Rockafellar [Roc64], substituting the compactness assumptions with convex analysis-type assumptions involving notions of relative interior and directions of recession.

Propositions 2.6.4 and 2.6.8(a) constitute variations of the corresponding minimax theorems of Rockafellar ([Roc70], Ths. 37.3 and 37.6), where instead of assuming that the level sets of t (and/or r) are compact, a stronger condition is assumed, namely that there is no nonzero common direction of recession of all the functions t_z, $z \in \mathrm{ri}(Z)$ [and/or r_x, $x \in \mathrm{ri}(X)$, respectively], but some of the other assumptions of Props. 2.6.4 and 2.6.8(a) are not made. The conclusions of Prop. 2.6.4 using the condition that there exists a vector $\bar{z} \in Z$ such that all the sets $\{x \in X \mid \phi(x,\bar{z}) \leq \gamma\}$, $\gamma \in \Re$, are compact, in place of our compactness of level sets assumption have been given by Borwein and Lewis ([BoL00], p. 96). An alternative proof of the Saddle Point Theorem (Prop. 2.6.9), fundamentally different from ours, is given by Hiriart-Urruty and Lemarechal [HiL93] (under more restrictive assumptions on ϕ). Propositions 2.6.5-2.6.7 and 2.6.8(b) seem to be new in the form given here.

EXERCISES

2.1

Let $f : \Re^n \mapsto \Re$ be a given function.

(a) Assume that f is convex. Show that a vector x^* is a global minimum of f if and only if it is a global minimum of f along any line passing through x^* [i.e., for all $d \in \Re^n$, the function $g : \Re \mapsto \Re$, defined by $g(\alpha) = f(x^* + \alpha d)$, has $\alpha^* = 0$ as its global minimum].

(b) Assume that f is not convex. Show that a vector x^* need not be a local minimum of f if it is a local minimum of f along every line passing through

x^*. *Hint:* Use the function $f : \Re^2 \mapsto \Re$ given by

$$f(x_1, x_2) = (x_2 - px_1^2)(x_2 - qx_1^2),$$

where p and q are scalars with $0 < p < q$, and $x^* = (0,0)$. Show that $f(y, my^2) < 0$ for $y \neq 0$ and m satisfying $p < m < q$, while $f(0,0) = 0$.

2.2 (Lipschitz Continuity of Convex Functions)

Let $f : \Re^n \mapsto \Re$ be a convex function and X be a bounded set in $\Re^n$. Show that f is Lipschitz continuous over X, i.e., there exists a positive scalar L such that

$$\big|f(x) - f(y)\big| \leq L\|x - y\|, \qquad \forall \ x, y \in X.$$

2.3 (Exact Penalty Functions)

Let $f : Y \mapsto \Re$ be a function defined on a subset Y of $\Re^n$. Assume that f is Lipschitz continuous with constant L, i.e.,

$$\big|f(x) - f(y)\big| \leq L\|x - y\|, \qquad \forall \ x, y \in Y.$$

Let also X be a nonempty closed subset of Y, and c be a scalar with $c > L$.

(a) Show that if x^* minimizes f over X, then x^* minimizes

$$F_c(x) = f(x) + c \inf_{y \in X} \|y - x\|$$

over Y.

(b) Show that if x^* minimizes $F_c(x)$ over Y, then $x^* \in X$, so that x^* minimizes f over X.

2.4 (Ekeland's Variational Principle [Eke74])

This exercise shows how ϵ-optimal solutions of optimization problems can be approximated by (exactly) optimal solutions of some other slightly perturbed problems. Let $f : \Re^n \mapsto (-\infty, \infty]$ be a closed proper function, and let $\bar{x} \in \Re^n$ be a vector such that

$$f(\bar{x}) \leq \inf_{x \in \Re^n} f(x) + \epsilon,$$

where $\epsilon > 0$. Then, for any $\delta > 0$, there exists a vector $\tilde{x} \in \Re^n$ such that

$$\|\bar{x} - \tilde{x}\| \leq \frac{\epsilon}{\delta}, \qquad f(\tilde{x}) \leq f(\bar{x}),$$

and $\tilde{x}$ is the unique optimal solution of the perturbed problem of minimizing $f(x) + \delta\|x - \tilde{x}\|$ over $\Re^n$.

2.5 (Approximate Minima of Convex Functions)

Let X be a closed convex subset of $\Re^n$, and let $f : \Re^n \mapsto (-\infty, \infty]$ be a closed convex function such that $X \cap \operatorname{dom}(f) \neq \emptyset$. Assume that f and X have no common nonzero direction of recession. Let X^* be the set of global minima of f over X (which is nonempty and compact by Prop. 2.3.2), and let $f^* = \inf_{x \in X} f(x)$. Show that:

(a) For every $\epsilon > 0$ there exists a $\delta > 0$ such that every vector $x \in X$ with $f(x) \leq f^* + \delta$ satisfies $\min_{x^* \in X^*} \|x - x^*\| \leq \epsilon$.

(b) If f is real-valued, for every $\delta > 0$ there exists a $\epsilon > 0$ such that every vector $x \in X$ with $\min_{x^* \in X^*} \|x - x^*\| \leq \epsilon$ satisfies $f(x) \leq f^* + \delta$.

(c) Every sequence $\{x_k\} \subset X$ satisfying $f(x_k) \to f^*$ is bounded and all its limit points belong to X^*.

2.6 (Directions Along Which a Function is Flat)

The purpose of the exercise is to provide refinements of results relating to set intersections and existence of optimal solutions (cf. Props. 1.5.6 and 2.3.4). Let $f : \Re^n \mapsto (-\infty, \infty]$ be a closed proper convex function, and let F_f be the set of all directions y such that for every $x \in \operatorname{dom}(f)$, the limit $\lim_{\alpha \to \infty} f(x + \alpha y)$ exists. We refer to F_f as the set of *directions along which f is flat*. Note that

$$L_f \subset F_f \subset R_f,$$

where L_f and R_f are the constancy space and recession cone of f, respectively. Let X be a subset of $\Re^n$ specified by linear inequality constraints, i.e.,

$$X = \{x \mid a_j'x \leq b_j, \ j = 1, \ldots, r\},$$

where a_j are vectors in $\Re^n$ and b_j are scalars. Assume that

$$R_X \cap F_f \subset L_f,$$

where R_X is the recession cone of X.

(a) Let

$$C_k = \{x \mid f(x) \leq w_k\},$$

where $\{w_k\}$ is a monotonically decreasing and convergent scalar sequence, and assume that $X \cap C_k \neq \emptyset$ for all k. Show that

$$X \cap \left(\cap_{k=0}^{\infty} C_k\right) \neq \emptyset.$$

(b) Show that if $\inf_{x \in X} f(x)$ is finite, the function f attains a minimum over the set X.

(c) Show by example that f need not attain a minimum over X if we just assume (as in Prop. 2.3.4) that $X \cap \operatorname{dom}(f) \neq \emptyset$.

2.7 (Bidirectionally Flat Functions)

The purpose of the exercise is to provide refinements of the results involving convex quadratic functions and relating to set intersections, closedness under linear transformations, existence of optimal solutions, and closedness under partial minimization [cf. Props. 1.5.7, 1.5.8(c), 1.5.9, 2.3.4, and 2.3.10].

Let $f : \Re^n \mapsto (-\infty, \infty]$ be a closed proper convex function, and let F_f be the set of directions along which f is flat (cf. Exercise 2.6). We say that f is *bidirectionally flat* if $L_f = F_f$ (i.e., if it is flat in some direction it must be flat, and hence constant, in the opposite direction). Note that every convex quadratic function is bidirectionally flat. More generally, a function of the form

$$f(x) = h(Ax) + c'x,$$

where A is an $m \times n$ matrix and $h : \Re^m \mapsto (-\infty, \infty]$ is a coercive closed proper convex function, is bidirectionally flat. In this case, we have

$$L_f = F_f = \{y \mid Ay = 0, \, c'y = 0\}.$$

Let $g_j : \Re^n \mapsto (-\infty, \infty]$, $j = 0, 1, \ldots, r$, be closed proper convex functions that are bidirectionally flat.

(a) Assume that each vector x such that $g_0(x) \le 0$ belongs to $\cap_{j=1}^r \mathrm{dom}(g_j)$, and that for some scalar sequence $\{w_k\}$ with $w_k \downarrow 0$, the set

$$C_k = \left\{x \mid g_0(x) \le w_k, \, g_j(x) \le 0, \, j = 1, \ldots, r\right\}$$

is nonempty for each k. Show that the intersection $\cap_{k=0}^\infty C_k$ is nonempty.

(b) Assume that each g_j, $j = 1, \ldots, r$, is real-valued and the set

$$C = \left\{x \mid g_j(x) \le 0, \, j = 1, \ldots, r\right\}$$

is nonempty. Show that for any $m \times n$ matrix A, the set AC is closed.

(c) Show that a closed proper convex function $f : \Re^n \mapsto (-\infty, \infty]$ that is bidirectionally flat attains a minimum over the set C of part (b), provided that $\inf_{x \in C} f(x)$ is finite.

(d) Let $F : \Re^{n+m} \mapsto (-\infty, \infty]$ be a function of the form

$$F(x, z) = \begin{cases} \overline{F}(x, z) & \text{if } (x, z) \in C, \\ \infty & \text{otherwise,} \end{cases}$$

where $\overline{F}$ is a bidirectionally flat real-valued convex function on $\Re^{n+m}$ and C is a subset of $\Re^{n+m}$ that is specified by r convex inequalities involving bidirectionally flat real-valued convex functions on $\Re^{n+m}$ [cf. part (b)]. Consider the function

$$p(x) = \inf_{z \in \Re^m} F(x, z).$$

Assume that there exists a vector $\overline{x} \in \Re^n$ such that $-\infty < p(\overline{x}) < \infty$. Show that p is convex, closed, and proper. Furthermore, for each $x \in \mathrm{dom}(p)$, the set of points that attain the infimum of $F(x, \cdot)$ over $\Re^m$ is nonempty.

2.8 (Minimization of Quasiconvex Functions)

We say that a function $f : \Re^n \mapsto (-\infty, \infty]$ is *quasiconvex* if all its level sets

$$V_\gamma = \{x \mid f(x) \le \gamma\}$$

are convex. Let X be a convex subset of $\Re^n$, let f be a quasiconvex function such that $X \cap \text{dom}(f) \ne \varnothing$, and denote $f^* = \inf_{x \in X} f(x)$.

(a) Assume that f is not constant on any line segment of X, i.e., we do not have $f(x) = c$ for some scalar c and all x in the line segment connecting any two distinct points of X. Show that every local minimum of f over X is also global.

(b) Assume that X is closed, and f is closed and proper. Let Γ be the set of all $\gamma > f^*$, and denote

$$R_f = \cap_{\gamma \in \Gamma} R_\gamma, \qquad L_f = \cap_{\gamma \in \Gamma} L_\gamma,$$

where R_γ and L_γ are the recession cone and the lineality space of V_γ, respectively. Use the line of proof of Prop. 2.3.4 and Exercise 2.7 to show that f attains a minimum over X if any one of the following conditions holds:

(1) $R_X \cap R_f = L_X \cap L_f$.

(2) $R_X \cap R_f \subset L_f$, and the set X is of the form

$$X = \{x \mid a_j' x \le b_j,\, j = 1, \ldots, r\},$$

where a_j are vectors in $\Re^n$ and b_j are scalars.

(3) $f^* > -\infty$, the set X is of the form

$$X = \{x \mid x' Q_j x + a_j' x + b_j \le 0,\ j = 1, \ldots, r\},$$

where Q_j are symmetric positive semidefinite $n \times n$ matrices, a_j are vectors in $\Re^n$, and b_j are scalars, and for some $\overline{\gamma} \in \Gamma$ and all $\gamma \in \Gamma$ with $\gamma \le \overline{\gamma}$, the level sets V_γ are of the form

$$V_\gamma = \{x \mid x' Q x + c' x + b(\gamma) \le 0\},$$

where Q is a symmetric positive semidefinite $n \times n$ matrix, c is a vector in $\Re^n$, and $b(\gamma)$ is a monotonically nondecreasing function of γ, such that the set $\{b(\gamma) \mid f^* < \gamma \le \overline{\gamma}\}$ is bounded.

(4) $f^* > -\infty$, the set X is of the form

$$X = \{x \mid g_j(x) \le 0,\ j = 1, \ldots, r\},$$

and for some $\overline{\gamma} \in \Gamma$ and all $\gamma \in \Gamma$ with $\gamma \le \overline{\gamma}$, the level sets V_γ are of the form

$$V_\gamma = \{x \mid g_0(x) + b(\gamma) \le 0\},$$

where $g_j,\ j = 0, 1, \ldots, r$, are real-valued, convex, and bidirectionally flat functions (cf. Exercise 2.7), and $b(\gamma)$ is a monotonically nondecreasing function of γ, such that the set $\{b(\gamma) \mid f^* < \gamma \le \overline{\gamma}\}$ is bounded.

2.9 (Partial Minimization)

(a) Let $f : \Re^n \mapsto [-\infty, \infty]$ be a function and consider the subset of $\Re^{n+1}$ given by

$$E_f = \big\{ (x, w) \mid f(x) < w \big\}.$$

Show that E_f is related to the epigraph of f as follows:

$$E_f \subset \mathrm{epi}(f) \subset \mathrm{cl}(E_f).$$

Show also that f is convex if and only if E_f is convex.

(b) Let $F : \Re^{m+n} \mapsto [-\infty, \infty]$ be a function and let

$$f(x) = \inf_{z \in \Re^m} F(x, z), \qquad x \in \Re^n.$$

Show that E_f is the projection of the set $\big\{ (x, z, w) \mid F(x, z) < w \big\}$ on the space of (x, w).

(c) Use parts (a) and (b) to show that convexity of F implies convexity of f (cf. Prop. 2.3.6).

2.10 (Partial Minimization of Nonconvex Functions)

Let $f : \Re^n \times \Re^m \mapsto (-\infty, \infty]$ be a closed proper function. Assume that f has the following property: for all $u^* \in \Re^m$ and $\gamma \in \Re$, there exists a neighborhood N of u^* such that the set $\big\{ (x, u) \mid u \in N, f(x, u) \leq \gamma \big\}$ is bounded, i.e., $f(x, u)$ is *level-bounded in x locally uniformly in u.* Let

$$p(u) = \inf_x f(x, u), \qquad P(u) = \arg\min_x f(x, u), \qquad u \in \Re^m.$$

(a) Show that the function p is closed and proper. Show also that for each $u \in \mathrm{dom}(p)$, the set $P(u)$ is nonempty and compact.

(b) Consider the following weaker assumption: for all $u \in \Re^m$ and $\gamma \in \Re$, the set $\big\{ x \mid f(x, u) \leq \gamma \big\}$ is bounded. Show that this assumption is not sufficient to guarantee the closedness of p.

(c) Let $\{u_k\}$ be a sequence such that $u_k \to u^*$ for some $u^* \in \mathrm{dom}(p)$, and assume that $p(u_k) \to p(u^*)$ (which holds when p is continuous at u^*). Let also $x_k \in P(u_k)$ for all k. Show that the sequence $\{x_k\}$ is bounded and all its limit points lie in $P(u^*)$.

(d) Show that a sufficient condition for p to be continuous at u^* is the existence of some $x^* \in P(u^*)$ such that $f(x^*, \cdot)$ is continuous at u^*.

2.11 (Projection on a Nonconvex Set)

Let C be a nonempty closed subset of $\Re^n$.

(a) Show that the distance function $d_C(x) : \Re^n \mapsto \Re$, defined by

$$d_C(x) = \inf_{w \in C} \|w - x\|,$$

is a continuous function of x.

(b) Show that the projection set $P_C(x)$, defined by

$$P_C(x) = \arg\min_{w \in C} \|w - x\|,$$

is nonempty and compact.

(c) Let $\{x_k\}$ be a sequence that converges to a vector x^* and let $\{w_k\}$ be a sequence with $w_k \in P_C(x_k)$. Show that the sequence $\{w_k\}$ is bounded and all its limit points belong to the set $P_C(x^*)$.

2.12 (Convergence of Penalty Methods [RoW98])

Let $f : \Re^n \mapsto (-\infty, \infty]$ be a closed proper function, let $F : \Re^n \mapsto \Re^m$ be a continuous function, and let $D \subset \Re^m$ be a nonempty closed set. Consider the problem

$$\text{minimize} \quad f(x)$$

$$\text{subject to } F(x) \in D.$$

Consider also the following approximation of this problem:

$$\text{minimize} \quad f(x) + \theta\big(F(x), c\big)$$

$$\text{subject to } x \in \Re^n, \tag{P_c}$$

with $c \in (0, \infty)$, where the function $\theta : \Re^m \times (0, \infty) \mapsto (-\infty, \infty]$ is lower semicontinuous, monotonically increasing in c for each $u \in \Re^m$, and satisfies

$$\lim_{c \to \infty} \theta(u, c) = \begin{cases} 0 & \text{if } u \in D, \\ \infty & \text{if } u \notin D. \end{cases}$$

Assume that for some $\bar{c} \in (0, \infty)$ sufficiently large, the level sets of the function $f(x) + \theta\big(F(x), \bar{c}\big)$ are bounded, and consider any sequence of parameter values $c_k \geq \bar{c}$ with $c_k \to \infty$. Show the following:

(a) The sequence of optimal values of the approximate problems (P_{c_k}) converges to the optimal value of the original problem.

(b) Any sequence $\{x_k\}$, where x_k is an optimal solution of the approximate problem (P_{c_k}), is bounded and each of its limit points is an optimal solution of the original problem.

2.13 (Approximation by Envelope Functions [RoW98])

Let $f : \Re^n \mapsto (-\infty, \infty]$ be a closed proper function. For a scalar $c > 0$, define the corresponding *envelope function* $e_c f$ and the *proximal mapping* $P_c f$ by

$$e_c f(x) = \inf_w \left\{ f(w) + \frac{1}{2c} \|w - x\|^2 \right\},$$

$$P_c f(x) = \arg\min_w \left\{ f(w) + \frac{1}{2c} \|w - x\|^2 \right\}.$$

Assume that there exists some $c > 0$ with $e_c f(x) > -\infty$ for some $x \in \Re^n$. Let c_f be the supremum of the set of all such c. Show the following:

(a) For every $c \in (0, c_f)$, the set $P_c f(x)$ is nonempty and compact, while the value $e_c f(x)$ is finite and depends continuously on (x, c) with

$$e_c f(x) \uparrow f(x) \text{ for all } x, \text{ as } c \downarrow 0.$$

(b) Let $\{w_k\}$ be a sequence such that $w_k \in P_{c_k} f(x_k)$ for some sequences $\{x_k\}$ and $\{c_k\}$ such that $x_k \to x^*$ and $c_k \to c^* \in (0, c_f)$. Then $\{w_k\}$ is bounded and all its limit points belong to the set $P_{c^*} f(x^*)$.

[*Note*: The approximation $e_c f$ is an underestimate of the function f, i.e., $e_c f(x) \le f(x)$ for all $x \in \Re^n$. Furthermore, $e_c f$ is a real-valued continuous function, whereas f itself may only be extended real-valued and lower semicontinuous.]

2.14 (Envelopes and Proximal Mappings under Convexity [RoW98])

Let $f : \Re^n \mapsto (-\infty, \infty]$ be a closed proper convex function. For $c > 0$, consider the envelope function $e_c f$ and the proximal mapping $P_c f$ (cf. Exercise 2.13). Show the following:

(a) The supremum of the set of all $c > 0$ such that $e_c f(x) > -\infty$ for some $x \in \Re^n$, is ∞.

(b) The proximal mapping $P_c f$ is single-valued and is continuous in the sense that $P_c f(x) \to P_{c^*} f(x^*)$ whenever $(x, c) \to (x^*, c^*)$ with $c^* > 0$.

(c) The envelope function $e_c f$ is convex and smooth, and its gradient is given by

$$\nabla e_c f(x) = \frac{1}{c} \big(x - P_c f(x) \big).$$

Note: The envelope function $e_c f$ is smooth, regardless of whether f is nonsmooth.

2.15

(a) Let C_1 be a convex set with nonempty interior and C_2 be a nonempty convex set that does not intersect the interior of C_1. Show that there exists a hyperplane such that one of the associated closed halfspaces contains C_2, and does not intersect the interior of C_1.

(b) Show by an example that we cannot replace interior with relative interior in the statement of part (a).

2.16

Let C be a nonempty convex set in $\Re^n$, and let M be a nonempty affine set in $\Re^n$. Show that $M \cap \mathrm{ri}(C) = \emptyset$ is a necessary and sufficient condition for the existence of a hyperplane H containing M, and such that $\mathrm{ri}(C)$ is contained in one of the open halfspaces associated with H.

2.17 (Strong Separation)

Let C_1 and C_2 be nonempty convex subsets of $\Re^n$, and let B denote the unit ball in $\Re^n$, $B = \{x \mid \|x\| \leq 1\}$. A hyperplane H is said to *separate strongly* C_1 and C_2 if there exists an $\epsilon > 0$ such that $C_1 + \epsilon B$ is contained in one of the open halfspaces associated with H and $C_2 + \epsilon B$ is contained in the other. Show that:

(a) The following three conditions are equivalent.

  (i) There exists a hyperplane separating strongly C_1 and C_2.

  (ii) There exists a vector $a \in \Re^n$ such that $\inf_{x \in C_1} a'x > \sup_{x \in C_2} a'x$.

  (iii) $\inf_{x_1 \in C_1, x_2 \in C_2} \|x_1 - x_2\| > 0$, i.e., $0 \notin \mathrm{cl}(C_2 - C_1)$.

(b) If C_1 and C_2 are disjoint, any one of the five conditions for strict separation, given in Prop. 2.4.3, implies that C_1 and C_2 can be strongly separated.

2.18

Let C_1 and C_2 be nonempty convex subsets of $\Re^n$ such that C_2 is a cone.

(a) Suppose that there exists a hyperplane that separates C_1 and C_2 properly. Show that there exists a hyperplane which separates C_1 and C_2 properly and passes through the origin.

(b) Suppose that there exists a hyperplane that separates C_1 and C_2 strictly. Show that there exists a hyperplane that passes through the origin such that one of the associated closed halfspaces contains the cone C_2 and does not intersect C_1.

2.19 (Separation Properties of Cones)

Define a *homogeneous halfspace* to be a closed halfspace associated with a hyperplane that passes through the origin. Show that:

- (a) A nonempty closed convex cone is the intersection of the homogeneous halfspaces that contain it.

- (b) The closure of the convex cone generated by a nonempty set X is the intersection of all the homogeneous halfspaces containing X.

2.20 (Convex System Alternatives)

Let $g_j : \Re^n \mapsto (-\infty, \infty]$, $j = 1, \ldots, r$, be closed proper convex functions, and let X be a nonempty closed convex set. Assume that one of the following four conditions holds:

- (1) $R_X \cap R_{g_1} \cap \cdots \cap R_{g_r} = L_X \cap L_{g_1} \cap \cdots \cap L_{g_r}$.

- (2) $R_X \cap R_{g_1} \cap \cdots \cap R_{g_r} \subset L_{g_1} \cap \cdots \cap L_{g_r}$ and X is specified by linear inequality constraints.

- (3) Each g_j is a convex quadratic function and X is specified by convex quadratic inequality constraints.

- (4) Each g_j is a convex bidirectionally flat function (see Exercise 2.7) and X is specified by convex bidirectionally flat functions.

Show that:

- (a) If there is no vector $x \in X$ such that

$$g_1(x) \leq 0, \ldots, g_r(x) \leq 0,$$

then there exist a positive scalar ϵ, and a vector $\mu \in \Re^r$ with $\mu \geq 0$, such that

$$\mu_1 g_1(x) + \cdots + \mu_r g_r(x) \geq \epsilon, \qquad \forall \, x \in X.$$

Hint: Show that under any one of the conditions (1)-(4), the set

$$C = \big\{ u \mid \text{there exists an } x \in X \text{ such that } g_j(x) \leq u_j, \, j = 1, \ldots, r \big\}$$

is closed, by viewing it as the projection of the set

$$\big\{ (x, u) \mid x \in X, \, g_j(x) \leq u_j, \, j = 1, \ldots, r \big\}$$

on the space of u. Furthermore, the origin does not belong to C, so it can be strictly separated from C by a hyperplane. The normal of this hyperplane provides the desired vector μ.

- (b) If for every $\epsilon > 0$, there exists a vector $x \in X$ such that

$$g_1(x) < \epsilon, \ldots, g_r(x) < \epsilon,$$

then there exists a vector $x \in X$ such that

$$g_1(x) \leq 0, \ldots, g_r(x) \leq 0.$$

Hint: Argue by contradiction and use part (a).

2.21

Let C be a nonempty closed convex subset of $\Re^{n+1}$ that contains no vertical lines. Show that C is equal to the intersection of the closed halfspaces that contain it and correspond to nonvertical hyperplanes.

2.22 (Min Common/Max Crossing Duality)

Consider the min common/max crossing framework, assuming that $w^* < \infty$.

(a) Assume that M is compact. Show that q^* is equal to the optimal value of the min common point problem corresponding to $\operatorname{conv}(M)$.

(b) Assume that M is closed and convex, and does not contain a halfline of the form $\{(x, w + \alpha) \mid \alpha \le 0\}$. Show that $\overline{M}$ is the epigraph of the function given by

$$f(x) = \inf\{w \mid (x, w) \in M\}, \qquad x \in \Re^n,$$

and that f is closed proper and convex.

(c) Assume that w^* is finite, and that $\overline{M}$ is convex and closed. Show that $q^* = w^*$.

2.23 (An Example of Lagrangian Duality)

Consider the problem

$$\text{minimize } f(x)$$
$$\text{subject to } x \in X, \qquad e_i' x = d_i, \quad i = 1, \dots, m,$$

where $f : \Re^n \mapsto \Re$ is a convex function, X is a convex set, and e_i and d_i are given vectors and scalars, respectively. Consider the min common/max crossing framework where M is the subset of $\Re^{m+1}$ given by

$$M = \left\{ \left(e_1' x - d_1, \dots, e_m' x - d_m, f(x) \right) \mid x \in X \right\}.$$

(a) Derive the corresponding max crossing problem.

(b) Show that the corresponding set $\overline{M}$ is convex.

(c) Show that if $w^* < \infty$ and X is compact, then $q^* = w^*$.

(d) Show that if $w^* < \infty$ and there exists a vector $\overline{x} \in \operatorname{ri}(X)$ such that $e_i' \overline{x} = d_i$ for all $i = 1, \dots, m$, then $q^* = w^*$ and the max crossing problem has an optimal solution.

2.24 (Saddle Points in Two Dimensions)

Consider a function ϕ of two real variables x and z taking values in compact intervals X and Z, respectively. Assume that for each $z \in Z$, the function $\phi(\cdot, z)$ is minimized over X at a unique point denoted $\hat{x}(z)$. Similarly, assume that for each $x \in X$, the function $\phi(x, \cdot)$ is maximized over Z at a unique point denoted $\hat{z}(x)$. Assume further that the functions $\hat{x}(z)$ and $\hat{z}(x)$ are continuous over Z and X, respectively. Show that ϕ has a saddle point (x^*, z^*). Use this to investigate the existence of saddle points of $\phi(x, z) = x^2 + z^2$ over $X = [0, 1]$ and $Z = [0, 1]$.

2.25 (Saddle Points of Quadratic Functions)

Consider a quadratic function $\phi : X \times Z \mapsto \Re$ of the form

$$\phi(x, z) = x'Qx + x'Dz - z'Rz,$$

where Q and R are symmetric positive semidefinite $n \times n$ and $m \times m$ matrices, respectively, D is some $n \times m$ matrix, and X and Z are subsets of $\Re^n$ and $\Re^m$, respectively. Derive conditions under which ϕ has at least one saddle point.

3

Polyhedral Convexity

Contents

3.1. Polar Cones . . . . p. 166
3.2. Polyhedral Cones and Polyhedral Sets . . . . p. 168
   3.2.1. Farkas' Lemma and Minkowski-Weyl Theorem . . p. 170
   3.2.2. Polyhedral Sets . . . . p. 175
   3.2.3. Polyhedral Functions . . . . p. 178
3.3. Extreme Points . . . . p. 180
   3.3.1. Extreme Points of Polyhedral Sets . . . . p. 183
3.4. Polyhedral Aspects of Optimization . . . . p. 186
   3.4.1. Linear Programming . . . . p. 188
   3.4.2. Integer Programming . . . . p. 189
3.5. Polyhedral Aspects of Duality . . . . p. 192
   3.5.1. Polyhedral Proper Separation . . . . p. 192
   3.5.2. Min Common/Max Crossing Duality . . . . . . p. 196
   3.5.3. Minimax Theory Under Polyhedral Assumptions . p. 199
   3.5.4. A Nonlinear Version of Farkas' Lemma . . . . . p. 203
   3.5.5. Convex Programming . . . . p. 208
3.6. Notes, Sources, and Exercises . . . . p. 210

In this chapter, we develop some basic results regarding the geometry of polyhedral sets. We start with a general duality relation between cones. We then specialize this relation to polyhedral cones, we discuss polyhedral sets and their extreme points, and we apply the results to linear and integer programming problems. Finally, we develop a specialized duality result based on the min common/max crossing framework and polyhedral assumptions, and we illustrate its use in minimax theory and constrained optimization.

3.1 POLAR CONES

Given a nonempty set C, the cone given by

$$C^* = \{y \mid y'x \le 0, \ \forall \ x \in C\},$$

is called the *polar cone* of C (see Fig. 3.1.1). Clearly, the polar cone C^*, being the intersection of a collection of closed halfspaces, is closed and convex (regardless of whether C is closed and/or convex). If C is a subspace, it can be seen that C^* is equal to the orthogonal subspace $C^\perp$. The following basic result generalizes the equality $C = (C^\perp)^\perp$, which holds in the case where C is a subspace.

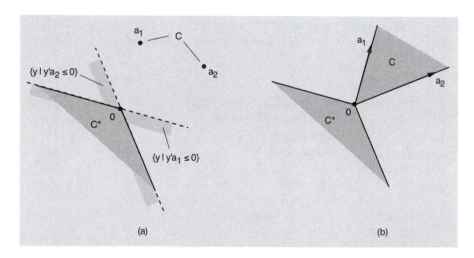

Figure 3.1.1. Illustration of the polar cone C^* of a subset C of $\Re^2$. In (a), C consists of just two points, a_1 and a_2, and C^* is the intersection of the two closed halfspaces $\{y \mid y'a_1 \le 0\}$ and $\{y \mid y'a_2 \le 0\}$. In (b), C is the convex cone

$$\{x \mid x = \mu_1 a_1 + \mu_2 a_2, \ \mu_1 \ge 0, \ \mu_2 \ge 0\},$$

and the polar cone C^* is the same as in case (a).

Proposition 3.1.1:

(a) For any nonempty set C, we have

$$C^* = \big(\mathrm{cl}(C)\big)^* = \big(\mathrm{conv}(C)\big)^* = \big(\mathrm{cone}(C)\big)^*.$$

(b) (*Polar Cone Theorem*) For any nonempty cone C, we have

$$(C^*)^* = \mathrm{cl}\big(\mathrm{conv}(C)\big).$$

In particular, if C is closed and convex, we have $(C^*)^* = C$.

Proof: (a) For any two sets X and Y with $X \supset Y$, we have $X^* \subset Y^*$, from which it follows that $\big(\mathrm{cl}(C)\big)^* \subset C^*$. Conversely, if $y \in C^*$, then $y'x_k \leq 0$ for all k and all sequences $\{x_k\} \subset C$, so that $y'x \leq 0$ for all $x \in \mathrm{cl}(C)$. Hence $y \in \big(\mathrm{cl}(C)\big)^*$, implying that $C^* \subset \big(\mathrm{cl}(C)\big)^*$.

Similarly, since $\mathrm{conv}(C) \supset C$, we have $\big(\mathrm{conv}(C)\big)^* \subset C^*$. Conversely, if $y \in C^*$, then $y'x \leq 0$ for all $x \in C$, so that $y'z \leq 0$ for all z that are convex combinations of vectors $x \in C$. Hence, $y \in \big(\mathrm{conv}(C)\big)^*$ and $C^* \subset \big(\mathrm{conv}(C)\big)^*$. A nearly identical argument also shows that $C^* = \big(\mathrm{cone}(C)\big)^*$.

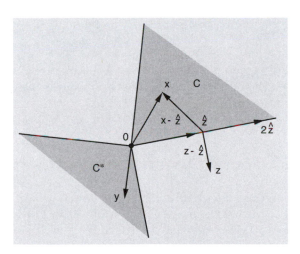

Figure 3.1.2. Illustration of the proof of the Polar Cone Theorem for the case where C is a closed and convex cone.

(b) We first show the result for the case where C is closed and convex. Indeed, if this is so, then for any $x \in C$, we have $x'y \leq 0$ for all $y \in C^*$, which implies that $x \in (C^*)^*$. Hence, $C \subset (C^*)^*$. To prove the reverse inclusion, choose any $z \in (C^*)^*$. Since C is closed, by the Projection Theorem (Prop. 2.2.1), there exists a unique projection of z on C, denoted $\hat{z}$, which satisfies

$$(z - \hat{z})'(x - \hat{z}) \leq 0, \qquad \forall \, x \in C;$$

see Fig. 3.1.2. By taking in the preceding relation $x = 0$ and $x = 2\hat{z}$ (which belong to C since C is a closed cone), it is seen that

$$(z - \hat{z})'\hat{z} = 0.$$

Combining the last two relations, we obtain $(z - \hat{z})'x \leq 0$ for all $x \in C$. Therefore, $(z - \hat{z}) \in C^*$, and since $z \in (C^*)^*$, we have $(z - \hat{z})'z \leq 0$, which when added to $-(z - \hat{z})'\hat{z} = 0$ yields $\|z - \hat{z}\|^2 \leq 0$. It follows that $z = \hat{z}$ and $z \in C$, implying that $(C^*)^* \subset C$.

    Using the result just shown for the case where C is closed and convex, it follows that

$$\left(\left(\mathrm{cl}\big(\mathrm{conv}(C)\big) \right)^* \right)^* = \mathrm{cl}\big(\mathrm{conv}(C)\big).$$

By using part (a), we have

$$C^* = \big(\mathrm{conv}(C)\big)^* = \left(\mathrm{cl}\big(\mathrm{conv}(C)\big) \right)^*.$$

By combining the above two relations, we obtain $(C^*)^* = \mathrm{cl}\big(\mathrm{conv}(C)\big)$. **Q.E.D.**

    The polarity relation between cones is a special case of a more general duality relation, the conjugacy relation between convex functions. We will discuss this relation in Section 7.1.

3.2   POLYHEDRAL CONES AND POLYHEDRAL SETS

We now focus on cones that have a polyhedral structure. We introduce two different ways to view such cones. Our first objective in this section is to show that these two views are related through the polarity relation and are in some sense equivalent. Subsequently, we will discuss characterizations of more general polyhedral sets and their extreme points.

    We say that a cone $C \subset \Re^n$ is *polyhedral*, if it has the form

$$C = \{x \mid a_j'x \leq 0, \, j = 1, \ldots, r\},$$

where $a_1, \ldots, a_r$ are some vectors in $\Re^n$, and r is a positive integer; see Fig. 3.2.1(a).

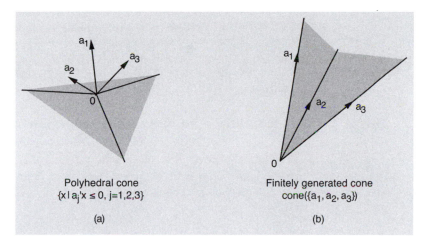

Polyhedral cone
$\{x \mid a_j'x \le 0, j=1,2,3\}$

(a)

Finitely generated cone
$\text{cone}(\{a_1, a_2, a_3\})$

(b)

Figure 3.2.1. (a) Polyhedral cone defined by the inequality constraints $a_j'x \le 0$, $j = 1, 2, 3$. (b) Cone generated by the vectors a_1, a_2, a_3.

We say that a cone $C \subset \Re^n$ is *finitely generated*, if it is generated by a finite set of vectors, i.e., if it has the form

$$C = \text{cone}(\{a_1, \ldots, a_r\}) = \left\{ x \ \middle| \ x = \sum_{j=1}^{r} \mu_j a_j, \ \mu_j \ge 0, \ j = 1, \ldots, r \right\},$$

where $a_1, \ldots, a_r$ are some vectors in $\Re^n$, and r is a positive integer; see Fig. 3.2.1(b).

Note that sets defined by linear equality constraints of the form $e_i'x = 0$ (as well as linear inequality constraints) are also polyhedral cones, since a linear equality constraint may be converted into two inequality constraints. In particular, if $e_1, \ldots, e_m$ and $a_1, \ldots, a_r$ are given vectors, the cone

$$\{x \mid e_i'x = 0, \ i = 1, \ldots, m, \ a_j'x \le 0, \ j = 1, \ldots, r\}$$

is polyhedral, since it can be written as

$$\{x \mid e_i'x \le 0, \ -e_i'x \le 0, \ i = 1, \ldots, m, \ a_j'x \le 0, \ j = 1, \ldots, r\}.$$

Using a related conversion, it can be seen that the cone

$$\left\{ x \ \middle| \ x = \sum_{i=1}^{m} \lambda_i e_i + \sum_{j=1}^{r} \mu_j a_j, \ \lambda_i \in \Re, \ i = 1, \ldots, m, \ \mu_j \ge 0, \ j = 1, \ldots, r \right\}$$

is finitely generated, since it can be written as

$$\left\{ x \ \middle| \ x = \sum_{i=1}^{m} \lambda_i^+ e_i + \sum_{i=1}^{m} \lambda_i^- (-e_i) + \sum_{j=1}^{r} \mu_j a_j, \right.$$

$$\left. \lambda_i^+ \ge 0, \ \lambda_i^- \ge 0, \ i = 1, \ldots, m, \ \mu_j \ge 0, \ j = 1, \ldots, r \right\}.$$

Since a subspace can be represented by linear equalities and also as the span of a finite set of vectors, it follows from the preceding conversions that a subspace is both a polyhedral cone and also a finitely generated cone. We will see shortly that this property generalizes, and that in fact a cone is polyhedral if and only if it is finitely generated.

3.2.1 Farkas' Lemma and Minkowski-Weyl Theorem

We will now discuss some fundamental results relating to polyhedral convexity. It turns out that polyhedral and finitely generated cones, defined by the same vectors $a_1, \ldots, a_r$, are polar to each other, i.e.,

$$\{x \mid a_j'x \leq 0,\, j = 1, \ldots, r\} = \text{cone}(\{a_1, \ldots, a_r\})^*.$$

This is the subject of part (a) of the following proposition, which also shows that $\text{cone}(\{a_1, \ldots, a_r\})$ is a closed set (this requires a nontrivial proof as will be shown below). By considering the polars of the cones in the two sides of the above relation, and by using the Polar Cone Theorem and the closedness property of $\text{cone}(\{a_1, \ldots, a_r\})$ just mentioned, it is seen that

$$\{x \mid a_j'x \leq 0,\, j = 1, \ldots, r\}^* = \text{cone}(\{a_1, \ldots, a_r\}).$$

This is Farkas' Lemma, a very useful result, given as part (b) of the following proposition (in somewhat more general terms). Finally, part (c) of the following proposition, the Minkowski-Weyl Theorem, shows that any polyhedral cone can be represented as a finitely generated cone and conversely.

Proposition 3.2.1:

(a) Let $a_1, \ldots, a_r$ be vectors in $\Re^n$. Then, the finitely generated cone

$$C = \text{cone}(\{a_1, \ldots, a_r\})$$

is closed and its polar cone is the polyhedral cone given by

$$C^* = \{y \mid a_j'y \leq 0,\, j = 1, \ldots, r\}.$$

(b) *(Farkas' Lemma)* Let $x, e_1, \ldots, e_m$, and $a_1, \ldots, a_r$ be vectors in $\Re^n$. We have $x'y \leq 0$ for all vectors $y \in \Re^n$ such that

$$y'e_i = 0, \quad \forall\, i = 1, \ldots, m, \qquad y'a_j \leq 0, \quad \forall\, j = 1, \ldots, r,$$

if and only if x can be expressed as

$$x = \sum_{i=1}^{m} \lambda_i e_i + \sum_{j=1}^{r} \mu_j a_j,$$

where λ_i and μ_j are some scalars with $\mu_j \geq 0$ for all j.

(c) (*Minkowski-Weyl Theorem*) A cone is polyhedral if and only if it is finitely generated.

Proof: (a) We first show that $C^* = \{y \mid a_j'y \leq 0, \ j = 1, \ldots, r\}$. If $a_j'y \leq 0$ for all j, then $x'y \leq 0$ for all $x \in C$, so $y \in C^*$ and

$$C^* \supset \{y \mid a_j'y \leq 0, \ j = 1, \ldots, r\}.$$

Conversely, if $y \in C^*$, i.e., if $x'y \leq 0$ for all $x \in C$, then, since $a_j \in C$, we have $a_j'y \leq 0$, for all j. Thus,

$$C^* \subset \{y \mid a_j'y \leq 0, \ j = 1, \ldots, r\}.$$

There remains to show that C is closed. We will give two proofs of this fact. The first proof is simpler and suffices for the purpose of showing Farkas' Lemma [part (b)]. The second proof is more complicated, but shows a stronger result, namely that C is polyhedral, and in fact provides an algorithm for constructing a polyhedral representation of the finitely generated cone C. Thus, the second proof not only shows that C is closed, but also proves half of the Minkowski-Weyl Theorem [part (c)].

The first proof that C is closed is based on induction on the number of vectors r. When $r = 1$, C is either $\{0\}$ (if $a_1 = 0$) or a halfline, and is therefore closed. Suppose that for some $r \geq 1$, all cones generated by at most r vectors are closed. We consider a cone of the form

$$C_{r+1} = \text{cone}(\{a_1, \ldots, a_{r+1}\})$$

and we show that it is also closed. In particular, let $\{x_k\}$ be a convergent sequence belonging to C_{r+1}, so that for all k, we have

$$x_k = \xi_k a_{r+1} + y_k,$$

where $\xi_k \geq 0$ and $y_k \in C_r$, where

$$C_r = \text{cone}(\{a_1, \ldots, a_r\}).$$

We will prove that the limit of $\{x_k\}$ belongs to C_{r+1}.

Without loss of generality, we may assume that $\|a_j\| = 1$ for all j, and that the negative of one of the vectors, say $-a_{r+1}$, does not belong to C_{r+1} (otherwise, the vectors $-a_1, \ldots, -a_{r+1}$ belong to C_{r+1}, so that C_{r+1} is the subspace spanned by $a_1, \ldots, a_{r+1}$ and is therefore closed). Let

$$\gamma = \min_{x \in C_r, \|x\|=1} a'_{r+1} x.$$

The set C_r is closed by the induction hypothesis, so the set $\{x \in C_r \mid \|x\| = 1\}$ is nonempty and compact. Therefore, by Weierstrass' Theorem, the minimum in the definition of γ is attained at some $x^* \in C_r$ with $\|x^*\| = 1$. Using the Schwarz Inequality, we have

$$\gamma = a'_{r+1} x^* \geq -\|a_{r+1}\| \cdot \|x^*\| = -1,$$

with equality if and only if $x^* = -a_{r+1}$. Because $x^* \in C_r$ and $-a_{r+1} \notin C_r$, equality cannot hold above, so that $\gamma > -1$.

Using the fact $\|a_{r+1}\| = 1$, we obtain

$$\begin{aligned}
\|x_k\|^2 &= \|\xi_k a_{r+1} + y_k\|^2 \\
&= \xi_k^2 + \|y_k\|^2 + 2\xi_k a'_{r+1} y_k \\
&\geq \xi_k^2 + \|y_k\|^2 + 2\gamma \xi_k \|y_k\| \\
&= \left(\xi_k - \|y_k\|\right)^2 + 2(1+\gamma)\xi_k \|y_k\|.
\end{aligned}$$

Since $\xi_k \geq 0$, $1 + \gamma > 0$, and $\{x_k\}$ converges, it follows that the sequences $\{\xi_k\}$ and $\{y_k\}$ are bounded and hence, they have limit points denoted by ξ and y, respectively. The limit of $\{x_k\}$ is

$$\lim_{k \to \infty} \left(\xi_k a_{r+1} + y_k\right) = \xi a_{r+1} + y,$$

which belongs to C_{r+1}, since $\xi \geq 0$ and $y \in C_r$ (by the closedness hypothesis on C_r). We conclude that C_{r+1} is closed, completing the proof.

We now give an alternative proof (due to Wets [Wet90]) that C is closed, by showing that it is polyhedral. The proof is constructive and also uses induction on the number of vectors r. When $r = 1$, there are two possibilities: (a) $a_1 = 0$, in which case C is the trivial subspace $\{0\}$, which is polyhedral; (b) $a_1 \neq 0$, in which case C is a closed halfline, which using the Polar Cone Theorem, is characterized as the set of vectors that are orthogonal to the subspace orthogonal to a_1 and also make a nonnegative inner product with a_1, i.e.,

$$C = \{x \mid b'_i x \leq 0, \ -b'_i x \leq 0, \ i = 1, \ldots, n-1, \ -a'_1 x \leq 0\},$$

where $b_1, \ldots, b_{n-1}$ are a set of basis vectors for the $(n-1)$-dimensional subspace that is orthogonal to the vector a_1. Thus, C is polyhedral in case (b) as well.

Assume that for some $r \geq 1$, a finitely generated cone of the form

$$C_r = \mathrm{cone}(\{a_1, \ldots, a_r\})$$

has a polyhedral representation

$$P_r = \{x \mid b'_j x \leq 0, \; j = 1, \ldots, m\}.$$

Let a_{r+1} be a given vector in $\Re^n$, and consider the set

$$C_{r+1} = \mathrm{cone}(\{a_1, \ldots, a_{r+1}\}).$$

We will show that C_{r+1} has a polyhedral representation.
Let

$$\beta_j = a'_{r+1} b_j, \qquad j = 1, \ldots, m,$$

and define the index sets

$$J^- = \{j \mid \beta_j < 0\}, \qquad J^0 = \{j \mid \beta_j = 0\}, \qquad J^+ = \{j \mid \beta_j > 0\}.$$

If $J^- \cup J^0 = \emptyset$, or equivalently $a'_{r+1} b_j > 0$ for all j, we see that $-a_{r+1}$ is an interior point of P_r. It follows that for any $x \in \Re^n$, we have

$$x - \mu_{r+1} a_{r+1} \in P_r$$

for a sufficiently large μ_{r+1}, implying that x belongs to C_{r+1}. Therefore $C_{r+1} = \Re^n$, so that C_{r+1} is a polyhedral set.
We may thus assume that $J^- \cup J^0 \neq \emptyset$. We will show that if $J^+ = \emptyset$ or $J^- = \emptyset$, the set C_{r+1} has the polyhedral representation

$$P_{r+1} = \{x \mid b'_j x \leq 0, \; j \in J^- \cup J^0\},$$

and if $J^+ \neq \emptyset$ and $J^- \neq \emptyset$, it has the polyhedral representation

$$P_{r+1} = \{x \mid b'_j x \leq 0, \; j \in J^- \cup J^0, \; b'_{l,k} x \leq 0, \; l \in J^+, \; k \in J^-\},$$

where

$$b_{l,k} = b_l - \frac{\beta_l}{\beta_k} b_k, \qquad \forall \; l \in J^+, \; \forall \; k \in J^-.$$

This will complete the induction.
Indeed, we have $C_{r+1} \subset P_{r+1}$ since by construction, the vectors $a_1, \ldots, a_{r+1}$ satisfy the inequalities defining P_{r+1}. To show the reverse inclusion, we fix a vector $x \in P_{r+1}$ and we verify that there exist $\mu_1, \ldots, \mu_{r+1} \geq 0$ such that $x = \sum_{j=1}^{r+1} \mu_j a_j$, or equivalently that there exists $\mu_{r+1} \geq 0$ such that

$$x - \mu_{r+1} a_{r+1} \in P_r.$$

We consider three cases.

(1) $J^+ = \emptyset$. Then, since $x \in P_{r+1}$, we have $b_j' x \leq 0$ for all $j \in J^- \cup J^0$, implying $x \in P_r$, so that $x - \mu_{r+1} a_{r+1} \in P_r$ for $\mu_{r+1} = 0$.

(2) $J^+ \neq \emptyset$, $J^0 \neq \emptyset$, and $J^- = \emptyset$. Then, since $x \in P_{r+1}$, we have $b_j' x \leq 0$ for all $j \in J^0$, so that

$$b_j'(x - \mu a_{r+1}) \leq 0, \qquad \forall \, j \in J^0, \ \forall \, \mu \geq 0,$$

$$b_j'(x - \mu a_{r+1}) \leq 0, \qquad \forall \, j \in J^+, \ \forall \, \mu \geq \max_{j \in J^+} \frac{b_j' x}{\beta_j}.$$

Hence, for μ_{r+1} sufficiently large, we have $x - \mu_{r+1} a_{r+1} \in P_r$.

(3) $J^+ \neq \emptyset$ and $J^- \neq \emptyset$. Then, since $x \in P_{r+1}$, we have $b_{l,k}' x \leq 0$ for all $l \in J^+$, $k \in J^-$, implying that

$$\frac{b_l' x}{\beta_l} \leq \frac{b_k' x}{\beta_k}, \qquad \forall \, l \in J^+, \ \forall \, k \in J^-.$$

Furthermore, for $x \in P_{r+1}$, there holds $b_k' x / \beta_k \geq 0$ for all $k \in J^-$, so that for μ_{r+1} satisfying

$$\max \left\{ 0, \max_{l \in J^+} \frac{b_l' x}{\beta_l} \right\} \leq \mu_{r+1} \leq \min_{k \in J^-} \frac{b_k' x}{\beta_k},$$

we have

$$b_j' x - \mu_{r+1} b_j' a_{r+1} \leq 0, \qquad \forall \, j \in J^0,$$

$$b_k' x - \mu_{r+1} b_k' a_{r+1} = \beta_k \left(\frac{b_k' x}{\beta_k} - \mu_{r+1} \right) \leq 0, \qquad \forall \, k \in J^-,$$

$$b_l' x - \mu_{r+1} b_l' a_{r+1} = \beta_l \left(\frac{b_l' x}{\beta_l} - \mu_{r+1} \right) \leq 0, \qquad \forall \, l \in J^+.$$

Hence, $b_j' x - \mu_{r+1} a_{r+1} \leq 0$ for all j, implying that $x - \mu_{r+1} a_{r+1} \in P_r$, and completing the proof.

(b) Define $a_{r+i} = e_i$ and $a_{r+m+i} = -e_i$, $i = 1, \ldots, m$. The result to be shown can be stated as

$$x \in P^* \qquad \Longleftrightarrow \qquad x \in C,$$

where

$$P = \{ y \mid y' a_j \leq 0, \ j = 1, \ldots, r + 2m \},$$

$$C = \mathrm{cone}(\{ a_1, \ldots, a_{r+2m} \}).$$

Since by part (a), $P = C^*$, and C is closed and convex, we have $P^* = (C^*)^* = C$ by the Polar Cone Theorem [Prop. 3.1.1(b)].

(c) We have already shown in the alternative proof of part (a) that a finitely generated cone is polyhedral, so there remains to show the converse. Let P be a polyhedral cone given by

$$P = \{x \mid a_j'x \leq 0, \, j = 1, \ldots, r\}.$$

By part (a),

$$P = \text{cone}(\{a_1, \ldots, a_r\})^*,$$

and by using also the closedness of finitely generated cones [proved in part (a)] and the Polar Cone Theorem, we have

$$P^* = \text{cone}(\{a_1, \ldots, a_r\}).$$

Thus, P^* is finitely generated and therefore polyhedral, so that by part (b), its polar $(P^*)^*$ is finitely generated. By the Polar Cone Theorem, $(P^*)^* = P$, implying that P is finitely generated.   **Q.E.D.**

Farkas' Lemma [Prop. 3.2.1(b)] plays an important role in various optimization analyses, and can be proved in several different ways. The preceding proof was obtained by using the Polar Cone Theorem and the closedness of finitely generated cones [Prop. 3.2.1(a), which was proved by two different methods]. Two more independent proofs of Farkas' Lemma will be given later. The first proof, given in Section 3.5, is based on the min common/max crossing duality developed in the preceding chapter (Section 2.5). In fact, this latter proof applies to a more general (nonlinear) version of Farkas' Lemma. The second proof, given in Section 5.4, is based on the Fritz John theory developed in Chapter 5 and ultimately relies on penalty function arguments.

3.2.2   Polyhedral Sets

A subset P of $\Re^n$ is said to be a *polyhedral set* (or *polyhedron*) if it is nonempty and has the form

$$P = \{x \mid a_j'x \leq b_j, \, j = 1, \ldots, r\},$$

where a_j are some vectors in $\Re^n$ and b_j are some scalars. A polyhedral set may also involve linear equalities, which may be converted into two linear inequalities. In particular, if $e_1, \ldots, e_m$ and $a_1, \ldots, a_r$ are some vectors in $\Re^n$, and $d_1, \ldots, d_m$ and $b_1, \ldots, b_r$ are some scalars, the set

$$\{x \mid e_i'x = d_i, \, i = 1, \ldots, m, \, a_j'x \leq b_j, \, j = 1, \ldots, r\}$$

is polyhedral, since it can be written as

$$\{x \mid e_i'x \leq d_i, \, -e_i'x \leq -d_i, \, i = 1, \ldots, m, \, a_j'x \leq b_j, \, j = 1, \ldots, r\}.$$

Note that any affine set can be described by linear equalities, so it is a polyhedral set.

The following is a fundamental result, showing that a polyhedral set can be represented as the sum of the convex hull of a finite set of points and a finitely generated cone (see Fig. 3.2.2). The result is proved by applying the Minkowski-Weyl Theorem to a polyhedral cone in $\Re^{n+1}$, which is derived from the given polyhedral set by a construction similar to the one used earlier in the proof of Caratheodory's Theorem (see Fig. 3.2.3).

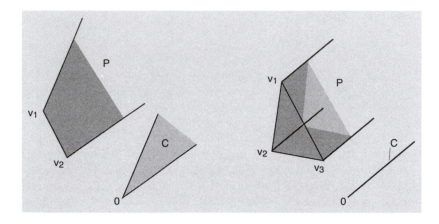

Figure 3.2.2. Examples of a Minkowski-Weyl representation of a two-dimensional and a three-dimensional polyhedral set P. It has the form

$$P = \text{conv}\big(\{v_1, \ldots, v_m\}\big) + C,$$

where $v_1, \ldots, v_m$ are vectors and C is a finitely generated cone.

Proposition 3.2.2: (Minkowski-Weyl Representation) A set P is polyhedral if and only if there is a nonempty finite set $\{v_1, \ldots, v_m\}$ and a finitely generated cone C such that $P = \text{conv}\big(\{v_1, \ldots, v_m\}\big) + C$, i.e.,

$$P = \left\{ x \ \bigg| \ x = \sum_{j=1}^{m} \mu_j v_j + y, \ \sum_{j=1}^{m} \mu_j = 1, \ \mu_j \ge 0, \ j = 1, \ldots, m, \ y \in C \right\}.$$

Proof: Assume that P is polyhedral. Then, it has the form

$$P = \{x \mid a_j'x \le b_j, \ j = 1, \ldots, r\},$$

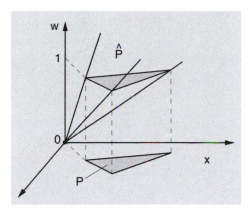

Figure 3.2.3. Illustration of the proof of the Minkowski-Weyl representation and the cone of $\Re^{n+1}$

$$\hat{P} = \left\{(x, w) \mid 0 \leq w,\ a_j'x \leq b_j w,\ j = 1, \ldots, r\right\},$$

derived from the polyhedral set

$$P = \left\{x \mid a_j'x \leq b_j,\ j = 1, \ldots, r\right\}.$$

for some vectors $a_j \in \Re^n$ and scalars b_j. Consider the polyhedral cone in $\Re^{n+1}$ given by

$$\hat{P} = \left\{(x, w) \mid 0 \leq w,\ a_j'x \leq b_j w,\ j = 1, \ldots, r\right\}$$

(see Fig. 3.2.3), and note that

$$P = \left\{x \mid (x, 1) \in \hat{P}\right\}.$$

By the Minkowski-Weyl Theorem [Prop. 3.2.1(c)], $\hat{P}$ is finitely generated, so it has the form

$$\hat{P} = \left\{(x, w) \ \middle| \ x = \sum_{j=1}^m \nu_j \tilde{v}_j,\ w = \sum_{j=1}^m \nu_j \tilde{d}_j,\ \nu_j \geq 0,\ j = 1, \ldots, m\right\},$$

for some vectors $\tilde{v}_j$ and scalars $\tilde{d}_j$. Since $w \geq 0$ for all vectors $(x, w) \in \hat{P}$, we see that $\tilde{d}_j \geq 0$ for all j. Let

$$J^+ = \{j \mid \tilde{d}_j > 0\}, \qquad J^0 = \{j \mid \tilde{d}_j = 0\}.$$

By replacing, for all $j \in J^+$, $\nu_j \tilde{d}_j$ with μ_j and $\tilde{v}_j / \tilde{d}_j$ with v_j, we obtain the equivalent description

$$\hat{P} = \left\{(x, w) \ \middle| \ x = \sum_{j \in J^+ \cup J^0} \mu_j v_j,\ w = \sum_{j \in J^+} \mu_j,\ \mu_j \geq 0,\ j \in J^+ \cup J^0\right\}.$$

$$(3.1)$$

Since $P = \{x \mid (x,1) \in \hat{P}\}$, we obtain

$$P = \left\{ x \;\middle|\; x = \sum_{j \in J^+ \cup J^0} \mu_j v_j, \; \sum_{j \in J^+} \mu_j = 1, \, \mu_j \geq 0, \, j \in J^+ \cup J^0 \right\}. \quad (3.2)$$

Thus, P is the vector sum of $\text{conv}(\{v_j \mid j \in J^+\})$ and the finitely generated cone

$$\left\{ \sum_{j \in J^0} \mu_j v_j \;\middle|\; \mu_j \geq 0, \, j \in J^0 \right\}.$$

To prove that the vector sum of $\text{conv}(\{v_1, \ldots, v_m\})$ and a finitely generated cone is a polyhedral set, we reverse the preceding argument: starting from Eq. (3.2), we express P as $\{x \mid (x,1) \in \hat{P}\}$, where $\hat{P}$ is the finitely generated cone of Eq. (3.1). We then use the Minkowski-Weyl Theorem to assert that this cone is polyhedral, and we finally construct a polyhedral set description using the equation $P = \{x \mid (x,1) \in \hat{P}\}$. **Q.E.D.**

As indicated by the examples of Fig. 3.2.2, the finitely generated cone C in a Minkowski-Weyl representation of a polyhedral set P is just the recession cone of P (see Exercise 3.8).

3.2.3 Polyhedral Functions

Polyhedral sets can also be used to define functions with polyhedral structure. In particular, we say that a function $f : \Re^n \mapsto (-\infty, \infty]$ is *polyhedral* if its epigraph is a polyhedral set in $\Re^{n+1}$; see Fig. 3.2.4. Note that a polyhedral function f is closed, convex, and proper [since $\text{epi}(f)$ is closed, convex, and nonempty (based on our convention that only nonempty sets can be polyhedral)]. The following proposition provides a useful representation of polyhedral functions.

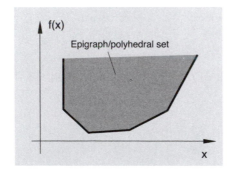

Figure 3.2.4. Illustration of a polyhedral function. By definition, the function must be proper, and its epigraph must be a polyhedral set.

Proposition 3.2.3: Let $f : \Re^n \mapsto (-\infty, \infty]$ be a convex function. Then f is polyhedral if and only if $\text{dom}(f)$ is a polyhedral set and

$$f(x) = \max_{j=1,\ldots,m} \{a_j' x + b_j\}, \qquad \forall\, x \in \text{dom}(f),$$

where a_j are vectors in $\Re^n$, b_j are scalars, and m is a positive integer.

Proof: If f has the representation given above, then $\text{epi}(f)$ is nonempty [since $\text{dom}(f)$ is polyhedral and hence nonempty], and can be written as

$$\text{epi}(f) = \{(x, w) \mid x \in \text{dom}(f)\} \cap \{(x, w) \mid a_j' x + b_j \leq w, j = 1, \ldots, m\}.$$

Since the two sets in the right-hand side above are polyhedral, their intersection, $\text{epi}(f)$, is also polyhedral (it follows easily from the definitions that if the intersection of two polyhedral sets is nonempty, then it is polyhedral; see also Exercise 3.10). Hence f is polyhedral.

Conversely, if f is polyhedral, its epigraph is a polyhedral set, and can be represented as

$$\{(x, w) \mid a_j' x + b_j \leq c_j w, j = 1, \ldots, r\},$$

where a_j are some vectors in $\Re^n$, and b_j and c_j are some scalars. Since for any $(x, w) \in \text{epi}(f)$, we have $(x, w + \gamma) \in \text{epi}(f)$ for all $\gamma \geq 0$, it follows that $c_j \geq 0$, so by normalizing if necessary, we may assume without loss of generality that either $c_j = 0$ or $c_j = 1$. If $c_j = 0$ for all j, then f would not be proper, contradicting the fact that a polyhedral function is proper. Hence we must have, for some m with $1 \leq m \leq r$, $c_j = 1$ for $j = 1, \ldots, m$, and $c_j = 0$ for $j = m + 1, \ldots, r$, i.e.,

$$\text{epi}(f) = \{(x, w) \mid a_j' x + b_j \leq w, j = 1, \ldots, m, \; a_j' x + b_j \leq 0, j = m+1, \ldots, r\}.$$

Thus the effective domain of f is the polyhedral set

$$\text{dom}(f) = \{x \mid a_j' x + b_j \leq 0, j = m + 1, \ldots, r\},$$

and we have

$$f(x) = \max_{j=1,\ldots,m} \{a_j' x + b_j\}, \qquad \forall\, x \in \text{dom}(f).$$

Q.E.D.

Some additional properties of polyhedral functions are discussed in Exercises 3.12, 3.13. In particular, under mild technical assumptions, common operations on polyhedral functions, such as convex combination, composition with a linear transformation, partial minimization, etc., preserve their polyhedral character.

3.3   EXTREME POINTS

A geometrically apparent property of a bounded polyhedral set in the plane is that it can be represented as the convex hull of a finite number of points, its "corner" points. This property can be generalized, as we show in this section, and has great analytical and algorithmic significance.

Given a nonempty convex set C, a vector $x \in C$ is said to be an *extreme point* of C if it does not lie strictly between the endpoints of any line segment contained in the set, i.e., if there do not exist vectors $y \in C$ and $z \in C$, with $y \neq x$ and $z \neq x$, and a scalar $\alpha \in (0,1)$ such that $x = \alpha y + (1 - \alpha)z$. It can be seen that an equivalent definition is that x cannot be expressed as a convex combination of some vectors of C, all of which are different from x.

It follows from the definition that an extreme point of a set that consists of more than one point must lie on the relative boundary of the set. As a result, a relatively open set has no extreme points, except in the special case where the set consists of a single point. As another consequence of the definition, a convex cone may have at most one extreme point, the origin. We will also show later in this section that a polyhedral set has at most a finite number of extreme points (possibly none). Figure 3.3.1 illustrates the extreme points of various types of sets.

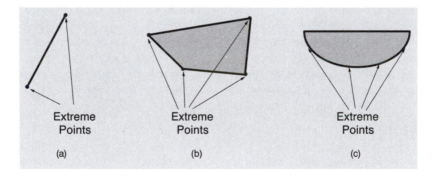

Figure 3.3.1. Illustration of extreme points of various convex sets in $\Re^2$. For the set in (c), the extreme points are the ones that lie on the circular arc.

The following proposition provides some intuition into the nature of extreme points.

Proposition 3.3.1: Let C be a nonempty convex subset of $\Re^n$.

(a) If a hyperplane H contains C in one of its closed halfspaces, then every extreme point of $C \cap H$ is also an extreme point of C.

(b) If C is closed, then C has at least one extreme point if and only if it does not contain a line, i.e., a set of the form $\{x + \alpha d \mid \alpha \in \Re\}$, where d is a nonzero vector in $\Re^n$ and x is some vector in C.

(c) (*Krein-Milman Theorem*) If C is compact, then C is equal to the convex hull of its extreme points.

Proof: (a) Let $\overline{x}$ be an extreme point of $C \cap H$, and assume, to arrive at a contradiction, that it is not an extreme point of C. Then, $\overline{x} = \alpha y + (1-\alpha)z$ for some $\alpha \in (0,1)$, and some $y \in C$ and $z \in C$, with $y \neq \overline{x}$ and $z \neq \overline{x}$ (see Fig. 3.3.2). Since $\overline{x} \in H$, the closed halfspace containing C is of the form $\{x \mid a'x \geq a'\overline{x}\}$, where $a \neq 0$, and H is of the form $\{x \mid a'x = a'\overline{x}\}$. Thus, we have $a'y \geq a'\overline{x}$ and $a'z \geq a'\overline{x}$, which in view of $\overline{x} = \alpha y + (1-\alpha)z$, implies that $a'y = a'\overline{x}$ and $a'z = a'\overline{x}$. Therefore, $y \in C \cap H$ and $z \in C \cap H$, contradicting the assumption that $\overline{x}$ is an extreme point of $C \cap H$.

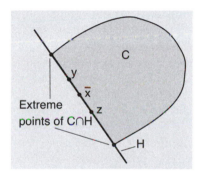

Figure 3.3.2. Construction used in the proof of Prop. 3.3.1(a).

(b) Assume, to arrive at a contradiction, that C has an extreme point x and contains a line $\{\overline{x} + \alpha d \mid \alpha \in \Re\}$, where $\overline{x} \in C$ and $d \neq 0$. Then, by the Recession Cone Theorem [Prop. 1.5.1(b)] and the closedness of C, both d and $-d$ are directions of recession of C, so the line $\{x + \alpha d \mid \alpha \in \Re\}$ belongs to C. This contradicts the fact that x is an extreme point.

Conversely, we use induction on the dimension of the space to show that if C does not contain a line, it must have an extreme point. This result is true for $\Re$, so assume it is true for $\Re^{n-1}$, where $n \geq 2$. We will show that any nonempty closed convex subset C of $\Re^n$, which does not contain a line, must have an extreme point. Since C does not contain a line, there must exist points $x \in C$ and $y \notin C$. The line segment connecting x and y intersects the relative boundary of C at some point $\overline{x}$, which belongs to C, since C is closed. Consider a hyperplane H passing through $\overline{x}$ and containing C in one of its closed halfspaces (see Fig. 3.3.3). By using a translation argument if necessary, we may assume without loss of generality that $\bar{x} = 0$. Then H is an $(n-1)$-dimensional subspace, so the set $C \cap H$ lies in an $(n-1)$-dimensional space and does not contain a line, since C does not. Hence, by the induction hypothesis, it must have an extreme point. By part (a), this extreme point must also be an extreme point of C.

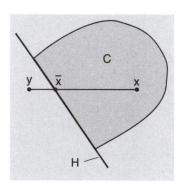

Figure 3.3.3. Construction used in the induction proof of Prop. 3.3.1(b) to show that if a closed convex set C does not contain a line, it must have an extreme point. Using the line segment connecting two points $x \in C$ and $y \notin C$, we obtain a relative boundary point $\bar{x}$ of C. The proof is then reduced to asserting the existence of an extreme point of the lower-dimensional set $C \cap H$, where H is a supporting hyperplane passing through $\bar{x}$, which is also an extreme point of C by Prop. 3.3.1(a).

(c) By convexity, C contains the convex hull of its extreme points. To show the reverse inclusion, we use induction on the dimension of the space. On the real line, a compact convex set C is a line segment whose endpoints are the extreme points of C, so every point in C is a convex combination of the two endpoints. Suppose now that every vector in a compact and convex subset of $\Re^{n-1}$ can be represented as a convex combination of extreme points of the set. We will show that the same is true for compact and convex subsets of $\Re^n$.

Let C be a compact and convex subset of $\Re^n$, and choose any $x \in C$. If x is the only point in C, it is an extreme point and we are done, so assume that $\bar{x}$ is another point in C, and consider the line that passes through x and $\bar{x}$. Since C is compact, the intersection of this line and C is a compact line segment whose endpoints, say x_1 and x_2, belong to the relative boundary of C. Let H_1 be a hyperplane that passes through x_1 and contains C in one of its closed halfspaces. Similarly, let H_2 be a hyperplane that passes through x_2 and contains C in one of its closed halfspaces (see Fig. 3.3.4). The intersections $C \cap H_1$ and $C \cap H_2$ are compact convex sets that lie in the hyperplanes H_1 and H_2, respectively. By viewing H_1 and H_2 as $(n-1)$-dimensional spaces, and by using the induction hypothesis, we see that each of the sets $C \cap H_1$ and $C \cap H_2$ is the convex hull of its extreme points. Hence, x_1 is a convex combination of some extreme points of $C \cap H_1$, and x_2 is a convex combination of some extreme points of $C \cap H_2$. By part (a), all the extreme points of $C \cap H_1$ and all the extreme points of $C \cap H_2$ are also extreme points of C, so both x_1 and x_2 are convex combinations of some extreme points of C. Since x lies in the line segment connecting x_1 and x_2, it follows that x is a convex combination of some extreme points of C, showing that C is contained in the convex hull of the extreme points of C.   **Q.E.D.**

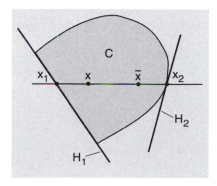

Figure 3.3.4. The main idea of the proof of the Krein-Milman Theorem, i.e., that any vector x of a convex and compact set C can be represented as a convex combination of extreme points of C. If $\bar{x}$ is another point in C, the points x_1 and x_2 shown can be represented as convex combinations of extreme points of the lower dimensional convex and compact sets $C \cap H_1$ and $C \cap H_2$, which are also extreme points of C by Prop. 3.3.1(a).

As an example of application of the preceding proposition, consider a nonempty polyhedral set of the form

$$\{x \mid Ax = b, \, x \geq 0\},$$

or of the form

$$\{x \mid Ax \leq b, \, x \geq 0\},$$

where A is an $m \times n$ matrix and b is a vector in $\Re^m$. Such polyhedra arise commonly in linear programming, and clearly do not contain a line. Hence, by Prop. 3.3.1(b), they always have at least one extreme point.

Note that the boundedness assumption is essential in the Krein-Milman Theorem. For example, a line is a closed convex set that has no extreme points, while a closed convex cone has at most one extreme point, the origin, but none of its other points can be generated as a convex combination of its extreme points.

3.3.1 Extreme Points of Polyhedral Sets

We now consider a polyhedral set and we characterize its extreme points.

Proposition 3.3.2: Let P be a polyhedral subset of $\Re^n$ that has a Minkowski-Weyl representation

$$P = \text{conv}\big(\{v_1, \ldots, v_m\}\big) + C,$$

where $v_1, \ldots, v_m$ are some vectors and C is a finitely generated cone. Then the set of extreme points of P is a subset of $\{v_1, \ldots, v_m\}$.

Proof: An extreme point $\bar{x}$ of P cannot be of the form $\bar{x} = \tilde{x} + y$, where $\tilde{x} \in \text{conv}(\{v_1, \dots, v_m\})$ and $y \neq 0$, $y \in C$, since in this case $\bar{x}$ would be the midpoint of the line segment connecting the distinct vectors $\tilde{x}$ and $\tilde{x} + 2y$. It follows that an extreme point of P must belong to $\text{conv}(\{v_1, \dots, v_m\})$, and therefore must be one of the vectors $v_1, \dots, v_m$, since otherwise this point would be expressible as a convex combination of $v_1, \dots, v_m$.   **Q.E.D.**

Note that the preceding proposition shows that a polyhedral set can have at most a finite number of extreme points. The following proposition gives a characterization of extreme points of polyhedral sets that is central in the theory of linear programming (see Fig. 3.3.5).

Proposition 3.3.3: Let P be a polyhedral subset of $\Re^n$.

(a) If P has the form

$$P = \{x \mid a_j'x \leq b_j, \, j = 1, \dots, r\},$$

where a_j are vectors in $\Re^n$ and b_j are scalars, respectively, then a vector $v \in P$ is an extreme point of P if and only if the set

$$A_v = \{a_j \mid a_j'v = b_j, \, j \in \{1, \dots, r\}\}$$

contains n linearly independent vectors.

(b) If P has the form

$$P = \{x \mid Ax = b, \, x \geq 0\},$$

where A is an $m \times n$ matrix and b is a vector in $\Re^m$, then a vector $v \in P$ is an extreme point of P if and only if the columns of A corresponding to the nonzero coordinates of v are linearly independent.

(c) If P has the form

$$P = \{x \mid Ax = b, \, c \leq x \leq d\},$$

where A is an $m \times n$ matrix, b is a vector in $\Re^m$, and c, d are vectors in $\Re^n$, then a vector $v \in P$ is an extreme point of P if and only if the columns of A corresponding to the coordinates of v that lie strictly between the corresponding coordinates of c and d are linearly independent.

Proof: (a) If the set A_v contains fewer than n linearly independent vectors,

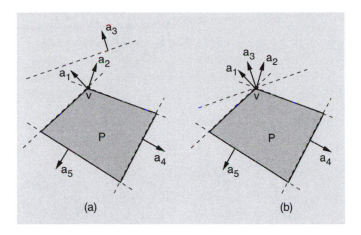

Figure 3.3.5. Illustration of extreme points of a 2-dimensional polyhedral set

$$P = \left\{ x \mid a_j' x \le b_j, \, j = 1, \ldots, r \right\},$$

[see Prop. 3.3.3(a)]. For a vector $v \in P$ to be an extreme point it is necessary and sufficient that the set of vectors

$$A_v = \left\{ a_j \mid a_j' v = b_j, \, j \in \{1, \ldots, r\} \right\}$$

contains $n = 2$ linearly independent vectors. In the case on the left-hand side, the set A_v consists of the two linearly independent vectors a_1 and a_2. In the case on the right-hand side, the set A_v consists of the three vectors a_1, a_2, a_3, any pair of which are linearly independent.

then the system of equations

$$a_j' w = 0, \qquad \forall \, a_j \in A_v$$

has a nonzero solution, call it $\overline{w}$. For sufficiently small $\gamma > 0$, we have $v + \gamma \overline{w} \in P$ and $v - \gamma \overline{w} \in P$, thus showing that v is not an extreme point. Thus, if v is an extreme point, A_v must contain n linearly independent vectors.

Conversely, assume that A_v contains a subset $\bar{A}_v$ consisting of n linearly independent vectors. Suppose that for some $y \in P$, $z \in P$, and $\alpha \in (0, 1)$, we have

$$v = \alpha y + (1 - \alpha) z.$$

Then, for all $a_j \in \bar{A}_v$, we have

$$b_j = a_j' v = \alpha a_j' y + (1 - \alpha) a_j' z \le \alpha b_j + (1 - \alpha) b_j = b_j.$$

Thus, v, y, and z are all solutions of the system of n linearly independent equations

$$a_j' w = b_j, \qquad \forall \, a_j \in \bar{A}_v.$$

Hence, $v = y = z$, implying that v is an extreme point of P.

(b) We write P in terms of inequalities in the equivalent form

$$P = \{x \mid Ax \le b, \ -Ax \le -b, \ -x \le 0\}.$$

Let R_v be the set of vectors of the form $(0, \ldots, 0, -1, 0, \ldots 0)$, where -1 is in a position corresponding to a zero coordinate of v. Consider the matrix $\bar{A}$, which is the same as A except that all the columns corresponding to the zero coordinates of v are set to zero. By applying the result of part (a) to the above inequality representation of P, we see that v is an extreme point if and only if the rows of A together with the vectors in R_v contain n linearly independent vectors. Equivalently, v is an extreme point if and only if $\bar{A}$ contains $n - k$ linearly independent rows, where k is the number of vectors in R_v. This is equivalent to the $n - k$ nonzero columns of A (corresponding to the nonzero coordinates of v) being linearly independent.

(c) The proof is essentially the same as the proof of part (b). **Q.E.D.**

3.4 POLYHEDRAL ASPECTS OF OPTIMIZATION

Polyhedral convexity plays a very important role in optimization. One reason is that many practical problems can be formulated in terms of polyhedral sets and functions. Another reason is that for polyhedral constraint sets and/or linear cost functions, it is often possible to show stronger optimization results than those available for general convex constraint sets and/or general cost functions. We have seen a few such instances so far. In particular:

(1) A linear or convex quadratic function that is bounded below over a nonempty polyhedral set P attains a minimum over P (see Prop. 2.3.4).

(2) A linear (or, more generally, concave) function $f : \Re^n \mapsto \Re$ that attains a minimum over a convex set C at some relative interior point of C must be constant over C (see Prop. 1.4.2). Thus, a linear function that is not constant over a convex constraint set C, can only attain its minimum at a relative boundary point of C.

In this section, we explore some further consequences of polyhedral convexity in optimization. One of the fundamental linear programming results is that if a linear function f attains a minimum over a polyhedral set C that has at least one extreme point, then f attains a minimum at some extreme point of C (as well as possibly at some other nonextreme points). This is a special case of a more general result, which holds when f is concave, and C is closed and convex.

Proposition 3.4.1: Let C be a closed convex subset of $\Re^n$ that has at least one extreme point. A concave function $f : C \mapsto \Re$ that attains a minimum over C attains the minimum at some extreme point of C.

Proof: Let x^* minimize f over C. If $x^* \in \mathrm{ri}(C)$ [see Fig. 3.4.1(a)], by Prop. 1.4.2, f must be constant over C, so it attains a minimum at an extreme point of C (since C has at least one extreme point by assumption). If $x^* \notin \mathrm{ri}(C)$, then by Prop. 2.4.5, there exists a hyperplane H_1 properly separating x^* and C. Since $x^* \in C$, H_1 must contain x^*, so by the proper separation property, H_1 cannot contain C, and it follows that the intersection $C \cap H_1$ has dimension smaller than the dimension of C.

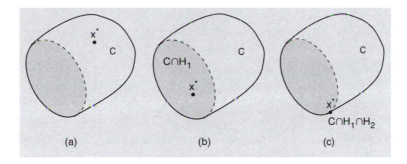

Figure 3.4.1. An outline of the argument of the proof of Prop. 3.4.1 for a 3-dimensional set C.

If $x^* \in \mathrm{ri}(C \cap H_1)$ [see Fig. 3.4.1(b)], then f must be constant over $C \cap H_1$, so it attains a minimum at an extreme point of $C \cap H_1$ [since C contains an extreme point, it does not contain a line by Prop. 3.3.1(b), and hence $C \cap H_1$ does not contain a line, which implies that $C \cap H_1$ has an extreme point]. By Prop. 3.3.1(a), this optimal extreme point is also an extreme point of C. If $x^* \notin \mathrm{ri}(C \cap H_1)$, there exists a hyperplane H_2 properly separating x^* and $C \cap H_1$. Again, since $x^* \in C \cap H_1$, H_2 contains x^*, so it cannot contain $C \cap H_1$, and it follows that the intersection $C \cap H_1 \cap H_2$ has dimension smaller than the dimension of $C \cap H_1$.

If $x^* \in \mathrm{ri}(C \cap H_1 \cap H_2)$ [see Fig. 3.4.1(c)], then f must be constant over $C \cap H_1 \cap H_2$, etc. Since with each new hyperplane, the dimension of the intersection of C with the generated hyperplanes is reduced, this process will be repeated at most n times, until x^* is a relative interior point of some set $C \cap H_1 \cap \cdots \cap H_k$, at which time an extreme point of $C \cap H_1 \cap \cdots \cap H_k$ will be obtained. Through a reverse argument, repeatedly applying Prop. 3.3.1(a), it follows that this extreme point is an extreme point of C. **Q.E.D.**

3.4.1   Linear Programming

We now specialize the preceding result to the case of a linear program, where the cost function f is linear.

Proposition 3.4.2:   (Fundamental Theorem of Linear Programming) Let P be a polyhedral set that has at least one extreme point. A linear function that is bounded below over P attains a minimum at some extreme point of P.

Proof: Since the cost function is bounded below over P, it attains a minimum by Prop. 2.3.4. The result now follows from Prop. 3.4.1.   **Q.E.D.**

Figure 3.4.2 illustrates the possibilities in the case of a linear programming problem. There are two cases:

(a) The constraint set P contains an extreme point [equivalently, by Prop. 3.3.1(b), P does not contain a line]. In this case, the linear cost function is either unbounded below over P or else it attains a minimum at an extreme point of P.

(b) The constraint set P does not contain an extreme point [equivalently, by Prop. 3.3.1(b), P contains a line]. In this case, L_P, the lineality space of P, is a subspace of dimension greater or equal to one. If the linear cost function f is bounded below over P, it attains a minimum by Prop. 2.3.4. However, because every direction in L_P must be a direction along which f is constant (otherwise f would be unbounded below over P), the set of minima is a polyhedral set whose lineality space is equal to L_P. Therefore, the set of minima must be unbounded.

Here is another interesting special case of Prop. 3.4.1.

Proposition 3.4.3: Let $f : C \mapsto \Re$ be a concave function, where C is a closed convex subset of $\Re^n$. Assume that for some $m \times n$ matrix A of rank n and some vector b in $\Re^m$, we have

$$Ax \geq b, \qquad \forall\, x \in C.$$

Then if f attains a minimum over C, it attains a minimum at some extreme point of C.

Proof: In view of Prop. 3.4.1, it suffices to show that C contains an ex-

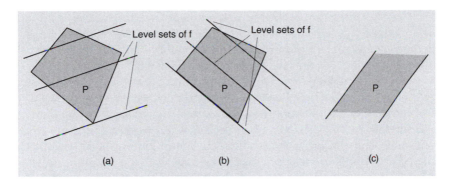

Figure 3.4.2. Illustration of the fundamental theorem of linear programming. In (a) and (b), the constraint set P has at least one extreme point and the linear cost function f is bounded over P. Then f either attains a minimum at a unique extreme point as in (a), or it attains a minimum at multiple extreme points as well as at an infinite number of nonextreme points as in (b). In (c), the constraint set has no extreme points because it contains a line [cf. Prop. 3.3.1(b)], and the linear cost function is either unbounded below over P or attains a minimum at an infinite set of (nonextreme) points whose lineality space is the same as the lineality space of P.

treme point, or equivalently by Prop. 3.3.1(b), that C does not contain a line. Assume, to obtain a contradiction, that C contains the line

$$L = \{\overline{x} + \lambda d \mid \lambda \in \Re\},$$

where $\overline{x} \in C$ and d is a nonzero vector. Since A has rank n, the vector Ad is nonzero, implying that the image

$$A \cdot L = \{A\overline{x} + \lambda Ad \mid \lambda \in \Re\}$$

is also a line. This contradicts the assumption $Ax \geq b$ for all $x \in C$. **Q.E.D.**

The theory of polyhedral cones and sets, and their extreme points can be used for the development of additional algorithmic or analytical aspects of linear programming, such as the simplex method and the corresponding duality theory. The simplex method is outside our scope, so we refer to standard textbooks, such as Dantzig [Dan63], Chvatal [Chv83], and Bertsimas and Tsitsiklis [BeT97]. Linear programming duality will be discussed in Chapter 6, within the broader context of convex programming duality.

3.4.2   Integer Programming

Many important optimization problems, in addition to the usual equality and inequality constraints, include the requirement that the optimization variables take integer values, such as 0 or 1. We refer to such problems as *integer programming* problems, and we note that they arise in a broad variety of practical settings, such as in scheduling, resource allocation, and engineering design, as well as in many combinatorial optimization contexts, such as matching and traveling salesman problems.

The methodology for solving an integer programming problem is very diverse, but an important subset of this methodology relies on the solution of a continuous optimization problem, called the *relaxed problem*, which is derived from the original by neglecting the integer constraints while maintaining all the other constraints. In many important situations the relaxed problem is a linear program that can be solved for an optimal solution, which is an extreme point, by a variety of algorithms, including the simplex method. A particularly fortuitous situation arises if this extreme point optimal solution happens to have integer components, since it will then solve optimally not just the relaxed problem, but also the original integer programming problem. Thus, polyhedra whose extreme points have integer components are of special significance. We will now characterize an important class of polyhedra with this property.

Consider a polyhedral set P of the form

$$P = \{x \mid Ax = b, \, c \le x \le d\},$$

where A is an $m \times n$ matrix, b is a vector in $\Re^m$, and c and d are vectors in $\Re^n$. We assume that all the components of the matrix A and the vectors b, c, and d are integer. We want to determine conditions under which the extreme points of P have integer components.

Let us say that a square matrix with integer components is *unimodular* if its determinant is 0, 1, or -1, and let us say that a rectangular matrix with integer components is *totally unimodular* if each of its square submatrices is unimodular. We have the following proposition.

Proposition 3.4.4: Let P be a polyhedral set of the form

$$P = \{x \mid Ax = b, \, c \le x \le d\},$$

where A is an $m \times n$ matrix, b is a vector in $\Re^m$, and c and d are vectors in $\Re^n$. Assume that all the components of the matrix A and the vectors b, c, and d are integer, and that the matrix A is totally unimodular. Then all the extreme points of P have integer components.

Proof: Let v be an extreme point of P. Consider the subset of indices

$$I = \{i \mid c_i < v_i < d_i\},$$

and without loss of generality, assume that

$$I = \{1, \ldots, \overline{m}\}$$

for some integer $\overline{m}$. Let $\overline{A}$ be the matrix consisting of the first $\overline{m}$ columns of A and let $\overline{v}$ be the vector consisting of the first $\overline{m}$ components of v. Note that each of the last $n - \overline{m}$ components of v is equal to either the corresponding component of c or to the corresponding component of d, which are integer. Thus the extreme point v has integer components if and only if the subvector $\overline{v}$ has integer components.

By Prop. 3.3.3, $\overline{A}$ has linearly independent columns, so $\overline{v}$ is the unique solution of the system of equations

$$\overline{A}y = \overline{b},$$

where $\overline{b}$ is equal to b minus the last $n - \overline{m}$ columns of A multiplied with the corresponding components of v (each of which is equal to either the corresponding component of c or the corresponding component of d, so that $\overline{b}$ has integer components). Equivalently, there exists an invertible $\overline{m} \times \overline{m}$ submatrix $\tilde{A}$ of $\overline{A}$ and a subvector $\tilde{b}$ of $\overline{b}$ with $\overline{m}$ components such that

$$\overline{v} = (\tilde{A})^{-1}\tilde{b}.$$

The components of $\overline{v}$ will be integer if we can guarantee that the components of the inverse $(\tilde{A})^{-1}$ are integer. By Cramer's rule for solving linear systems of equations, each of the components of the inverse of a matrix is a fraction with a sum of products of the components of the matrix in the numerator and the determinant of the matrix in the denominator. Since by hypothesis, A is totally unimodular, the invertible submatrix $\tilde{A}$ is unimodular, and its determinant is equal to either 1 or -1. Hence, $(\tilde{A})^{-1}$ has integer components, and it follows that $\overline{v}$ (and hence also the extreme point v) has integer components.   **Q.E.D.**

The total unimodularity of the matrix A in the above proposition can be verified in a number of important special cases, some of which are discussed in Exercises 3.22-3.25. The most important such case arises in network optimization problems, where A is the, so-called, *arc incidence matrix* of a given directed graph (we refer to network optimization books such as Rockafellar [Roc84] or Bertsekas [Ber98] for a detailed discussion of these problems and their broad applications). Here A has a row for each node and a column for each arc of the graph. The component corresponding to the ith row and a given arc is a 1 if the arc is outgoing from node i, is a -1 if the arc is incoming to i, and is a 0 otherwise. Then, it can be shown, by using induction on the dimension of the submatrix, that the determinant of each square submatrix of A is 0, 1, or -1. Thus, in linear network optimization where the only constraints, other than upper and lower bounds on the variables, are equality constraints corresponding to an arc incidence matrix, integer extreme point optimal solutions can be found, assuming a feasible solution exists.

3.5 POLYHEDRAL ASPECTS OF DUALITY

Polyhedral convexity plays a special role in duality theory. In particular, it turns out that some of the important duality theorems can be enhanced in the presence of polyhedral convexity assumptions. In this section, we first show a min common/max crossing duality theorem that relies on polyhedral structure. We then use this theorem to prove a corresponding minimax theorem, as well as a nonlinear version of Farkas' Lemma. These results will in turn be revisited in Chapters 6 and 7, when we discuss duality in more detail.

We first introduce a special type of proper hyperplane separation that involves polyhedral convexity assumptions.

3.5.1 Polyhedral Proper Separation

As shown in Prop. 2.4.5, two nonempty convex sets C and P such that

$$\mathrm{ri}(C) \cap \mathrm{ri}(P) = \emptyset$$

can be properly separated, i.e., can be separated by a hyperplane that does not contain both C and P. The following proposition shows that if P is polyhedral and the slightly stronger condition

$$\mathrm{ri}(C) \cap P = \emptyset$$

holds, then there exists a properly separating hyperplane satisfying the extra restriction that it does not contain the nonpolyhedral set C (rather than just the milder requirement that it does not contain either C or P); see Fig. 3.5.1.

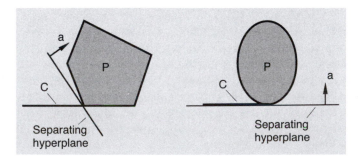

Figure 3.5.1. Illustration of the special proper separation property of a convex set C and a polyhedral set P, under the condition $\mathrm{ri}(C) \cap P = \emptyset$. In the figure on the left, the separating hyperplane can be chosen so that it does not contain C. If P is not polyhedral, as in the figure on the right, this may not be possible.

> **Proposition 3.5.1: (Polyhedral Proper Separation Theorem)**
> Let C and P be two nonempty convex subsets of $\Re^n$ such that P is polyhedral. There exists a hyperplane that separates C and P, and does not contain C if and only if
>
> $$\mathrm{ri}(C) \cap P = \emptyset.$$

Proof: First, as a general observation, we recall from our discussion of proper separation in Section 2.4 that given a convex set X and a hyperplane H that contains X in one of its closed halfspaces, we have

$$X \subset H \qquad \text{if and only if} \qquad \mathrm{ri}(X) \cap H \neq \emptyset. \tag{3.3}$$

We will use this relation in the subsequent proof.

Assume that there exists a hyperplane H that separates C and P, and does not contain C. Then, by Eq. (3.3), H cannot contain a point in $\mathrm{ri}(C)$, and since H separates C and P, we must have $\mathrm{ri}(C) \cap P = \emptyset$.

Conversely, assume that $\mathrm{ri}(C) \cap P = \emptyset$. We will show that there exists a separating hyperplane that does not contain C. Denote

$$D = P \cap \mathrm{aff}(C).$$

If $D = \emptyset$, then since $\mathrm{aff}(C)$ is closed and P is polyhedral, the Strict Separation Theorem [cf. Prop. 2.4.3 under condition (4)] applies and shows that there exists a hyperplane H that separates $\mathrm{aff}(C)$ and P strictly, and hence does not contain C.

We may thus assume that $D \neq \emptyset$. The idea now is to first construct a hyperplane that properly separates C and D, and then extend this hyperplane so that it suitably separates C and P. [If C had nonempty interior, the proof would be much simpler, since then $\mathrm{aff}(C) = \Re^n$ and $D = P$.]

By assumption, we have $\mathrm{ri}(C) \cap P = \emptyset$ implying that

$$\mathrm{ri}(C) \cap \mathrm{ri}(D) \subset \mathrm{ri}(C) \cap \big(P \cap \mathrm{aff}(C)\big) = \big(\mathrm{ri}(C) \cap P\big) \cap \mathrm{aff}(C) = \emptyset.$$

Hence, by Prop. 2.4.6, there exists a hyperplane H that properly separates C and D. Furthermore, H does not contain C, since if it would, H would also contain $\mathrm{aff}(C)$ and hence also D, contradicting the proper separation property. Thus, C is contained in one of the closed halfspaces of H, but not in both. Let $\overline{C}$ be the intersection of $\mathrm{aff}(C)$ and the closed halfspace of H that contains C; see Fig. 3.5.2. Note that H does not contain $\overline{C}$ (since H does not contain C), and by Eq. (3.3), we have $H \cap \mathrm{ri}(\overline{C}) = \emptyset$, implying that

$$P \cap \mathrm{ri}(\overline{C}) = \emptyset,$$

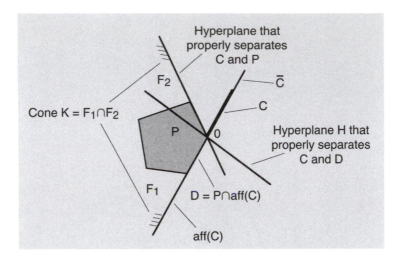

Figure 3.5.2. Illustration of the proof of Prop. 3.5.1 in the case where $D = P \cap \operatorname{aff}(C) \neq \emptyset$. The figure shows the construction of a hyperplane that properly separates C and P, and does not contain C, starting from the hyperplane H that properly separates C and D. In this two-dimensional example we have $M = \{0\}$, so $K = \operatorname{cone}(P) + M = \operatorname{cone}(P)$.

since P and $\overline{C}$ lie in the opposite closed halfspaces of H.

If $P \cap \overline{C} = \emptyset$, then by using again the Strict Separation Theorem [cf. Prop. 2.4.3 under condition (4)], we can construct a hyperplane that strictly separates P and $\overline{C}$. This hyperplane also strictly separates P and C, and we are done. We thus assume that $P \cap \overline{C} \neq \emptyset$, and by using a translation argument if necessary, we assume that

$$0 \in P \cap \overline{C},$$

as indicated in Fig. 3.5.2. The polyhedral set P can be represented as the intersection of halfspaces $\{x \mid a_j'x \le b_j\}$ with $b_j \ge 0$ (since $0 \in P$) and with $b_j = 0$ for at least one j (since otherwise 0 would be in the interior of P, which is impossible since $0 \in H$ and P lies in a closed halfspace of H). Thus, we have

$$P = \{x \mid a_j'x \le 0, \ j = 1, \ldots, m\} \cap \{x \mid a_j'x \le b_j, \ j = m+1, \ldots, \overline{m}\},$$

for some integers $m \ge 1$ and $\overline{m} \ge m$, vectors a_j, and scalars $b_j > 0$.

Let M be the relative boundary of $\overline{C}$, i.e.,

$$M = H \cap \operatorname{aff}(C),$$

and consider the cone

$$K = \{x \mid a_j'x \le 0, \ j = 1, \ldots, m\} + M.$$

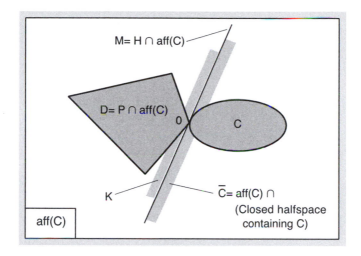

Figure 3.5.3. Illustration of the construction of the cone $K = \text{cone}(P) + M$ in the proof of Prop. 3.5.1.

Note that $K = \text{cone}(P) + M$ (see Figs. 3.5.2 and 3.5.3).

We claim that $K \cap \text{ri}(\overline{C}) = \emptyset$. The proof is by contradiction. If there exists $\overline{x} \in K \cap \text{ri}(\overline{C})$, then $\overline{x}$ can be expressed as $\overline{x} = \alpha w + v$ for some $\alpha > 0$, $w \in P$, and $v \in M$ [since $K = \text{cone}(P) + M$ and $0 \in P$], so that $(\overline{x}/\alpha) - (v/\alpha) \in P$. On the other hand, since $\overline{x} \in \text{ri}(\overline{C})$, $0 \in \overline{C}$, and M is a subset of the lineality space of $\overline{C}$ [and hence also of the lineality space of $\text{ri}(\overline{C})$], all vectors of the form $\overline{\alpha}\overline{x} + \overline{v}$, with $\overline{\alpha} > 0$ and $\overline{v} \in M$, belong to $\text{ri}(\overline{C})$, so in particular the vector $(\overline{x}/\alpha) - (v/\alpha)$ also belongs to $\text{ri}(\overline{C})$. This is a contradiction since $P \cap \text{ri}(\overline{C}) = \emptyset$, and it follows that $K \cap \text{ri}(\overline{C}) = \emptyset$.

The cone K is polyhedral (since it is the vector sum of two polyhedral sets), so it is the intersection of some closed halfspaces $F_1, \ldots, F_r$ that pass through 0 (cf. Fig. 3.5.2). Since $K = \text{cone}(P) + M$, each of these closed halfspaces contains M, the relative boundary of the set $\overline{C}$, and furthermore $\overline{C}$ is the closed half of an affine set. It follows that if any of the closed halfspaces $F_1, \ldots, F_r$ contains a vector in $\text{ri}(\overline{C})$, then that closed halfspace entirely contains $\overline{C}$. Hence, since K does not contain any point in $\text{ri}(\overline{C})$, at least one of $F_1, \ldots, F_r$, say F_1, does not contain any point in $\text{ri}(\overline{C})$ (cf. Fig. 3.5.2). Therefore, the hyperplane corresponding to F_1 contains no points of $\text{ri}(\overline{C})$, and hence also no points of $\text{ri}(C)$. Thus, this hyperplane does not contain C, while separating K and C. Since K contains P, this hyperplane also separates P and C. **Q.E.D.**

Note that in the preceding proof, it is essential to introduce M, the relative boundary of the set $\overline{C}$, and to define $K = \text{cone}(P) + M$. If instead we define $K = \text{cone}(P)$, then the corresponding halfspaces $F_1, \ldots, F_r$ may all intersect $\text{ri}(\overline{C})$, and the proof argument fails (see Fig. 3.5.3).

The preceding polyhedral proper separation theorem will now be used to derive another type of min common/max crossing theorem, which relies on polyhedral assumptions. The latter theorem will then be used to prove a corresponding minimax theorem and a nonlinear version of Farkas' Lemma.

3.5.2 Min Common/Max Crossing Duality

In the following proposition, the set M of the min common/max crossing framework of Section 2.5 has a partially polyhedral structure. The most interesting special case is when, in the notation of the proposition, V is the epigraph of a convex function f, in which case M has the same closure as the epigraph of the function $p(u) = \inf_{Ax-b-Bu \in P} f(x)$; compare with the proof of the subsequent Nonlinear Farkas' Lemma (Prop. 3.5.4).

Proposition 3.5.2: (Min Common/Max Crossing Theorem III) Consider the min common and max crossing problems, and assume the following:

(1) The set M is defined in terms of a convex set $V \subset \Re^{m+1}$, an $r \times m$ matrix A, an $r \times n$ matrix B, a vector b in $\Re^r$, and a polyhedral cone $P \subset \Re^r$ as follows:

$$M = \big\{(u, w) \mid u \in \Re^n, \text{ and there is a vector } (x, w) \in V$$
$$\text{such that } Ax - b - Bu \in P\big\}.$$

(2) There exists a vector $(\overline{x}, \overline{w})$ in the relative interior of V such that $A\overline{x} - b \in P$.

Then $q^* = w^*$ and there exists a vector μ in the polar cone P^* such that $q(B'\mu) = q^*$.

Proof: If $w^* = -\infty$, then the conclusion holds since, by the Weak Duality Theorem (Prop. 2.5.3), we have $q^* \le w^*$, so that $q^* = w^* = -\infty$, and $q(B'\mu) = q^*$ for all μ, including the vector $\mu = 0$, which belongs to P^*. We may thus assume that $-\infty < w^*$, which also implies w^* is finite, since the min common problem has a feasible solution in view of the assumptions (1) and (2). Consider the convex subsets of $\Re^{m+1}$ defined by

$$C_1 = \big\{(x, v) \mid \text{there is a vector } (x, w) \in V \text{ such that } v > w\big\},$$

$$C_2 = \big\{(x, w^*) \mid Ax - b \in P\big\}$$

(cf. Fig. 3.5.4). The set C_1 is nonempty since $(\overline{x}, v) \in C_1$ for all $v > \overline{w}$, while the set C_2 is nonempty since $(\overline{x}, w^*) \in C_2$. Finally, C_1 and C_2 are disjoint. To see this, note that

$$w^* = \inf_{(x,w)\in V,\, Ax-b\in P} w, \tag{3.4}$$

and if $(x, w^*) \in C_1 \cap C_2$, by the definition of C_2, we must have $Ax - b \in P$, while by the definition of C_1, we must have $w^* > w$ for some w with $(x, w) \in V$, contradicting Eq. (3.4).

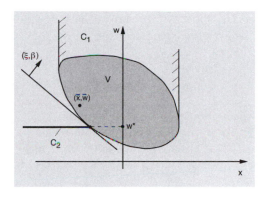

Figure 3.5.4. Illustration of the sets

$$C_1 = \big\{ (x, v) \mid \text{there is a vector } (x, w) \in V \text{ such that } v > w \big\},$$

$$C_2 = \big\{ (x, w^*) \mid Ax - b \in P \big\},$$

and the hyperplane separating them in the proof of Prop. 3.5.2. Note that since $(\overline{x}, \overline{w}) \in \mathrm{ri}(V)$, all the vectors $(\overline{x}, v)$ with $v > \overline{w}$ belong to $\mathrm{ri}(C_1)$.

Since $C_1 \cap C_2 = \varnothing$ and C_2 is polyhedral, by Prop. 3.5.1, there exists a hyperplane that separates C_1 and C_2, and does not contain C_1, i.e., a vector (ξ, β) such that

$$\beta w^* + \xi'z \le \beta v + \xi'x, \qquad \forall\, (x,v) \in C_1,\ \forall\, z \text{ such that } Az - b \in P, \tag{3.5}$$

$$\inf_{(x,v)\in C_1} \big\{ \beta v + \xi'x \big\} < \sup_{(x,v)\in C_1} \big\{ \beta v + \xi'x \big\}. \tag{3.6}$$

If $\beta = 0$, then from Eq. (3.5), we have

$$\xi'\overline{x} \le \sup_{Az-b\in P} \xi'z \le \inf_{(x,v)\in C_1} \xi'x \le \xi'\overline{x}.$$

Thus, equality holds throughout in the preceding relation, implying that all the vectors $(\overline{x}, v)$ with $v > \overline{w}$ minimize the linear function $(\xi, 0)'(x, v)$ over the set C_1. Since $(\overline{x}, \overline{w})$ is a relative interior point of V, all these vectors are relative interior points of C_1 (cf. Fig. 3.5.4). It follows, by Prop. 1.4.2, that the linear function $(\xi, 0)'(x, v)$ is constant over C_1. This, however, contradicts Eq. (3.6). Therefore, we must have $\beta \ne 0$.

By using Eq. (3.5) with $z = \bar{x}$ and $v > \bar{w}$, we obtain $\beta w^* + \xi'\bar{x} \leq \beta v + \xi'\bar{x}$, or $\beta w^* \leq \beta v$. Since $w^* \leq \bar{w} < v$ and $\beta \neq 0$, it follows that $\beta > 0$, and by normalizing (ξ, β) if necessary, we may assume that $\beta = 1$. Thus, from Eq. (3.5) and the definition of C_1, we have

$$\sup_{Az-b\in P} \{w^* + \xi'z\} \leq \inf_{(x,w)\in V} \{w + \xi'x\}. \tag{3.7}$$

Let $\{y \mid Dy \leq 0\}$ be a representation of the polyhedral cone P in terms of a matrix D. Then the maximization problem on the left-hand side of Eq. (3.7) involves the linear program

$$\begin{aligned} &\text{maximize} \quad \xi'z \\ &\text{subject to} \quad DAz - Db \leq 0. \end{aligned} \tag{3.8}$$

Since the minimization problem in the right-hand side of Eq. (3.7) is feasible by assumption, the linear program (3.8) is bounded (as well as feasible) and therefore, by Prop. 2.3.4, has an optimal solution, which is denoted by z^*. Let c_j' be the rows of DA, let $(Db)_j$ denote the corresponding components of Db, and let

$$J = \{j \mid c_j'z^* = (Db)_j\}.$$

If $J = \emptyset$, then z^* lies in the interior of the constraint set of problem (3.8), so we must have $\xi = 0$. If $J \neq \emptyset$ and y is such that $c_j'y \leq 0$ for all $j \in J$, then there is a small enough $\epsilon > 0$ such that $A(z^* + \epsilon y) - b \in P$, and the optimality of z^* implies that $\xi'(z^* + \epsilon y) \leq \xi'z^*$ or $\xi'y \leq 0$. Hence, by Farkas' Lemma [Prop. 3.2.1(b)], there exist scalars $\zeta_j \geq 0$, $j \in J$, such that

$$\xi = \sum_{j\in J} \zeta_j c_j.$$

Thus, by defining $\zeta_j = 0$ for $j \notin J$, we see that for the vector ζ, we have

$$\xi = A'D'\zeta, \qquad \zeta'(DAz^* - Db) = 0.$$

Let $\mu = D'\zeta$, and note that since $\zeta \geq 0$, by Farkas' Lemma, we have $\mu \in P^*$. Furthermore, the preceding relations can be written as

$$\xi = A'\mu, \qquad \mu'(Az^* - b) = 0,$$

from which we obtain

$$\xi'z^* = \mu'Az^* = \mu'b.$$

Thus, from Eq. (3.7) and the equalities $\xi = A'\mu$ and $\xi'z^* = \mu'b$, we have $w^* + \mu'b \leq \inf_{(x,w)\in V}\{w + \mu'Ax\}$ or equivalently,

$$w^* \leq \inf_{(x,w)\in V} \{w + \mu'(Ax - b)\}.$$

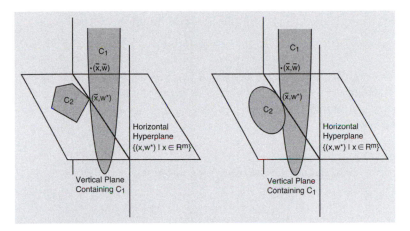

Figure 3.5.5. Illustration of the need for the polyhedral assumption on P, which implies that C_2 is polyhedral in the separation argument of the proof of Prop. 3.5.2. In the figure on the left, C_2 is polyhedral and there exists a hyperplane separating C_1 and C_2 but not containing C_1; this hyperplane can be taken to be nonvertical thanks to the relative interior condition (3). In the figure on the right, C_2 is not polyhedral and all hyperplanes separating C_1 and C_2 must contain C_1. As a result, all these hyperplanes must be vertical, the argument that $\beta \neq 0$ based on Eqs. (3.5) and (3.6) fails, and the proof of the proposition breaks down.

Since $\mu \in P^*$, we have $\mu'(Ax - b) \leq \mu'Bu$ for all (x, u) such that $u \in \Re^n$ and $Ax - b - Bu \in P$, so that

$$\inf_{(x,w)\in V} \big\{ w + \mu'(Ax - b) \big\} \leq \inf_{\substack{(x,w)\in V,\, u\in\Re^n \\ Ax-b-Bu\in P}} \big\{ w + \mu'(Ax - b) \big\}$$

$$\leq \inf_{\substack{(x,w)\in V,\, u\in\Re^n \\ Ax-b-Bu\in P}} \big\{ w + \mu'Bu \big\}$$

$$= \inf_{(u,w)\in M} \big\{ w + \mu'Bu \big\}$$

$$= q(B'\mu)$$

$$\leq q^*.$$

By combining the preceding relations, we obtain $w^* \leq q(B'\mu) \leq q^*$. On the other hand, by the weak duality relation, we have $q^* \leq w^*$, so that $q(B'\mu) = q^* = w^*$.   **Q.E.D.**

Figure 3.5.5 illustrates how the preceding proposition breaks down if P is not assumed polyhedral, even when the relative interior condition (3) is satisfied.

3.5.3   Minimax Theory Under Polyhedral Assumptions

Similar to Section 2.6, we can prove a minimax theorem by applying the preceding result to a case where the set M in the min common/max crossing framework is defined as the epigraph of the function

$$p(u) = \inf_{x \in X} \sup_{z \in Z} \{\phi(x, z) - u'z\}.$$

We thus obtain the following minimax theorem, which applies to the case where x and z are linearly coupled. We state the assumptions of the theorem with as much generality as possible, and we discuss several simpler special cases later.

Proposition 3.5.3: (Minimax Theorem III) Let $\phi : X \times Z \mapsto \Re$ be a function of the form

$$\phi(x, z) = f(x) + z'Qx - h(z),$$

where X and Z are convex subsets of $\Re^n$ and $\Re^m$, respectively, Q is an $m \times n$ matrix, $f : X \mapsto \Re$ is a convex function, and $h : Z \mapsto \Re$ is a closed convex function. Consider the function

$$h^*(\zeta) = \sup_{z \in Z} \{z'\zeta - h(z)\}, \qquad \zeta \in \Re^m$$

and assume the following:

(1) X is the intersection of a polyhedron P_1 and a convex set C_1, and f can be extended to a real-valued convex function over C_1 [i.e., there exists a convex function $\overline{f} : C_1 \mapsto \Re$ such that $f(x) = \overline{f}(x)$ for all $x \in X$].

(2) $\text{dom}(h^*)$ is the intersection of a polyhedron P_2 and a convex set C_2, and h^* can be extended to a real-valued convex function over C_2 [i.e., there exists a convex function $\overline{h}^* : C_2 \mapsto \Re$ such that $h^*(\zeta) = \overline{h}^*(\zeta)$ for all $\zeta \in \text{dom}(h^*)$].

(3) The sets $Q \cdot (X \cap \text{ri}(C_1))$ and $\text{dom}(h^*) \cap \text{ri}(C_2)$ have nonempty intersection.

Then the minimax equality holds, i.e.,

$$\sup_{z \in Z} \inf_{x \in X} \phi(x, z) = \inf_{x \in X} \sup_{z \in Z} \phi(x, z),$$

and the supremum over Z in the left-hand side above is attained by some vector $z \in Z$.

Proof: The function $p(u) = \inf_{x \in X} \sup_{z \in Z} \{\phi(x, z) - u'z\}$ is given by

$$p(u) = \inf_{x \in X} \sup_{z \in Z} \{f(x) + z'Qx - h(z) - u'z\}$$

$$= \inf_{x \in X} \left\{ f(x) + \sup_{z \in Z} \left\{ z'(Qx - u) - h(z) \right\} \right\}$$

$$= \inf_{x \in X, \ (Qx-u) \in \text{dom}(h^*)} \left\{ f(x) + h^*(Qx - u) \right\}.$$

Because f is convex, and h is convex and closed, Lemmas 2.6.1 and 2.6.2, which connect the min common/max crossing framework with minimax problems apply. Thus, if we can use the Min Common/Max Crossing Theorem III in conjunction with the set

$$M = \text{epi}(p),$$

the minimax equality will be proved, and the supremum over $z \in Z$ will be attained. We will thus show that the assumptions of the Min Common/Max Crossing Theorem III are satisfied under the assumptions of the present proposition.

Let P_1 and P_2 be represented in terms of linear inequalities as

$$P_1 = \{x \mid a_j' x - b_j \leq 0, \ j = 1, \ldots, r_1\},$$

and

$$P_2 = \{\zeta \mid c_k' \zeta - d_k \leq 0, \ k = 1, \ldots, r_2\}.$$

We have from the preceding expressions

$$p(u) = \inf_{\substack{x \in C_1, \ \zeta \in C_2 \\ A(x,\zeta) - b - Bu \leq 0}} \left\{ f(x) + h^*(\zeta) \right\},$$

where $A(x, \zeta) - b - Bu \leq 0$ is a matrix representation of the inequalities

$$a_j' x - b_j \leq 0, \ j = 1, \ldots, r_1, \qquad c_k' \zeta - d_k \leq 0, \ k = 1, \ldots, r_2,$$

$$\zeta - Qx + u = 0,$$

for suitable matrices A and B, and vector b. Consider the convex set

$$V = \left\{ (x, \zeta, w) \mid x \in C_1, \ \zeta \in C_2, \ w \in \Re, \ \overline{f}(x) + \overline{h}^*(\zeta) \leq w \right\},$$

where $\overline{f}$ and $\overline{h}^*$ are real-valued convex functions, which are the extended versions of f and h^* over C_1 and C_2, respectively [cf. conditions (1) and (2)]. Thus, the epigraph of p has the form

$$\text{epi}(p) = \left\{ (u, w) \mid \text{there is a vector } (x, \zeta, w) \text{ such that} \right.$$

$$\left. x \in C_1, \ \zeta \in C_2, \ A(x, \zeta) - b - Bu \leq 0, \ \overline{f}(x) + \overline{h}^*(\zeta) \leq w \right\},$$

so it can be written as

$$\text{epi}(p) = \big\{(u, w) \mid \text{ there is a vector } (x, \zeta, w) \in V$$
$$\text{such that } A(x, \zeta) - b - Bu \leq 0\big\}.$$

Therefore, epi(p) has the appropriate form for the application of the Min Common/Max Crossing Theorem III with the polyhedral cone P being the nonpositive orthant.

Finally, consider condition (3), which in conjunction with conditions (1) and (2), says that there exists a vector $\overline{x} \in P_1 \cap \text{ri}(C_1)$ such that $Q\overline{x} \in P_2 \cap \text{ri}(C_2)$. By letting $\overline{\zeta} = Q\overline{x}$, this holds if and only if there exist $\overline{x} \in \text{ri}(C_1)$ and $\overline{\zeta} \in \text{ri}(C_2)$ such that

$$\overline{x} \in P_1, \qquad \overline{\zeta} \in P_2, \qquad \overline{\zeta} = Q\overline{x},$$

or, in view of the definition of A and b,

$$A(\overline{x}, \overline{\zeta}) - b \leq 0.$$

Thus, given conditions (1) and (2), condition (3) is equivalent to the existence of $\overline{x} \in \text{ri}(C_1)$ and $\overline{\zeta} \in \text{ri}(C_2)$ such that $A(\overline{x}, \overline{\zeta}) - b \leq 0$.

On the other hand, since V is the epigraph of the function $\overline{f}(x) + \overline{h}^*(\zeta)$, whose domain is $C_1 \times C_2$, we have

$$\text{ri}(V) = \big\{(x, \zeta, w) \mid x \in \text{ri}(C_1), \, \zeta \in \text{ri}(C_2), \, w \in \Re, \, \overline{f}(x) + \overline{h}^*(\zeta) < w\big\}.$$

Therefore, condition (3) is equivalent to the second assumption of the Min Common/Max Crossing Theorem III, i.e., that there exists a vector of the form $(\overline{x}, \overline{\zeta}, \overline{w})$ in ri(V) such that $A(\overline{x}, \overline{\zeta}) - b \leq 0$. Thus, all the assumptions needed for application of the Min Common/Max Crossing Theorem III are satisfied, and the proof is complete. **Q.E.D.**

There are several simpler special cases of the above proposition. In particular, its assumptions (1)-(3) are satisfied if one of the following two conditions holds:

(a) There exists a vector $\overline{x} \in \text{ri}(X)$ such that $Q\overline{x} \in \text{dom}(h^*)$, and the function h^* is polyhedral. [Take $C_1 = X$, $C_2 = \Re^m$, and $P_1 = \Re^n$, $P_2 = \text{dom}(h^*)$ in Prop. 3.5.3.]

(b) There exists a vector $\overline{x} \in \text{ri}(X)$ such that $Q\overline{x} \in \text{ri}\big(\text{dom}(h^*)\big)$. [Take $C_1 = X$, $C_2 = \text{dom}(h^*)$, and $P_1 = \Re^n$, $P_2 = \Re^m$ in Prop. 3.5.3.]

Another interesting special case arises when the nonempty level sets $\big\{z \in Z \mid h(z) \leq \gamma\big\}$ have no nonzero directions of recession. Then the supremum in the definition of h^* is attained, so h^* is real-valued and $\text{dom}(h^*) = \Re^m$. In this case, condition (b) above is automatically satisfied, and the conclusions of Prop. 3.5.3 hold. The following example, which was also given as Example 2.6.1, illustrates the need for the conditions of the proposition.

Example 3.5.1:

Let x be two-dimensional, let z be scalar, and let

$$X = \big\{(x_1, x_2) \mid x \geq 0\big\}, \qquad Z = \{z \in \Re \mid z \geq 0\}.$$

Consider the function

$$\phi(x, z) = e^{-\sqrt{x_1 x_2}} + zx_1,$$

which can be written as $\phi(x, z) = f(x) + z'Qx - h(z)$, with

$$f(x) = e^{-\sqrt{x_1 x_2}}, \qquad Q = [1, 0], \qquad h(z) = 0.$$

We have

$$0 = \sup_{z \geq 0} \inf_{x \geq 0} \phi(x, z) < \inf_{x \geq 0} \sup_{z \geq 0} \phi(x, z) = 1,$$

as shown in Example 2.6.1. Here the convexity assumptions of Prop. 3.5.3 are satisfied, and h^* is given by

$$h^*(\zeta) = \sup_{z > 0} z\zeta = \begin{cases} 0 & \text{if } \zeta \leq 0, \\ \infty & \text{otherwise,} \end{cases}$$

so it is a polyhedral function. On the other hand, the conditions of Prop. 3.5.3 are not satisfied because $\text{ri}(X) = \big\{(x_1, x_2) \mid x_1 > 0, \, x_2 > 0\big\}$ and $\text{dom}(h^*) = \{\zeta \mid \zeta \leq 0\}$, so $Q\overline{x} \notin \text{dom}(h^*)$ for all $\overline{x} \in \text{ri}(X)$.

We finally note that a minimax problem involving a function ϕ of the form

$$\phi(x, z) = f(x) + z'Qx - h(z)$$

is closely related to conjugate functions and Fenchel duality, as will be discussed in Chapter 7. In fact the conditions of Prop. 3.5.3, which guarantee that the minimax equality holds, have counterparts, which guarantee that there is no duality gap in the Fenchel duality context. Within this context, we will also see that the function h^* is the conjugate convex function of h, and that the polyhedral assumption on h^* in the preceding proposition is satisfied if Z is a polyhedral set and h is a polyhedral function.

3.5.4   A Nonlinear Version of Farkas' Lemma

We will now use the min common/max crossing duality framework to prove the following nonlinear version of Farkas' Lemma [if $C = \Re^n$ and the functions f and g_j are linear in the statement below, we obtain the version of Farkas' Lemma that we have already given in Prop. 3.2.1(b)]. Note that part (a) of the lemma corresponds to the interior point condition of the Min Common/Max Crossing Theorem II of Section 2.5. Part (b) corresponds to the alternative polyhedral/relative interior condition of the Min

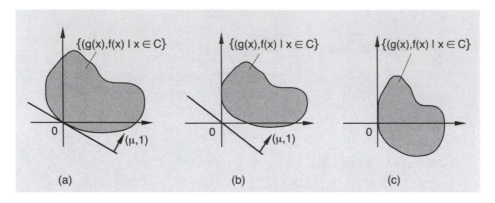

Figure 3.5.6. Geometrical interpretation of the Nonlinear Farkas' Lemma. Assuming that for all $x \in C$ with $g(x) \leq 0$, we have $f(x) \geq 0$, and other conditions, the lemma asserts the existence of a nonvertical hyperplane in $\Re^{r+1}$, with normal $(\mu, 1)$, that passes through the origin and contains the set

$$\left\{ \big(g(x), f(x)\big) \mid x \in C \right\}$$

in its positive halfspace. Figures (a) and (b) show examples where such a hyperplane exists, and figure (c) shows an example where it does not.

Common/Max Crossing Theorem III of this section. Figure 3.5.6 interprets the lemma geometrically, and suggests its strong connection with the min common/max crossing framework.

Proposition 3.5.4: (Nonlinear Farkas' Lemma) Let C be a nonempty convex subset of $\Re^n$, and let $f : C \mapsto \Re$ and $g_j : C \mapsto \Re$, $j = 1, \ldots, r$, be convex functions. Consider the set F given by

$$F = \left\{ x \in C \mid g(x) \leq 0 \right\},$$

where $g(x) = \big(g_1(x), \ldots, g_r(x)\big)$, and assume that

$$f(x) \geq 0, \qquad \forall \, x \in F. \tag{3.9}$$

Consider the subset Q^* of $\Re^r$ given by

$$Q^* = \left\{ \mu \mid \mu \geq 0, \; f(x) + \mu'g(x) \geq 0, \, \forall \, x \in C \right\}.$$

Then:

(a) Q^* is nonempty and compact if and only if there exists a vector $\overline{x} \in C$ such that

$$g_j(\overline{x}) < 0, \qquad \forall\, j = 1, \ldots, r.$$

(b) Q^* is nonempty if the functions g_j, $j = 1, \ldots, r$, are affine, and F contains a relative interior point of C.

Proof: (a) Assume that there exists a vector $\overline{x} \in C$ such that $g(\overline{x}) < 0$. We will apply the Min Common/Max Crossing Theorem II (Prop. 2.5.5) to the subset of $\Re^{r+1}$ given by

$$M = \big\{(u, w) \mid \text{there exists } x \in C \text{ such that } g(x) \leq u,\ f(x) \leq w\big\}$$

(cf. Fig. 3.5.7). To this end, we verify that the assumptions of the theorem are satisfied for the above choice of M.

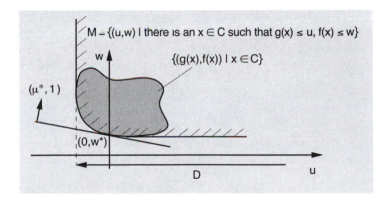

Figure 3.5.7. Illustration of the sets used in the min common/max crossing framework of the proof of Prop. 3.5.4(a). We have

$$M = \overline{M} = \big\{(u, w) \mid \text{there exists } x \in C \text{ such that } g(x) \leq u,\ f(x) \leq w\big\}$$

and

$$D = \big\{u \mid \text{there exists } w \in \Re \text{ such that } (u, w) \in \overline{M}\big\}$$
$$= \big\{u \mid \text{there exists } x \in C \text{ such that } g(x) \leq u\big\}.$$

In particular, we will show that:

(i) The optimal value w^* of the corresponding min common problem,

$$w^* = \inf\big\{w \mid (0, w) \in M\big\},$$

is such that $-\infty < w^*$.

(ii) The set

$$\overline{M} = \big\{(u, w) \mid \text{there exists } \overline{w} \text{ with } \overline{w} \le w \text{ and } (u, \overline{w}) \in M\big\},$$

is convex. (Note here that $\overline{M} = M$.)

(iii) The set

$$D = \big\{u \mid \text{there exists } w \in \Re \text{ such that } (u, w) \in \overline{M}\big\}$$

contains the origin in its interior.

To show (i), note that since $f(x) \ge 0$ for all $x \in F$, we have $w \ge 0$ for all $(0, w) \in M$, so that $w^* \ge 0$.

To show (iii), note that the set D can also be written as

$$D = \big\{u \mid \text{ there exists } x \in C \text{ such that } g(x) \le u\big\}.$$

If $g(\overline{x}) < 0$ for some $\overline{x} \in C$, then since D contains the set $g(\overline{x}) + \{u \mid u \ge 0\}$, we have $0 \in \text{int}(D)$.

There remains to show (ii), i.e., that the set $\overline{M}$ is convex. Since $\overline{M} = M$, we will prove that M is convex. To this end, we consider vectors $(u, w) \in M$ and $(\tilde{u}, \tilde{w}) \in M$, and we show that their convex combinations lie in M. By the definition of M, for some $x \in C$ and $\tilde{x} \in C$, we have

$$f(x) \le w, \qquad g_j(x) \le u_j, \quad \forall\, j = 1, \ldots, r,$$

$$f(\tilde{x}) \le \tilde{w}, \qquad g_j(\tilde{x}) \le \tilde{u}_j, \quad \forall\, j = 1, \ldots, r.$$

For any $\alpha \in [0, 1]$, we multiply these relations with α and $1-\alpha$, respectively, and add them. By using the convexity of f and g_j for all j, we obtain

$$f\big(\alpha x + (1 - \alpha)\tilde{x}\big) \le \alpha f(x) + (1 - \alpha)f(\tilde{x}) \le \alpha w + (1 - \alpha)\tilde{w},$$

$$g_j\big(\alpha x + (1-\alpha)\tilde{x}\big) \le \alpha g_j(x) + (1-\alpha)g_j(\tilde{x}) \le \alpha u_j + (1-\alpha)\tilde{u}_j, \quad \forall\, j = 1, \ldots, r.$$

By convexity of C, we have $\alpha x + (1 - \alpha)\tilde{x} \in C$ for all $\alpha \in [0, 1]$, so the preceding inequalities imply that the convex combination $\alpha(u, w) + (1 - \alpha)(\tilde{u}, \tilde{w})$ belongs to M, showing that M is convex.

Thus all the assumptions of the Min Common/Max Crossing Theorem II hold, and by the conclusions of the theorem, we have $w^* = \sup_\mu q(\mu)$, where

$$q(\mu) = \inf_{(u,w) \in M} \{w + \mu'u\}.$$

Furthermore, the optimal solution set $\tilde{Q} = \{\mu \mid q(\mu) \ge w^*\}$ is nonempty and compact. Using the definition of M, it can be seen that

$$q(\mu) = \begin{cases} \inf_{x \in C}\{f(x) + \mu'g(x)\} & \text{if } \mu \ge 0, \\ -\infty & \text{otherwise.} \end{cases}$$

From the definition of Q^*, we have

$$Q^* = \{\mu \mid \mu \geq 0, \ f(x) + \mu'g(x) \geq 0, \ \forall \ x \in C\} = \{\mu \mid q(\mu) \geq 0\},$$

so Q^* and $\tilde{Q}$ are level sets of the proper convex function $-q$, which is closed by Prop. 2.5.2. Therefore, since $\tilde{Q}$ is nonempty and compact, Q^* is compact (cf. Prop. 2.3.1). Furthermore, Q^* is nonempty since $Q^* \supset \tilde{Q}$.

Conversely, assuming that Q^* is nonempty and compact, we will show that there exists a vector $\bar{x} \in C$ such that $g(\bar{x}) < 0$. Indeed, if this were not so, then 0 would not be an interior point of the set D. Since D is convex, there exists a hyperplane that passes through 0 and contains D in its positive halfspace, i.e., there is a nonzero vector $\nu \in \Re^r$ such that $\nu'u \geq 0$ for all $u \in D$. From the definition of D, it follows that $\nu \geq 0$. Since $g(x) \in D$ for all $x \in C$, we obtain

$$\nu'g(x) \geq 0, \qquad \forall \ x \in C.$$

Thus, for any $\mu \in Q^*$, we have

$$f(x) + (\mu + \gamma\nu)'g(x) \geq 0, \qquad \forall \ x \subset C, \ \forall \ \gamma \geq 0.$$

Since we also have $\nu \geq 0$, it follows that $(\mu + \gamma\nu) \in Q^*$ for all $\gamma \geq 0$, which contradicts the boundedness of Q^*.

(b) We apply the Min Common/Max Crossing Theorem III (cf. Prop. 3.5.2), with the polyhedral cone P in that theorem being equal to the nonpositive orthant, i.e., $P = \{u \mid u \leq 0\}$, and with the matrix B being the identity matrix. The assumptions of the theorem for this choice of P and B are:

(i) The subset M of $\Re^{r+1}$ is defined in terms of a convex set V of $\Re^{n+1}$, an $r \times n$ matrix A, and a vector b in $\Re^r$ as follows:

$$M = \{(u, w) \mid \text{there is a vector } (x, w) \in V \text{ such that } Ax - b \leq u\}.$$

(ii) There exists a vector $(\bar{x}, \bar{w})$ in the relative interior of V such that $A\bar{x} - b \leq 0$.

Let the affine functions g_j have the form

$$g_j(x) = a_j'x - b_j,$$

where a_j are some vectors in $\Re^n$ and b_j are some scalars. We choose the matrix A to have as rows the vectors a_j', and the vector b to be equal to $(b_1, \ldots, b_r)'$. We also choose the convex set V to be

$$V = \{(x, w) \mid x \in C, \ f(x) \leq w\}.$$

To prove that (ii) holds, note that by our assumptions, there exists a vector $\overline{x}$ in $F \cap \mathrm{ri}(C)$, i.e., $\overline{x} \in \mathrm{ri}(C)$ and $A\overline{x} - b \leq 0$. Then, the vector $(\overline{x}, \overline{w})$ with $\overline{w} > f(\overline{x})$ belongs to $\mathrm{ri}(V)$. Hence, all the assumptions of the Min Common/Max Crossing Theorem III are satisfied, and by using this theorem, similar to the proof of part (a), we have

$$\inf_{x \in C} \big\{ f(x) + \mu^{*\prime} g(x) \big\} = q(\mu^*) = q^* = w^*$$

for some $\mu^* \geq 0$. Since $w^* \geq 0$, it follows that $f(x) + \mu^{*\prime} g(x) \geq 0$ for all $x \in C$. **Q.E.D.**

Figure 3.5.8 illustrates the Nonlinear Farkas' Lemma for the case of a one-dimensional function and a single inequality constraint. The existence of a vector $\overline{x} \in C$ such that $g(\overline{x}) < 0$ [part (a) of the lemma] is known as the *Slater condition*, and will be reencountered in Chapters 5 and 6.

3.5.5 Convex Programming

The results of this section are very useful for the analysis of convex optimization problems, as we will see in Chapter 6, where these problems will be discussed in greater detail. As a preview of that analysis, let us indicate how the Nonlinear Farkas' Lemma can be applied to the problem

$$\begin{aligned} \text{minimize} \quad & f(x) \\ \text{subject to} \quad & x \in C, \quad g_j(x) \leq 0, \ j = 1, \ldots, r, \end{aligned} \tag{3.10}$$

where C is a nonempty convex subset of $\Re^n$, and $f : C \mapsto \Re$ and $g_j : C \mapsto \Re$, $j = 1, \ldots, r$, are convex functions. Assume that the feasible set

$$F = \big\{ x \in C \mid g_j(x) \leq 0, \ j = 1, \ldots, r \big\}$$

is nonempty and that the optimal value

$$f^* = \inf_{x \in F} f(x)$$

is finite. Then, we have

$$0 \leq f(x) - f^*, \qquad \forall \, x \in F,$$

so by replacing $f(x)$ by $f(x) - f^*$ and by applying the Nonlinear Farkas' Lemma [assuming that the assumption of part (a) or part (b) of the lemma holds], it follows that there exists a vector $\mu^* \geq 0$ such that

$$f^* \leq f(x) + \mu^{*\prime} g(x), \qquad \forall \, x \in C.$$

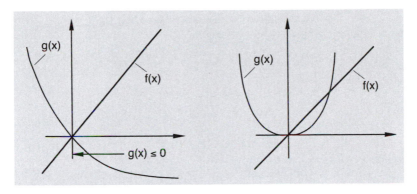

Figure 3.5.8. Two examples illustrating the Nonlinear Farkas' Lemma for the case of a one-dimensional function and a single inequality constraint. In both examples, we have

$$C = \Re, \qquad f(x) = x.$$

In the example on the left, g is given by

$$g(x) = e^{-x} - 1,$$

while in the example on the right, g is given by

$$g(x) = x^2.$$

In both examples, we have $f(x) \geq 0$ for all x such that $g(x) \leq 0$.

In the example on the left, the Slater condition [the existence of a vector $\overline{x} \in C$ such that $g(\overline{x}) < 0$; cf. part (a) of the lemma] is satisfied, and for $\mu^* = 1$, we have

$$f(x) + \mu^* g(x) = x + e^{-x} - 1 \geq 0, \qquad \forall \, x \in \Re.$$

In the example on the right, this condition, as well as the assumption of part (b), are violated, and it can be seen that for every $\mu^* \geq 0$, the function

$$f(x) + \mu^* g(x) = x + \mu^* x^2$$

takes negative values for x negative and sufficiently close to 0.

Thus, since $F \subset C$, and we have $\mu^{*\prime} g(x) \leq 0$ for all $x \in F$, the preceding equation yields

$$f^* \leq \inf_{x \in C} \big\{ f(x) + \mu^{*\prime} g(x) \big\} \leq \inf_{x \in F} f(x) = f^*.$$

Thus equality holds throughout in the above relation, and we have

$$f^* = \inf_{x \in C} \big\{ f(x) + \mu^{*\prime} g(x) \big\}. \tag{3.11}$$

Vectors $\mu^* \geq 0$ satisfying the above equation will be called *geometric multipliers* in Chapter 6. We will see there that the geometric multipliers are the optimal solutions to a dual problem associated with the original problem (3.10) (see also the discussion of Section 2.6).

An important property of geometric multipliers is that they relate to optimal solutions of the original problem (3.10). If x^* is such an optimal solution, we have

$$f(x^*) = f^* \leq f(x^*) + \mu^{*\prime} g(x^*) \leq f(x^*),$$

where the first inequality follows from Eq. (3.11), and the second inequality holds since $\mu_j^* g_j(x^*) \leq 0$ for all j. We thus obtain the necessary condition that an optimal solution x^* also minimizes $f(x) + \mu^{*\prime} g(x)$ over C, while satisfying $\mu_j^* g_j(x^*) = 0$ for all j:

$$x^* \in \arg\min_{x \in C} \{ f(x) + \mu^{*\prime} g(x) \}, \qquad \mu_j^* g_j(x^*) = 0, \quad j = 1, \dots, r.$$

Vectors $\mu^* \geq 0$ satisfying the above condition will be called *Lagrange multipliers* associated with the optimal solution x^*, and will be discussed within a more general context in Chapter 5.

3.6 NOTES, SOURCES, AND EXERCISES

The theory of polar cones is attributed to Voronoi [Vor08], and to Steinitz [Ste13], [Ste14], [Ste16]. The study of linear inequalities and polyhedral cones originated with Farkas and Minkowski near the end of the 19th century. The Minkowski-Weyl Theorem given here is due to Weyl [Wey35]. The Krein-Milman Theorem was first proved by Minkowski [Min11], but it is named after the infinite dimensional generalization by Krein and Milman [KrM40].

An extensive source on polyhedral convexity and its applications to integer programming and combinatorial optimization is Schrijver [Sch86], which also contains a detailed historical account of the early work on the subject. The books by Nemhauser and Wolsey [NeW88], Cook, Cunningham, Pulleyblank, and Schrijver [CCP98], and Wolsey [Wol98] also discuss integer programming and combinatorial optimization. The Polyhedral Proper Separation Theorem (Prop. 3.5.1) is due to Rockafellar ([Roc70], Th. 20.2).

The third min common/max crossing duality theorem (Prop. 3.5.2) was given in Bertsekas, Nedić, and Ozdaglar [BNO02]. The line of development of the corresponding minimax theorem (Prop. 3.5.3) is new. The theorem itself is closely related to Fenchel duality theory (see Chapter 7 and the references given there).

Several versions of the Nonlinear Farkas' Lemma are available in the literature, dating to Fan, Glicksberg, and Hoffman [FGH57] (who showed the result under the Slater condition), and including Berge and Ghouila-Houri [BeG62], and Rockafellar [Roc70].

E X E R C I S E S

3.1 (Cone Decomposition Theorem)

Let C be a nonempty closed convex cone in $\Re^n$ and let x be a vector in $\Re^n$. Show that:

(a) $\hat{x}$ is the projection of x on C if and only if

$$\hat{x} \in C, \qquad (x - \hat{x})'\hat{x} = 0, \qquad x - \hat{x} \in C^*.$$

(b) The following two statements are equivalent:

    (i) x_1 and x_2 are the projections of x on C and C^*, respectively.

    (ii) $x = x_1 + x_2$ with $x_1 \in C$, $x_2 \in C^*$, and $x_1' x_2 = 0$.

3.2

Let C be a nonempty closed convex cone in $\Re^n$ and let a be a vector in $\Re^n$. Show that for any positive scalars β and γ, we have

$$\max_{\|x\| \le \beta,\, x \in C} a'x \le \gamma \qquad \text{if and only if} \qquad a \in C^* + \big\{x \mid \|x\| \le \gamma/\beta\big\}.$$

(This may be viewed as an "approximate" version of the Polar Cone Theorem.)

3.3

Let C be a nonempty cone in $\Re^n$. Show that

$$L_{C^*} = \big(\mathrm{aff}(C)\big)^{\perp},$$

$$\dim(C) + \dim(L_{C^*}) = n,$$

$$\dim(C^*) + \dim\big(L_{\mathrm{conv}(C)}\big) \le \dim(C^*) + \dim\big(L_{cl(\mathrm{conv}(C))}\big) = n,$$

where L_X denotes the lineality space of a convex set X.

3.4 (Polar Cone Operations)

Show the following:

(a) For any nonempty cones $C_i \subset \Re^{n_i}$, $i = 1, \ldots, m$, we have

$$(C_1 \times \cdots \times C_m)^* = C_1^* \times \cdots \times C_m^*.$$

(b) For any collection of nonempty cones $\{C_i \mid i \in I\}$, we have

$$\left(\cup_{i \in I} C_i\right)^* = \cap_{i \in I} C_i^*.$$

(c) For any two nonempty cones C_1 and C_2, we have

$$(C_1 + C_2)^* = C_1^* \cap C_2^*.$$

(d) For any two nonempty closed convex cones C_1 and C_2, we have

$$(C_1 \cap C_2)^* = \text{cl}(C_1^* + C_2^*).$$

Furthermore, if $\text{ri}(C_1) \cap \text{ri}(C_2) \neq \emptyset$, then the cone $C_1^* + C_2^*$ is closed and the closure operation in the preceding relation can be omitted.

(e) Consider the following cones in $\Re^3$

$$C_1 = \left\{(x_1, x_2, x_3) \mid x_1^2 + x_2^2 \leq x_3^2, \ x_3 \leq 0\right\},$$

$$C_2 = \left\{(x_1, x_2, x_3) \mid x_2 = -x_3\right\}.$$

Verify that $\text{ri}(C_1) \cap \text{ri}(C_2) = \emptyset$, $(1, 1, 1) \in (C_1 \cap C_2)^*$, and $(1, 1, 1) \notin C_1^* + C_2^*$, thus showing that the closure operation in the relation of part (c) may not be omitted when $\text{ri}(C_1) \cap \text{ri}(C_2) = \emptyset$.

3.5 (Linear Transformations and Polar Cones)

Let C be a nonempty cone in $\Re^n$, K be a nonempty closed convex cone in $\Re^m$, and A be a linear transformation from $\Re^n$ to $\Re^m$. Show that

$$(AC)^* = (A')^{-1} \cdot C^*, \qquad \left(A^{-1} \cdot K\right)^* = \text{cl}(A' K^*).$$

Show also that if $\text{ri}(K) \cap R(A) \neq \emptyset$, then the cone $A' K^*$ is closed and $(A')^{-1}$ and the closure operation in the above relation can be omitted.

3.6 (Pointed Cones and Bases)

Let C be a closed convex cone in $\Re^n$. We say that C is a *pointed cone* if $C \cap (-C) = \{0\}$. A convex set $D \subset \Re^n$ is said to be a *base* for C if $C = \text{cone}(D)$ and $0 \notin \text{cl}(D)$. Show that the following properties are equivalent:

(a) C is a pointed cone.

(b) $\text{cl}(C^* - C^*) = \Re^n$.

(c) $C^* - C^* = \Re^n$.

(d) C^* has nonempty interior.

(e) There exist a nonzero vector $\hat{x} \in \Re^n$ and a positive scalar δ such that $\hat{x}'x \geq \delta\|x\|$ for all $x \in C$.

(f) C has a bounded base.

Hint: Use Exercise 3.4 to show the implications (a) $\Rightarrow$ (b) $\Rightarrow$ (c) $\Rightarrow$ (d) $\Rightarrow$ (e) $\Rightarrow$ (f) $\Rightarrow$ (a).

3.7

Show that a closed convex cone is polyhedral if and only if its polar cone is polyhedral.

3.8

Let P be a polyhedral set in $\Re^n$, with a Minkowski-Weyl Representation

$$P = \left\{ x \ \middle| \ x = \sum_{j=1}^{m} \mu_j v_j + y, \ \sum_{j=1}^{m} \mu_j = 1, \ \mu_j \geq 0, \ j = 1, \ldots, m, \ y \in C \right\},$$

where $v_1, \ldots, v_m$ are some vectors in $\Re^n$ and C is a finitely generated cone in $\Re^n$ (cf. Prop. 3.2.2). Show that:

(a) The recession cone of P is equal to C.

(b) Each extreme point of P is equal to some vector v_i that cannot be represented as a convex combination of the remaining vectors v_j, $j \neq i$.

3.9 (Polyhedral Cones and Sets under Linear Transformations)

(a) Show that the image and the inverse image of a polyhedral cone under a linear transformation are polyhedral cones.

(b) Show that the image and the inverse image of a polyhedral set under a linear transformation are polyhedral sets.

3.10

Show the following:

    (a) For polyhedral cones $C_i \subset \Re^{n_i}$, $i = 1, \ldots, m$, the Cartesian product $C_1 \times \cdots \times C_m$ is a polyhedral cone.

    (b) For polyhedral cones $C_i \subset \Re^n$, $i = 1 \ldots, m$, the intersection $\cap_{i=1}^m C_i$ and the vector sum $\sum_{i=1}^m C_i$ are polyhedral cones.

    (c) For polyhedral sets $P_i \subset \Re^{n_i}$, $i = 1, \ldots, m$, the Cartesian product $P_1 \times \cdots \times P_m$ is a polyhedral set.

    (d) For polyhedral sets $P_i \subset \Re^n$, $i = 1 \ldots, m$, the intersection $\cap_{i=1}^m P_i$ and the vector sum $\sum_{i=1}^m P_i$ are polyhedral sets.

Hint: In part (b) and in part (d), for the case of the vector sum, use Exercise 3.9.

3.11

Show that if P is a polyhedral set in $\Re^n$ containing the origin, then $\text{cone}(P)$ is a polyhedral cone. Give an example showing that if P does not contain the origin, then $\text{cone}(P)$ may not be a polyhedral cone.

3.12 (Properties of Polyhedral Functions)

Show the following:

    (a) The sum of two polyhedral functions f_1 and f_2, such that $\text{dom}(f_1) \cap \text{dom}(f_2) \neq \emptyset$, is a polyhedral function.

    (b) If A is a matrix and g is a polyhedral function such that $\text{dom}(g)$ contains a point in the range of A, the function f given by $f(x) = g(Ax)$ is polyhedral.

3.13 (Partial Minimization of Polyhedral Functions)

Let $F : \Re^{n+m} \mapsto (-\infty, \infty]$ be a polyhedral function. Show that the function f obtained by the partial minimization

$$f(x) = \inf_{z \in \Re^m} F(x, z), \qquad x \in \Re^n,$$

has a polyhedral epigraph, and is therefore polyhedral under the additional assumption $f(x) > -\infty$ for all $x \in \Re^n$. *Hint*: Use the following relation, shown at the end of Section 2.3:

$$P\big(\text{epi}(F)\big) \subset \text{epi}(f) \subset \text{cl}\Big(P\big(\text{epi}(F)\big)\Big),$$

where $P(\cdot)$ denotes projection on the space of (x, w), i.e., $P(x, z, w) = (x, w)$.

3.14 (Existence of Minima of Polyhedral Functions)

Let P be a polyhedral set in $\Re^n$, and let $f : \Re^n \mapsto (-\infty, \infty]$ be a polyhedral function such that $P \cap \mathrm{dom}(f) \neq \emptyset$. Show that the set of minima of f over P is nonempty if and only if $\inf_{x \in P} f(x)$ is finite. *Hint*: Use Prop. 3.2.3 to replace the problem of minimizing f over P with an equivalent linear program.

3.15 (Existence of Solutions of Quadratic Nonconvex Programs [FrW56])

This exercise is an extension of Prop. 2.3.4 to the case where the quadratic cost may not be convex. Let $f : \Re^n \mapsto \Re$ be a quadratic function of the form

$$f(x) = x'Qx + c'x,$$

where Q is a symmetric $n \times n$ matrix and c is a vector in $\Re^n$, and let X be a polyhedral set. Use the Minkowski-Weyl representation of X to show that the following are equivalent:

(a) f attains a minimum over X.

(b) $f^* = \inf_{x \in X} f(x) > -\infty$.

(c) For all y such that $Ay \leq 0$, we have either $y'Qy > 0$, or else $y \in N(Q)$ and $c'y \geq 0$.

3.16

Let P be a polyhedral set in $\Re^n$ of the form

$$P = \big\{ x \mid a_j'x \leq b_j,\, j = 1, \ldots, r \big\},$$

where a_j are some vectors in $\Re^n$ and b_j are some scalars. Show that P has an extreme point if and only if the set of vectors $\{a_j \mid j = 1, \ldots, r\}$ contains a subset of n linearly independent vectors.

3.17

Let C be a nonempty convex subset of $\Re^n$, and let A be an $m \times n$ matrix with linearly independent columns. Show that a vector $x \in C$ is an extreme point of C if and only if Ax is an extreme point of the image AC. Show by example that if the columns of A are linearly dependent, then Ax can be an extreme point of AC, for some non-extreme point x of C.

3.18

Show by example that the set of extreme points of a nonempty compact set need not be closed. *Hint*: Consider a line segment $C_1 = \big\{(x_1, x_2, x_3) \mid x_1 = 0, x_2 = 0, -1 \le x_3 \le 1\big\}$ and a circular disk $C_2 = \big\{(x_1, x_2, x_3) \mid (x_1 - 1)^2 + x_2^2 \le 1, \ x_3 = 0\big\}$, and verify that the set $\operatorname{conv}(C_1 \cup C_2)$ is compact, while its set of extreme points is not closed.

3.19

Show that a nonempty compact convex set is polyhedral if and only if it has a finite number of extreme points. Give an example showing that the assertion fails if compactness of the set is replaced by the weaker assumption that the set is closed and does not contain a line.

3.20 (Faces)

Let P be a polyhedral set. For any hyperplane H that passes through a boundary point of P and contains P in one of its halfspaces, we say that the set $F = P \cap H$ is a *face* of P. Show the following:

    (a) Each face is a polyhedral set.

    (b) Each extreme point of P, viewed as a singleton set, is a face.

    (c) If P is not an affine set, there is a face of P whose dimension is $\dim(P) - 1$.

    (d) The number of distinct faces of P is finite.

3.21 (Isomorphic Polyhedral Sets)

Let P and Q be polyhedral sets in $\Re^n$ and $\Re^m$, respectively. We say that P and Q are *isomorphic* if there exist affine functions $f : P \mapsto Q$ and $g : Q \mapsto P$ such that

$$x = g\big(f(x)\big), \quad \forall\, x \in P, \qquad y = f\big(g(y)\big), \quad \forall\, y \in Q.$$

    (a) Show that if P and Q are isomorphic, then their extreme points are in one-to-one correspondence.

    (b) Let A be an $r \times n$ matrix and b be a vector in $\Re^r$, and let

$$P = \{x \in \Re^n \mid Ax \le b, \ x \ge 0\},$$

$$Q = \big\{(x, z) \in \Re^{n+r} \mid Ax + z = b, \ x \ge 0, \ z \ge 0\big\}.$$

    Show that P and Q are isomorphic.

3.22 (Unimodularity I)

Let A be an $n \times n$ invertible matrix with integer entries. Show that A is unimodular if and only if the solution of the system $Ax = b$ has integer components for every vector $b \in \Re^n$ with integer components. *Hint*: To prove that A is unimodular when the given property holds, use the system $Ax = u_i$, where u_i is the ith unit vector, to show that A^{-1} has integer components, and then use the equality $\det(A) \cdot \det(A^{-1}) = 1$. To prove the converse, use Cramer's rule.

3.23 (Unimodularity II)

Let A be an $m \times n$ matrix.

(a) Show that A is totally unimodular if and only if its transpose A' is totally unimodular.

(b) Show that A is totally unimodular if and only if every subset J of $\{1, \ldots, n\}$ can be partitioned into two subsets J_1 and J_2 such that

$$\left| \sum_{j \in J_1} a_{ij} - \sum_{j \subset J_2} a_{ij} \right| \leq 1, \qquad \forall\, i = 1, \ldots, m.$$

3.24 (Unimodularity III)

Show that a matrix A is totally unimodular if one of the following holds:

(a) The entries of A are -1, 0, or 1, and there are exactly one 1 and exactly one -1 in each of its columns.

(b) The entries of A are 0 or 1, and in each of its columns, the entries that are equal to 1 appear consecutively.

3.25 (Unimodularity IV)

Let A be a matrix with entries -1, 0, or 1, and exactly two nonzero entries in each of its columns. Show that A is totally unimodular if and only if the rows of A can be divided into two subsets such that for each column the following hold: if the two nonzero entries in the column have the same sign, their rows are in different subsets, and if they have the opposite sign, their rows are in the same subset.

3.26 (Gordan's Theorem of the Alternative [Gor73])

Let $a_1, \ldots, a_r$ be vectors in $\Re^n$.

(a) Show that exactly one of the following two conditions holds:

(i) There exists a vector $x \in \Re^n$ such that

$$a_1' x < 0, \ldots, a_r' x < 0.$$

(ii) There exists a vector $\mu \in \Re^r$ such that $\mu \neq 0$, $\mu \geq 0$, and

$$\mu_1 a_1 + \cdots + \mu_r a_r = 0.$$

(b) Show that an equivalent statement of part (a) is the following: a polyhedral cone has nonempty interior if and only if its polar cone does not contain a line, i.e., a set of the form $\{x + \alpha z \mid \alpha \in \Re\}$, where x lies in the polar cone and z is a nonzero vector. (*Note*: This statement is a special case of Exercise 3.3.)

3.27 (Linear System Alternatives)

Let $a_1, \ldots, a_r$ be vectors in $\Re^n$ and let $b_1, \ldots b_r$ be scalars. Show that exactly one of the following two conditions holds:

(i) There exists a vector $x \in \Re^n$ such that

$$a_1' x \leq b_1, \ldots, a_r' x \leq b_r.$$

(ii) There exists a vector $\mu \in \Re^r$ such that $\mu \geq 0$ and

$$\mu_1 a_1 + \cdots + \mu_r a_r = 0, \qquad \mu_1 b_1 + \cdots + \mu_r b_r < 0.$$

3.28 (Convex System Alternatives [FGH57])

Let $f_i : C \mapsto (-\infty, \infty]$, $i = 1, \ldots, r$, be convex functions, where C is a nonempty convex subset of $\Re^n$ such that $\mathrm{ri}(C) \subset \mathrm{dom}(f_i)$ for all i. Show that exactly one of the following two conditions holds:

(i) There exists a vector $x \in C$ such that

$$f_1(x) < 0, \ldots, f_r(x) < 0.$$

(ii) There exists a vector $\mu \in \Re^r$ such that $\mu \neq 0$, $\mu \geq 0$, and

$$\mu_1 f_1(x) + \cdots + \mu_r f_r(x) \geq 0, \qquad \forall \, x \in C.$$

3.29 (Convex-Affine System Alternatives)

Let $f_i : C \mapsto (-\infty, \infty]$, $i = 1, \ldots, \bar{r}$, be convex functions, where C is a convex set in $\Re^n$ such that $\operatorname{ri}(C) \subset \operatorname{dom}(f_i)$ for all $i = 1, \ldots, \bar{r}$. Let $f_i : C \mapsto \Re$, $i = \bar{r} + 1, \ldots, r$, be affine functions such that the system

$$f_{\bar{r}+1}(x) \leq 0, \ldots, f_r(x) \leq 0$$

has a solution $\bar{x} \in \operatorname{ri}(C)$. Show that exactly one of the following two conditions holds:

(i) There exists a vector $x \in C$ such that

$$f_1(x) < 0, \ldots, f_{\bar{r}}(x) < 0, \qquad f_{\bar{r}+1}(x) \leq 0, \ldots, f_r(x) \leq 0.$$

(ii) There exists a vector $\mu \in \Re^r$ such that not all $\mu_1, \ldots, \mu_{\bar{r}}$ are zero, $\mu \geq 0$, and

$$\mu_1 f_1(x) + \cdots + \mu_r f_r(x) \geq 0, \qquad \forall \, x \in C.$$

3.30 (Elementary Vectors [Roc69])

Given a vector $z = (z_1, \ldots, z_n)$ in $\Re^n$, the *support* of z is the set of indices $\{j \mid z_j \neq 0\}$. We say that a nonzero vector z of a subspace S of $\Re^n$ is *elementary* if there is no vector $\bar{z} \neq 0$ in S that has smaller support than z, i.e., for all nonzero $\bar{z} \in S$, $\{j \mid \bar{z}_j \neq 0\}$ is not a strict subset of $\{j \mid z_j \neq 0\}$. Show that:

(a) Two elementary vectors with the same support are scalar multiples of each other.

(b) For every nonzero vector y, there exists an elementary vector with support contained in the support of y.

(c) (*Conformal Realization Theorem*) We say that a vector x is in *harmony* with a vector z if

$$x_j z_j \geq 0, \qquad \forall \, j = 1, \ldots, n.$$

Show that every nonzero vector x of a subspace S can be written in the form

$$x = z^1 + \ldots + z^m,$$

where $z^1, \ldots, z^m$ are elementary vectors of S, and each of them is in harmony with x and has support contained in the support of x. *Note:* Among other subjects, this result finds significant application in network optimization algorithms (see Rockafellar [Roc69] and Bertsekas [Ber98]).

3.31 (Combinatorial Separation Theorem [Cam68], [Roc69])

Let S be a subspace of $\Re^n$. Consider a set B that is a Cartesian product of n nonempty intervals, and is such that $B \cap S^\perp = \varnothing$ (by an interval, we mean a convex set of scalars, which may be open, closed, or neither open nor closed.) Show that there exists an elementary vector z of S (cf. Exercise 3.30) such that

$$t'z < 0, \qquad \forall\, t \in B,$$

i.e., a hyperplane that separates B and $S^\perp$, and does not contain any point of B. *Note*: There are two points here: (1) The set B need not be closed, as required for application of the Strict Separation Theorem (cf. Prop. 2.4.3), and (2) the hyperplane normal can be one of the elementary vectors of S (not just any vector of S). For application of this result in duality theory for network optimization and monotropic programming, see Rockafellar [Roc84] and Bertsekas [Ber98].

3.32 (Tucker's Complementarity Theorem)

(a) Let S be a subspace of $\Re^n$. Show that there exist disjoint index sets I and $\bar{I}$ with $I \cup \bar{I} = \{1, \ldots, n\}$, and vectors $x \in S$ and $y \in S^\perp$ such that

$$x_i > 0, \quad \forall\, i \in I, \qquad x_i = 0, \quad \forall\, i \in \bar{I},$$

$$y_i = 0, \quad \forall\, i \in I, \qquad y_i > 0, \quad \forall\, i \in \bar{I}.$$

Furthermore, the index sets I and $\bar{I}$ with this property are unique. In addition, we have

$$x_i = 0, \qquad \forall\, i \in \bar{I}, \quad \forall\, x \in S \text{ with } x \geq 0,$$

$$y_i = 0, \qquad \forall\, i \in I, \quad \forall\, y \in S^\perp \text{ with } y \geq 0.$$

Hint: Use a hyperplane separation argument based on Exercise 3.31.

(b) Let A be an $m \times n$ matrix and let b be a vector in $\Re^n$. Assume that the set $F = \{x \mid Ax = b,\ x \geq 0\}$ is nonempty. Apply part (a) to the subspace

$$S = \big\{ (x, w) \mid Ax - bw = 0,\ x \in \Re^n,\ w \in \Re \big\},$$

and show that there exist disjoint index sets I and $\bar{I}$ with $I \cup \bar{I} = \{1, \ldots, n\}$, and vectors $x \in F$ and $z \in \Re^m$ such that $b'z = 0$ and

$$x_i > 0, \quad \forall\, i \in I, \qquad x_i = 0, \quad \forall\, i \in \bar{I},$$

$$y_i = 0, \quad \forall\, i \in I, \qquad y_i > 0, \quad \forall\, i \in \bar{I},$$

where $y = A'z$. *Note*: A special choice of A and b yields an important result, which relates optimal primal and dual solutions in linear programming: the Goldman-Tucker Complementarity Theorem (see the exercises of Chapter 6).

4

Subgradients and Constrained Optimization

Contents

4.1. Directional Derivatives . . . . p. 222
4.2. Subgradients and Subdifferentials . . . . p. 227
4.3. ϵ-Subgradients . . . . p. 235
4.4. Subgradients of Extended Real-Valued Functions . . . p. 241
4.5. Directional Derivative of the Max Function . . . . . . p. 245
4.6. Conical Approximations . . . . p. 248
4.7. Optimality Conditions . . . . p. 255
4.8. Notes, Sources, and Exercises . . . . p. 261

Much of optimization theory revolves around comparing the value of the cost function at a given point with its values at neighboring points. This motivates an analysis that uses derivatives, but such an analysis breaks down when the cost function is nondifferentiable. Fortunately, however, it turns out that in the case of a convex cost function there is a convenient substitute, the notion of directional differentiability and the related notion of a subgradient.

In this chapter, we develop the theory of subgradients in Sections 4.1-4.5, and we discuss its application to optimality conditions for constrained optimization in Section 4.7. We discuss both convex and nonconvex problems, and in order to treat the latter, we need the machinery of conical approximations to constraint sets, which we develop in Section 4.6.

4.1 DIRECTIONAL DERIVATIVES

Convex sets and functions can be characterized in many ways by their behavior along lines. For example, a set is convex if and only if its intersection with any line is convex, a convex set is bounded if and only if its intersection with every line is bounded, a function is convex if and only if it is convex along any line, and a convex function is coercive if and only if it is coercive along any line. Similarly, it turns out that the differentiability properties of a convex function are determined by the corresponding properties along lines. With this in mind, we first consider convex functions of a single variable.

One-Dimensional Convex Functions

Let $f : I \mapsto \Re$ be a convex function, where I is an interval, i.e., a convex set of scalars, which may be open, closed, or neither open nor closed. The infimum and the supremum of I are referred to as its *left and right end points*, respectively. Note that since we allow the possibility that I is unbounded above or below (or both), the end points may be equal to $-\infty$ or ∞. The convexity of f implies the following important inequality

$$\frac{f(y) - f(x)}{y - x} \le \frac{f(z) - f(x)}{z - x} \le \frac{f(z) - f(y)}{z - y}, \qquad (4.1)$$

which holds for all $x, y, z \in I$ such that $x < y < z$, and is illustrated in Fig. 4.1.1 [see also Exercise 1.6(a)]. To see this, note that, using the definition of a convex function [cf. Eq. (1.3)], we obtain

$$f(y) \le \left(\frac{y - x}{z - x} \right) f(z) + \left(\frac{z - y}{z - x} \right) f(x)$$

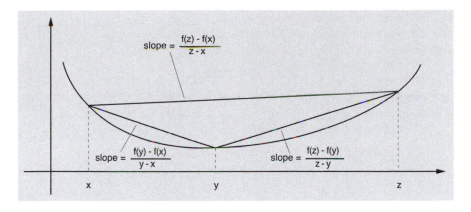

Figure 4.1.1. Illustration of the inequalities (4.1). The rate of change of the function f is nondecreasing in its argument.

and either of the desired inequalities follows by rearranging terms.

For any $x \in I$ that is not equal to the right end point, and for any $\alpha > 0$ such that $x + \alpha \in I$, we define

$$s^+(x, \alpha) = \frac{f(x + \alpha) - f(x)}{\alpha}.$$

Let $0 < \alpha \le \alpha'$. We use the first inequality in Eq. (4.1) with $y = x + \alpha$ and $z = x + \alpha'$ to obtain $s^+(x, \alpha) \le s^+(x, \alpha')$. Therefore, $s^+(x, \alpha)$ is a nondecreasing function of α, and as α decreases to zero, $s^+(x, \alpha)$ either converges to some real number or decreases to $-\infty$. Based on this fact, we define the *right derivative* of f at x to be

$$f^+(x) = \lim_{\alpha \downarrow 0} \frac{f(x + \alpha) - f(x)}{\alpha} = \inf_{\alpha > 0} \frac{f(x + \alpha) - f(x)}{\alpha}. \qquad (4.2)$$

Similarly, for any $x \in I$ that is not equal to the left end point, and for any $\alpha > 0$ such that $x - \alpha \in I$, we define

$$s^-(x, \alpha) = \frac{f(x) - f(x - \alpha)}{\alpha}.$$

By a symmetrical argument, it can be seen that $s^-(x, \alpha)$ is a nonincreasing function of α, and as α decreases to zero, $s^-(x, \alpha)$ either converges to some real number or increases to ∞. We define the *left derivative* of f at x to be

$$f^-(x) = \lim_{\alpha \downarrow 0} \frac{f(x) - f(x - \alpha)}{\alpha} = \sup_{\alpha > 0} \frac{f(x) - f(x - \alpha)}{\alpha}.$$

In the case where the end points of the domain I of f belong to I, we define for completeness $f^-(a) = -\infty$ and $f^+(b) = \infty$, where a denotes the left endpoint and b denotes the right endpoint. The basic facts about the differentiability properties of one-dimensional convex functions can be easily visualized, and are given in the following proposition.

Proposition 4.1.1: Let I be an interval of real numbers, whose left and right end points are denoted by a and b, respectively, and let $f : I \mapsto \Re$ be a convex function.

(a) We have $f^-(x) \le f^+(x)$ for every $x \in I$.

(b) If x belongs to the interior of I, then $f^+(x)$ and $f^-(x)$ are finite.

(c) If $x, z \in I$ and $x < z$, then $f^+(x) \le f^-(z)$.

(d) The functions $f^-, f^+ : I \mapsto [-\infty, +\infty]$ are nondecreasing.

(e) The function f^+ (respectively, f^-) is right– (respectively, left–) continuous at every interior point of I. Furthermore, if $a \in I$ (respectively, $b \in I$) and f is continuous at a (respectively, b), then f^+ (respectively, f^-) is right– (respectively, left–) continuous at a (respectively, b).

(f) The function $f^+ : I \mapsto (-\infty, \infty]$ [respectively, $f^- : I \mapsto [-\infty, \infty)$] is upper (respectively, lower) semicontinuous at every $x \in I$.

Proof: (a) If x is an end point of I, the result follows, since if $x = a$ then $f^-(x) = -\infty$, and if $x = b$ then $f^+(x) = \infty$. Assume that x is an interior point of I. Then we let $\alpha > 0$, and we use Eq. (4.1), with x, y, z replaced by $x - \alpha, x, x + \alpha$, respectively, to obtain $s^-(x, \alpha) \le s^+(x, \alpha)$. Taking the limit as α decreases to zero, we obtain $f^-(x) \le f^+(x)$.

(b) Let x belong to the interior of I and let $\alpha > 0$ be such that $x - \alpha \in I$. Then $f^-(x) \ge s^-(x, \alpha) > -\infty$. Similarly, we obtain $f^+(x) < \infty$. Part (a) then implies that $f^+(x)$ and $f^-(x)$ are finite.

(c) We use Eq. (4.1), with $y = (z + x)/2$, to obtain $s^+\big(x, (z - x)/2\big) \le s^-\big(z, (z - x)/2\big)$. The result then follows because $f^+(x) \le s^+\big(x, (z-x)/2\big)$ and $s^-\big(z, (z - x)/2\big) \le f^-(z)$.

(d) This follows by combining parts (a) and (c).

(e) Fix some $x \in I$, $x \ne b$, and some positive δ and α such that $x + \delta + \alpha < b$. We allow x to be equal to a, in which case f is assumed to be continuous at a. We have $f^+(x + \delta) \le s^+(x + \delta, \alpha)$. We take the limit, as δ decreases to zero, to obtain $\lim_{\delta \downarrow 0} f^+(x + \delta) \le s^+(x, \alpha)$. We have used here the fact that $s^+(x, \alpha)$ is a continuous function of x, which is a consequence of the continuity of f (Prop. 1.4.6). We now let α decrease to zero to obtain $\lim_{\delta \downarrow 0} f^+(x + \delta) \le f^+(x)$. The reverse inequality also holds because f^+ is nondecreasing by part (d) and this proves the right–continuity of f^+. The proof for f^- is similar.

(f) This follows from parts (a), (d), (e), and the defining property of semi-continuity (Definition 1.1.5).   **Q.E.D.**

Multi-Dimensional Convex Functions

We will now discuss notions of directional differentiability of multidimensional real-valued functions (the extended real-valued case will be discussed later). Recall from Section 1.1.4 that the directional derivative of a function $f : \Re^n \mapsto \Re$ at a point $x \in \Re^n$ in the direction $y \in \Re^n$ is given by

$$f'(x; y) = \lim_{\alpha \downarrow 0} \frac{f(x + \alpha y) - f(x)}{\alpha}, \tag{4.3}$$

provided that the limit exists, in which case we say that f is directionally differentiable at x in the direction y, and we call $f'(x; y)$ the *directional derivative of f at x in the direction y*. We say that f is *directionally differentiable at x* if it is directionally differentiable at x in all directions. Recall also from Section 1.1.4 that f is differentiable at x if it is directionally differentiable at x and $f'(x; y)$ is linear, as a function of y, of the form

$$f'(x; y) = \nabla f(x)'y,$$

where $\nabla f(x)$ is the gradient of f at x.

An important property is that if f is convex, it is directionally differentiable at all points. This is a consequence of the directional differentiability of scalar convex functions, as can be seen from the relation

$$f'(x; y) = \lim_{\alpha \downarrow 0} \frac{f(x + \alpha y) - f(x)}{\alpha} = \lim_{\alpha \downarrow 0} \frac{F_y(\alpha) - F_y(0)}{\alpha} = F_y^+(0),$$

where $F_y^+(0)$ is the right derivative of the convex scalar function

$$F_y(\alpha) = f(x + \alpha y)$$

at $\alpha = 0$. Note that the above calculation also shows that the left derivative $F_y^-(0)$ of F_y is equal to $-f'(x; -y)$ and, by using Prop. 4.1.1(a), we obtain $F_y^-(0) \le F_y^+(0)$, or equivalently,

$$-f'(x; -y) \le f'(x; y), \qquad \forall \, y \in \Re^n. \tag{4.4}$$

Note also that for a convex function, an equivalent definition of the directional derivative is

$$f'(x; y) = \inf_{\alpha > 0} \frac{f(x + \alpha y) - f(x)}{\alpha}; \tag{4.5}$$

[cf. Eq. (4.2)]. The exercises elaborate further on the properties of directional derivatives for real-valued convex functions, and their extension to extended real-valued functions.

The following proposition generalizes the upper semicontinuity property of right derivatives of scalar convex functions [Prop. 4.1.1(f)], and shows that if f is differentiable, then its gradient is continuous over $\Re^n$.

Proposition 4.1.2: Let $f : \Re^n \mapsto \Re$ be a convex function, and let $\{f_k\}$ be a sequence of convex functions $f_k : \Re^n \mapsto \Re$ with the property that $\lim_{k\to\infty} f_k(x_k) = f(x)$ for every $x \in \Re^n$ and every sequence $\{x_k\}$ that converges to x. Then, for any $x \in \Re^n$ and $y \in \Re^n$, and any sequences $\{x_k\}$ and $\{y_k\}$ converging to x and y, respectively, we have

$$\limsup_{k\to\infty} f_k'(x_k; y_k) \le f'(x; y).$$

Furthermore, if f is differentiable over $\Re^n$, then it is continuously differentiable over $\Re^n$.

Proof: From the definition of directional derivative [cf. Eq. (4.3)], it follows that for any $\epsilon > 0$, there exists an $\alpha > 0$ such that

$$\frac{f(x + \alpha y) - f(x)}{\alpha} < f'(x; y) + \epsilon.$$

Hence, for all sufficiently large k, we have, using also Eq. (4.5),

$$f_k'(x_k; y_k) \le \frac{f_k(x_k + \alpha y_k) - f_k(x_k)}{\alpha} < f'(x; y) + \epsilon,$$

so by taking the limit as $k \to \infty$,

$$\limsup_{k\to\infty} f_k'(x_k; y_k) \le f'(x; y) + \epsilon.$$

Since this is true for all $\epsilon > 0$, we obtain $\limsup_{k\to\infty} f_k'(x_k; y_k) \le f'(x; y)$.

If f is differentiable at all $x \in \Re^n$, then using the continuity of f and the part of the proposition just proved, we have for every sequence $\{x_k\}$ converging to x and every $y \in \Re^n$,

$$\limsup_{k\to\infty} \nabla f(x_k)'y = \limsup_{k\to\infty} f'(x_k; y) \le f'(x; y) = \nabla f(x)'y.$$

By replacing y with $-y$ in the preceding argument, we obtain

$$-\liminf_{k\to\infty} \nabla f(x_k)'y = \limsup_{k\to\infty} \big(-\nabla f(x_k)'y\big) \le -\nabla f(x)'y.$$

Therefore, we have $\nabla f(x_k)'y \to \nabla f(x)'y$ for every y, which implies that $\nabla f(x_k) \to \nabla f(x)$. Hence, $\nabla f(\cdot)$ is continuous. **Q.E.D.**

As a special case, when $f_k = f$ for all k, Prop. 4.1.2 shows that $f'(x; y)$ is upper semicontinuous as a function of (x, y).

4.2 SUBGRADIENTS AND SUBDIFFERENTIALS

Let $f : \Re^n \mapsto \Re$ be a convex function. We say that a vector $d \in \Re^n$ is a *subgradient* of f at a point $x \in \Re^n$ if

$$f(z) \geq f(x) + (z - x)'d, \qquad \forall\, z \in \Re^n. \tag{4.6}$$

If instead f is a concave function, we say that d is a subgradient of f at x if $-d$ is a subgradient of the convex function $-f$ at x. The set of all subgradients of a convex (or concave) function f at $x \in \Re^n$ is called the *subdifferential of f at x*, and is denoted by $\partial f(x)$.

A subgradient admits an intuitive geometrical interpretation: it can be identified with a nonvertical supporting hyperplane to the epigraph of f at $\big(x, f(x)\big)$, as illustrated in Fig. 4.2.1. Such a hyperplane provides a linear approximation to the function f, which is an underestimate of f if f is convex and an overestimate of f if f is concave. Figure 4.2.2 provides some examples of subdifferentials.

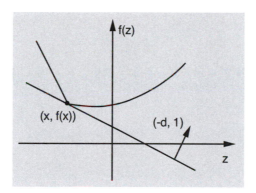

Figure 4.2.1. Illustration of a subgradient of a convex function f. The defining relation (4.6) can be written as

$$f(z) - z'd \geq f(x) - x'd, \qquad \forall\, z \in \Re^n.$$

Equivalently, d is a subgradient of f at x if and only if the hyperplane in $\Re^{n+1}$ that has normal $(-d, 1)$ and passes through $\big(x, f(x)\big)$ supports the epigraph of f, as shown in the figure.

The question of existence of a nonvertical hyperplane supporting the epigraph of f can be addressed within the min common/max crossing framework of Section 2.5, with a special choice of the set M in that framework. This is the basis for the proof of the following proposition.

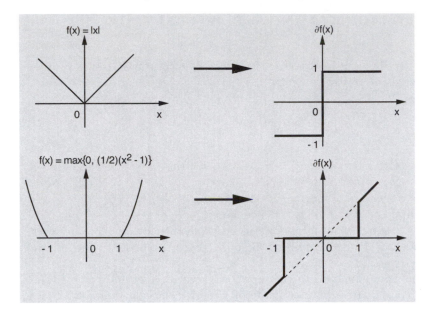

Figure 4.2.2. The subdifferential of some scalar convex functions as a function of the argument x.

Proposition 4.2.1: Let $f : \Re^n \mapsto \Re$ be a convex function. The subdifferential $\partial f(x)$ is nonempty, convex, and compact for all $x \in \Re^n$.

Proof: Let the set M in the min common/max crossing framework of Section 2.5 be

$$M = \big\{(u, w) \mid u \in \Re^n,\; f(x + u) \le w\big\}. \tag{4.7}$$

Then the min common value is

$$w^* = \inf_{(0,w)\in M} w = f(x),$$

and the max crossing value is

$$q^* = \sup_{\mu} q(\mu),$$

where

$$q(\mu) = \inf_{(u,w)\in M} \{w + \mu'u\}.$$

Thus, for a given μ, we have $w^* = q^* = q(\mu)$ if and only if

$$f(x) = \inf_{(u,w)\in M} \{w + \mu'u\},$$

or equivalently, if and only if

$$f(x) \leq f(x+u) + \mu'u, \qquad \forall \, u \in \Re^n.$$

It follows that the set of optimal solutions of the max crossing problem is precisely $-\partial f(x)$. By using the Min Common/Max Crossing Theorem II (Prop. 2.5.5), it follows that if the set

$$D = \big\{ u \mid \text{there exists } w \in \Re \text{ with } (u,w) \in M \big\}$$

contains the origin in its interior, then the set of optimal solutions of the max crossing problem is nonempty, convex, and compact. Since f is real-valued over $\Re^n$, D is equal to $\Re^n$, so it contains the origin in its interior, and the result follows.   **Q.E.D.**

The directional derivative and the subdifferential of a convex function are closely linked. To see this, note that the subgradient inequality (4.6) is equivalent to

$$\frac{f(x+\alpha y) - f(x)}{\alpha} \geq y'd, \qquad \forall \, y \in \Re^n, \, \forall \, \alpha > 0.$$

Since the quotient on the left above decreases monotonically to $f'(x;y)$ as $\alpha \downarrow 0$, we conclude that the subgradient inequality (4.6) is equivalent to $f'(x;y) \geq y'd$ for all $y \in \Re^n$. Therefore, we obtain

$$d \in \partial f(x) \qquad \Longleftrightarrow \qquad f'(x;y) \geq y'd, \qquad \forall \, y \in \Re^n, \qquad (4.8)$$

and it follows that

$$f'(x;y) \geq \max_{d \in \partial f(x)} y'd.$$

The following proposition shows that in fact equality holds in the above relation.

Proposition 4.2.2: Let $f : \Re^n \mapsto \Re$ be a convex function. For every $x \in \Re^n$, we have

$$f'(x;y) = \max_{d \in \partial f(x)} y'd, \qquad \forall \, y \in \Re^n. \qquad (4.9)$$

In particular, f is differentiable at x with gradient $\nabla f(x)$ if and only if it has $\nabla f(x)$ as its unique subgradient at x.

Proof: We have already shown that $f'(x;y) \geq \max_{d \in \partial f(x)} y'd$ for all $y \in \Re^n$ [cf. Eq. (4.8)]. To show the reverse inequality, we take any x and y in $\Re^n$, and we consider the subset of $\Re^{n+1}$

$$C_1 = \big\{ (z,w) \mid f(z) < w \big\},$$

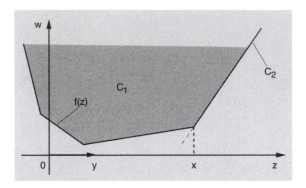

Figure 4.2.3. Illustration of the sets C_1 and C_2 used in the hyperplane separation argument of the proof of Prop. 4.2.2.

and the half-line

$$C_2 = \big\{(z, w) \mid z = x + \alpha y, \ w = f(x) + \alpha f'(x; y), \ \alpha \geq 0\big\};$$

see Fig. 4.2.3. Using the definition of directional derivative and the convexity of f, it follows that these two sets are nonempty, convex, and disjoint. Thus we can use the Separating Hyperplane Theorem (Prop. 2.4.2), to assert the existence of a nonzero vector $(\mu, \gamma) \in \Re^{n+1}$ such that

$$\gamma w + \mu' z \geq \gamma\big(f(x) + \alpha f'(x; y)\big) + \mu'(x + \alpha y), \quad \forall \, \alpha \geq 0, \ z \in \Re^n, \ w > f(z). \tag{4.10}$$

We cannot have $\gamma < 0$ since then the left-hand side above could be made arbitrarily small by choosing w sufficiently large. Also if $\gamma = 0$, then Eq. (4.10) implies that $\mu = 0$, which is a contradiction. Therefore, $\gamma > 0$ and by dividing with γ in Eq. (4.10) if necessary, we may assume that $\gamma = 1$, i.e.,

$$w + (z - x)'\mu \geq f(x) + \alpha f'(x; y) + \alpha y'\mu, \quad \forall \, \alpha \geq 0, \ z \in \Re^n, \ w > f(z). \tag{4.11}$$

By setting $\alpha = 0$ in the above relation and by taking the limit as $w \downarrow f(z)$, we obtain

$$f(z) \geq f(x) - (z - x)'\mu, \quad \forall \, z \in \Re^n,$$

implying that $-\mu \in \partial f(x)$. By setting $z = x$ and $\alpha = 1$ in Eq. (4.11), and by taking the limit as $w \downarrow f(x)$, we obtain $-y'\mu \geq f'(x; y)$, which implies that

$$\max_{d \in \partial f(x)} y'd \geq f'(x; y),$$

and completes the proof of Eq. (4.9).

From the definition of directional derivative, we see that f is differentiable at x if and only if the directional derivative $f'(x; y)$ is a linear function of the form $f'(x; y) = \nabla f(x)'y$. Thus, from Eq. (4.9), f is differentiable at x if and only if it has $\nabla f(x)$ as its unique subgradient at x. **Q.E.D.**

The following proposition provides some important boundedness and continuity properties of the subdifferential.

Proposition 4.2.3: Let $f : \Re^n \mapsto \Re$ be a convex function.

(a) If X is a bounded set, then the set $\cup_{x \in X} \partial f(x)$ is bounded.

(b) If a sequence $\{x_k\}$ converges to a vector $x \in \Re^n$ and $d_k \in \partial f(x_k)$ for all k, then the sequence $\{d_k\}$ is bounded and each of its limit points is a subgradient of f at x.

Proof: (a) Assume the contrary, i.e., that there exists a sequence $\{x_k\} \subset X$, and an unbounded sequence $\{d_k\}$ with $d_k \in \partial f(x_k)$ for all k. Without loss of generality, we assume that $d_k \neq 0$ for all k, and we denote $y_k = d_k/\|d_k\|$. Since both $\{x_k\}$ and $\{y_k\}$ are bounded, they must contain convergent subsequences. We assume without loss of generality that $\{x_k\}$ converges to some x and $\{y_k\}$ converges to some y. Since $d_k \in \partial f(x_k)$, we have

$$f(x_k + y_k) - f(x_k) \geq d_k' y_k = \|d_k\|.$$

Since $\{x_k\}$ and $\{y_k\}$ converge, by the continuity of f (cf. Prop. 1.4.6), the left-hand side above is bounded. This implies that the right-hand side is bounded, thereby contradicting the unboundedness of $\{d_k\}$.

(b) By Prop. 4.2.2, we have

$$y'd_k \leq f'(x_k; y), \qquad \forall \, y \in \Re^n.$$

By part (a), the sequence $\{d_k\}$ is bounded, so let d be a limit point of $\{d_k\}$. By taking limit along the relevant subsequence in the above relation and by using Prop. 4.1.2, it follows that

$$y'd \leq \limsup_{k \to \infty} f'(x_k; y) \leq f'(x; y), \qquad \forall \, y \in \Re^n.$$

Therefore, by Eq. (4.8), we have $d \in \partial f(x)$.   **Q.E.D.**

The subdifferential of the sum of convex functions is obtained as the vector sum of the corresponding subdifferentials, as shown in the following proposition.

Proposition 4.2.4: Let $f_j : \Re^n \mapsto \Re$, $j = 1, \ldots, m$, be convex functions and let $f = f_1 + \cdots + f_m$. Then

$$\partial f(x) = \partial f_1(x) + \cdots + \partial f_m(x).$$

Proof: It will suffice to prove the result for the case where $f = f_1 + f_2$. If $d_1 \in \partial f_1(x)$ and $d_2 \in \partial f_2(x)$, then from the subgradient inequality (4.6), we have

$$f_1(z) \geq f_1(x) + (z - x)'d_1, \qquad \forall\, z \in \Re^n,$$

$$f_2(z) \geq f_2(x) + (z - x)'d_2, \qquad \forall\, z \in \Re^n,$$

so by adding, we obtain

$$f(z) \geq f(x) + (z - x)'(d_1 + d_2), \qquad \forall\, z \in \Re^n.$$

Hence, $d_1 + d_2 \in \partial f(x)$, implying that $\partial f_1(x) + \partial f_2(x) \subset \partial f(x)$.

To prove the reverse inclusion, assume to arrive at a contradiction, that there exists a $d \in \partial f(x)$ such that $d \notin \partial f_1(x) + \partial f_2(x)$. Since by Prop. 4.2.1, the sets $\partial f_1(x)$ and $\partial f_2(x)$ are compact, the set $\partial f_1(x) + \partial f_2(x)$ is compact (cf. Prop. 1.5.4), and by the Strict Separation Theorem (Prop. 2.4.3), there exists a hyperplane strictly separating d from $\partial f_1(x) + \partial f_2(x)$, i.e., a vector y and a scalar b such that

$$y'(d_1 + d_2) < b < y'd, \qquad \forall\, d_1 \in \partial f_1(x),\ \forall\, d_2 \in \partial f_2(x).$$

Therefore,

$$\sup_{d_1 \in \partial f_1(x)} y'd_1 + \sup_{d_2 \in \partial f_2(x)} y'd_2 < y'd,$$

and by Prop. 4.2.2,

$$f_1'(x; y) + f_2'(x; y) < y'd.$$

By using the definition of directional derivative, we have $f_1'(x; y) + f_2'(x; y) = f'(x; y)$, so that

$$f'(x; y) < y'd,$$

which contradicts the assumption $d \in \partial f(x)$, in view of Prop. 4.2.2. **Q.E.D.**

Finally, we present some versions of the chain rule for directional derivatives and subgradients.

Proposition 4.2.5: (Chain Rule)

(a) Let $f : \Re^m \mapsto \Re$ be a convex function, and let A be an $m \times n$ matrix. Then, the subdifferential of the function F, defined by

$$F(x) = f(Ax),$$

is given by

$$\partial F(x) = A'\partial f(Ax) = \{A'g \mid g \in \partial f(Ax)\}.$$

(b) Let $f : \Re^n \mapsto \Re$ be a convex function and let $g : \Re \mapsto \Re$ be a smooth scalar function. Then the function F, defined by

$$F(x) = g\big(f(x)\big),$$

is directionally differentiable at all x, and its directional derivative is given by

$$F'(x;y) = \nabla g\big(f(x)\big)f'(x;y), \qquad \forall\, x \in \Re^n, \ \forall\, y \in \Re^n. \quad (4.12)$$

Furthermore, if g is convex and monotonically nondecreasing, then F is convex and its subdifferential is given by

$$\partial F(x) = \nabla g\big(f(x)\big)\partial f(x), \qquad \forall\, x \in \Re^n. \quad (4.13)$$

Proof: (a) From the definition of directional derivative, it can be seen that

$$F'(x;y) = f'(Ax; Ay), \qquad \forall\, y \in \Re^n.$$

Let $g \in \partial f(Ax)$. Then by Prop. 4.2.2, we have

$$g'z \leq f'(Ax; z), \qquad \forall\, z \in \Re^m,$$

and in particular,

$$g'Ay \leq f'(Ax; Ay), \qquad \forall\, y \in \Re^n,$$

or

$$(A'g)'y \leq F'(x;y), \qquad \forall\, y \in \Re^n.$$

Hence, by Eq. (4.8), we have $A'g \in \partial F(x)$, so that $A'\partial f(Ax) \subset \partial F(x)$.

To prove the reverse inclusion, assume to arrive at a contradiction, that there exists a $d \in \partial F(x)$ such that $d \notin A'\partial f(Ax)$. Since by Prop.

4.2.1, the set $\partial f(Ax)$ is compact, the set $A'\partial f(Ax)$ is also compact [cf. Prop. 1.1.9(d)], and by the Strict Separation Theorem (Prop. 2.4.3), there exists a hyperplane strictly separating d from $A'\partial f(Ax)$, i.e., a vector y and a scalar c such that

$$y'(A'g) < c < y'd, \qquad \forall\ g \in \partial f(Ax).$$

From this we obtain

$$\max_{g\in\partial f(Ax)} (Ay)'g < y'd,$$

or, by using Prop. 4.2.2,

$$f'(Ax; Ay) < y'd.$$

Since $f'(Ax; Ay) = F'(x; y)$, it follows that

$$F'(x; y) < y'd,$$

which contradicts the assumption $d \in \partial F(x)$, in view of Prop. 4.2.2.

(b) We have $F'(x; y) = \lim_{\alpha\downarrow 0} s(x, y, \alpha)$, provided the limit exists, where

$$s(x, y, \alpha) = \frac{F(x + \alpha y) - F(x)}{\alpha} = \frac{g\big(f(x + \alpha y)\big) - g\big(f(x)\big)}{\alpha}. \qquad (4.14)$$

From the convexity of f it follows that there are three possibilities (see Exercise 1.6): (1) for some $\bar{\alpha} > 0$, $f(x + \alpha y) = f(x)$ for all $\alpha \in (0, \bar{\alpha}]$, (2) for some $\bar{\alpha} > 0$, $f(x + \alpha y) > f(x)$ for all $\alpha \in (0, \bar{\alpha}]$, (3) for some $\bar{\alpha} > 0$, $f(x + \alpha y) < f(x)$ for all $\alpha \in (0, \bar{\alpha}]$.

In case (1), from Eq. (4.14), we have $\lim_{\alpha\downarrow 0} s(x, y, \alpha) = 0$ and $F'(x; y) = f'(x; y) = 0$, so Eq. (4.12) holds. In case (2), Eq. (4.14) is written as

$$s(x, y, \alpha) = \frac{f(x + \alpha y) - f(x)}{\alpha} \cdot \frac{g\big(f(x + \alpha y)\big) - g\big(f(x)\big)}{f(x + \alpha y) - f(x)}.$$

As $\alpha \downarrow 0$, we have $f(x + \alpha y) \to f(x)$, so the preceding equation yields

$$F'(x; y) = \lim_{\alpha\downarrow 0} s(x, y, \alpha) = \nabla g\big(f(x)\big)f'(x; y).$$

The proof for case (3) is similar.

If g is convex and monotonically nondecreasing, then F is convex (see Exercise 1.4). To obtain the equation for the subdifferential of F, we note that by Eq. (4.8), $d \in \partial F(x)$ if and only if $y'd \le F'(x; y)$ for all $y \in \Re^n$, or equivalently (from what has already been shown)

$$y'd \le \nabla g\big(f(x)\big)f'(x; y), \qquad \forall\ y \in \Re^n.$$

If $\nabla g\big(f(x)\big) = 0$, this relation yields $d = 0$, so $\partial F(x) = \{0\}$ and the desired Eq. (4.13) holds. If $\nabla g\big(f(x)\big) \neq 0$, we have $\nabla g\big(f(x)\big) > 0$ by the monotonicity of g, so we obtain

$$y' \frac{d}{\nabla g(f(x))} \leq f'(x; y), \qquad \forall\, y \in \Re^n,$$

which, by Eq. (4.8), is equivalent to $d/\nabla g\big(f(x)\big) \in \partial f(x)$. Thus, we have shown that $d \in \partial F(x)$ if and only if $d/\nabla g\big(f(x)\big) \in \partial f(x)$, which proves the desired Eq. (4.13).   **Q.E.D.**

4.3 ε-SUBGRADIENTS

We now consider a notion of approximate subgradient. Given a convex function $f : \Re^n \mapsto \Re$ and a positive scalar ϵ, we say that a vector $d \in \Re^n$ is an *ε-subgradient* of f at a point $x \in \Re^n$ if

$$f(z) \geq f(x) + (z - x)'d - \epsilon, \qquad \forall\, z \in \Re^n. \tag{4.15}$$

If instead f is a concave function, we say that d is an ε-subgradient of f at x if $-d$ is a subgradient of the convex function $-f$ at x.

To interpret geometrically an ε-subgradient, note that the defining relation (4.15) can be written as

$$f(z) - z'd \geq \big(f(x) - \epsilon\big) - x'd, \qquad \forall\, z \in \Re^n.$$

Thus d is an ε-subgradient at x if and only if the epigraph of f is contained in the positive halfspace corresponding to the hyperplane in $\Re^{n+1}$ that has normal $(-d, 1)$ and passes through $\big(x, f(x) - \epsilon\big)$, as illustrated in Fig. 4.3.1.

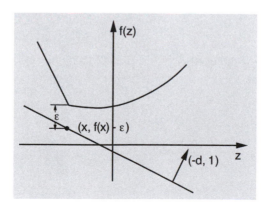

Figure 4.3.1. Illustration of an ε-subgradient of a convex function f.

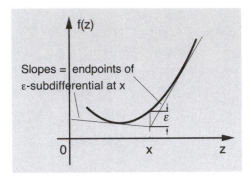

Figure 4.3.2. Illustration of the ϵ-subdifferential $\partial_\epsilon f(x)$ of a convex one-dimensional function $f : \Re \mapsto \Re$. The ϵ-subdifferential is a bounded interval, and corresponds to the set of slopes indicated in the figure. Its left endpoint is the left slope shown, given by

$$f_\epsilon^-(x) = \sup_{\delta < 0} \frac{f(x + \delta) - f(x) + \epsilon}{\delta},$$

and its right endpoint is the right slope shown, given by

$$f_\epsilon^+(x) = \inf_{\delta > 0} \frac{f(x + \delta) - f(x) + \epsilon}{\delta}.$$

The set of all ϵ-subgradients of a convex (or concave) function f at $x \in \Re^n$ is called the *ϵ-subdifferential* of f at x, and is denoted by $\partial_\epsilon f(x)$. The ϵ-subdifferential is illustrated geometrically in Fig. 4.3.2.

The ϵ-subdifferential finds several applications in nondifferentiable optimization, particularly in the context of computational methods. Some of its important properties are given in the following proposition.

Proposition 4.3.1: Let $f : \Re^n \mapsto \Re$ be a convex function and let ϵ be a positive scalar. For every $x \in \Re^n$, the following hold:

(a) The ϵ-subdifferential $\partial_\epsilon f(x)$ is a nonempty, convex, and compact set, and for all $y \in \Re^n$ there holds

$$\inf_{\alpha > 0} \frac{f(x + \alpha y) - f(x) + \epsilon}{\alpha} = \max_{d \in \partial_\epsilon f(x)} y'd.$$

(b) We have $0 \in \partial_\epsilon f(x)$ if and only if

$$f(x) \le \inf_{z \in \Re^n} f(z) + \epsilon.$$

(c) If a direction y is such that $y'd < 0$ for all $d \in \partial_\epsilon f(x)$, then

$$\inf_{\alpha > 0} f(x + \alpha y) < f(x) - \epsilon.$$

(d) If $0 \notin \partial_\epsilon f(x)$, then the direction $y = -\overline{d}$, where

$$\overline{d} = \arg \min_{d \in \partial_\epsilon f(x)} \|d\|,$$

satisfies $y'd < 0$ for all $d \in \partial_\epsilon f(x)$.

(e) If f is equal to the sum $f_1 + \cdots + f_m$ of convex functions f_j : $\Re^n \mapsto \Re$, $j = 1, \ldots, m$, then

$$\partial_\epsilon f(x) \subset \partial_\epsilon f_1(x) + \cdots + \partial_\epsilon f_m(x) \subset \partial_{m\epsilon} f(x).$$

Proof: (a) We have

$$d \in \partial_\epsilon f(x) \quad \Longleftrightarrow \quad f(x + \alpha y) \geq f(x) + \alpha y'd - \epsilon, \quad \forall \, \alpha > 0, \, \forall \, y \in \Re^n.$$
$$(4.16)$$

Hence

$$d \in \partial_\epsilon f(x) \quad \Longleftrightarrow \quad \inf_{\alpha > 0} \frac{f(x + \alpha y) - f(x) + \epsilon}{\alpha} \geq y'd, \quad \forall \, y \in \Re^n. \quad (4.17)$$

It follows that $\partial_\epsilon f(x)$ is the intersection of the closed halfspaces

$$\left\{ d \ \Big| \ \inf_{\alpha > 0} \frac{f(x + \alpha y) - f(x) + \epsilon}{\alpha} \geq y'd \right\},$$

as y ranges over $\Re^n$. Hence, $\partial_\epsilon f(x)$ is closed and convex. To show that $\partial_\epsilon f(x)$ is also bounded, suppose to arrive at a contradiction that there is a sequence $\{d_k\} \subset \partial_\epsilon f(x)$ with $\|d_k\| \to \infty$. Let $y_k = \frac{d_k}{\|d_k\|}$. Then, from Eq. (4.16), we have for $\alpha = 1$,

$$f(x + y_k) \geq f(x) + \|d_k\| - \epsilon, \qquad \forall \, k,$$

so it follows that $f(x + y_k) \to \infty$. This is a contradiction since f is convex and hence continuous, so it is bounded on any bounded set. Thus $\partial_\epsilon f(x)$ is bounded.

To show that $\partial_\epsilon f(x)$ is nonempty and satisfies

$$\inf_{\alpha > 0} \frac{f(x + \alpha y) - f(x) + \epsilon}{\alpha} = \max_{d \in \partial_\epsilon f(x)} y'd, \qquad \forall \, y \in \Re^n,$$

we argue similar to the proof of Prop. 4.2.2. Consider the subset of $\Re^{n+1}$

$$C_1 = \{(z, w) \mid f(z) < w\},$$

and the half-line

$$C_2 = \left\{(z, w) \;\middle|\; w = f(x) - \epsilon + \beta \inf_{\alpha > 0} \frac{f(x + \alpha y) - f(x) + \epsilon}{\alpha}, \right.$$
$$\left. z = x + \beta y, \beta \geq 0 \right\}.$$

These sets are nonempty and convex. They are also disjoint, since we have for all $(z, w) \in C_2$,

$$\begin{aligned} w &= f(x) - \epsilon + \beta \inf_{\alpha > 0} \frac{f(x + \alpha y) - f(x) + \epsilon}{\alpha} \\ &\leq f(x) - \epsilon + \beta \frac{f(x + \beta y) - f(x) + \epsilon}{\beta} \\ &= f(x + \beta y) \\ &= f(z). \end{aligned}$$

Hence, there exists a hyperplane separating them, i.e., a vector $(\mu, \gamma) \neq (0, 0)$ such that for all $\beta \geq 0$, $z \in \Re^n$, and $w > f(z)$,

$$\gamma w + \mu' z \geq \gamma \left(f(x) - \epsilon + \beta \inf_{\alpha > 0} \frac{f(x + \alpha y) - f(x) + \epsilon}{\alpha} \right) + \mu'(x + \beta y).$$
$$(4.18)$$

We cannot have $\gamma < 0$, since then the left-hand side above could be made arbitrarily small by choosing w to be sufficiently large. Also, if $\gamma = 0$, Eq. (4.18) implies that $\mu = 0$, contradicting the fact that $(\mu, \gamma) \neq (0, 0)$. Therefore, $\gamma > 0$ and after dividing Eq. (4.18) by γ, we obtain for all $\beta \geq 0$, $z \in \Re^n$, and $w > f(z)$

$$w + \left(\frac{\mu}{\gamma}\right)' (z - x) \geq f(x) - \epsilon + \beta \inf_{\alpha > 0} \frac{f(x + \alpha y) - f(x) + \epsilon}{\alpha} + \beta \left(\frac{\mu}{\gamma}\right)' y.$$
$$(4.19)$$

Taking the limit above as $w \downarrow f(z)$ and setting $\beta = 0$, we obtain

$$f(z) \geq f(x) - \epsilon + \left(-\frac{\mu}{\gamma}\right)' (z - x), \qquad \forall z \in \Re^n.$$

Hence, $-\mu/\gamma$ belongs to $\partial_\epsilon f(x)$, showing that $\partial_\epsilon f(x)$ is nonempty. Also by taking $z = x$ in Eq. (4.19), and by letting $w \downarrow f(x)$ and by dividing with β, we obtain

$$-\frac{\mu'}{\gamma} y \geq -\frac{\epsilon}{\beta} + \inf_{\alpha > 0} \frac{f(x + \alpha y) - f(x) + \epsilon}{\alpha}.$$

Since β can be chosen as large as desired, we see that

$$-\frac{\mu'}{\gamma}y \geq \inf_{\alpha>0} \frac{f(x+\alpha y)-f(x)+\epsilon}{\alpha}.$$

Combining this relation with Eq. (4.17), we obtain

$$\max_{d\in\partial_\epsilon f(x)} d'y = \inf_{\alpha>0} \frac{f(x+\alpha y)-f(x)-\epsilon}{\alpha}.$$

(b) By definition, $0 \in \partial_\epsilon f(x)$ if and only if $f(z) \geq f(x)-\epsilon$ for all $z \in \Re^n$, which is equivalent to $\inf_{z\in\Re^n} f(z) \geq f(x)-\epsilon$.

(c) Assume that a direction y is such that

$$\max_{d\in\partial_\epsilon f(x)} y'd < 0, \tag{4.20}$$

while $\inf_{\alpha>0} f(x+\alpha y) \geq f(x)-\epsilon$. Then $f(x+\alpha y)-f(x) \geq -\epsilon$ for all $\alpha > 0$, or equivalently

$$\frac{f(x+\alpha y)-f(x)+\epsilon}{\alpha} \geq 0, \qquad \forall\, \alpha > 0.$$

Consequently, using part (a), we have

$$\max_{d\in\partial_\epsilon f(x)} y'd = \inf_{\alpha>0} \frac{f(x+\alpha y)-f(x)+\epsilon}{\alpha} \geq 0,$$

which contradicts Eq. (4.20).

(d) The vector $\bar{d}$ is the projection of the zero vector on the convex and compact set $\partial_\epsilon f(x)$. If $0 \notin \partial_\epsilon f(x)$, we have $\|\bar{d}\| > 0$, while by the Projection Theorem (Prop. 2.2.1),

$$(d-\bar{d})'\bar{d} \geq 0, \qquad \forall\, d \in \partial_\epsilon f(x),$$

implying that

$$d'\bar{d} \geq \|\bar{d}\|^2 > 0, \qquad \forall\, d \in \partial_\epsilon f(x).$$

(e) It will suffice to prove the result for the case where $f = f_1 + f_2$. If $d_1 \in \partial_\epsilon f_1(x)$ and $d_2 \in \partial_\epsilon f_2(x)$, then from Eq. (4.16), we have

$$f_1(x+\alpha y) \geq f_1(x)+\alpha y'd_1 - \epsilon, \qquad \forall\, \alpha > 0,\ \forall\, y \in \Re^n,$$

$$f_2(x+\alpha y) \geq f_2(x)+\alpha y'd_2 - \epsilon, \qquad \forall\, \alpha > 0,\ \forall\, y \in \Re^n,$$

so by adding, we obtain

$$f(x+\alpha y) \geq f(x)+\alpha y'(d_1+d_2) - 2\epsilon, \qquad \forall\, \alpha > 0,\ \forall\, y \in \Re^n.$$

Hence from Eq. (4.16), we have $d_1 + d_2 \in \partial_{2\epsilon} f(x)$, implying that $\partial_\epsilon f_1(x) + \partial_\epsilon f_2(x) \subset \partial_{2\epsilon} f(x)$.

To prove that $\partial_\epsilon f(x) \subset \partial_\epsilon f_1(x) + \partial_\epsilon f_2(x)$, we use an argument similar to the one of the proof of Prop. 4.2.4. Assume to arrive at a contradiction, that there exists a $d \in \partial_\epsilon f(x)$ such that $d \notin \partial_\epsilon f_1(x) + \partial_\epsilon f_2(x)$. Since by part (a), the sets $\partial_\epsilon f_1(x)$ and $\partial_\epsilon f_2(x)$ are compact, the set $\partial_\epsilon f_1(x) + \partial_\epsilon f_2(x)$ is compact (cf. Prop. 1.5.4), and by the Strict Separation Theorem (Prop. 2.4.3), there exists a hyperplane strictly separating d from $\partial_\epsilon f_1(x) + \partial_\epsilon f_2(x)$, i.e., a vector y and a scalar b such that

$$y'(d_1 + d_2) < b < y'd, \qquad \forall\, d_1 \in \partial_\epsilon f_1(x),\ \forall\, d_2 \in \partial_\epsilon f_2(x).$$

From this we obtain

$$\sup_{d_1 \in \partial_\epsilon f_1(x)} y'd_1 + \sup_{d_2 \in \partial_\epsilon f_2(x)} y'd_2 < y'd,$$

and by part (a),

$$\inf_{\alpha > 0} \frac{f_1(x + \alpha y) - f_1(x) + \epsilon}{\alpha} + \inf_{\alpha > 0} \frac{f_2(x + \alpha y) - f_2(x) + \epsilon}{\alpha} < y'd.$$

Let α_1 and α_2 be positive scalars such that

$$\frac{f_1(x + \alpha_1 y) - f_1(x) + \epsilon}{\alpha_1} + \frac{f_2(x + \alpha_2 y) - f_2(x) + \epsilon}{\alpha_2} < y'd, \qquad (4.21)$$

and let

$$\overline{\alpha} = \frac{1}{1/\alpha_1 + 1/\alpha_2}.$$

As a consequence of the convexity of f_j, $j = 1, 2$, the ratio $\left(f_j(x + \alpha y) - f_j(x)\right)/\alpha$ is monotonically nondecreasing in α. Thus, since $\alpha_j \geq \overline{\alpha}$, we have

$$\frac{f_j(x + \alpha_j y) - f_j(x)}{\alpha_j} \geq \frac{f_j(x + \overline{\alpha} y) - f_j(x)}{\overline{\alpha}}, \qquad j = 1, 2,$$

and Eq. (4.21) together with the definition of $\overline{\alpha}$ yields

$$\begin{aligned}
y'd &> \frac{f_1(x + \alpha_1 y) - f_1(x) + \epsilon}{\alpha_1} + \frac{f_2(x + \alpha_2 y) - f_2(x) + \epsilon}{\alpha_2} \\
&\geq \frac{\epsilon}{\overline{\alpha}} + \frac{f_1(x + \overline{\alpha} y) - f_1(x)}{\overline{\alpha}} + \frac{f_2(x + \overline{\alpha} y) - f_2(x)}{\overline{\alpha}} \\
&\geq \inf_{\alpha > 0} \frac{f(x + \alpha y) - f(x) + \epsilon}{\alpha}.
\end{aligned}$$

This contradicts Eq. (4.17), thus implying that $\partial_\epsilon f(x) \subset \partial_\epsilon f_1(x) + \partial_\epsilon f_2(x)$.
Q.E.D.

ϵ-Descent Algorithm

Parts (b)-(d) of Prop. 4.3.1 contain the elements of an important iterative algorithm (called the ϵ-*descent algorithm* and introduced by Bertsekas and Mitter [BeM71], [BeM73]) for minimizing convex functions to within a tolerance of ϵ. In this algorithm, at the current iterate x, we check whether $0 \in \partial_\epsilon f(x)$. If this is so, then by Prop. 4.3.1(b), x is an ϵ-optimal solution. If not, then by going along the direction opposite to the vector of minimum norm in $\partial_\epsilon f(x)$, we are guaranteed a cost improvement of at least ϵ. Thus, the algorithm is guaranteed to find an ϵ-optimal solution, assuming that f is bounded below, and to yield a sequence $\{x_k\}$ with $f(x_k) \to -\infty$, if f is unbounded below.

One may use Prop. 4.3.1(e) to approximate $\partial_\epsilon f(x)$ in cases where f is the sum of convex functions whose ϵ-subdifferential is easily computed. In particular, by using $\partial_\epsilon f_1(x) + \cdots + \partial_\epsilon f_m(x)$ in place of (the potentially harder to compute) $\partial_\epsilon f(x)$, the ϵ-descent algorithm yields a cost improvement of at least ϵ at all points x where $f(x) > \inf_{z \in \Re^n} f(z) + m\epsilon$. This idea finds application in several contexts, including monotropic programming and network optimization (see Rockafellar [Roc84] or Bertsekas [Ber98]). An implementation of the ϵ-descent algorithm (due to Lemaréchal [Lem74], [Lem75]), which does not require the computation of the full ϵ-subdifferential, is given in Exercise 4.17.

4.4 SUBGRADIENTS OF EXTENDED REAL-VALUED FUNCTIONS

We have focused so far on real-valued convex functions $f : \Re^n \mapsto \Re$, which are defined over the entire space $\Re^n$ and are convex over $\Re^n$. The notion of a subdifferential and a subgradient of an extended real-valued proper convex function $f : \Re^n \mapsto (-\infty, \infty]$ can be developed along similar lines. In particular, a vector d is a subgradient of f at a point $x \in \mathrm{dom}(f)$ if the subgradient inequality holds, i.e.,

$$f(z) \geq f(x) + (z - x)'d, \qquad \forall \, z \in \Re^n.$$

If the extended real-valued function $f : \Re^n \mapsto [-\infty, \infty)$ is concave, d is said to be a subgradient of f at a point x with $f(x) > -\infty$ if $-d$ is a subgradient of the convex function $-f$ at x.

The subdifferential $\partial f(x)$ is the set of all subgradients of the convex (or concave) function f at x. By convention, $\partial f(x)$ is considered empty for all $x \notin \mathrm{dom}(f)$. Note that contrary to the case of real-valued functions, $\partial f(x)$ may be empty, or unbounded. For example, the extended real-valued convex function given by

$$f(x) = \begin{cases} -\sqrt{x} & \text{if } 0 \leq x \leq 1, \\ \infty & \text{otherwise,} \end{cases}$$

has the subdifferential

$$\partial f(x) = \begin{cases} -\frac{1}{2\sqrt{x}} & \text{if } 0 < x < 1, \\ [-1/2, \infty) & \text{if } x = 1, \\ \varnothing & \text{if } x \leq 0 \text{ or } 1 < x. \end{cases}$$

Thus, $\partial f(x)$ can be empty or unbounded at points x within dom(f) (as in the cases $x = 0$ and $x = 1$, respectively, of the above example).

The nonemptiness of the subdifferential $\partial f(x)$ is closely related to existence of a nonvertical hyperplane supporting the epigraph of f at the point $(x, f(x))$, as shown in the following proposition.

Proposition 4.4.1: Let $f : \Re^n \mapsto (-\infty, \infty]$ be a proper convex function and let x be a vector in dom(f). Then $\partial f(x)$ is closed and convex, and it is nonempty if and only if there exists a nonvertical hyperplane that supports epi(f) at the point $(x, f(x))$.

Proof: Clearly, $\partial f(x)$ is closed and convex, since it is defined as the intersection of the collection of closed halfspaces $\{\mu \mid f(x) + \mu'(z - x) \leq f(z)\}$ as z ranges over $\Re^n$.

Suppose that there exists a nonvertical hyperplane supporting epi(f) at $(x, f(x))$, i.e., there exists a vector $(d, \beta) \in \Re^{n+1}$ such that $\beta \neq 0$ and

$$d'x + \beta f(x) \leq d'z + \beta w, \qquad \forall \, (z, w) \in \text{epi}(f). \tag{4.22}$$

We must have $\beta \geq 0$, since otherwise the preceding relation would be contradicted for sufficiently large w. Thus, since $\beta \neq 0$, we must have $\beta > 0$. By dividing with β, and by setting $w = f(z)$ and $\overline{d} = d/\beta$ in Eq. (4.22), we obtain

$$\overline{d}'x + f(x) \leq \overline{d}'z + f(z), \qquad \forall \, z \in \text{dom}(f),$$

or equivalently,

$$f(z) \geq f(x) + (-\overline{d})'(z - x), \qquad \forall \, z \in \text{dom}(f).$$

Since $f(z) = \infty$ for all $z \notin \text{dom}(f)$, the preceding inequality holds for all $z \in \Re^n$, so that $(-\overline{d}) \in \partial f(x)$, and $\partial f(x)$ is nonempty.

The converse is a direct consequence of the definition of the subgradient (compare with Fig. 4.2.1). In particular, if $d \in \partial f(x)$, we have

$$f(z) \geq f(x) + d'(z - x), \qquad \forall \, z \in \Re^n.$$

By restricting z to be in dom(f) and by rearranging terms, we see that

$$f(z) - d'z \geq f(x) - d'x, \qquad \forall \, z \in \text{dom}(f).$$

Since $z \in \text{dom}(f)$ and $f(z) \leq w$ for all $(z, w) \in \text{epi}(f)$, the preceding relation yields

$$w - d'z \geq f(x) - d'x, \qquad \forall \, (z, w) \in \text{epi}(f),$$

showing that the hyperplane with normal $(-d, 1)$ supports $\text{epi}(f)$ at $(x, f(x))$. **Q.E.D.**

As in the case of real-valued functions (cf. Prop. 4.2.1), the question of existence of a nonvertical hyperplane supporting the epigraph of a proper convex function $f : \Re^n \mapsto (-\infty, \infty]$ at $(x, f(x))$ can be addressed within the min common/max crossing framework of Section 2.5, with a special choice of the set M in that framework. We thus obtain the following proposition.

Proposition 4.4.2: Let $f : \Re^n \mapsto (-\infty, \infty]$ be a proper convex function. For every x in the relative interior of $\text{dom}(f)$, the subdifferential $\partial f(x)$ has the form
$$\partial f(x) = S^\perp + G,$$
where S is the subspace that is parallel to the affine hull of $\text{dom}(f)$, and G is a nonempty and compact set. Furthermore, $\partial f(x)$ is nonempty and compact if and only if x is in the interior of $\text{dom}(f)$.

Proof: Let the set M in the min common/max crossing framework of Section 2.5 be

$$M = \big\{ (u, w) \mid u \in \Re^n, \, f(x + u) \leq w \big\}. \tag{4.23}$$

Then the min common value is $w^* = f(x)$ and the max crossing value is $q^* = \sup_\mu q(\mu)$, where $q(\mu) = \inf_{(u,w) \in M} \{w + \mu'u\}$. For a given μ, we have $w^* = q^* = q(\mu)$ if and only if

$$f(x) = \inf_{(u,w) \in M} \{w + \mu'u\},$$

or equivalently, if and only if

$$f(x) \leq f(x + u) + \mu'u, \qquad \forall \, u \in \Re^n.$$

Thus, the set of optimal solutions of the max crossing problem is precisely $-\partial f(x)$. Furthermore, we have $\overline{M} = M$ and

$$D = \big\{ u \mid \text{there exists } w \in \Re \text{ with } (u, w) \in \overline{M} \big\} = \text{dom}(f) - \{x\}.$$

With the preceding identifications, the result follows from the Min Common/Max Crossing Theorem II (Prop. 2.5.5).   **Q.E.D.**

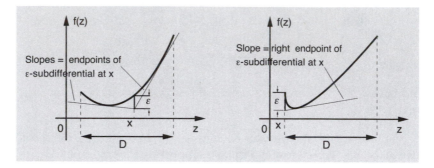

Figure 4.4.1. Illustration of the ϵ-subdifferential $\partial_\epsilon f(x)$ of a one-dimensional function $f : \Re \mapsto (-\infty, \infty]$, which is convex and has as effective domain an interval D. The ϵ-subdifferential corresponds to the set of slopes indicated in the figure. Note that $\partial_\epsilon f(x)$ is nonempty at all $x \in D$. Its left endpoint is

$$f_\epsilon^-(x) = \begin{cases} \sup_{\delta<0,\, x+\delta\in D} \frac{f(x+\delta)-f(x)+\epsilon}{\delta} & \text{if } \inf D < x, \\ -\infty & \text{if } \inf D = x, \end{cases}$$

and its right endpoint is

$$f_\epsilon^+(x) = \begin{cases} \inf_{\delta>0,\, x+\delta\in D} \frac{f(x+\delta)-f(x)+\epsilon}{\delta} & \text{if } x < \sup D, \\ \infty & \text{if } x = \sup D. \end{cases}$$

Note that these endpoints can be $-\infty$ (as in the figure on the right) or ∞. For $\epsilon = 0$, the above formulas also give the endpoints of the subdifferential $\partial f(x)$. Note also that while $\partial f(x)$ is nonempty for all x in the interior of D, it may be empty for x at the relative boundary of D (as in the figure on the right).

There is also a notion of ϵ-subgradient for extended real-valued functions. In particular, a vector d is an ϵ-subgradient of a proper convex function $f : \Re^n \mapsto (-\infty, \infty]$ at a point $x \in \text{dom}(f)$ if

$$f(z) \geq f(x) + (z-x)'d - \epsilon, \qquad \forall\, z \in \Re^n.$$

The ϵ-subdifferential $\partial_\epsilon f(x)$ is the set of all ϵ-subgradients of f, and by convention, $\partial_\epsilon f(x) = \emptyset$ for $x \notin \text{dom}(f)$. Figure 4.4.1 illustrates the definition of $\partial_\epsilon f(x)$ for the case of a one-dimensional function f. The figure indicates that if f is closed, then $\partial_\epsilon f(x)$ is nonempty at all points of $\text{dom}(f)$. This can be shown for multi-dimensional functions f as well.

One can provide generalized versions of the results of Props. 4.2.2-4.3.1 within the context of extended real-valued convex functions, but with appropriate adjustments and additional assumptions to deal with cases where $\partial f(x)$ may be empty or noncompact. Some of these generalizations are discussed in the exercises.

4.5 DIRECTIONAL DERIVATIVE OF THE MAX FUNCTION

As mentioned in Section 2.6, the max function $f(x) = \max_{z \in Z} \phi(x, z)$ arises in a variety of interesting contexts, including duality. It is therefore important to characterize this function, and the following proposition gives the directional derivative and the subdifferential of f for the case where $\phi(\cdot, z)$ is convex for all $z \in Z$.

Proposition 4.5.1: (Danskin's Theorem) Let Z be a compact subset of $\Re^m$, and let $\phi : \Re^n \times Z \mapsto \Re$ be continuous and such that $\phi(\cdot, z) : \Re^n \mapsto \Re$ is convex for each $z \in Z$.

(a) The function $f : \Re^n \mapsto \Re$ given by

$$f(x) = \max_{z \in Z} \phi(x, z) \qquad (4.24)$$

is convex and has directional derivative given by

$$f'(x; y) = \max_{z \in Z(x)} \phi'(x, z; y), \qquad (4.25)$$

where $\phi'(x, z; y)$ is the directional derivative of the function $\phi(\cdot, z)$ at x in the direction y, and $Z(x)$ is the set of maximizing points in Eq. (4.24)

$$Z(x) = \left\{ \overline{z} \;\middle|\; \phi(x, \overline{z}) = \max_{z \in Z} \phi(x, z) \right\}.$$

In particular, if $Z(x)$ consists of a unique point $\overline{z}$ and $\phi(\cdot, \overline{z})$ is differentiable at x, then f is differentiable at x, and $\nabla f(x) = \nabla_x \phi(x, \overline{z})$, where $\nabla_x \phi(x, \overline{z})$ is the vector with components

$$\frac{\partial \phi(x, \overline{z})}{\partial x_i}, \qquad i = 1, \ldots, n.$$

(b) If $\phi(\cdot, z)$ is differentiable for all $z \in Z$ and $\nabla_x \phi(x, \cdot)$ is continuous on Z for each x, then

$$\partial f(x) = \mathrm{conv}\{\nabla_x \phi(x, z) \mid z \in Z(x)\}, \qquad \forall\, x \in \Re^n.$$

Proof: (a) The convexity of f has been established in Prop. 1.2.4(c). We note that since ϕ is continuous and Z is compact, the set $Z(x)$ is nonempty by Weierstrass' Theorem (Prop. 2.1.1) and f takes real values. For any $z \in Z(x)$, $y \in \Re^n$, and $\alpha > 0$, we use the definition of f to write

$$\frac{f(x + \alpha y) - f(x)}{\alpha} \geq \frac{\phi(x + \alpha y, z) - \phi(x, z)}{\alpha}.$$

Taking the limit as α decreases to zero, we obtain $f'(x; y) \geq \phi'(x, z; y)$. Since this is true for every $z \in Z(x)$, we conclude that

$$f'(x; y) \geq \sup_{z \in Z(x)} \phi'(x, z; y), \qquad \forall\, y \in \Re^n. \tag{4.26}$$

To prove the reverse inequality and that the supremum in the right-hand side of the above inequality is attained, consider a sequence $\{\alpha_k\}$ with $\alpha_k \downarrow 0$ and let $x_k = x + \alpha_k y$. For each k, let z_k be a vector in $Z(x_k)$. Since $\{z_k\}$ belongs to the compact set Z, it has a subsequence converging to some $\overline{z} \in Z$. Without loss of generality, we assume that the entire sequence $\{z_k\}$ converges to $\overline{z}$. We have

$$\phi(x_k, z_k) \geq \phi(x_k, z), \qquad \forall\, z \in Z,$$

so by taking the limit as $k \to \infty$ and by using the continuity of ϕ, we obtain

$$\phi(x, \overline{z}) \geq \phi(x, z), \qquad \forall\, z \in Z.$$

Therefore, $\overline{z} \in Z(x)$. We now have

$$
\begin{aligned}
f'(x; y) &\leq \frac{f(x + \alpha_k y) - f(x)}{\alpha_k} \\
&= \frac{\phi(x + \alpha_k y, z_k) - \phi(x, \overline{z})}{\alpha_k} \\
&\leq \frac{\phi(x + \alpha_k y, z_k) - \phi(x, z_k)}{\alpha_k} \\
&= -\frac{\phi(x + \alpha_k y + \alpha_k(-y), z_k) - \phi(x + \alpha_k y, z_k)}{\alpha_k} \\
&\leq -\phi'(x + \alpha_k y, z_k; -y) \\
&\leq \phi'(x + \alpha_k y, z_k; y),
\end{aligned}
\tag{4.27}
$$

where the last inequality follows from Eq. (4.4). By letting $k \to \infty$, we obtain

$$f'(x; y) \leq \limsup_{k \to \infty} \phi'(x + \alpha_k y, z_k; y) \leq \phi'(x, \overline{z}; y),$$

where the right inequality in the preceding relation follows from Prop. 4.1.2 with $f_k(\cdot) = \phi(\cdot, z_k)$, $x_k = x + \alpha_k y$, and $y_k = y$. This relation together with inequality (4.26) proves Eq. (4.25).

For the last statement of part (a), if $Z(x)$ consists of the unique point $\overline{z}$, Eq. (4.25) and the differentiability assumption on ϕ yield

$$f'(x; y) = \phi'(x, \overline{z}; y) = y' \nabla_x \phi(x, \overline{z}), \qquad \forall \, y \in \Re^n,$$

which implies that f is differentiable at x and $\nabla f(x) = \nabla_x \phi(x, \overline{z})$.

(b) By part (a), we have

$$f'(x; y) = \max_{z \in Z(x)} \nabla_x \phi(x, z)' y,$$

while by Prop. 4.2.2,

$$f'(x; y) = \max_{z \in \partial f(x)} d' y.$$

For all $\overline{z} \in Z(x)$ and $y \in \Re^n$, we have

$$\begin{aligned}
f(y) &= \max_{z \in Z} \phi(y, z) \\
&\geq \phi(y, \overline{z}) \\
&\geq \phi(x, \overline{z}) + \nabla_x \phi(x, \overline{z})'(y - x) \\
&= f(x) + \nabla_x \phi(x, \overline{z})'(y - x).
\end{aligned}$$

Therefore, $\nabla_x \phi(x, \overline{z})$ is a subgradient of f at x, implying that

$$\operatorname{conv}\big\{\nabla_x \phi(x, z) \mid z \in Z(x)\big\} \subset \partial f(x).$$

To prove the reverse inclusion, we use a hyperplane separation argument. Since $\phi(x, \cdot)$ is continuous and Z is compact, from Weierstrass' Theorem it follows that $Z(x)$ is compact, so since $\nabla_x \phi(x, \cdot)$ is continuous, the set $\big\{\nabla_x \phi(x, z) \mid z \in Z(x)\big\}$ is compact. By Prop. 1.3.2, it follows that $\operatorname{conv}\big\{\nabla_x \phi(x, z) \mid z \in Z(x)\big\}$ is compact. If $d \in \partial f(x)$ while $d \notin \operatorname{conv}\big\{\nabla_x \phi(x, z) \mid z \in Z(x)\big\}$, by the Strict Separation Theorem (Prop. 2.4.3), there exist $y \neq 0$ and $\gamma \in \Re$, such that

$$d' y > \gamma > \nabla_x \phi(x, z)' y, \qquad \forall \, z \in Z(x).$$

Therefore, we have

$$d' y > \max_{z \in Z(x)} \nabla_x \phi(x, z)' y = f'(x; y),$$

contradicting Prop. 4.2.2. Thus, $\partial f(x) \subset \operatorname{conv}\big\{\nabla_x \phi(x, z) \mid z \in Z(x)\big\}$ and the proof is complete.   **Q.E.D.**

A simple but important application of the above proposition is when Z is a finite set and ϕ is linear in x for all z. In this case, $f(x)$ can be represented as

$$f(x) = \max\{a_1' x + b_1, \ldots, a_m' x + b_m\},$$

where $a_1, \ldots, a_m$ are some vectors in $\Re^n$ and $b_1, \ldots, b_m$ are some scalars. Then, using Prop. 4.5.1(b), it follows that $\partial f(x)$ is a polyhedral set. It is the convex hull of the set of all vectors a_i for which the above maximum is attained:

$$a_i' x + b_i = \max\{a_1' x + b_1, \ldots, a_m' x + b_m\}.$$

4.6 CONICAL APPROXIMATIONS

The analysis of a constrained optimization problem often revolves around the behavior of the cost function along directions leading away from a local minimum to some neighboring feasible points. For example, suppose that x^* minimizes a smooth function $f : \Re^n \mapsto \Re$ over a convex set X. Then, for any $x \in X$, the function

$$g(\alpha) = f\big(x^* + \alpha(x - x^*)\big)$$

is minimized over $\alpha \in [0, 1]$ at $\alpha = 0$, so $\nabla g(\alpha)\big|_{\alpha=0} = \nabla f(x^*)'(x - x^*) \geq 0$. We thus obtain the necessary optimality condition

$$\nabla f(x^*)'(x - x^*) \geq 0, \qquad \forall \ x \in X,$$

which expresses the fact that the slope of f along any direction leading from the minimum x^* to any other vector $x \in X$ should be nonnegative.

The preceding argument can be used, even when X is not convex, to show that $\nabla f(x^*)'y \geq 0$ for any direction y such that $x^* + \alpha y \in X$ for α in some nontrivial interval $[0, \overline{\alpha}]$. Directions having this property or other related properties constitute cones that can be viewed as approximations to the constraint set, locally near a point of interest. We will introduce two such cones, and in the next section, we will discuss their use in connection with optimality conditions.

Definition 4.6.1: Given a subset X of $\Re^n$ and a vector x in X, a vector $y \in \Re^n$ is said to be a *feasible direction* of X at x if there exists an $\overline{\alpha} > 0$ such that $x + \alpha y \in X$ for all $\alpha \in [0, \overline{\alpha}]$. The set of all feasible directions of X at x is denoted by $F_X(x)$.

Clearly, $F_X(x)$ is a cone containing the origin. It can be seen that if X is convex, $F_X(x)$ consists of all vectors of the form $\alpha(\overline{x} - x)$ with $\alpha > 0$ and $\overline{x} \in X$. However, when X is nonconvex, $F_X(x)$ may not provide interesting information about the local structure of the set X near the point x. For example, often there is no nonzero feasible direction at x when X is nonconvex [think of the set $X = \big\{x \mid h(x) = 0\big\}$, where $h : \Re^n \mapsto \Re$ is a nonlinear function]. The next definition introduces a cone that provides information on the local structure of X even when there are no feasible directions other than zero.

Definition 4.6.2: Given a subset X of $\Re^n$ and a vector x in X, a vector $y \in \Re^n$ is said to be a *tangent* of X at x if either $y = 0$ or there exists a sequence $\{x_k\} \subset X$ such that $x_k \neq x$ for all k and

$$x_k \longrightarrow x, \qquad \frac{x_k - x}{\|x_k - x\|} \rightarrow \frac{y}{\|y\|}.$$

The set of all tangents of X at x is called the *tangent cone* of X at x, and is denoted by $T_X(x)$.

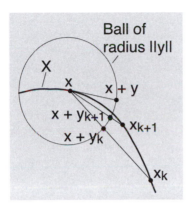

Figure 4.6.2. Illustration of a tangent y at a vector $x \in X$. There is a sequence $\{x_k\} \subset X$ that converges to x and is such that the normalized direction sequence $(x_k - x)/\|x_k - x\|$ converges to $y/\|y\|$, the normalized direction of y, or equivalently, the sequence

$$y_k = \frac{\|y\|(x_k - x)}{\|x_k - x\|}$$

illustrated in the figure converges to y

It is clear that $T_X(x)$ is a cone, thereby justifying the name "tangent cone." According to the definition, a nonzero vector y is a tangent at x if it is possible to approach x with a feasible sequence $\{x_k\}$ such that the normalized direction sequence $(x_k - x)/\|x_k - x\|$ converges to $y/\|y\|$, the normalized direction of y (see Fig. 4.6.2). The following proposition provides an equivalent definition of a tangent, which is occasionally more convenient for analysis.

Proposition 4.6.1: Let X be a subset of $\Re^n$ and let x be a vector in X. A vector y is a tangent of X at x if and only if there exists a sequence $\{x_k\} \subset X$ with $x_k \to x$, and a positive scalar sequence $\{\alpha_k\}$ such that $\alpha_k \to 0$ and $(x_k - x)/\alpha_k \to y$.

Proof: Let y be a tangent of X at x. If $y = 0$, take $x_k = x$ and $\alpha_k = 1/k$ for all k. If $y \neq 0$, take $\{x_k\}$ to be the sequence in the definition of a tangent, and take $\alpha_k = \|x_k - x\|/\|y\|$.

Conversely, suppose that y is such that there exist sequences $\{x_k\}$ and $\{\alpha_k\}$ with the given properties. If $y = 0$ then y is a tangent. If $y \neq 0$, since $\alpha_k \to 0$ and $(x_k - x)/\alpha_k \to y$, we clearly have $x_k \to x$ and

$$\frac{x_k - x}{\|x_k - x\|} = \frac{(x_k - x)/\alpha_k}{\|(x_k - x)/\alpha_k\|} \rightarrow \frac{y}{\|y\|}.$$

It follows that y is a tangent of X at x.   **Q.E.D.**

Figure 4.6.3 illustrates with examples the cones $F_X(x)$ and $T_X(x)$, and the following proposition establishes some of their properties.

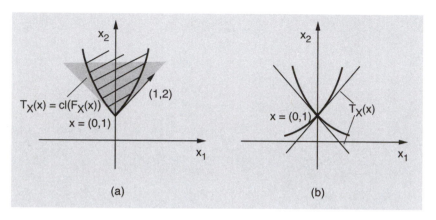

Figure 4.6.3. Examples of the cones $F_X(x)$ and $T_X(x)$ of a set X at the vector $x = (0, 1)$. In (a), we have

$$X = \left\{(x_1, x_2) \mid (x_1 + 1)^2 - x_2 \leq 0, \ (x_1 - 1)^2 - x_2 \leq 0\right\}.$$

Here the set X is convex and the tangent cone $T_X(x)$ is equal to the closure of the cone of feasible directions $F_X(x)$. Note, however, that the vectors $(1, 2)$ and $(-1, 2)$ belong to $T_X(x)$ and also to the closure of $F_X(x)$, but are not feasible directions. In (b), we have

$$X = \left\{(x_1, x_2) \mid \left((x_1 + 1)^2 - x_2\right)\left((x_1 - 1)^2 - x_2\right) = 0\right\}.$$

Here the set X is nonconvex, and $T_X(x)$ is closed but not convex. On the other hand, $F_X(x)$ consists of just the origin.

Proposition 4.6.2: Let X be a subset of $\Re^n$ and let x be a vector in X. The following hold regarding the cone of feasible directions $F_X(x)$ and the tangent cone $T_X(x)$.

(a) $T_X(x)$ is a closed cone.

(b) $\text{cl}\big(F_X(x)\big) \subset T_X(x)$.

(c) If X is convex, then $F_X(x)$ and $T_X(x)$ are convex, and we have

$$\text{cl}\big(F_X(x)\big) = T_X(x).$$

Proof: (a) Let $\{y_k\}$ be a sequence in $T_X(x)$ that converges to some $y \in \Re^n$. We will show that $y \in T_X(x)$. If $y = 0$, then $y \in T_X(x)$, so assume that $y \neq 0$. By the definition of a tangent, for every k there is a sequence $\{x_k^i\} \subset X$ with $x_k^i \neq x$ such that

$$\lim_{i \to \infty} x_k^i = x, \qquad \lim_{i \to \infty} \frac{x_k^i - x}{\|x_k^i - x\|} = \frac{y_k}{\|y_k\|}.$$

For each k, choose an index i_k such that $i_1 < i_2 < \ldots < i_k$ and

$$\lim_{k \to \infty} x_k^{i_k} = x, \qquad \lim_{k \to \infty} \left\| \frac{x_k^{i_k} - x}{\|x_k^{i_k} - x\|} - \frac{y_k}{\|y_k\|} \right\| = 0.$$

For all k, we have

$$\left\| \frac{x_k^{i_k} - x}{\|x_k^{i_k} - x\|} - \frac{y}{\|y\|} \right\| \leq \left\| \frac{x_k^{i_k} - x}{\|x_k^{i_k} - x\|} - \frac{y_k}{\|y_k\|} \right\| + \left\| \frac{y_k}{\|y_k\|} - \frac{y}{\|y\|} \right\|,$$

and by combining the preceding two relations with the fact $y_k \to y$, we obtain

$$\lim_{k \to \infty} x_k^{i_k} = x, \qquad \lim_{k \to \infty} \left\| \frac{x_k^{i_k} - x}{\|x_k^{i_k} - x\|} - \frac{y}{\|y\|} \right\| = 0.$$

Hence $y \in T_X(x)$, thereby showing that $T_X(x)$ is closed.

(b) It is easy to see, using Prop. 4.6.1, that every feasible direction is a tangent, so $F_X(x) \subset T_X(x)$. Since by part (a), $T_X(x)$ is closed, the result follows.

(c) Since X is convex, the set $F_X(x)$ consists of all vectors of the form $\alpha(\bar{x} - x)$ with $\alpha > 0$ and $\bar{x} \in X$. To show that $F_X(x)$ is convex, let y_1 and y_2 be vectors in $F_X(x)$, so that $y_1 = \alpha_1(\bar{x}_1 - x)$ and $y_2 = \alpha_2(\bar{x}_2 - x)$, with $\alpha_1 > 0$, $\alpha_2 > 0$, and $\bar{x}_1, \bar{x}_2 \in X$. Let also γ_1 and γ_2 be nonnegative scalars with $\gamma_1 + \gamma_2 = 1$. We will show that $\gamma_1 y_1 + \gamma_2 y_2$ belongs to $F_X(x)$. Indeed, we have

$$\gamma_1 y_1 + \gamma_2 y_2 = \gamma_1 \alpha_1 (\bar{x}_1 - x) + \gamma_2 \alpha_2 (\bar{x}_2 - x)$$

$$= (\gamma_1 \alpha_1 + \gamma_2 \alpha_2) \left(\frac{\gamma_1 \alpha_1}{\gamma_1 \alpha_1 + \gamma_2 \alpha_2} \bar{x}_1 + \frac{\gamma_2 \alpha_2}{\gamma_1 \alpha_1 + \gamma_2 \alpha_2} \bar{x}_2 - x \right)$$

$$= \alpha(\bar{x} - x),$$

where

$$\alpha = \gamma_1 \alpha_1 + \gamma_2 \alpha_2, \qquad \bar{x} = \frac{\gamma_1 \alpha_1}{\gamma_1 \alpha_1 + \gamma_2 \alpha_2} \bar{x}_1 + \frac{\gamma_2 \alpha_2}{\gamma_1 \alpha_1 + \gamma_2 \alpha_2} \bar{x}_2.$$

By the convexity of X, $\bar{x}$ belongs to X, and since $\gamma_1 y_1 + \gamma_2 y_2 = \alpha(\bar{x} - x)$, it follows that $\gamma_1 y_1 + \gamma_2 y_2 \in F_X(x)$. Therefore, $F_X(x)$ is convex.

Convexity of $T_X(x)$ will follow from the convexity of $F_X(x)$ once we show that $\mathrm{cl}\big(F_X(x)\big) = T_X(x)$, since the closure of a convex set is convex [Prop. 1.2.1(d)]. In view of part (b), it will suffice to show that $T_X(x) \subset \mathrm{cl}\big(F_X(x)\big)$. Let $y \in T_X(x)$ and, using Prop. 4.6.1, let $\{x_k\}$ be a sequence in X and $\{\alpha_k\}$ be a positive scalar sequence such that $\alpha_k \to 0$ and $(x_k - x)/\alpha_k \to y$. Since X is a convex set, the direction $(x_k - x)/\alpha_k$ is feasible at x for all k. Hence $y \in \mathrm{cl}\big(F_X(x)\big)$, and it follows that $T_X(x) \subset \mathrm{cl}\big(F_X(x)\big)$.   **Q.E.D.**

Note that if X is polyhedral, it can be seen that the cones $F_X(x)$ and $T_X(x)$ are polyhedral and hence closed, so by Prop. 4.6.2(c), they are equal. This is essentially the only case of interest where $F_X(x) = T_X(x)$ for all $x \in X$. Nonetheless, when X is convex, $F_X(x)$ is as useful as $T_X(x)$ [since its closure yields $T_X(x)$], and is often easier to work with. For example, an important class of iterative methods for minimizing a smooth cost function over a convex set X, called *methods of feasible directions*, involves the calculation of a feasible direction of X at the current iterate, and a movement along that direction by an amount that improves the cost function value (see, for example, Bertsekas [Ber99a], Chapter 2).

The Normal Cone

In addition to the cone of feasible directions and the tangent cone, there is one more conical approximation that is of special interest for the optimization topics covered in this book.

Definition 4.6.3: Given a subset X of $\Re^n$ and a vector $x \in X$, a vector $z \in \Re^n$ is said to be a *normal* of X at x if there exist sequences $\{x_k\} \subset X$ and $\{z_k\}$ with

$$x_k \to x, \qquad z_k \to z, \qquad z_k \in T_X(x_k)^*, \quad \forall\, k.$$

The set of all normals of X at x is called the *normal cone* of X at x, and is denoted by $N_X(x)$. If

$$N_X(x) = T_X(x)^*,$$

X is said to be *regular* at x.

It can be seen that $N_X(x)$ is a closed cone containing $T_X(x)^*$, but it need not be convex like $T_X(x)^*$, so it may strictly contain $T_X(x)^*$ (see the examples of Fig. 4.6.4). The notion of regularity will prove to be of special interest in the Lagrange multiplier theory of Chapter 5. An important

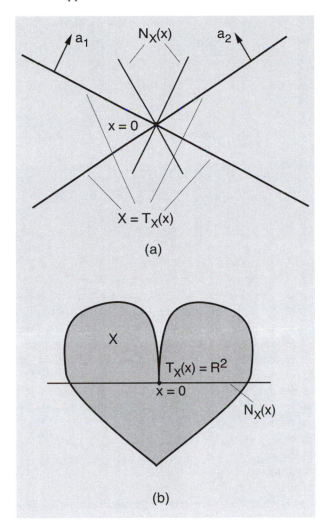

Figure 4.6.4. Examples of normal cones. In the case of figure (a), the set X is the union of two lines passing through the origin:

$$X = \left\{ x \mid (a_1'x)(a_2'x) = 0 \right\}.$$

For $x = 0$, we have $T_X(x) = X$, $T_X(x)^* = \{0\}$, while $N_X(x)$ is the nonconvex set consisting of the two lines of vectors that are collinear to either a_1 or a_2. Thus X is not regular at $x = 0$. At all other vectors $x \in X$, we have regularity with $T_X(x)^*$ and $N_X(x)$ equal to either the line of vectors that are collinear to a_1 or the line of vectors that are collinear to a_2.

In the case of figure (b), the set X is regular at all points except at $x = 0$, where we have $T_X(0) = \Re^2$, $T_X(0)^* = \{0\}$, while $N_X(0)$ is equal to the horizontal axis.

consequence of convexity of X is that it implies regularity, as shown in the following proposition.

Proposition 4.6.3: Let X be a nonempty convex subset of $\Re^n$. Then for all $x \in X$, we have

$$z \in T_X(x)^* \qquad \text{if and only if} \qquad z'(\overline{x} - x) \leq 0, \quad \forall\, \overline{x} \in X. \quad (4.28)$$

Furthermore, X is regular at all $x \in X$. In particular, we have

$$T_X(x)^* = N_X(x), \qquad T_X(x) = N_X(x)^*.$$

Proof: Since $F_X(x) \subset T_X(x)$ and $(\overline{x} - x) \in F_X(x)$ for all $\overline{x} \in X$, it follows that if $z \in T_X(x)^*$, then $z'(\overline{x} - x) \leq 0$ for all $\overline{x} \in X$. Conversely, let z be such that $z'(\overline{x} - x) \leq 0$ for all $\overline{x} \in X$, and to arrive at a contradiction, assume that $z \notin T_X(x)^*$. Then there exists some $y \in T_X(x)$ such that $z'y > 0$. Since $\mathrm{cl}\big(F_X(x)\big) = T_X(x)$ [cf. Prop. 4.6.2(c)], there exists a sequence $\{y_k\} \subset F_X(x)$ such that $y_k \to y$, so that for all k, we have $y_k = \alpha_k(x_k - x)$ for some $\alpha_k > 0$ and $x_k \in X$. Since $z'y > 0$, we have $\alpha_k z'(x_k - x) > 0$ for large enough k, which is a contradiction.

To show regularity of X at x, note that if $x \in X$ and $z \in N_X(x)$, there exist sequences $\{x_k\} \subset X$ and $\{z_k\}$ such that $x_k \to x$, $z_k \to z$, and $z_k \in T_X(x_k)^*$. By Eq. (4.28), we must have $z_k'(\overline{x} - x_k) \leq 0$ for all $\overline{x} \in X$. Taking the limit as $k \to \infty$, we obtain $z'(\overline{x} - x) \leq 0$ for all $\overline{x} \in X$, which by Eq. (4.28), implies that $z \in T_X(x)^*$. Thus, we have $N_X(x) \subset T_X(x)^*$. Since the reverse inclusion always holds, it follows that $T_X(x)^* = N_X(x)$, i.e., X is regular at x.

Finally, note that the convexity of X implies that $T_X(x)$ is a closed convex cone [cf. Prop. 4.6.2(a) and (c)], so by using the relation $T_X(x)^* = N_X(x)$ and the Polar Cone Theorem (Prop. 3.1.1), we obtain $T_X(x) = N_X(x)^*$. **Q.E.D.**

Note that convexity of $T_X(x)$ does not imply regularity of X at x, as the example of Fig. 4.6.4(b) shows. However, it can be shown that if X is closed and regular at x, then $T_X(x)$ is equal to the polar of $N_X(x)$:

$$T_X(x) = N_X(x)^* \qquad \text{if } X \text{ is closed and regular at } x.$$

This result, which will not be needed in our development, requires an elaborate proof (see Rockafellar and Wets [RoW98], p. 221). One interesting implication of the result is that for a closed set X, regularity at x implies that $T_X(x)$ is convex.

4.7 OPTIMALITY CONDITIONS

In this section we derive constrained optimality conditions for three types of problems, respectively involving:

(a) A smooth cost function and an arbitrary constraint set.

(b) A convex, possibly nonsmooth, cost function and a convex constraint set.

(c) A cost function consisting of the sum of a smooth function and a convex function, and an arbitrary constraint set.

The necessary conditions of this section relating to a local minimum can also be shown to hold if the cost function is defined and is continuously differentiable within just an open set containing the local minimum, instead of being smooth. The proofs are essentially identical to those given here.

When the constraint set is nonconvex, the tangent cone can be used as a suitable approximation to the constraint set, as illustrated in the following basic necessary condition for local optimality.

Proposition 4.7.1: Let $f : \Re^n \mapsto \Re$ be a smooth function, and let x^* be a local minimum of f over a subset X of $\Re^n$. Then

$$\nabla f(x^*)'y \geq 0, \qquad \forall \, y \in T_X(x^*).$$

If X is convex, this condition can be equivalently written as

$$\nabla f(x^*)'(x - x^*) \geq 0, \qquad \forall \, x \in X,$$

and in the case where $X = \Re^n$, it reduces to $\nabla f(x^*) = 0$.

Proof: Let y be a nonzero tangent of X at x^*. Then, there exist a sequence $\{\xi_k\} \subset \Re$ and a sequence $\{x_k\} \subset X$ such that $x_k \neq x^*$ for all k,

$$\xi_k \to 0, \qquad x_k \to x^*,$$

and

$$\frac{x_k - x^*}{\|x_k - x^*\|} = \frac{y}{\|y\|} + \xi_k.$$

By the Mean Value Theorem (Prop. 1.1.12), we have for all k,

$$f(x_k) = f(x^*) + \nabla f(\tilde{x}_k)'(x_k - x^*),$$

where $\tilde{x}_k$ is a vector that lies on the line segment joining x_k and x^*. Combining the last two equations, we obtain

$$f(x_k) = f(x^*) + \frac{\|x_k - x^*\|}{\|y\|} \nabla f(\tilde{x}_k)'y_k, \tag{4.29}$$

where

$$y_k = y + \|y\|\xi_k.$$

If $\nabla f(x^*)'y < 0$, since $\tilde{x}_k \to x^*$ and $y_k \to y$, it follows that for all sufficiently large k, $\nabla f(\tilde{x}_k)'y_k < 0$ and [from Eq. (4.29)] $f(x_k) < f(x^*)$. This contradicts the local optimality of x^*.

When X is convex, by Prop. 4.6.2, we have $\mathrm{cl}(F_X(x)) = T_X(x)$. Thus the condition already shown, $\nabla f(x^*)'y \geq 0$ for all $y \in T_X(x^*)$, can be written as

$$\nabla f(x^*)'y \geq 0, \qquad \forall\, y \in \mathrm{cl}(F_X(x)),$$

which in turn is equivalent to

$$\nabla f(x^*)'(x - x^*) \geq 0, \qquad \forall\, x \in X.$$

If $X = \Re^n$, by setting $x = x^* + e_i$ and $x = x^* - e_i$, where e_i is the ith unit vector (all components are 0 except for the ith, which is 1), we obtain $\partial f(x^*)/\partial x_i = 0$ for all $i = 1, \ldots, n$, so that $\nabla f(x^*) = 0$. **Q.E.D.**

A direction y for which $\nabla f(x^*)'y < 0$ may be viewed as a *descent direction* of f at x^*, in the sense that we have (by the definition of a gradient)

$$f(x^* + \alpha y) = f(x^*) + \alpha \nabla f(x^*)'y + o(\alpha) < f(x^*)$$

for all sufficiently small but positive α. Thus, Prop. 4.7.1 says that *if x^* is a local minimum, there is no descent direction within the tangent cone* $T_X(x^*)$.

Note that the necessary condition of Prop. 4.7.1 can equivalently be written as

$$-\nabla f(x^*) \in T_X(x^*)^*$$

(see Fig. 4.7.1). There is an interesting converse of this result, namely that given any vector $z \in T_X(x^*)^*$, there exists a smooth function f such that $-\nabla f(x^*) = z$ and x^* is a local minimum of f over X (see Gould and Tolle [GoT71] for the original version of this result, and Rockafellar and Wets [RoW98] for a refinement, given here as Exercise 4.24).

We now use directional derivatives to provide a necessary and sufficient condition for optimality in the problem of minimizing a (possibly nondifferentiable) convex function $f : \Re^n \mapsto \Re$ over a convex subset X of $\Re^n$. It can be seen that x^* is a global minimum of f over X if and only if

$$f'(x^*; x - x^*) \geq 0, \qquad \forall\, x \in X.$$

This follows using the definition (4.3) of directional derivative and the fact that the difference quotient

$$\frac{f(x^* + \alpha(x - x^*)) - f(x^*)}{\alpha}$$

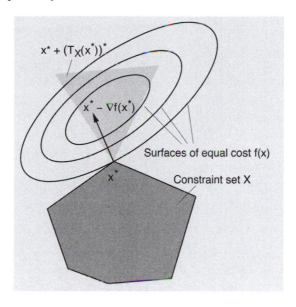

Figure 4.7.1. Illustration of the necessary optimality condition

$$-\nabla f(x^*) \in T_X(x^*)^*$$

for x^* to be a local minimum of f over X.

is a monotonically nondecreasing function of α. This leads to the following optimality condition.

Proposition 4.7.2: Let $f : \Re^n \mapsto \Re$ be a convex function. A vector x^* minimizes f over a convex subset X of $\Re^n$ if and only if there exists a subgradient $d \in \partial f(x^*)$ such that

$$d'(x - x^*) \geq 0, \qquad \forall \, x \in X.$$

Equivalently, x^* minimizes f over a convex subset X of $\Re^n$ if and only if

$$0 \in \partial f(x^*) + T_X(x^*)^*.$$

Proof: Suppose that for some $d \in \partial f(x^*)$ and all $x \in X$, we have $d'(x - x^*) \geq 0$. Then, since from the definition of a subgradient we have $f(x) - f(x^*) \geq d'(x - x^*)$ for all $x \in X$, we obtain $f(x) - f(x^*) \geq 0$ for all $x \in X$, so x^* minimizes f over X.

Conversely, suppose that x^* minimizes f over X. Then, since f is convex and hence continuous, x^* minimizes f over the closure of X, and we have $f'(x^*; x - x^*) \geq 0$ for all $x \in \text{cl}(X)$. By Prop. 4.2.2, we have

$$f'(x^*; x - x^*) = \sup_{d \in \partial f(x^*)} d'(x - x^*),$$

so it follows that

$$\sup_{d \in \partial f(x^*)} d'(x - x^*) \geq 0, \qquad \forall\, x \in \text{cl}(X).$$

Therefore,

$$\inf_{x \in \text{cl}(X) \cap \{z | \|z - x^*\| \leq 1\}} \ \sup_{d \in \partial f(x^*)} d'(x - x^*) = 0.$$

In view of the compactness and convexity of the sets over which the infimum and the supremum are taken above, we can apply the Saddle Point Theorem (Prop. 2.6.9) to interchange infimum and supremum. We conclude that

$$\sup_{d \in \partial f(x^*)} \ \inf_{x \in \text{cl}(X) \cap \{z | \|z - x^*\| \leq 1\}} d'(x - x^*) = 0,$$

and that the supremum is attained by some $d \in \partial f(x^*)$. Thus, we have $d'(x - x^*) \geq 0$ for all $x \in \text{cl}(X) \cap \{z \mid \|z - x^*\| \leq 1\}$. Since X is convex, this implies that

$$d'(x - x^*) \geq 0, \qquad \forall\, x \in X.$$

The final statement in the proposition follows from the convexity of X, which implies that $T_X(x^*)^*$ is the set of all z such that $z'(x - x^*) \leq 0$ for all $x \in X$ (cf. Props. 4.6.2 and 4.6.3). **Q.E.D.**

Note that the above proposition generalizes the optimality condition of Prop. 4.7.1 for the case where f is convex and smooth:

$$\nabla f(x^*)'(x - x^*) \geq 0, \qquad \forall\, x \in X.$$

In the special case where $X = \Re^n$, we obtain a basic necessary and sufficient condition for unconstrained optimality of x^*:

$$0 \in \partial f(x^*).$$

This optimality condition is also evident from the subgradient inequality (4.6).

We finally extend the optimality conditions of Props. 4.7.1 and 4.7.2 to the case where the cost function involves a (possibly nonconvex) smooth component and a convex (possibly nondifferentiable) component.

Proposition 4.7.3: Let x^* be a local minimum of a function $f : \Re^n \mapsto \Re$ over a subset X of $\Re^n$. Assume that the tangent cone $T_X(x^*)$ is convex, and that f has the form

$$f(x) = f_1(x) + f_2(x),$$

where f_1 is convex and f_2 is smooth. Then

$$-\nabla f_2(x^*) \in \partial f_1(x^*) + T_X(x^*)^*.$$

Proof: The proof is analogous to the one of Prop. 4.7.1. Let y be a nonzero tangent of X at x^*. Then there exists a sequence $\{\xi_k\}$ and a sequence $\{x_k\} \subset X$ such that $x_k \neq x^*$ for all k,

$$\xi_k \to 0, \qquad x_k \to x^*,$$

and

$$\frac{x_k - x^*}{\|x_k - x^*\|} = \frac{y}{\|y\|} + \xi_k.$$

We write this equation as

$$x_k - x^* = \frac{\|x_k - x^*\|}{\|y\|} y_k, \tag{4.30}$$

where

$$y_k = y + \|y\|\xi_k.$$

By the convexity of f_1, we have

$$-f_1'(x_k; x_k - x^*) \leq f_1'(x_k; x^* - x_k)$$

and

$$f_1(x_k) + f_1'(x_k; x^* - x_k) \leq f_1(x^*),$$

and by adding these inequalities, we obtain

$$f_1(x_k) \leq f_1(x^*) + f_1'(x_k; x_k - x^*), \qquad \forall\, k.$$

Also, by the Mean Value Theorem (Prop. 1.1.12), we have

$$f_2(x_k) = f_2(x^*) + \nabla f_2(\tilde{x}_k)'(x_k - x^*), \qquad \forall\, k,$$

where $\tilde{x}_k$ is a vector on the line segment joining x_k and x^*. By adding the last two relations, we see that

$$f(x_k) \leq f(x^*) + f_1'(x_k; x_k - x^*) + \nabla f_2(\tilde{x}_k)'(x_k - x^*), \qquad \forall\, k,$$

so using Eq. (4.30), we obtain

$$f(x_k) \leq f(x^*) + \frac{\|x_k - x^*\|}{\|y\|} \big(f_1'(x_k; y_k) + \nabla f_2(\tilde{x}_k)'y_k\big), \qquad \forall\, k. \quad (4.31)$$

We now show by contradiction that $f_1'(x^*; y) + \nabla f_2(x^*)'y \geq 0$. Indeed, assume that $f_1'(x^*; y) + \nabla f_2(x^*)'y < 0$. Then since $\tilde{x}_k \to x^*$, $y_k \to y$, and f_2 is smooth, it follows using also Prop. 4.1.2, that for sufficiently large k,

$$f_1'(x_k; y_k) + \nabla f_2(\tilde{x}_k)'y_k < 0,$$

and [from Eq. (4.31)] $f(x_k) < f(x^*)$. This contradicts the local optimality of x^*.

We have thus shown that

$$f_1'(x^*; y) + \nabla f_2(x^*)'y \geq 0, \qquad \forall\, y \in T_X(x^*),$$

or equivalently, by Prop. 4.2.2,

$$\max_{d \in \partial f_1(x^*)} d'y + \nabla f_2(x^*)'y \geq 0, \qquad \forall\, y \in T_X(x^*),$$

or equivalently, since $T_X(x^*)$ is a cone,

$$\min_{\substack{\|y\| \leq 1 \\ y \in T_X(x^*)}} \max_{d \in \partial f_1(x^*)} \big(d + \nabla f_2(x^*)\big)'y = 0.$$

We can now apply the Saddle Point Theorem in the above equation (cf. Prop. 2.6.9; the convexity/concavity and compactness assumptions of that proposition are satisfied). We can thus assert that there exists a $d \in \partial f_1(x^*)$ such that

$$\min_{\substack{\|y\| \leq 1 \\ y \in T_X(x^*)}} \big(d + \nabla f_2(x^*)\big)'y = 0.$$

This implies that

$$-\big(d + \nabla f_2(x^*)\big) \in T_X(x^*)^*,$$

which in turn is equivalent to $-\nabla f_2(x^*) \in \partial f_1(x^*) + T_X(x^*)^*$. **Q.E.D.**

Note that in the special case where $f_1(x) \equiv 0$, the preceding proposition yields Prop. 4.7.1. The convexity assumption on $T_X(x^*)$ is unnecessary in this case, but is essential in general. As an example, consider the subset $X = \big\{(x_1, x_2) \mid x_1 x_2 = 0\big\}$ of $\Re^2$. Then $T_X(0)^* = \{0\}$, and any convex nondifferentiable function f for which $x^* = 0$ is a global minimum over X, but $x^* = 0$ is not an unconstrained global minimum [so that $0 \notin \partial f(0)$] violates the necessary condition of Prop. 4.7.3.

In the special case where $f_2(x) \equiv 0$ and X is convex, Prop. 4.7.3 yields the necessity part of Prop. 4.7.2. More generally, when X is convex, an equivalent statement of Prop. 4.7.3 is that if x^* is a local minimum of f over X, there exists a subgradient $d \in \partial f_1(x^*)$ such that

$$\big(d + \nabla f_2(x^*)\big)'(x - x^*) \geq 0, \qquad \forall\, x \in X.$$

This is because for a convex set X, we have $z \in T_X(x^*)^*$ if and only if $z'(x - x^*) \leq 0$ for all $x \in X$ (cf. Prop. 4.6.3).

4.8 NOTES, SOURCES, AND EXERCISES

Subdifferential theory was systematically developed in Fenchel's 1951 lecture notes [Fen51]. There are some interesting results relating to differentiable convex functions, which we have not covered. For example, the set of points D where a convex function $f : \Re^n \mapsto \Re$ is differentiable is a dense set, and its complement has measure zero. Furthermore, the gradient mapping is continuous over D. These and other related results can be found in Rockafellar [Roc70].

The notion of a tangent and the tangent cone originated with Bouligand [Bou30], [Bou32]. The normal cone was introduced by Mordukhovich [Mor76]. Related conical approximations are central in the work of Clarke on nonsmooth analysis [Cla83]. Our definition of regularity is related to the definition of "regularity in the sense of Clarke" used by Rockafellar and Wets [RoW98]: the two definitions coincide in the case where X is locally closed at x, i.e., there is a closed sphere S centered at x such that $X \cap S$ is closed. The book by Rockafellar and Wets [RoW98] contains a lot of material on conical approximations. See also Aubin and Frankowska [AuF90], and Borwein and Lewis [BoL00].

E X E R C I S E S

4.1 (Directional Derivative of Extended Real-Valued Functions)

Let $f : \Re^n \mapsto (-\infty, \infty]$ be a convex function, and let x be a vector in $\mathrm{dom}(f)$. Define

$$f'(x; y) = \inf_{\alpha > 0} \frac{f(x + \alpha y) - f(x)}{\alpha}, \qquad y \in \Re^n.$$

Show the following:

(a) $f'(x; \lambda y) = \lambda f'(x; y)$ for all $\lambda \geq 0$ and $y \in \Re^n$.

(b) $f'(x; \cdot)$ is a convex function.

(c) $-f'(x; -y) \leq f'(x; y)$ for all $y \in \Re^n$.

(d) If $\mathrm{dom}(f) = \Re^n$, then the level set $\{y \mid f'(x; y) \leq 0\}$ is a closed convex cone and its polar is given by

$$\Big(\{y \mid f'(x; y) \leq 0\}\Big)^* = \mathrm{cl}\Big(\mathrm{cone}\big(\partial f(x)\big)\Big).$$

4.2 (Chain Rule for Directional Derivatives)

Let $f : \Re^n \mapsto \Re^m$ and $g : \Re^m \mapsto \Re$ be some functions, and let x be a vector in $\Re^n$. Suppose that all the components of f and g are directionally differentiable at x, and that g is such that for all $w \in \Re^m$,

$$g'(y; w) = \lim_{\alpha \downarrow 0, \ z \to w} \frac{g(y + \alpha z) - g(y)}{\alpha}.$$

Then, the composite function $F(x) = g(f(x))$ is directionally differentiable at x and the following chain rule holds:

$$F'(x; d) = g'\big(f(x); f'(x; d)\big), \qquad \forall \, d \in \Re^n.$$

4.3

Let $f : \Re^n \mapsto \Re$ be a convex function. Show that a vector $d \in \Re^n$ is a subgradient of f at x if and only if the function $d'y - f(y)$ attains its maximum at $y = x$.

4.4

Show that:

(a) For the function $f(x) = \|x\|$, we have

$$\partial f(x) = \begin{cases} \{x/\|x\|\} & \text{if } x \neq 0, \\ \{d \mid \|d\| \leq 1\} & \text{if } x = 0. \end{cases}$$

(b) For a nonempty convex subset C of $\Re^n$ and the function $f : \Re^n \mapsto (-\infty, \infty]$ given by

$$f(x) = \begin{cases} 0 & \text{if } x \in C, \\ \infty & \text{if } x \notin C, \end{cases}$$

we have

$$\partial f(x) = \begin{cases} N_C(x) & \text{if } x \in C, \\ \varnothing & \text{if } x \notin C. \end{cases}$$

4.5

Show that for a scalar convex function $f : \Re \mapsto \Re$, we have

$$\partial f(x) = \big\{d \mid f^-(x) \leq d \leq f^+(x)\big\}, \qquad \forall \, x \in \Re.$$

4.6

Let $f : \Re^n \mapsto \Re$ be a convex function, and let x and y be given vectors in $\Re^n$. Consider the scalar function $\varphi : \Re \mapsto \Re$ defined by $\varphi(t) = f\big(tx + (1 - t)y\big)$ for all $t \in \Re$, and show that

$$\partial\varphi(t) = \big\{ (x - y)'d \,\big|\, d \in \partial f\big(tx + (1 - t)y\big) \big\}, \qquad \forall\, t \in \Re.$$

Hint: Apply the Chain Rule [Prop. 4.2.5(a)].

4.7

Let $f : \Re^n \mapsto \Re$ be a convex function, and let X be a nonempty bounded subset of $\Re^n$. Show that f is Lipschitz continuous over X, i.e., that there exists a scalar L such that

$$\big|f(x) - f(y)\big| \le L \,\|x - y\|, \qquad \forall\, x, y \in X.$$

Show also that

$$f'(x; y) \le L \,\|y\|, \qquad \forall\, x \in X, \quad \forall\, y \in \Re^n.$$

Hint: Use the boundedness property of the subdifferential (Prop. 4.2.3).

4.8 (Nonemptiness of Subdifferential)

Let $f : \Re^n \mapsto (-\infty, \infty]$ be a proper convex function, and let x be a vector in $\mathrm{dom}(f)$. Show that $\partial f(x)$ is nonempty if and only if $f'(x; z - x)$ is finite for all $z \in \mathrm{dom}(f)$.

4.9 (Subdifferential of Sum of Extended Real-Valued Functions)

This exercise is a refinement of Prop. 4.2.4. Let $f_i : \Re^n \mapsto (-\infty, \infty]$, $i = 1, \ldots, m$, be convex functions, and let $f = f_1 + \cdots + f_m$. Show that

$$\partial f_1(x) + \cdots + \partial f_m(x) \subset \partial f(x), \qquad \forall\, x \in \Re^n.$$

Furthermore, if

$$\cap_{i=1}^m \mathrm{ri}\big(\mathrm{dom}(f_i)\big) \ne \varnothing,$$

then

$$\partial f_1(x) + \cdots + \partial f_m(x) = \partial f(x), \qquad \forall\, x \in \Re^n.$$

In addition, if the functions f_i, $i = r + 1, \ldots, m$, are polyhedral, the preceding relation holds under the weaker assumption that

$$\Big(\cap_{i=1}^r \mathrm{ri}\big(\mathrm{dom}(f_i)\big) \Big) \cap \big(\cap_{i=r+1}^m \mathrm{dom}(f_i) \big) \ne \varnothing, \qquad \forall\, x \in \Re^n.$$

4.10 (Chain Rule for Extended Real-Valued Functions)

This exercise is a refinement of Prop. 4.2.5(a). Let $f : \Re^m \mapsto (-\infty, \infty]$ be a convex function, and let A be an $m \times n$ matrix. Assume that the range of A contains a point in the relative interior of $\text{dom}(f)$. Then, the subdifferential of the function F, defined by

$$F(x) = f(Ax),$$

is given by

$$\partial F(x) = A' \partial f(Ax).$$

4.11

Let $f : \Re^n \mapsto \Re$ be a convex function, and let X be a bounded subset of $\Re^n$. Show that for all $\epsilon > 0$, the set $\cup_{x \in X} \partial_\epsilon f(x)$ is bounded.

4.12

Let $f : \Re^n \mapsto \Re$ be a convex function. Show that for all $x \in \Re^n$, we have

$$\cap_{\epsilon > 0} \partial_\epsilon f(x) = \partial f(x).$$

4.13 (Continuity Properties of ϵ-Subdifferential [Nur77])

Let $f : \Re^n \mapsto \Re$ be a convex function and let ϵ be a positive scalar. Show that for every $x \in \Re^n$, the following hold:

(a) If a sequence $\{x_k\}$ converges to x and $d_k \in \partial_\epsilon f(x_k)$ for all k, then the sequence $\{d_k\}$ is bounded and each of its limit points is an ϵ-subgradient of f at x.

(b) If $d \in \partial_\epsilon f(x)$, then for every sequence $\{x_k\}$ converging to x, there exists a subsequence $\{d_k\}_K$ converging to d with $d_k \in \partial_\epsilon f(x_k)$ for all $k \in K$.

4.14 (Subgradient Mean Value Theorem)

(a) *Scalar Case*: Let $\varphi : \Re \mapsto \Re$ be a scalar convex function, and let a and b be scalars with $a < b$. Show that there exists a scalar $t^* \in (a, b)$ such that

$$\frac{\varphi(b) - \varphi(a)}{b - a} \in \partial \varphi(t^*).$$

Hint: Show that the scalar convex function

$$g(t) = \varphi(t) - \varphi(a) - \frac{\varphi(b) - \varphi(a)}{b - a}(t - a)$$

has a minimum $t^* \in (a, b)$, and use the optimality condition $0 \in \partial g(t^*)$.

(b) *Vector Case:*   Let $f : \Re^n \mapsto \Re$ be a convex function, and let x and y be vectors in $\Re^n$. Show that there exist a scalar $\alpha \in (0, 1)$ and a subgradient $d \in \partial f\big(\alpha x + (1 - \alpha)y\big)$ such that

$$f(y) = f(x) + d'(y - x).$$

Hint: Apply part (a) to the scalar function $\varphi(t) = f\big(tx + (1 - t)y\big)$, $t \in \Re$.

4.15 (Steepest Descent Direction of a Convex Function)

Let $f : \Re^n \mapsto \Re$ be a convex function and let x be a vector in $\Re^n$. Show that a vector $\bar{d}$ is the vector of minimum norm in $\partial f(x)$ if and only if either $\bar{d} = 0$ or else $\bar{d}/\|\bar{d}\|$ minimizes $f'(x; d)$ over all d with $\|d\| \le 1$.

4.16 (Generating Descent Directions of Convex Functions)

This exercise provides a method for generating a descent direction in circumstances where obtaining a single subgradient is relatively easy.

Let $f : \Re^n \mapsto \Re$ be a convex function, and let x be a fixed vector in $\Re^n$. A vector $d \in \Re^n$ is said to be a *descent direction* of f at x if the corresponding directional derivative of f satisfies

$$f'(x; d) < 0.$$

Assume that x does not minimize f, and let g_1 be a subgradient of f at x. For $k = 2, 3, \ldots$, let w_k be the vector of minimum norm in the convex hull of $g_1, \ldots, g_{k-1}$,

$$w_k = \arg \min_{g \in \mathrm{conv}\{g_1, \ldots, g_{k-1}\}} \|g\|.$$

If $-w_k$ is a descent direction of f at x, then stop; else let g_k be a vector in $\partial f(x)$ such that

$$g_k' w_k = \min_{g \in \partial f(x)} g' w_k.$$

Show that this process terminates in a finite number of steps with a descent direction of f at x. *Hint*: If $-w_k$ is not a descent direction, then $g_i' w_k \ge \|w_k\|^2 \ge \|g^*\|^2 > 0$ for all $i = 1, \ldots, k - 1$, where g^* is the subgradient of f at x with minimum norm, while at the same time $g_k' w_k \le 0$. Consider a limit point of $\big\{(w_k, g_k)\big\}$.

4.17 (Generating ϵ-Descent Directions of Convex Functions [Lem74])

This exercise shows how the procedure of Exercise 4.16 can be modified so that it generates an ϵ-descent direction.

Let $f : \Re^n \mapsto \Re$ be a convex function, and let x be a fixed vector in $\Re^n$. Assume that x is not an ϵ-minimizer of the function f, i.e., $f(x) > \inf_{z \in \Re^n} f(z) + \epsilon$, and let g_1 be an ϵ-subgradient of f at x. For $k = 2, 3, \ldots$, let w_k be the vector of minimum norm in the convex hull of $g_1, \ldots, g_{k-1}$,

$$ w_k = \arg \min_{g \in \text{conv}\{g_1, \ldots, g_{k-1}\}} \|g\|. $$

By a search along the direction $-w_k$, determine whether there exists a scalar $\overline{\alpha}$ such that $f(x - \overline{\alpha} w_k) < f(x) - \epsilon$. If such an $\overline{\alpha}$ exists, then stop ($-w_k$ is an ϵ-descent direction of f at x); otherwise, let g_k be a vector in $\partial_\epsilon f(x)$ such that

$$ g_k' w_k = \min_{g \in \partial_\epsilon f(x)} g' w_k. $$

Show that this process will terminate in a finite number of steps with an ϵ-descent direction of f at x.

4.18

For the following subsets C of $\Re^n$, specify the tangent cone and the normal cone at every point of C.

(a) C is the unit ball.

(b) C is a subspace.

(c) C is a closed halfspace, i.e., $C = \{x \mid a'x \leq b\}$ for a nonzero vector $a \in \Re^n$ and a scalar b.

(d) $C = \{x \mid x_i \geq 0, \ i \in I\}$ with $I \subset \{1, \ldots, n\}$.

4.19

Let C be a convex subset of $\Re^n$, and let x be a vector in C. Show that the following properties are equivalent:

(a) x lies in the relative interior of C.

(b) $T_C(x)$ is a subspace.

(c) $N_C(x)$ is a subspace.

4.20 (Tangent and Normal Cones of Affine Sets)

Let A be an $m \times n$ matrix and b be a vector in $\Re^m$. Show that the tangent cone and the normal cone of the set $\{x \mid Ax = b\}$ at any of its points are the null space of A and the range space of A', respectively.

4.21 (Tangent and Normal Cones of Level Sets)

Let $f : \Re^n \mapsto \Re$ be a convex function, and let x be a vector in $\Re^n$ such that the level set $\{z \mid f(z) \leq f(x)\}$ is nonempty. Show that the tangent cone and the normal cone of the level set $\{z \mid f(z) \leq f(x)\}$ at the point x coincide with $\{y \mid f'(x; y) \leq 0\}$ and $\mathrm{cl}\big(\mathrm{cone}(\partial f(x))\big)$, respectively. Furthermore, if x does not minimize f over $\Re^n$, the closure operation is unnecessary.

4.22

Let $C_i \subset \Re^{n_i}$, $i = 1, \ldots, m$, be convex sets and let $x_i \in C_i$ for all i. Show that

$$T_{C_1 \times \cdots \times C_m}(x_1, \ldots, x_m) = T_{C_1}(x_1) \times \cdots \times T_{C_m}(x_m),$$

$$N_{C_1 \times \cdots \times C_m}(x_1, \ldots, x_m) = N_{C_1}(x_1) \times \cdots \times N_{C_m}(x_m).$$

4.23 (Tangent and Normal Cone Relations)

Let C_1, C_2, and C be nonempty convex subsets of $\Re^n$. Show the following properties:

(a) We have

$$N_{C_1 \cap C_2}(x) \supset N_{C_1}(x) + N_{C_2}(x), \qquad \forall\, x \in C_1 \cap C_2,$$

$$T_{C_1 \cap C_2}(x) \subset T_{C_1}(x) \cap T_{C_2}(x), \qquad \forall\, x \in C_1 \cap C_2.$$

Furthermore, if $\mathrm{ri}(C_1) \cap \mathrm{ri}(C_2)$ is nonempty, the preceding relations hold with equality. This is also true if $\mathrm{ri}(C_1) \cap C_2$ is nonempty and the set C_2 is polyhedral.

(b) For $x_1 \in C_1$ and $x_2 \in C_2$, we have

$$N_{C_1 + C_2}(x_1 + x_2) = N_{C_1}(x_1) \cap N_{C_2}(x_2),$$

$$T_{C_1 + C_2}(x_1 + x_2) = \mathrm{cl}\big(T_{C_1}(x_1) + T_{C_2}(x_2)\big).$$

(c) For an $m \times n$ matrix A and any $x \in C$, we have

$$N_{AC}(Ax) = (A')^{-1} \cdot N_C(x), \qquad T_{AC}(Ax) = \mathrm{cl}\big(A \cdot T_C(x)\big).$$

4.24 [GoT71], [RoW98]

Let C be a subset of $\Re^n$ and let $x^* \in C$. Show that for every $y \in T_C(x^*)^*$ there is a smooth function f with $-\nabla f(x^*) = y$, and such that x^* is the unique global minimum of f over C.

4.25

Let C_1, C_2, and C_3 be nonempty closed subsets of $\Re^n$. Consider the problem of finding a triangle with minimum perimeter that has one vertex on each of the sets, i.e., the problem of minimizing $\|x_1 - x_2\| + \|x_2 - x_3\| + \|x_3 - x_1\|$ subject to $x_i \in C_i$, $i = 1, 2, 3$, and the additional condition that x_1, x_2, and x_3 do not lie on the same line. Show that if (x_1^*, x_2^*, x_3^*) defines an optimal triangle, there exists a vector z^* in the triangle such that

$$(z^* - x_i^*) \in T_{C_i}(x_i^*)^*, \qquad i = 1, 2, 3.$$

4.26

Consider the problem of minimizing a convex function $f : \Re^n \mapsto \Re$ over the polyhedral set

$$X = \{x \mid a_j' x \le b_j, \, j = 1, \ldots, r\}.$$

Show that x^* is an optimal solution if and only if there exist scalars $\mu_1^*, \ldots, \mu_r^*$ such that

(i) $\mu_j^* \ge 0$ for all j, and $\mu_j^* = 0$ for all j such that $a_j' x^* < b_j$.

(ii) $0 \in \partial f(x^*) + \sum_{j=1}^r \mu_j^* a_j$.

Hint: Characterize the cone $T_X(x^*)^*$, and use Prop. 4.7.2 and Farkas' Lemma.

4.27 (Quasiregularity)

Let $f : \Re^n \mapsto \Re$ be a smooth function, let X be a subset of $\Re^n$, and let x^* be a local minimum of f over X. Denote $D(x) = \{y \mid \nabla f(x)'y < 0\}$.

(a) Show that $D(x^*) \cap T_X(x^*) = \emptyset$.

(b) Suppose that X has the form

$$X = \{x \mid h_1(x) = 0, \ldots, h_m(x) = 0, \, g_1(x) \le 0, \ldots, g_r(x) \le 0\},$$

where the functions $h_i : \Re^n \mapsto \Re$, $i = 1, \ldots, m$, and $g_j : \Re^n \mapsto \Re$, $j = 1, \ldots, r$, are smooth. For any $x \in X$ consider the cone

$$V(x) = \{y \mid \nabla h_i(x)'y = 0, \, i = 1, \ldots, m, \, \nabla g_j(x)'y \le 0, \, j \in A(x)\},$$

where $A(x) = \{j \mid g_j(x) = 0\}$. Show that $T_X(x) \subset V(x)$.

(c) Use Farkas' Lemma and part (a) to show that if $T_X(x^*) = V(x^*)$, then there exist scalars $\lambda_1^*, \ldots, \lambda_m^*$ and $\mu_1^*, \ldots, \mu_r^*$, satisfying

$$\nabla f(x^*) + \sum_{i=1}^m \lambda_i^* \nabla h_i(x^*) + \sum_{j=1}^r \mu_j^* \nabla g_j(x^*) = 0,$$

$$\mu_j^* \ge 0, \quad \forall \, j = 1, \ldots, r, \qquad \mu_j^* = 0, \quad \forall \, j \in A(x^*).$$

Note: The condition $T_X(x^*) = V(x^*)$ is called *quasiregularity* at x^*, and will be discussed further in Chapter 5.

5

Lagrange Multipliers

Contents

5.1. Introduction to Lagrange Multipliers . . . . p. 270
5.2. Enhanced Fritz John Optimality Conditions . . . . . p. 281
5.3. Informative Lagrange Multipliers . . . . p. 288
    5.3.1. Sensitivity . . . . p. 297
    5.3.2. Alternative Lagrange Multipliers . . . . p. 299
5.4. Pseudonormality and Constraint Qualifications . . . . p. 302
5.5. Exact Penalty Functions . . . . p. 313
5.6. Using the Extended Representation . . . . p. 319
5.7. Extensions Under Convexity Assumptions . . . . . . p. 324
5.8. Notes, Sources, and Exercises . . . . p. 335

The optimality conditions developed in Section 4.7 apply to any type of constraint set. On the other hand, the constraint set of an optimization problem is usually specified in terms of equality and inequality constraints. If we take into account this structure, we can obtain more refined optimality conditions, involving *Lagrange multipliers*. These multipliers facilitate the characterization of optimal solutions and often play an important role in computational methods. They are also central in duality theory, as will be discussed in Chapter 6.

In this chapter, we provide a new treatment of Lagrange multiplier theory. It is simple and it is also more powerful than the classical treatments. In particular, it establishes new necessary conditions that non-trivially supplement the classical Karush-Kuhn-Tucker conditions. It also deals effectively with problems where in addition to equality and inequality constraints, there is an additional abstract set constraint.

The starting point for our development is an enhanced version of the classical Fritz John necessary conditions, which is given in Section 5.2. In Section 5.3, we provide a classification of different types of Lagrange multipliers, and in Section 5.4, we introduce the notion of pseudonormality, which unifies and expands the conditions that guarantee the existence of Lagrange multipliers. In the final sections of the chapter, we discuss various topics related to Lagrange multipliers, including their connection with exact penalty functions.

5.1 INTRODUCTION TO LAGRANGE MULTIPLIERS

Let us provide some orientation by first considering a problem with equality constraints of the form

$$\text{minimize} \quad f(x)$$
$$\text{subject to} \quad h_i(x) = 0, \qquad i = 1, \ldots, m.$$

We assume that $f : \Re^n \mapsto \Re$ and $h_i : \Re^n \mapsto \Re$, $i = 1, \ldots, m$, are smooth functions. The basic Lagrange multiplier theorem for this problem states that, under appropriate assumptions, for a given local minimum x^*, there exist scalars $\lambda_1^*, \ldots, \lambda_m^*$, called *Lagrange multipliers*, such that

$$\nabla f(x^*) + \sum_{i=1}^m \lambda_i^* \nabla h_i(x^*) = 0. \tag{5.1}$$

This condition can be viewed as n equations, which together with the m constraints $h_i(x^*) = 0$, form a system of $n + m$ equations with $n + m$ unknowns, the vector x^* and the multipliers λ_i^*. Thus, a constrained optimization problem can be "transformed" into a problem of solving a system of nonlinear equations. This is the traditional rationale for the importance of Lagrange multipliers.

To see why Lagrange multipliers may ordinarily be expected to exist, consider the case where the functions h_i are affine, so that

$$h_i(x) = a_i'x - b_i, \qquad i = 1, \dots, m,$$

for some vectors a_i and scalars b_i. Then it can be seen that the tangent cone of the constraint set at a local minimum x^* is

$$T(x^*) = \{y \mid a_i'y = 0, \ i = 1, \dots, m\},$$

and according to the necessary conditions of Section 4.7, we have

$$-\nabla f(x^*) \in T(x^*)^*.$$

Since $T(x^*)$ is the nullspace of the $m \times n$ matrix having $a_1', \dots, a_m'$ as rows, $T(x^*)^*$ is the range space of the matrix having $a_1, \dots, a_m$ as columns. It follows that $-\nabla f(x^*)$ can be expressed as a linear combination of the vectors $a_1, \dots, a_m$, so that for some scalars λ_i^*, we have

$$\nabla f(x^*) + \sum_{i=1}^{m} \lambda_i^* a_i = 0.$$

In the general case where the constraint functions h_i are nonlinear, the argument above would work if we could guarantee that the tangent cone $T(x^*)$ can be represented as

$$T(x^*) = \{y \mid \nabla h_i(x^*)'y = 0, \ i = 1, \dots, m\}. \tag{5.2}$$

Unfortunately, additional assumptions are needed to guarantee the validity of this representation. One such assumption, called *Linear Independence Constraint Qualification* (LICQ), is that the gradients $\nabla h_i(x^*)$ are linearly independent; this will be shown in Section 5.4. Thus, when x^* satisfies the LICQ, there exist Lagrange multipliers. However, in general the equality (5.2) may fail, and there may not exist Lagrange multipliers, as shown by the following example.

Example 5.1.1: (A Problem with no Lagrange Multipliers)

Consider the problem of minimizing

$$f(x) = x_1 + x_2$$

subject to the two constraints

$$h_1(x) = (x_1 + 1)^2 + x_2^2 - 1 = 0, \qquad h_2(x) = (x_1 - 2)^2 + x_2^2 - 4 = 0.$$

The geometry of this problem is illustrated in Fig. 5.1.1. It can be seen that at the local minimum $x^* = (0,0)$ (the only feasible solution), the cost gradient $\nabla f(x^*) = (1,1)$ cannot be expressed as a linear combination of the constraint gradients $\nabla h_1(x^*) = (2,0)$ and $\nabla h_2(x^*) = (-4,0)$. Thus the Lagrange multiplier condition

$$\nabla f(x^*) + \lambda_1^* \nabla h_1(x^*) + \lambda_2^* \nabla h_2(x^*) = 0$$

cannot hold for any λ_1^* and λ_2^*. The difficulty here is that $T(x^*) = \{0\}$, while the set $\{y \mid \nabla h_1(x^*)'y = 0, \ \nabla h_2(x^*)'y = 0\}$ is equal to $\{(0,\gamma) \mid \gamma \in \Re\}$, so the characterization (5.2) fails.

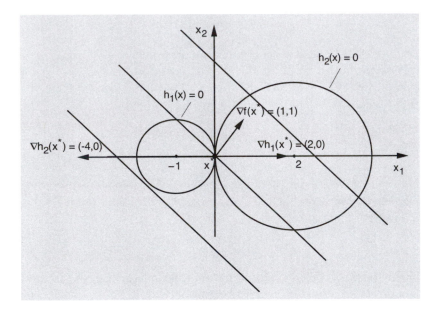

Figure 5.1.1. Illustration of how Lagrange multipliers may not exist (cf. Example 5.1.1). The problem is

$$\text{minimize} \quad f(x) = x_1 + x_2$$
$$\text{subject to} \quad h_1(x) = (x_1 + 1)^2 + x_2^2 - 1 = 0,$$
$$h_2(x) = (x_1 - 2)^2 + x_2^2 - 4 = 0,$$

with a local minimum $x^* = (0,0)$ (the only feasible solution). Here, $\nabla f(x^*)$ cannot be expressed as a linear combination of $\nabla h_1(x^*)$ and $\nabla h_2(x^*)$, so there are no Lagrange multipliers.

Fritz John Conditions

In summary, based on the preceding discussion, there are two possibilities at a local minimum x^*:

(a) The gradients $\nabla h_i(x^*)$ are linearly independent. Then, there exist scalars/Lagrange multipliers $\lambda_1^*, \ldots, \lambda_m^*$ such that

$$\nabla f(x^*) + \sum_{i=1}^{m} \lambda_i^* \nabla h_i(x^*) = 0.$$

(b) The gradients $\nabla h_i(x^*)$ are linearly dependent, so there exist scalars $\lambda_1^*, \ldots, \lambda_m^*$, not all equal to 0, such that

$$\sum_{i=1}^{m} \lambda_i^* \nabla h_i(x^*) = 0.$$

These two possibilities can be lumped into a single condition: at a local minimum x^*, there exist scalars $\mu_0, \lambda_1, \ldots, \lambda_m$, not all equal to 0, such that $\mu_0 \geq 0$ and

$$\mu_0 \nabla f(x^*) + \sum_{i=1}^{m} \lambda_i \nabla h_i(x^*) = 0. \tag{5.3}$$

Possibility (a) corresponds to the case where $\mu_0 > 0$, in which case the scalars $\lambda_i^* = \lambda_i / \mu_0$ are Lagrange multipliers. Possibility (b) corresponds to the case where $\mu_0 = 0$, in which case the condition (5.3) provides no information regarding the existence of Lagrange multipliers.

Necessary conditions that involve a multiplier μ_0 for the cost gradient, such as the one of Eq. (5.3), are known as *Fritz John necessary conditions*, and were first given by John [Joh48]. These conditions can be extended to inequality-constrained problems, and they hold without any further assumptions on x^* (such as the LICQ). However, this extra generality comes at a price, because the issue of whether the cost multiplier μ_0 can be taken to be positive is left unresolved. Unfortunately, asserting that $\mu_0 > 0$ is nontrivial under some commonly used assumptions, and for this reason, traditionally Fritz John conditions in their classical form have played a somewhat peripheral role in the development of Lagrange multiplier theory. Nonetheless, the Fritz John conditions, when properly strengthened, can provide a simple and powerful line of analysis, as we will see in the next section.

Sensitivity

Lagrange multipliers frequently have an interesting interpretation in specific practical contexts. In economic applications they can often be interpreted as prices, while in other problems they represent quantities with concrete physical meaning. It turns out that within our mathematical framework, they can be viewed as rates of change of the optimal cost as the level of constraint changes. This is fairly easy to show for the case of linear and independent constraints, as indicated in Fig. 5.1.2.

When the constraints are nonlinear, the sensitivity interpretation of Lagrange multipliers is valid, provided some assumptions are satisfied. Typically, these assumptions include the linear independence of the constraint gradients, but also additional conditions involving second derivatives (see Exercises 5.1-5.3, or a nonlinear programming textbook, such as Bertsekas [Ber99a]).

Inequality Constraints

The preceding discussion can be extended to the case where there are both equality and inequality constraints. Consider for example the case of linear inequality constraints:

$$\text{minimize} \ \ f(x)$$
$$\text{subject to} \ \ a_j' x \le b_j, \qquad j = 1, \dots, r.$$

The tangent cone of the constraint set at a local minimum x^* is given by

$$T(x^*) = \{y \mid a_j' y \le 0, \, j \in A(x^*)\},$$

where $A(x^*)$ is the set of indices of the *active* constraints, i.e., those constraints that are satisfied as equations at x^*,

$$A(x^*) = \{j \mid a_j' x^* = 0\}.$$

The necessary condition $-\nabla f(x^*) \in T(x^*)^*$ can be shown to be equivalent to the existence of scalars μ_j^* such that

$$\nabla f(x^*) + \sum_{j=1}^{r} \mu_j^* a_j = 0,$$

$$\mu_j^* \ge 0, \quad \forall \, j = 1, \dots, r, \qquad \mu_j^* = 0, \quad \forall \, j \in A(x^*).$$

This follows from the characterization of the cone $T(x^*)^*$ as the cone generated by the vectors a_j, $j \in A(x^*)$ (Farkas' Lemma). The above necessary conditions, properly generalized for nonlinear constraints, are known as the *Karush-Kuhn-Tucker conditions.*

Similar to the equality-constrained case, the typical method to ascertain existence of Lagrange multipliers for the general nonlinearly-constrained problem

$$\text{minimize} \ \ f(x)$$
$$\text{subject to} \ \ h_i(x) = 0, \ \ i = 1, \dots, m, \ \ g_j(x) \le 0, \ \ j = 1, \dots, r, \tag{5.4}$$

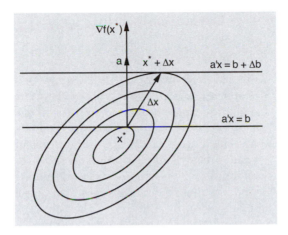

Figure 5.1.2. Sensitivity interpretation of Lagrange multipliers. Consider first a problem involving a single linear constraint,

$$\text{minimize } f(x)$$

$$\text{subject to } a'x = b,$$

where $a \neq 0$. Let x^* be a local minimum and λ^* be the unique corresponding Lagrange multiplier. If the level of constraint b is changed to $b + \Delta b$, the minimum x^* will change to $x^* + \Delta x$. Since $b + \Delta b = a'(x^* + \Delta x) = a'x^* + a'\Delta x = b + a'\Delta x$, we see that the variations Δx and Δb are related by

$$a'\Delta x = \Delta b.$$

Using the Lagrange multiplier condition $\nabla f(x^*) = -\lambda^* a$, the corresponding cost change can be written as

$$\Delta \text{cost} = f(x^* + \Delta x) - f(x^*) = \nabla f(x^*)'\Delta x + o(\|\Delta x\|) = -\lambda^* a'\Delta x + o(\|\Delta x\|).$$

By combining the above two relations, we obtain $\Delta \text{cost} = -\lambda^* \Delta b + o(\|\Delta x\|)$, so up to first order we have

$$\lambda^* = -\frac{\Delta \text{cost}}{\Delta b}.$$

Thus, the Lagrange multiplier λ^* gives the rate of optimal cost decrease as the level of constraint increases.

In the case of multiple constraints $a_i'x = b_i$, $i = 1, \ldots, m$, the preceding argument can be appropriately modified. In particular, assuming that the a_i are linearly independent, the system $a_i'x = b_i + \Delta b_i$ has a solution $x + \Delta x$ for any set for changes Δb_i, and we have

$$\Delta \text{cost} = f(x^* + \Delta x) - f(x^*) = \nabla f(x^*)'\Delta x + o(\|\Delta x\|) = -\sum_{i=1}^{m} \lambda_i^* a_i'\Delta x + o(\|\Delta x\|),$$

and $a_i'\Delta x = \Delta b_i$ for all i, so we obtain $\Delta \text{cost} = -\sum_{i=1}^{m} \lambda_i^* \Delta b_i + o(\|\Delta x\|)$.

is to assume structure which guarantees that the tangent cone has the form

$$T(x^*) = \{y \mid \nabla h_i(x^*)'y = 0, \ i = 1, \ldots, m, \nabla g_j(x^*)'y \leq 0, \ j \in A(x^*)\};$$
(5.5)

a condition known as *quasiregularity* (see also Exercise 4.27). This is the classical line of development of Lagrange multiplier theory: easily verifiable conditions implying quasiregularity (known as *constraint qualifications*) are established, and the existence of Lagrange multipliers is then inferred using Farkas' Lemma. This line of analysis is insightful and intuitive, but has traditionally required fairly complex proofs to show that specific constraint qualifications imply the quasiregularity condition (5.5). A more serious difficulty, however, is that the analysis based on quasiregularity does not extend to the case where, in addition to the equality and inequality constraints, there is an additional abstract set constraint $x \in X$.

Uniqueness of Lagrange Multipliers

For a given local minimum, the set of Lagrange multiplier vectors, call it M, may contain more than one element. In the case of equality constraints only, M contains a single element if and only if the gradients $\nabla h_1(x^*), \ldots, \nabla h_m(x^*)$ are linearly independent. In fact, in this case, if M is nonempty, it is an affine set, which is a translation of the nullspace of the $n \times m$ matrix

$$\nabla h(x^*) = \begin{bmatrix} \nabla h_1(x^*) & \cdots & \nabla h_m(x^*) \end{bmatrix}.$$

For the mixed equality/inequality constrained case of problem (5.4), it can be seen that the set of Lagrange multipliers

$$M = \left\{ (\lambda, \mu) \mid \nabla f(x^*) + \sum_{i=1}^{m} \lambda_i \nabla h_i(x^*) + \sum_{j=1}^{r} \mu_j \nabla g_j(x^*) = 0, \right.$$

$$\left. \mu \geq 0, \ \mu_j = 0, \ j \in A(x^*) \right\}$$

is a polyhedral set, which (if nonempty) may be bounded or unbounded.

What kind of sensitivity information do Lagrange multipliers convey when they are not unique? The classical theory does not offer satisfactory answers to this question. The following example and the accompanying Fig. 5.1.3 indicate that while some of the Lagrange multipliers are special and provide some sensitivity information, other Lagrange multipliers are meaningless from a sensitivity point of view.

Example 5.1.2:

Assume that there are no equality constraints, that f and g_j are affine functions,

$$f(x) = c'x, \qquad g_j(x) = a_j'x - b_j, \quad j = 1, \ldots, r,$$

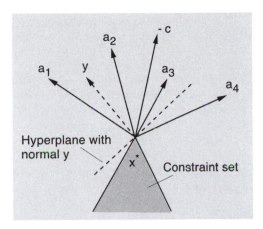

Figure 5.1.3. Illustration of Lagrange multipliers for the case of a two-dimensional problem with linear cost and four linear inequality constraints (cf. Example 5.1.2). We assume that $-c$ and a_1, a_2, a_3, a_4 lie in the same halfspace, and $-c$ lies "between" a_1 and a_2 on one side, and a_3 and a_4 on the other side, as shown in the figure. For $(\mu_1, \mu_2, \mu_3, \mu_4)$ to be a Lagrange multiplier vector, we must have $\mu_j \geq 0$ for all j, and

$$-c = \mu_1 a_1 + \mu_2 a_2 + \mu_3 a_3 + \mu_4 a_4.$$

There exist exactly four Lagrange multiplier vectors with two positive multipliers/components and the other two components equal to zero. It can be seen that it is impossible to improve the cost by moving away from x^*, while violating the constraints with positive multipliers but not violating any constraints with zero multipliers. For example, when

$$\mu_1 = 0, \quad \mu_2 > 0, \quad \mu_3 > 0, \quad \mu_4 = 0,$$

moving from x^* along a direction y such that $c'y < 0$ will violate either the 1st or the 4th inequality constraint. To see this, note that to improve the cost while violating only the constraints with positive multipliers, the direction y must be the normal of a hyperplane that contains the vectors $-c$, a_2, and a_3 in one of its closed halfspaces, and the vectors a_1 and a_4 in the complementary subspace (see the figure).

There also exist an infinite number of Lagrange multiplier vectors where all four multipliers are positive, or any three multipliers are positive and the fourth is zero. For any one of these vectors, it is possible to find a direction y such that $c'y < 0$ and moving from x^* in the direction of y will violate only the constraints that have positive multipliers.

and that all inequality constraints are active at a minimum x^* (cf. Fig. 5.1.3). The Lagrange multiplier vectors are the vectors $\mu^* = (\mu_1^*, \ldots, \mu_r^*)$ such that $\mu^* \geq 0$ and

$$c + \mu_1^* a_1 + \cdots + \mu_r^* a_r = 0.$$

In a naive extension of the classical sensitivity interpretation, a positive (or zero) multiplier μ_j would indicate that the jth inequality is "significant" (or "insignificant") in that the cost can be improved (or cannot be improved) through its violation. Figure 5.1.3, however, indicates that for each j, one can find a Lagrange multiplier vector where $\mu_j > 0$ and another Lagrange multiplier vector where $\mu_j = 0$. Furthermore, one can find Lagrange multiplier vectors where the signs of the multipliers are "wrong" in the sense that it is impossible to improve the cost while violating only constraints with positive multipliers. Thus the classical sensitivity interpretation may fail when there are multiple Lagrange multipliers.

Exact Penalty Functions

An important analytical and algorithmic technique in nonlinear programming involves the use of penalty functions, whereby the equality and inequality constraints are discarded, and simultaneously some terms are added in the cost function that penalize the violation of the constraints. An important example for the mixed equality and inequality constraint problem (5.4) is the quadratic penalty function

$$Q_c(x) = f(x) + \frac{c}{2}\left(\sum_{i=1}^{m}\left(h_i(x)\right)^2 + \sum_{j=1}^{r}\left(g_j^+(x)\right)^2\right),$$

where c is a positive penalty parameter, and we use the notation

$$g_j^+(x) = \max\{0, g_j(x)\}.$$

We may expect that by minimizing $Q_{c^k}(x)$ over $x \in \Re^n$ for a sequence $\{c^k\}$ of penalty parameters with $c^k \to \infty$, we will obtain in the limit a solution of the original problem. Indeed, convergence of this type can generically be shown, and it turns out that typically a Lagrange multiplier vector can also be simultaneously obtained (assuming such a vector exists); see e.g., Bertsekas [Ber82], [Ber99a]. We will use these convergence ideas in various proofs, starting with the next section.

The quadratic penalty function is not exact in the sense that a local minimum x^* of the constrained minimization problem is typically not a local minimum of $Q_c(x)$ for any value of c. A different type of penalty function is given by

$$F_c(x) = f(x) + c\left(\sum_{i=1}^{m}|h_i(x)| + \sum_{j=1}^{r}g_j^+(x)\right),$$

where c is a positive penalty parameter. It can be shown that x^* is typically a local minimum of F_c, provided c is larger than some threshold value. The conditions under which this is true, as well as the threshold value for c bear an intimate connection with Lagrange multipliers. Figure 5.1.4 illustrates this connection for the case of a problem with a single inequality constraint. In particular, it can be seen that if the penalty parameter exceeds the value of a Lagrange multiplier, x^* usually (though not always) is also a local minimum of F_c.

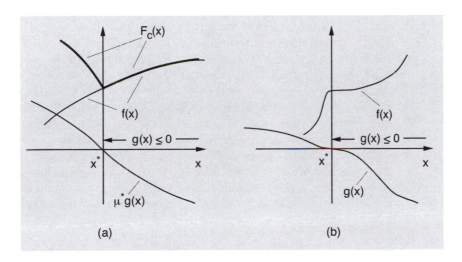

Figure 5.1.4. Illustration of an exact penalty function for the case of a one-dimensional problem with a single inequality constraint. Figure (a) illustrates that, typically, if c is sufficiently large, then x^* is also a local minimum of $F_c(x) = f(x) + cg^+(x)$. (The threshold value for c in this figure is $c > \mu^*$, where μ^* is a Lagrange multiplier, and this observation can be generalized, as will be seen in Section 7.3.) Figure (b) illustrates an exceptional case where the penalty function is not exact, even though there exists a Lagrange multiplier. In this case, $\nabla g(x^*) = 0$, thus violating the condition of constraint gradient linear independence (we will show later that one condition guaranteeing the exactness of F_c is that x^* is a strict local minimum and that the constraint gradients at x^* are linearly independent). Here we have $\nabla f(x^*) = \nabla g(x^*) = 0$, so any nonnegative scalar μ^* is a Lagrange multiplier. However, it is possible that $F_c(x)$ does not have a local minimum at x^* for any $c > 0$ (if for example, as x^* decreases, the downward order of growth rate of f exceeds the upward order of growth of g).

Our Treatment of Lagrange Multipliers

Our development of Lagrange multiplier theory in this chapter differs from the classical treatment in a number of ways:

(a) The optimality conditions of the Lagrange multiplier type that we will develop are sharper than the classical Karush-Kuhn-Tucker conditions (they include extra conditions, which may narrow down the field of candidate local minima). They are also more general in that they apply when in addition to the equality and inequality constraints, there is an abstract set constraint. To achieve this level of generality, we will bring to bear some of the theory of nonsmooth analysis that we developed in Chapter 4. We will find that the notion of *regularity* of the abstract constraint set, introduced in Section 4.6, provides the critical distinction between problems that do and do not admit a satisfactory Lagrange multiplier theory.

(b) We will simplify the proofs of various Lagrange multiplier theorems by introducing the notion of *pseudonormality*, which serves as a connecting link between the major constraint qualifications and the existence of Lagrange multipliers. In the case of equality constraints only, pseudonormality is a property of the constraint set that is implied by either one of the two major constraint qualifications mentioned in the beginning of this section: the LICQ and the linearity of the constraint functions. In fact for equality constraints only, pseudonormality is not much different than the union of these two constraint qualifications. However, pseudonormality is an effective unifying property that is meaningful in the case of inequality constraints and an additional set constraint, where the classical proof arguments based on quasiregularity fail.

(c) We will see in Section 5.3 that there may be several different types of Lagrange multipliers for a given problem, which can be characterized in terms of their sensitivity properties and the information they provide regarding the significance of the corresponding constraints. We show that one particular Lagrange multiplier vector, the one that has minimum norm, has nice sensitivity properties in that it characterizes the direction of steepest rate of improvement of the cost function for a given level of constraint violation. Along that direction, the equality and inequality constraints are violated consistently with the signs of the corresponding multipliers.

(d) We will make a connection between pseudonormality, the existence of Lagrange multipliers, and the unconstrained minimization of the penalty function

$$F_c(x) = f(x) + c \left(\sum_{i=1}^{m} |h_i(x)| + \sum_{j=1}^{r} g_j^+(x) \right),$$

where c is a positive penalty parameter. In particular, we will show in Section 5.5 that pseudonormality implies that strict local minima of the general nonlinear programming problem (5.4) are also unconstrained local minima of F_c when c is large enough. We will also show that this in turn implies the existence of Lagrange multipliers.

Much of our analysis is based on an enhanced set of Fritz John necessary conditions that are introduced in the next section.

5.2 ENHANCED FRITZ JOHN OPTIMALITY CONDITIONS

Throughout this chapter we will focus on optimization problems of the form

$$\text{minimize} \quad f(x)$$
$$\text{subject to} \quad x \in C, \tag{5.6}$$

where the constraint set C consists of equality and inequality constraints as well as an additional abstract set constraint X:

$$C = X \cap \{x \mid h_1(x) = 0, \ldots, h_m(x) = 0\} \cap \{x \mid g_1(x) \le 0, \ldots, g_r(x) \le 0\}. \tag{5.7}$$

Except for Section 5.7, where we introduce some convexity conditions, we assume in this chapter that f, h_i, and g_j are smooth functions from $\Re^n$ to $\Re$, and X is a nonempty closed set.

We will use frequently the tangent and normal cone of X at any $x \in X$, which are denoted by $T_X(x)$ and $N_X(x)$, respectively. We recall from Section 4.6 that X is said to be *regular* at a point $x \in X$ if

$$N_X(x) = T_X(x)^*.$$

Furthermore, if X is convex, then it is regular at all $x \in X$ (cf. Prop. 4.6.3).

Definition 5.2.1: We say that the constraint set C, as represented by Eq. (5.7), *admits Lagrange multipliers* at a point $x^* \in C$ if for every smooth cost function f for which x^* is a local minimum of problem (5.6) there exist vectors $\lambda^* = (\lambda_1^*, \ldots, \lambda_m^*)$ and $\mu^* = (\mu_1^*, \ldots, \mu_r^*)$ that satisfy the following conditions:

$$\left(\nabla f(x^*) + \sum_{i=1}^m \lambda_i^* \nabla h_i(x^*) + \sum_{j=1}^r \mu_j^* \nabla g_j(x^*) \right)' y \ge 0, \quad \forall\, y \in T_X(x^*), \tag{5.8}$$

$$\mu_j^* \ge 0, \qquad \forall\, j = 1, \ldots, r, \tag{5.9}$$

$$\mu_j^* = 0, \qquad \forall\, j \notin A(x^*), \tag{5.10}$$

where $A(x^*) = \{j \mid g_j(x^*) = 0\}$ is the index set of inequality constraints that are active at x^*. A pair (λ^*, μ^*) satisfying Eqs. (5.8)-(5.10) is called a *Lagrange multiplier vector corresponding to f and x^**.

When there is no danger of confusion, we refer to (λ^*, μ^*) simply as a *Lagrange multiplier vector* or a *Lagrange multiplier*, without reference to the corresponding local minimum x^* and the function f. Figure 5.2.1

illustrates the definition of a Lagrange multiplier. Condition (5.10) is referred to as the *complementary slackness* condition (CS for short). Note
that from Eq. (5.8), it follows that the set of Lagrange multiplier vectors
corresponding to a given f and x^* is the intersection of a collection of closed
half spaces [one for each $y \in T_X(x^*)$], so it is a (possibly empty) closed and
convex set.

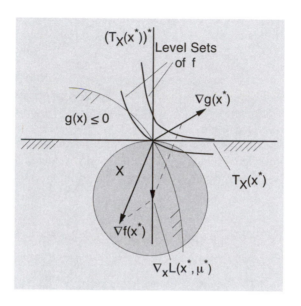

Figure 5.2.1. Illustration of a Lagrange multiplier for the case of a single inequality constraint and the spherical set X shown in the figure. The tangent cone
$T_X(x^*)$ is a closed halfspace and its polar $T_X(x^*)^*$ is the halfline shown in the
figure. There is a unique Lagrange multiplier μ^*, and it is such that

$$-\nabla_x L(x^*, \mu^*) \in T_X(x^*)^*.$$

Note that for this example, we have $\mu^* > 0$, and that there is a sequence $\{x^k\} \subset X$
that converges to x^* and is such that $f(x^k) < f(x^*)$ and $g(x^k) > 0$ for all k,
consistently with condition (iv) of the subsequent Prop. 5.2.1.

The condition (5.8), referred to as the *Lagrangian stationarity condition*, is consistent with the Lagrange multiplier theory outlined in the
preceding section for the case where $X = \Re^n$. It can be viewed as the necessary condition for x^* to be a local minimum of the Lagrangian function

$$L(x, \lambda^*, \mu^*) = f(x) + \sum_{i=1}^{m} \lambda_i^* h_i(x) + \sum_{j=1}^{r} \mu_j^* g_j(x)$$

over $x \in X$ (cf. Prop. 4.7.2). When X is a convex set, Eq. (5.8) is equivalent to

$$\left(\nabla f(x^*) + \sum_{i=1}^{m} \lambda_i^* \nabla h_i(x^*) + \sum_{j=1}^{r} \mu_j^* \nabla g_j(x^*)\right)'(x - x^*) \geq 0, \qquad \forall \ x \in X.$$

$$(5.11)$$

This is because when X is convex, $T_X(x^*)$ is equal to the closure of the set of feasible directions at x^* [cf. Prop. 4.6.2(c)], which is in turn equal to the closure of the set of vectors of the form $\alpha(x - x^*)$, where $\alpha > 0$ and $x \in X$. If $X = \Re^n$, Eq. (5.11) becomes

$$\nabla f(x^*) + \sum_{i=1}^{m} \lambda_i^* \nabla h_i(x^*) + \sum_{j=1}^{r} \mu_j^* \nabla g_j(x^*) = 0.$$

This equation, the nonnegativity condition (5.9), and the CS condition (5.10) comprise the classical Karush-Kuhn-Tucker conditions.

Note that the Lagrangian stationarity condition,

$$\left(\nabla f(x^*) + \sum_{i=1}^{m} \lambda_i^* \nabla h_i(x^*) + \sum_{j=1}^{r} \mu_j^* \nabla g_j(x^*)\right)' y \geq 0, \qquad \forall \ y \in T_X(x^*),$$

can equivalently be written as

$$-\left(\nabla f(x^*) + \sum_{i=1}^{m} \lambda_i^* \nabla h_i(x^*) + \sum_{j=1}^{r} \mu_j^* \nabla g_j(x^*)\right) \in T_X(x^*)^*.$$

Thus, the negative gradient of the Lagrangian function must lie in the polar of the tangent cone $T_X(x^*)^*$, and therefore must also lie in the normal cone $N_X(x^*)$ [since $T_X(x^*)^* \subset N_X(x^*)$, with equality if X is regular at x^*]. The following proposition revolves around this form of the Lagrangian stationarity condition.

The proposition enhances the classical Fritz John optimality conditions, and forms the basis for enhancing the Karush-Kuhn-Tucker conditions as well. It asserts the existence of multipliers corresponding to a local minimum x^*, including a multiplier μ_0^* for the cost function gradient. These multipliers have standard properties [conditions (i)-(iii) below], but they also have a special nonstandard property [condition (iv) below]. This condition implies that *by violating the constraints corresponding to nonzero multipliers, we can improve the optimal cost* (the remaining constraints, may also need to be violated, but the degree of their violation is arbitrarily small relative to the other constraints).

Proposition 5.2.1: Let x^* be a local minimum of problem (5.6)-(5.7). Then there exist scalars μ_0^*, $\lambda_1^*, \ldots, \lambda_m^*$, $\mu_1^*, \ldots, \mu_r^*$, satisfying the following conditions:

(i) $-\left(\mu_0^* \nabla f(x^*) + \sum_{i=1}^m \lambda_i^* \nabla h_i(x^*) + \sum_{j=1}^r \mu_j^* \nabla g_j(x^*)\right) \in N_X(x^*).$

(ii) $\mu_j^* \geq 0$ for all $j = 0, 1, \ldots, r$.

(iii) $\mu_0^*, \lambda_1^*, \ldots, \lambda_m^*, \mu_1^*, \ldots, \mu_r^*$ are not all equal to 0.

(iv) If the index set $I \cup J$ is nonempty, where

$$I = \{i \mid \lambda_i^* \neq 0\}, \qquad J = \{j \neq 0 \mid \mu_j^* > 0\},$$

there exists a sequence $\{x^k\} \subset X$ that converges to x^* and is such that for all k,

$$f(x^k) < f(x^*),$$

$$\lambda_i^* h_i(x^k) > 0, \ \ \forall \, i \in I, \qquad \mu_j^* g_j(x^k) > 0, \ \ \forall \, j \in J, \qquad (5.12)$$

$$|h_i(x^k)| = o(w(x^k)), \ \forall \, i \notin I, \qquad g_j^+(x^k) = o(w(x^k)), \ \forall \, j \notin J, \qquad (5.13)$$

where $g^+(x) = \max\{0, g_j(x)\}$ and

$$w(x) = \min\left\{ \min_{i \in I} |h_i(x)|, \ \min_{j \in J} g_j^+(x) \right\}. \qquad (5.14)$$

Proof: We use a quadratic penalty function approach. For each $k = 1, 2, \ldots$, let

$$F^k(x) = f(x) + \frac{k}{2} \sum_{i=1}^m \big(h_i(x)\big)^2 + \frac{k}{2} \sum_{j=1}^r \big(g_j^+(x)\big)^2 + \frac{1}{2}\|x - x^*\|^2,$$

and consider the "penalized" problem

$$\text{minimize} \ \ F^k(x)$$
$$\text{subject to} \ \ x \in X \cap S,$$

where

$$S = \{x \mid \|x - x^*\| \leq \epsilon\},$$

and ϵ is a positive scalar such that $f(x^*) \leq f(x)$ for all feasible x with $x \in S$. Since $X \cap S$ is compact, by Weierstrass' Theorem, there exists an optimal solution x^k of the above problem. We have

$$F^k(x^k) \leq F^k(x^*), \qquad \forall \, k,$$

which can be written as

$$f(x^k) + \frac{k}{2} \sum_{i=1}^{m} \left(h_i(x^k)\right)^2 + \frac{k}{2} \sum_{j=1}^{r} \left(g_j^+(x^k)\right)^2 + \frac{1}{2} \|x^k - x^*\|^2 \le f(x^*). \quad (5.15)$$

Since $f(x^k)$ is bounded over $X \cap S$, it follows that

$$\lim_{k \to \infty} |h_i(x^k)| = 0, \quad i = 1, \ldots, m, \qquad \lim_{k \to \infty} g_j^+(x^k) = 0, \quad j = 1, \ldots, r;$$

otherwise the left-hand side of Eq. (5.15) would tend to ∞ as $k \to \infty$. Therefore, every limit point $\bar{x}$ of $\{x^k\}$ is feasible, i.e., $\bar{x} \in C$. Furthermore, Eq. (5.15) yields $f(x^k) + (1/2)\|x^k - x^*\|^2 \le f(x^*)$ for all k, so by taking the limit as $k \to \infty$, we obtain

$$f(\bar{x}) + \tfrac{1}{2}\|\bar{x} - x^*\|^2 \le f(x^*).$$

Since $\bar{x} \in S$ and $\bar{x}$ is feasible, we have $f(x^*) \le f(\bar{x})$, which when combined with the preceding inequality yields $\|\bar{x} - x^*\| = 0$ so that $\bar{x} = x^*$. Thus the sequence $\{x^k\}$ converges to x^*, and it follows that x^k is an interior point of the closed sphere S for all k greater than some $\bar{k}$.

For $k \ge \bar{k}$, the optimality condition of Prop. 4.7.2,

$$-\nabla F^k(x^k) \in T_X(x^k)^*,$$

can be written [by using the formula $\nabla \left(g_j^+(x)\right)^2 = 2g_j^+(x)\nabla g_j(x)$] as

$$-\left(\nabla f(x^k) + \sum_{i=1}^{m} \xi_i^k \nabla h_i(x^k) + \sum_{j=1}^{r} \zeta_j^k \nabla g_j(x^k) + (x^k - x^*)\right) \in T_X(x^k)^*,$$
$$(5.16)$$

where

$$\xi_i^k = kh_i(x^k), \qquad \zeta_j^k = kg_j^+(x^k). \quad (5.17)$$

Denote

$$\delta^k = \sqrt{1 + \sum_{i=1}^{m} (\xi_i^k)^2 + \sum_{j=1}^{r} (\zeta_j^k)^2}, \quad (5.18)$$

$$\mu_0^k = \frac{1}{\delta^k}, \qquad \lambda_i^k = \frac{\xi_i^k}{\delta^k}, \quad i = 1, \ldots, m, \qquad \mu_j^k = \frac{\zeta_j^k}{\delta^k}, \quad j = 1, \ldots, r.$$
$$(5.19)$$

Then by dividing Eq. (5.16) with δ^k, we obtain for all $k \ge \bar{k}$,

$$-\left(\mu_0^k \nabla f(x^k) + \sum_{i=1}^{m} \lambda_i^k \nabla h_i(x^k) + \sum_{j=1}^{r} \mu_j^k \nabla g_j(x^k) + \frac{1}{\delta^k}(x^k - x^*)\right) \in T_X(x^k)^*$$
$$(5.20)$$

Since by construction we have

$$(\mu_0^k)^2 + \sum_{i=1}^{m}(\lambda_i^k)^2 + \sum_{j=1}^{r}(\mu_j^k)^2 = 1, \tag{5.21}$$

the sequence $\{\mu_0^k, \lambda_1^k, \ldots, \lambda_m^k, \mu_1^k, \ldots, \mu_r^k\}$ is bounded and must contain a subsequence that converges to some limit $\{\mu_0^*, \lambda_1^*, \ldots, \lambda_m^*, \mu_1^*, \ldots, \mu_r^*\}$.

From Eq. (5.20), the relation $\delta^k \geq 1$, and the defining property of the normal cone $N_X(x^*)$ [$x^k \to x^*$, $z^k \to z^*$, and $z^k \in T_X(x^k)^*$ for all k, imply that $z^* \in N_X(x^*)$], we see that μ_0^*, λ_i^*, and μ_j^* must satisfy condition (i). From Eqs. (5.17) and (5.19), μ_0^* and μ_j^* must satisfy condition (ii), and from Eq. (5.21), μ_0^*, λ_i^*, and μ_j^* must satisfy condition (iii).

Finally, to show that condition (iv) is satisfied, assume that $I \cup J$ is nonempty, and note that for all sufficiently large k within the index set $\mathcal{K}$ of the convergent subsequence, we must have $\lambda_i^* \lambda_i^k > 0$ for all $i \in I$ and $\mu_j^* \mu_j^k > 0$ for all $j \in J$. Therefore, for these k, from Eqs. (5.17) and (5.19), we must have $\lambda_i^* h_i(x^k) > 0$ for all $i \in I$ and $\mu_j^* g_j(x^k) > 0$ for all $j \in J$, while from Eq. (5.15), we have $f(x^k) < f(x^*)$ for k sufficiently large (the case where $x^k = x^*$ for infinitely many k is excluded by the assumption that $I \cup J$ is nonempty). Furthermore, in view of the definitions (5.17)-(5.19) of λ_i^k and μ_j^k, and the definition (5.14) of $w(x^k)$, the conditions

$$|h_i(x^k)| = o\big(w(x^k)\big), \qquad \forall\, i \notin I,$$

and

$$g_j^+(x^k) \leq o\big(w(x^k)\big), \qquad \forall\, j \notin J,$$

are equivalent to

$$|\lambda_i^k| = o\left(\min\left\{\min_{i \in I}|\lambda_i^k|, \min_{j \in J}\mu_j^k\right\}\right), \qquad \forall\, i \notin I,$$

and

$$\mu_j^k \leq o\left(\min\left\{\min_{i \in I}|\lambda_i^k|, \min_{j \in J}\mu_j^k\right\}\right), \qquad \forall\, j \notin J,$$

respectively, so they hold for $k \in \mathcal{K}$. This proves condition (iv). **Q.E.D.**

Note that condition (iv) of Prop. 5.2.1 is stronger than the CS condition (5.10): if $\mu_j^* > 0$, then according to condition (iv), the corresponding jth inequality constraint must be violated arbitrarily close to x^* [cf. Eq. (5.12)], implying that $g_j(x^*) = 0$. For ease of reference, we refer to condition (iv) as the *complementary violation condition* (CV for short).† This

---

† This term is in analogy with "complementary slackness," which is the condition that for all j, $\mu_j^* > 0$ implies $g_j(x^*) = 0$. Thus "complementary violation" reflects the condition that for all j, $\mu_j^* > 0$ implies $g_j(x) > 0$ for some x arbitrarily close to x^*.

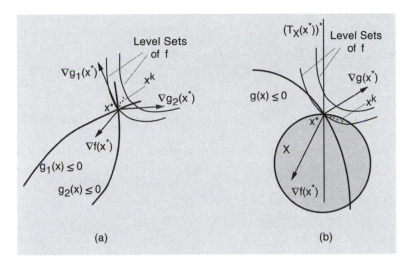

Figure 5.2.2. Illustration of the CV condition. In the example of (a), where $X = \Re^2$, the triplets that satisfy the enhanced Fritz John conditions are the positive multiples of a unique vector of the form $(1, \mu_1^*, \mu_2^*)$ where $\mu_1^* > 0$ and $\mu_2^* > 0$. It is possible to reduce the cost function while violating both constraints by approaching x^* along the sequence $\{x^k\}$ shown. The example of (b) is the one of Fig. 5.2.1, and is similar to the one of (a), except that X is the shaded sphere shown rather than $X = \Re^2$.

condition is illustrated in the examples of Fig. 5.2.2. It will turn out to be of crucial significance in the following sections.

If X is regular at x^*, i.e., $N_X(x^*) = T_X(x^*)^*$, condition (i) of Prop. 5.2.1 becomes

$$-\left(\mu_0^* \nabla f(x^*) + \sum_{i=1}^{m} \lambda_i^* \nabla h_i(x^*) + \sum_{j=1}^{r} \mu_j^* \nabla g_j(x^*)\right) \in T_X(x^*)^*,$$

or equivalently

$$\left(\mu_0^* \nabla f(x^*) + \sum_{i=1}^{m} \lambda_i^* \nabla h_i(x^*) + \sum_{j=1}^{r} \mu_j^* \nabla g_j(x^*)\right)' y \geq 0, \qquad \forall\, y \in T_X(x^*).$$

If in addition, the scalar μ_0^* is strictly positive, then by normalization we can choose $\mu_0^* = 1$, and condition (i) of Prop. 5.2.1 becomes equivalent to the Lagrangian stationarity condition (5.8). Thus, if X is regular at x^* and $\mu_0^* = 1$, the vector $(\lambda^*, \mu^*) = \{\lambda_1^*, \ldots, \lambda_m^*, \mu_1^*, \ldots, \mu_r^*\}$ is a Lagrange multiplier that satisfies the CV condition (which is stronger than the CS condition, as mentioned earlier).

As an example, if there is no abstract set constraint ($X = \Re^n$), and the gradients $\nabla h_i(x^*)$, $i = 1, \ldots, m$, and $\nabla g_j(x^*)$, $j \in A(x^*)$, are linearly independent, we cannot have $\mu_0^* = 0$, since then condition (i) of Prop. 5.2.1 would be violated. It follows that there exists a Lagrange multiplier vector, which in this case is unique in view of the linear independence assumption. We thus obtain the Lagrange multiplier theorem alluded to in Section 5.1. This is a classical result, found in almost all nonlinear programming textbooks, but it is obtained here in a stronger form, which includes the assertion that the multipliers satisfy the CV condition in place of the CS condition.

To illustrate the use of the generalized Fritz John conditions of Prop. 5.2.1 and the CV condition in particular, consider the following example.

Example 5.2.1:

Suppose that we convert the problem $\min_{h(x)=0} f(x)$, involving a single equality constraint, to the inequality constrained problem

$$\text{minimize} \quad f(x)$$
$$\text{subject to} \quad h(x) \leq 0, \quad -h(x) \leq 0.$$

The Fritz John conditions, in their classical form, assert the existence of nonnegative $\mu_0^*, \lambda^+, \lambda^-$, not all zero, such that

$$\mu_0^* \nabla f(x^*) + \lambda^+ \nabla h(x^*) - \lambda^- \nabla h(x^*) = 0. \tag{5.22}$$

The candidate multipliers that satisfy the above condition as well as the CS condition $\lambda^+ h(x^*) = \lambda^- h(x^*) = 0$, include those of the form $\mu_0^* = 0$ and $\lambda^+ = \lambda^- > 0$, which provide no relevant information about the problem. However, these multipliers fail the stronger CV condition of Prop. 5.2.1, showing that if $\mu_0^* = 0$, we must have either $\lambda^+ \neq 0$ and $\lambda^- = 0$, or $\lambda^+ = 0$ and $\lambda^- \neq 0$. Assuming $\nabla h(x^*) \neq 0$, this violates Eq. (5.22), so it follows that $\mu_0^* > 0$. Thus, by dividing Eq. (5.22) with μ_0^*, we recover the familiar first order condition $\nabla f(x^*) + \lambda^* \nabla h(x^*) = 0$ with $\lambda^* = (\lambda^+ - \lambda^-)/\mu_0^*$, under the assumption $\nabla h(x^*) \neq 0$. Note that this deduction would not have been possible without the CV condition.

We will further explore the CV condition as a vehicle for characterizing Lagrange multipliers in the next section. Subsequently, in Section 5.4, we will derive conditions guaranteeing that one can take $\mu_0^* = 1$ in Prop. 5.2.1, by using the notion of pseudonormality, which is motivated by the CV condition.

5.3   INFORMATIVE LAGRANGE MULTIPLIERS

The Lagrange multipliers whose existence is guaranteed by Prop. 5.2.1 (assuming that $\mu_0^* = 1$) are special: they satisfy the stronger CV condition in place of the CS condition. These multipliers *provide a significant amount of sensitivity information by indicating which constraints to violate in order to effect a cost reduction.* In view of this interpretation, we refer to a Lagrange multiplier vector (λ^*, μ^*) that satisfies, in addition to Eqs. (5.8)-(5.10), the CV condition [condition (iv) of Prop. 5.2.1] as being *informative*.

The salient property of informative Lagrange multipliers is consistent with the classical sensitivity interpretation of a Lagrange multiplier as the rate of cost reduction when the corresponding constraint is violated. Here we are not making enough assumptions for this stronger type of sensitivity interpretation to be valid. Yet it is remarkable that with hardly any assumptions, at least one informative Lagrange multiplier vector exists if X is regular and we can guarantee that we can take $\mu_0^* = 1$ in Prop. 5.2.1. In fact we will show shortly a stronger and more definitive property: *if the tangent cone $T_X(x^*)$ is convex and there exists at least one Lagrange multiplier vector, there exists one that is informative.*

An informative Lagrange multiplier vector is useful, among other things, if one is interested in identifying redundant constraints. Given such a vector, one may simply discard the constraints whose multipliers are 0 and check to see whether x^* is still a local minimum. While there is no general guarantee that this will be true, in many cases it will be; for example, as we will show in Chapter 6, in the special case where f and X are convex, the g_j are convex, and the h_i are affine, x^* is guaranteed to be a global minimum, even after the constraints whose multipliers are 0 are discarded.

If discarding constraints is of analytical or computational interest, and constraints whose Lagrange multipliers are equal to zero can be discarded as indicated above, we are motivated to find multiplier vectors that have a minimal number of nonzero components (a minimal support). We call such Lagrange multiplier vectors *minimal*, and we define them as having support $I \cup J$ that does not strictly contain the support of any other Lagrange multiplier vector.

Minimal Lagrange multipliers are not necessarily informative. For example, think of the case where some of the constraints are duplicates of others. Then in a minimal Lagrange multiplier vector, at most one of each set of duplicate constraints can have a nonzero multiplier, while in an informative Lagrange multiplier vector, either all or none of these duplicate constraints will have a nonzero multiplier. Another interesting example is provided by Fig. 5.1.3. There are four minimal Lagrange multiplier vectors in this example, and each of them has two nonzero components. However, none of these multipliers is informative. In fact all the other Lagrange multiplier vectors, those involving three out of four, or all four components positive, are informative. Thus, in the example of Fig. 5.1.3, the sets of minimal and informative Lagrange multipliers are nonempty and disjoint.

Nonetheless, minimal Lagrange multipliers turn out to be informative after the constraints corresponding to zero multipliers are neglected, as can be inferred by the subsequent Prop. 5.3.1. In particular, let us say that a Lagrange multiplier (λ^*, μ^*) is *strong* if in addition to Eqs. (5.8)-(5.10), it satisfies the condition

(iv′) If the set $I \cup J$ is nonempty, where $I = \{i \mid \lambda_i^* \neq 0\}$ and $J = \{j \neq 0 \mid \mu_j^* > 0\}$, there exists a sequence $\{x^k\} \subset X$ that converges to x^* and is such that for all k,

$$f(x^k) < f(x^*), \quad \lambda_i^* h_i(x^k) > 0, \ \forall\, i \in I, \quad \mu_j^* g_j(x^k) > 0, \ \forall\, j \in J. \tag{5.23}$$

This condition resembles the CV condition, but is weaker in that it makes no provision for negligibly small violation of the constraints corresponding to zero multipliers, as per Eqs. (5.13) and (5.14). As a result, informative Lagrange multipliers are also strong, but not reversely.

The following proposition, illustrated in Fig. 5.3.1, clarifies the relationships between different types of Lagrange multipliers, and shows the existence of an informative multiplier, assuming there exists at least one Lagrange multiplier. For this, the tangent cone $T_X(x^*)$ should be convex, which is true in particular if X is convex. In fact regularity of X at x^* guarantees that $T_X(x^*)$ is convex, as mentioned in Section 4.6.

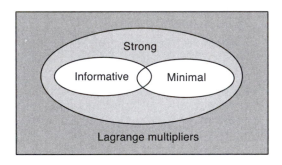

Figure 5.3.1. Relations of different types of Lagrange multipliers, assuming that the tangent cone $T_X(x^*)$ is convex (this is true in particular if X is regular at x^*).

Proposition 5.3.1: Let x^* be a local minimum of problem (5.6)-(5.7). Assume that the tangent cone $T_X(x^*)$ is convex and that the set of Lagrange multipliers is nonempty. Then:

(a) The set of informative Lagrange multiplier vectors is nonempty, and in fact the Lagrange multiplier vector that has minimum norm is informative.

(b) Each minimal Lagrange multiplier vector is strong.

Proof: (a) We summarize the essence of the proof argument in the following lemma.

Lemma 5.3.1: Let N be a closed convex cone in $\Re^n$, and let $a_0, \ldots, a_r$ be given vectors in $\Re^n$. Suppose that the set

$$M = \left\{ \mu \geq 0 \;\middle|\; -\left(a_0 + \sum_{j=1}^{r} \mu_j a_j \right) \in N \right\}$$

is nonempty. Then there exists a sequence $\{d^k\} \subset N^*$ such that

$$a_0' d^k \to -\|\mu^*\|^2, \tag{5.24}$$

$$(a_j' d^k)^+ \to \mu_j^*, \qquad j = 1, \ldots, r, \tag{5.25}$$

where $(a_j' d^k)^+ = \max\{0, a_j' d^k\}$ and μ^* is the vector of minimum norm in M. Furthermore, we have

$$-\tfrac{1}{2}\|\mu^*\|^2 - \inf_{d \in N^*} \left\{ a_0' d + \tfrac{1}{2} \sum_{j=1}^{r} \left((a_j' d)^+ \right)^2 \right\}$$
$$= \lim_{k \to \infty} \left\{ a_0' d^k + \tfrac{1}{2} \sum_{j=1}^{r} \left((a_j' d^k)^+ \right)^2 \right\}. \tag{5.26}$$

In addition, if the problem

$$\text{minimize} \quad a_0' d + \tfrac{1}{2} \sum_{j=1}^{r} \left((a_j' d)^+ \right)^2$$
$$\text{subject to} \quad d \in N^*, \tag{5.27}$$

has an optimal solution, denoted d^*, we have

$$a_0' d^* = -\|\mu^*\|^2, \qquad (a_j' d^*)^+ = \mu_j^*, \quad j = 1, \ldots, r. \tag{5.28}$$

Proof: We first note that the set M is closed and convex, so that there exists a vector μ^* of minimum norm on M. For any $\gamma \geq 0$, consider the function

$$L_\gamma(d, \mu) = \left(a_0 + \sum_{j=1}^{r} \mu_j a_j \right)' d + \frac{\gamma}{2}\|d\|^2 - \frac{1}{2}\|\mu\|^2. \tag{5.29}$$

Our proof will revolve around saddle point properties of the convex/concave function L_0, but to derive these properties, we will work with its γ-perturbed and *coercive* version L_γ for $\gamma > 0$, and then take the limit as $\gamma \to 0$.

From the Saddle Point Theorem (Prop. 2.6.9), for all $\gamma > 0$, the coercive convex/concave quadratic function L_γ has a unique saddle point, denoted (d^γ, μ^γ), over $d \in N^*$ and $\mu \geq 0$. We will characterize this saddle point by using the quadratic nature of L_γ. In particular, the maximization over $\mu \geq 0$ when $d = d^\gamma$ yields

$$\mu_j^\gamma = (a_j' d^\gamma)^+, \qquad j = 1, \ldots, r; \tag{5.30}$$

[to maximize $\mu_j a_j' d^\gamma - (1/2)\mu_j^2$ subject to the constraint $\mu_j \geq 0$, we calculate the unconstrained maximum, which is $a_j' d^\gamma$, and if it is negative we set it to 0, so that the maximum subject to $\mu_j \geq 0$ is attained for $\mu_j = (a_j' d^\gamma)^+$].

The minimization over $d \in N^*$ when $\mu = \mu^\gamma$ is equivalent to projecting the vector

$$s^\gamma = -\frac{a_0 + \sum_{j=1}^r \mu_j^\gamma a_j}{\gamma}$$

on the closed convex cone N^*, since by adding and subtracting $-(\gamma/2)\|s^\gamma\|^2$, we can write $L_\gamma(d, \mu^\gamma)$ as

$$L_\gamma(d, \mu^\gamma) = \frac{\gamma}{2} \|d - s^\gamma\|^2 - \frac{\gamma}{2}\|s^\gamma\|^2 - \frac{1}{2}\|\mu^\gamma\|^2. \tag{5.31}$$

It follows that

$$d^\gamma = P_{N^*}(s^\gamma), \tag{5.32}$$

where $P_{N^*}(\cdot)$ denotes projection on N^*. Furthermore, using the Projection Theorem and the fact that N^* is a closed and convex cone, we have

$$\|s^\gamma\|^2 - \|d^\gamma - s^\gamma\|^2 = \|d^\gamma\|^2 = \|P_{N^*}(s^\gamma)\|^2,$$

so Eq. (5.31) yields

$$L_\gamma(d^\gamma, \mu^\gamma) = -\frac{\gamma}{2} \|P_{N^*}(s^\gamma)\|^2 - \frac{1}{2}\|\mu^\gamma\|^2, \tag{5.33}$$

or equivalently using the definition of s^γ,

$$L_\gamma(d^\gamma, \mu^\gamma) = -\frac{\left\|P_{N^*}\left(-\left(a_0 + \sum_{j=1}^r \mu_j^\gamma a_j\right)\right)\right\|^2}{2\gamma} - \frac{1}{2}\|\mu^\gamma\|^2. \tag{5.34}$$

We will use the preceding facts to show that $\mu^\gamma \to \mu^*$ as $\gamma \to 0$, while d^γ yields the desired sequence d^k, satisfying Eqs. (5.24)-(5.26). Furthermore, we will show that if d^* is an optimal solution of problem (5.27), then (d^*, μ^*) is a saddle point of the function L_0.

We note that

$$\inf_{d \in N^*} L_0(d, \mu) = \begin{cases} -\frac{1}{2}\|\mu\|^2 & \text{if } \mu \in M, \\ -\infty & \text{otherwise}. \end{cases}$$

Since μ^* is the vector of minimum norm in M, we obtain for all $\gamma > 0$,

$$
\begin{aligned}
-\tfrac{1}{2}\|\mu^*\|^2 &= \sup_{\mu \geq 0} \inf_{d \in N^*} L_0(d, \mu) \\
&\leq \inf_{d \in N^*} \sup_{\mu \geq 0} L_0(d, \mu) \\
&\leq \inf_{d \in N^*} \sup_{\mu \geq 0} L_\gamma(d, \mu) \\
&= L_\gamma(d^\gamma, \mu^\gamma).
\end{aligned}
\tag{5.35}
$$

Combining Eqs. (5.34) and (5.35), we obtain

$$
L_\gamma(d^\gamma, \mu^\gamma) = -\frac{\left\| P_{N^*}\left(-\left(a_0 + \sum_{j=1}^r \mu_j^\gamma a_j\right)\right)\right\|^2}{2\gamma} - \frac{1}{2}\|\mu^\gamma\|^2 \geq -\frac{1}{2}\|\mu^*\|^2.
\tag{5.36}
$$

From this, we see that $\|\mu^\gamma\| \leq \|\mu^*\|$, so that μ^γ remains bounded as $\gamma \to 0$. By taking the limit in Eq. (5.36) as $\gamma \to 0$, we see that

$$
\lim_{\gamma \to 0} P_{N^*}\left(-\left(a_0 + \sum_{j=1}^r \mu_j^\gamma a_j\right)\right) = 0,
$$

so (by the continuity of the projection operation) any limit point of μ^γ, call it $\overline{\mu}$, satisfies $P_{N^*}\left(-\left(a_0 + \sum_{j=1}^r \overline{\mu}_j a_j\right)\right) = 0$, or $-\left(a_0 + \sum_{j=1}^r \overline{\mu}_j a_j\right) \in N$. Since $\mu^\gamma \geq 0$, it follows that $\overline{\mu} \geq 0$, so $\overline{\mu} \in M$. We also have $\|\overline{\mu}\| \leq \|\mu^*\|$ (since $\|\mu^\gamma\| \leq \|\mu^*\|$), so by using the minimum norm property of μ^*, we conclude that any limit point $\overline{\mu}$ of μ^γ must be equal to μ^*, i.e., $\mu^\gamma \to \mu^*$. From Eq. (5.36) we then obtain

$$
L_\gamma(d^\gamma, \mu^\gamma) \to -\tfrac{1}{2}\|\mu^*\|^2,
\tag{5.37}
$$

while from Eqs. (5.32) and (5.33), we have

$$
\frac{\gamma}{2}\|d^\gamma\|^2 \to 0.
\tag{5.38}
$$

We also have

$$
\begin{aligned}
L_\gamma(d^\gamma, \mu^\gamma) &= \sup_{\mu \geq 0} L_\gamma(d^\gamma, \mu) \\
&= a_0' d^\gamma + \frac{\gamma}{2}\|d^\gamma\|^2 + \frac{1}{2}\sum_{j=1}^r \left((a_j' d^\gamma)^+\right)^2 \\
&= a_0' d^\gamma + \frac{\gamma}{2}\|d^\gamma\|^2 + \frac{1}{2}\|\mu^\gamma\|^2.
\end{aligned}
\tag{5.39}
$$

Taking the limit as $\gamma \to 0$ and using the fact $\mu^\gamma \to \mu^*$ and $(\gamma/2)\|d^\gamma\|^2 \to 0$ [cf. Eq. (5.38)], it follows that

$$\lim_{\gamma \to 0} L_\gamma(d^\gamma, \mu^\gamma) = \lim_{\gamma \to 0} a_0' d^\gamma + \tfrac{1}{2}\|\mu^*\|^2.$$

Combining this equation with Eq. (5.37), we obtain

$$\lim_{\gamma \to 0} a_0' d^\gamma = -\|\mu^*\|^2,$$

which together with the fact $a_j' d^\gamma = \mu_j^\gamma \to \mu_j^*$ shown earlier, proves Eqs. (5.24) and (5.25).

The maximum of L_0 over $\mu \geq 0$ for fixed d is attained at $\mu_j = (a_j' d)^+$ [compare with Eq. (5.30)], so that

$$\sup_{\mu \geq 0} L_0(d, \mu) = a_0' d + \tfrac{1}{2} \sum_{j=1}^r \left((a_j' d)^+\right)^2.$$

Hence

$$\inf_{d \in N^*} \sup_{\mu \geq 0} L_0(d, \mu) = \inf_{d \in N^*} \left\{ a_0' d + \tfrac{1}{2} \sum_{j=1}^r \left((a_j' d)^+\right)^2 \right\}. \qquad (5.40)$$

Combining this with the equation

$$\inf_{d \in N^*} \sup_{\mu \geq 0} L_0(d, \mu) = \lim_{\gamma \to 0} L_\gamma(d^\gamma, \mu^\gamma) = -\tfrac{1}{2}\|\mu^*\|^2$$

[cf. Eqs. (5.35) and (5.37)], and Eqs. (5.38) and (5.39), we obtain

$$\begin{aligned} -\tfrac{1}{2}\|\mu^*\|^2 &= \inf_{d \in N^*} \left\{ a_0' d + \tfrac{1}{2} \sum_{j=1}^r \left((a_j' d)^+\right)^2 \right\} \\ &= \lim_{\gamma \to 0} \left\{ a_0' d^\gamma + \tfrac{1}{2} \sum_{j=1}^r \left((a_j' d^\gamma)^+\right)^2 \right\} \end{aligned}$$

[cf. Eq. (5.26)].

Finally, if d^* attains the infimum in the right-hand side above, from Eqs. (5.35), (5.37), and (5.40), we see that (d^*, μ^*) is a saddle point of L_0, and that

$$a_0' d^* = -\|\mu^*\|^2, \qquad (a_j' d^*)^+ = \mu_j^*, \quad j = 1, \ldots, r.$$

Q.E.D.

We now return to the proof of Prop. 5.3.1(a). For simplicity, we assume that all the constraints are inequalities that are active at x^* (equality constraints can be handled by conversion to two inequalities, and inactive inequality constraints are inconsequential, since the subsequent analysis focuses in a small neighborhood of x^*, within which these constraints remain inactive). We will use Lemma 5.3.1 with the following identifications:

$$N = T_X(x^*)^*, \qquad a_0 = \nabla f(x^*), \qquad a_j = \nabla g_j(x^*), \quad j = 1, \ldots, r,$$

$$M = \text{ set of Lagrange multipliers},$$

$$\mu^* = \text{ Lagrange multiplier of minimum norm}.$$

If $\mu^* = 0$, then μ^* is an informative Lagrange multiplier and we are done. We may thus assume that $\mu^* \neq 0$, and that the set

$$J^* = \{j \mid \mu_j^* > 0\}$$

is nonempty.

We note that since $T_X(x^*)$ is assumed convex, we have $N^* = T_X(x^*)$. Let $\{\epsilon^k\}$ be a sequence with $0 < \epsilon^k < 1$ for all k, and $\epsilon^k \to 0$. By Lemma 5.3.1 [cf. (5.24) and (5.25)], for each k, we can select a vector $d^k \in T_X(x^*)$ such that

$$\nabla f(x^*)'d^k \leq -(1 - \epsilon^k)\|\mu^*\|^2, \tag{5.41}$$

$$\nabla g_j(x^*)'d^k \geq (1 - \epsilon^k)\mu_j^*, \qquad \forall\, j \in J^*, \tag{5.42}$$

$$\nabla g_j(x^*)'d^k \leq \epsilon^k \min_{l \in J^*} \mu_l^*, \qquad \forall\, j \notin J^*. \tag{5.43}$$

Since $d^k \in T_X(x^*)$, for each k we can select a sequence $\{x^{k,t}\} \subset X$ such that $x^{k,t} \neq x^*$ for all t and

$$\lim_{t \to \infty} x^{k,t} = x^*, \qquad \lim_{t \to \infty} \frac{x^{k,t} - x^*}{\|x^{k,t} - x^*\|} = d^k. \tag{5.44}$$

Denote

$$\xi^{k,t} = \frac{x^{k,t} - x^*}{\|x^{k,t} - x^*\|} - d^k.$$

Using a first order expansion for the cost function f, we have for each k and t,

$$f(x^{k,t}) - f(x^*) = \nabla f(x^*)'(x^{k,t} - x^*) + o(\|x^{k,t} - x^*\|)$$
$$= \nabla f(x^*)'\, (d^k + \xi^{k,t})\, \|x^{k,t} - x^*\| + o(\|x^{k,t} - x^*\|)$$
$$= \|x^{k,t} - x^*\| \left(\nabla f(x^*)'d^k + \nabla f(x^*)'\xi^{k,t} + \frac{o(\|x^{k,t} - x^*\|)}{\|x^{k,t} - x^*\|} \right),$$

or

$$\frac{f(x^{k,t}) - f(x^*)}{\|x^{k,t} - x^*\|} = \nabla f(x^*)'d^k + \nabla f(x^*)'\xi^{k,t} + \frac{o(\|x^{k,t} - x^*\|)}{\|x^{k,t} - x^*\|}. \tag{5.45}$$

Similarly, we have for each k and t,

$$
\begin{aligned}
g_j(x^{k,t}) - g_j(x^*) &= \nabla g_j(x^*)'(x^{k,t} - x^*) + o(\|x^{k,t} - x^*\|) \\
&= \nabla g_j(x^*)'(d^k + \xi^{k,t})\|x^{k,t} - x^*\| + o(\|x^{k,t} - x^*\|) \\
&= \|x^{k,t} - x^*\| \left(\nabla g_j(x^*)'d^k + \nabla g_j(x^*)'\xi^{k,t} + \frac{o(\|x^{k,t} - x^*\|)}{\|x^{k,t} - x^*\|} \right),
\end{aligned}
$$

or, using also the fact $g_j(x^*) = 0$ for all $j = 1, \ldots, r$,

$$
\frac{g_j(x^{k,t})}{\|x^{k,t} - x^*\|} = \nabla g_j(x^*)'d^k + \nabla g_j(x^*)'\xi^{k,t} + \frac{o(\|x^{k,t} - x^*\|)}{\|x^{k,t} - x^*\|}. \tag{5.46}
$$

Using Eqs. (5.41)-(5.46), and the fact

$$
\lim_{t \to \infty} \xi^{k,t} = 0,
$$

we can select for each k, an index t_k such that for all $k = 1, 2, \ldots$, we have,

$$
\|x^{k,t_k} - x^*\| \le \epsilon^k,
$$

$$
\frac{f(x^{k,t_k}) - f(x^*)}{\|x^{k,t_k} - x^*\|} \le \frac{1}{2}\nabla f(x^*)'d^k \le -\frac{1 - \epsilon^k}{2}\|\mu^*\|^2,
$$

$$
\frac{g_j(x^{k,t_k})}{\|x^{k,t_k} - x^*\|} \ge \frac{1}{2}\nabla g_j(x^*)'d^k \ge \frac{1 - \epsilon^k}{2}\mu_j^*, \qquad \forall\, j \in J^*,
$$

$$
\frac{g_j(x^{k,t_k})}{\|x^{k,t_k} - x^*\|} \le |\nabla g_j(x^*)'d^k| + \epsilon^k \min_{l \in J^*} \mu_l^* \le 2\epsilon^k \min_{l \in J^*} \mu_l^*, \qquad \forall\, j \notin J^*.
$$

From these relations, we see that the sequence $\{x^{k,t_k}\}$ can be used to establish the CV condition for μ^*, so μ^* is an informative Lagrange multiplier.

(b) We summarize the essence of the proof argument of this part in the following lemma.

Lemma 5.3.2: Let N be a closed convex cone in $\Re^n$, and let $a_0, \ldots, a_r$ be given vectors in $\Re^n$. Suppose that the set

$$
M = \left\{ \mu \ge 0 \;\middle|\; -\left(a_0 + \sum_{j=1}^r \mu_j a_j \right) \in N \right\}
$$

is nonempty. Among index sets $J \subset \{1, \ldots, r\}$ such that for some $\mu \in M$ we have $J = \{j \mid \mu_j > 0\}$, let $\overline{J} \subset \{1, \ldots, r\}$ have a minimal number of elements. Then if $\overline{J}$ is nonempty, there exists a vector $\overline{d} \in N^*$ such that

$$
a_0'\overline{d} < 0, \qquad a_j'\overline{d} > 0, \quad \forall\, j \in \overline{J}. \tag{5.47}
$$

Proof: We apply Lemma 5.3.1 with the vectors $a_1, \ldots, a_r$ replaced by the vectors a_j, $j \in \overline{J}$. The subset of M given by

$$\overline{M} = \left\{ \mu \geq 0 \; \middle| \; -\left(a_0 + \sum_{j \in \overline{J}} \mu_j a_j \right) \in N, \; \mu_j = 0, \; \forall \, j \notin \overline{J} \right\}$$

is nonempty by assumption. Let $\overline{\mu}$ be the vector of minimum norm in the closed and convex set $\overline{M}$. Since $\overline{J}$ has a minimal number of indices, we must have $\overline{\mu}_j > 0$ for all $j \in \overline{J}$. If $\overline{J}$ is nonempty, Lemma 5.3.1 implies that there exists a $\overline{d} \in N^*$ such that Eq. (5.47) holds.   **Q.E.D.**

Given Lemma 5.3.2, the proof of Prop. 5.3.1(b) is very similar to the corresponding part of the proof of Prop. 5.3.1(a).   **Q.E.D.**

5.3.1   Sensitivity

Let us consider now the special direction d^* that appears in Lemma 5.3.1, and is a solution of problem (5.27) (assuming that this problem has an optimal solution). Let us note that *this problem is guaranteed to have at least one solution when N^* is a polyhedral cone.* This is because problem (5.27) can be written as

$$\text{minimize} \quad a_0' d + \tfrac{1}{2} \sum_{j=1}^{r} z_j^2$$
$$\text{subject to} \quad d \in N^*, \quad 0 \leq z_j, \quad a_j' d \leq z_j, \; j = 1, \ldots, r,$$

where z_j are auxiliary variables. Thus, if N^* is polyhedral, then problem (5.27) is a convex quadratic program with a cost function that is bounded below by Eq. (5.26), and as shown in Prop. 2.3.4, it has an optimal solution. An important context where this is relevant is when $X = \Re^n$ in which case $N_X(x^*)^* = T_X(x^*) = \Re^n$, or more generally when X is polyhedral, in which case $T_X(x^*)$ is polyhedral.

It can also be proved that problem (5.27) has an optimal solution if there exists a vector $\overline{\mu} \in M$ such that the vector

$$-\left(a_0 + \sum_{j=1}^{r} \overline{\mu}_j a_j \right)$$

lies in the relative interior of the cone N. This result, which is due to Xin Chen (private communication), will be shown in Section 6.4, after we develop more fully the theory of Lagrange multipliers for convex problems.

Assuming now that $T_X(x^*)$ is polyhedral [or more generally, that problem (5.27) has an optimal solution], the line of proof of Prop. 5.3.1(a) [combine Eqs. (5.45) and (5.46)] can be used to show that if the Lagrange multiplier of minimum norm, denoted by (λ^*, μ^*), is nonzero, there exists a sequence $\{x^k\} \subset X$, corresponding to the vector $d^* \in T_X(x^*)$ of Eq. (5.28) such that

$$f(x^k) = f(x^*) - \sum_{i=1}^{m} \lambda_i^* h_i(x^k) - \sum_{j=1}^{r} \mu_j^* g_j(x^k) + o(\|x^k - x^*\|). \qquad (5.48)$$

Furthermore, the vector d^* solves problem (5.27), from which it can be seen that d^* also solves the problem

minimize $\nabla f(x^*)'d$

subject to $\displaystyle\sum_{i=1}^{m} \left(\nabla h_i(x^*)'d\right)^2 + \sum_{j \in A(x^*)} \left((\nabla g_j(x^*)'d)^+\right)^2 = \overline{\beta}, \quad d \in T_X(x^*),$

where $\overline{\beta}$ is given by

$$\overline{\beta} = \sum_{i=1}^{m} \left(\nabla h_i(x^*)'d^*\right)^2 + \sum_{j \in A(x^*)} \left((\nabla g_j(x^*)'d^*)^+\right)^2.$$

More generally, it can be seen that for any positive scalar β, a positive multiple of d^* solves the problem

minimize $\nabla f(x^*)'d$

subject to $\displaystyle\sum_{i=1}^{m} \left(\nabla h_i(x^*)'d\right)^2 + \sum_{j \in A(x^*)} ((\nabla g_j(x^*)'d)^+)^2 = \beta, \quad d \in T_X(x^*).$

Thus, d^* is the tangent direction that maximizes the cost function improvement (calculated up to first order) for a given value of the norm of the constraint violation (calculated up to first order). From Eq. (5.48), this first order cost improvement is equal to

$$\sum_{i=1}^{m} \lambda_i^* h_i(x^k) + \sum_{j=1}^{r} \mu_j^* g_j(x^k).$$

Thus, the minimum norm multipliers λ_i^* and μ_j^* express the rate of improvement per unit constraint violation, along the maximum improvement (or steepest descent) direction d^*. This is consistent with the traditional sensitivity interpretation of Lagrange multipliers.

5.3.2 Alternative Lagrange Multipliers

Finally, let us make the connection with another treatment of Lagrange multipliers. Consider vectors $\lambda^* = (\lambda_1^*, \ldots, \lambda_m^*)$ and $\mu^* = (\mu_1^*, \ldots, \mu_r^*)$ that satisfy the conditions

$$-\left(\nabla f(x^*) + \sum_{i=1}^m \lambda_i^* \nabla h_i(x^*) + \sum_{j=1}^r \mu_j^* \nabla g_j(x^*)\right) \in N_X(x^*), \qquad (5.49)$$

$$\mu_j^* \geq 0, \quad \forall\, j = 1, \ldots, r, \qquad \mu_j^* = 0, \quad \forall\, j \notin A(x^*). \qquad (5.50)$$

Such vectors are called "Lagrange multipliers" by Rockafellar [Roc93], but here we will refer to them as *R-multipliers*, to distinguish them from Lagrange multipliers as we have defined them [cf. Eqs. (5.8)-(5.10)].

When X is regular at x^*, the sets of Lagrange multipliers and R-multipliers coincide. In general, however, the set of Lagrange multipliers is a (possibly strict) subset of the set of R-multipliers, since $T_X(x^*)^* \subset N_X(x^*)$ with inequality holding when X is not regular at x^*. Note that multipliers satisfying the enhanced Fritz John conditions of Prop. 5.2.1 with $\mu_0^* = 1$ are R-multipliers, and they still have the extra sensitivity-like property embodied in the CV condition.

However, R-multipliers have a serious conceptual limitation. If X is not regular at x^*, an R-multiplier may be such that the Lagrangian function has a negative slope along some tangent directions of X at x^*. This is in sharp contrast with Lagrange multipliers, whose salient defining property is that they render the slope of the Lagrangian function nonnegative along all tangent directions. Note also that the existence of R-multipliers does not guarantee the existence of Lagrange multipliers. Furthermore, even if Lagrange multipliers exist, none of them may be informative or strong. Thus regularity of X at the given local minimum is the property that separates problems that possess a satisfactory Lagrange multiplier theory and problems that do not.

The following examples illustrate the preceding discussion.

Example 5.3.1:

In this 2-dimensional example, there exists an R-multiplier for every smooth cost function f, but the constraint set does not admit Lagrange multipliers. Let X be the subset of $\Re^2$ given by

$$X = \big\{(x_1, x_2) \mid (x_2 - x_1)(x_2 + x_1) = 0\big\},$$

and let there be a single equality constraint

$$h(x) = x_2 = 0$$

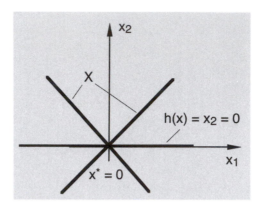

Figure 5.3.2. Constraints of Example 5.3.1. Here,

$$X = \big\{(x_1, x_2) \mid (x_2 - x_1)(x_2 + x_1) = 0\big\},$$

and X is not regular at $x^* = (0,0)$, since we have $T_X(x^*) = X$, $T_X(x^*)^* = \{0\}$, but $N_X(x^*) = X$. For

$$h(x) = x_2 = 0,$$

the constraint set does not admit Lagrange multipliers at x^*, yet there exist R-multipliers for every smooth cost function, since for any f, there exists a λ^* such that $-\big(\nabla f(x^*) + \lambda^* \nabla h(x^*)\big)$ belongs to $N_X(x^*)$. Except if this λ^* is a Lagrange multiplier, the Lagrangian function $L(\cdot, \lambda^*)$ has negative slope along some of the tangent directions of X at x^*.

(see Fig. 5.3.2). There is only one feasible point $x^* = (0,0)$, which is optimal for any cost function f. Here we have $T_X(x^*) = X$ and $T_X(x^*)^* = \{0\}$, so for λ^* to be a Lagrange multiplier, we must have

$$\nabla f(x^*) + \lambda^* \begin{bmatrix} 0 \\ 1 \end{bmatrix} = 0.$$

Thus, there exists a Lagrange multiplier if and only if $\partial f(x^*)/\partial x_1 = 0$. It follows that the constraint set does not admit Lagrange multipliers at x^*. On the other hand, it can be seen that we have

$$N_X(x^*) = X,$$

and that there exists an R-multiplier for every smooth cost function f.

Example 5.3.2:

In this 2-dimensional example, there exists a Lagrange multiplier for every smooth cost function f. However, for some cost functions, the set of Lagrange multipliers is a strict subset of the set of R-multipliers. There are two

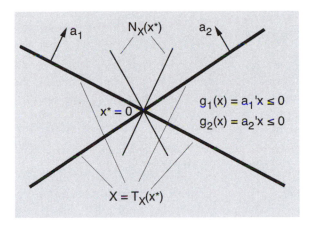

Figure 5.3.3. Constraints of Example 5.3.2. We have

$$T_X(x^*) = X = \left\{ x \mid (a_1'x)(a_2'x) = 0 \right\}$$

and $N_X(x^*)$ is the nonconvex set consisting of the two rays of vectors that are colinear to either a_1 or a_2.

linear constraints $a_1'x \leq 0$ and $a_2'x \leq 0$ with the vectors a_1 and a_2 linearly independent. The set X is the (nonconvex) cone

$$X = \left\{ x \mid (a_1'x)(a_2'x) = 0 \right\}.$$

Consider the vector $x^* = (0,0)$. Here $T_X(x^*) = X$ and $T_X(x^*)^* = \{0\}$. However, it can be seen that $N_X(x^*)$ consists of the two rays of vectors that are colinear to either a_1 or a_2:

$$N_X(x^*) = \{\gamma a_1 \mid \gamma \in \Re\} \cup \{\gamma a_2 \mid \gamma \in \Re\}$$

(see Fig. 5.3.3).

Because $N_X(x^*) \neq T_X(x^*)^*$, X is not regular at x^*. Furthermore, $T_X(x^*)$ and $N_X(x^*)$ are not convex. For any f for which x^* is a local minimum, there exists a unique Lagrange multiplier (μ_1^*, μ_2^*) satisfying Eqs. (5.8)-(5.10). The scalars μ_1^*, μ_2^* are determined from the requirement

$$\nabla f(x^*) + \mu_1^* a_1 + \mu_2^* a_2 = 0. \tag{5.51}$$

Except in the cases where $\nabla f(x^*)$ is equal to 0 or to $-a_1$ or to $-a_2$, we have $\mu_1^* > 0$ and $\mu_2^* > 0$, but the Lagrange multiplier (μ_1^*, μ_2^*) is neither informative nor strong, because there is no $x \in X$ that simultaneously violates both inequality constraints. The R-multipliers here are the vectors (μ_1^*, μ_2^*) such that $\nabla f(x^*) + \mu_1^* a_1 + \mu_2^* a_2$ is either equal to a multiple of a_1 or to a multiple of a_2. Except for the Lagrange multipliers, which satisfy Eq. (5.51), all other R-multipliers are such that the Lagrangian function $L(\cdot, \mu^*)$ has negative slope along some of the tangent directions of X at x^*.

5.4 PSEUDONORMALITY AND CONSTRAINT QUALIFICATIONS

In this section, we use the enhanced Fritz-John conditions of Section 5.2 to derive conditions on the constraint functions h_i and g_j, and the set X, which guarantee the existence of Lagrange multipliers for *any* smooth cost function f. Several such conditions, known as *constraint qualifications*, have been developed and have been the subject of extensive research.

In the standard line of analysis, the existence of Lagrange multipliers is proved separately for each constraint qualification. We will follow a simpler and more insightful approach. In particular, we will first introduce a general property of the constraint set, called *pseudonormality*, which implies the existence of Lagrange multipliers, based on the enhanced Fritz-John conditions, because it guarantees that the cost multiplier μ_0^* of Prop. 5.2.1 cannot be 0. We will then show that the most popular constraint qualifications, together with some new ones, are special cases of pseudonormality.

Definition 5.4.1: We say that a feasible vector x^* of problem (5.6)-(5.7) is *pseudonormal* if one cannot find scalars $\lambda_1, \ldots, \lambda_m, \mu_1, \ldots, \mu_r$, and a sequence $\{x^k\} \subset X$ such that:

(i) $-\left(\sum_{i=1}^m \lambda_i \nabla h_i(x^*) + \sum_{j=1}^r \mu_j \nabla g_j(x^*)\right) \in N_X(x^*)$.

(ii) $\mu_j \geq 0$ for all $j = 1, \ldots, r$, and $\mu_j = 0$ for all $j \notin A(x^*)$, where

$$A(x^*) = \{j \mid g_j(x^*)\}.$$

(iii) $\{x^k\}$ converges to x^* and

$$\sum_{i=1}^m \lambda_i h_i(x^k) + \sum_{j=1}^r \mu_j g_j(x^k) > 0, \qquad \forall\, k. \tag{5.52}$$

The salient point of the above definition is that if x^* is a pseudonormal local minimum, the enhanced Fritz John conditions of Prop. 5.2.1 cannot be satisfied with $\mu_0^* = 0$, so that μ_0^* can be taken equal to 1. Then, if X is regular at x^* so that $N_X(x^*) = T_X(x^*)^*$, the vector $(\lambda_1^*, \ldots, \lambda_m^*, \mu_1^*, \ldots, \mu_r^*)$ obtained from the enhanced Fritz John conditions is an informative Lagrange multiplier.

To get some insight into pseudonormality, consider the case where $X = \Re^n$ and there are no inequality constraints. Then it can be seen that x^* is pseudonormal if and only if one of the following two conditions holds:

(1) The gradients $\nabla h_i(x^*)$, $i = 1, \ldots, m$, are linearly independent.

(2) For every nonzero $\lambda = (\lambda_1, \ldots, \lambda_m)$ such that $\sum_{i=1}^{m} \lambda_i \nabla h_i(x^*) = 0$, the hyperplane through the origin with normal λ contains all vectors $h(x)$ for x within some sphere centered at x^* (see Fig. 5.4.1).

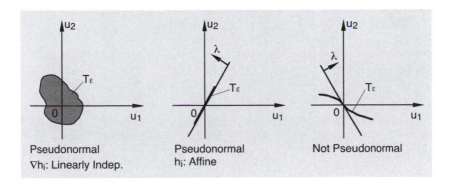

Figure 5.4.1. Geometrical interpretation of pseudonormality of x^* in the case where $X = \Re^n$ and there are no inequality constraints. Consider the set

$$T_\epsilon = \Big\{ h(x) \mid \|x - x^*\| < \epsilon \Big\}$$

for a small positive scalar ϵ. The vector x^* is pseudonormal if and only if either (1) the gradients $\nabla h_i(x^*)$, $i = 1, \ldots, m$, are linearly independent (in this case, it follows from the Implicit Function Theorem that the origin is an interior point of T_ϵ), or (2) for every nonzero λ satisfying $\sum_{i=1}^{m} \lambda_i \nabla h_i(x^*) = 0$, there is a small enough ϵ, such that the set T_ϵ lies in the hyperplane through the origin whose normal is λ.

Note that condition (2) above is satisfied in particular if the functions h_i are affine. To see this, note that if $h_i(x) = a_i'x - b_i$ and $\sum_{i=1}^{m} \lambda_i a_i = 0$, then we have $\lambda'h(x) = \sum_{i=1}^{m} \lambda_i a_i'(x - x^*) = 0$. It can also be argued that the only interesting special case where condition (2) is satisfied is when the functions h_i are affine (see Fig. 5.4.1). Thus, for the case of equality constraints only and $X = \Re^n$, pseudonormality is not much more general than the union of the two major constraint qualifications that guarantee that the constraint set admits Lagrange multipliers (cf. the discussion in the beginning of Section 5.1).

Figure 5.4.2 illustrates pseudonormality for the case where there are inequality constraints (but no abstract set constraint). Clearly, from the definition of pseudonormality, we have that x^* is pseudonormal if the gradients $\nabla g_j(x^*)$, $j \in A(x^*)$, are linearly independent. It can also be seen from the figure that x^* is pseudonormal if the functions g_j, $j \in A(x^*)$, are concave. There are also other interesting situations where pseudonormality holds, and they will be discussed and analyzed shortly.

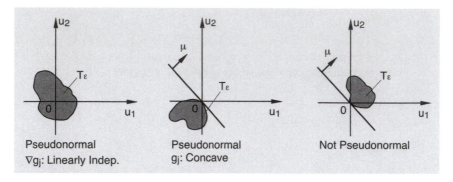

Figure 5.4.2. Geometrical interpretation of pseudonormality of x^* in the case where $X = \Re^n$, and all the constraints are inequalities that are active $[A(x^*) = \{1, \ldots, r\}]$. Consider the set

$$T_\epsilon = \left\{ g(x) \mid \|x - x^*\| < \epsilon \right\}$$

for a small positive scalar ϵ. The vector x^* is pseudonormal if and only if either (1) the gradients $\nabla g_i(x^*)$, $j = 1, \ldots, r$, are linearly independent (in this case, it follows from the Implicit Function Theorem that the origin is an interior point of T_ϵ), or (2) for every nonzero $\mu \geq 0$ satisfying $\sum_{j=1}^r \mu_j \nabla g_j(x^*) = 0$, there is a small enough ϵ, such that the set T_ϵ does not cross into the positive open halfspace of the hyperplane through the origin whose normal is μ.

If the functions g_j are concave, the condition $\sum_{j=1}^r \mu_j \nabla g_j(x^*) = 0$ implies that x^* maximizes $\mu' g(x)$, so that

$$\mu' g(x) \leq \mu' g(x^*) = 0, \qquad \forall\, x \in \Re^n.$$

Therefore, as illustrated in the figure, the set T_ϵ does not cross into the positive open halfspace of the hyperplane through the origin whose normal is μ. This implies that x^* is pseudonormal.

For the case where there is an abstract constraint set $(X \neq \Re^n)$, we will provide a geometric interpretation of pseudonormality in Section 5.7, after we specialize the enhanced Fritz John conditions to the case where X and the functions g_j are convex, and the functions h_i are affine.

We now introduce some constraint qualifications (CQ for short), which will be shown to imply pseudonormality of x^* and hence also existence of informative Lagrange multipliers (assuming also regularity of X at x^*).

> **CQ1:** $X = \Re^n$ and x^* satisfies the LICQ, i.e., the equality constraint gradients $\nabla h_i(x^*)$, $i = 1, \ldots, m$, and the active inequality constraint gradients $\nabla g_j(x^*)$, $j \in A(x^*)$, are linearly independent.

CQ2: $X = \Re^n$, the equality constraint gradients $\nabla h_i(x^*)$, $i = 1, \ldots, m$, are linearly independent, and there exists a $y \in \Re^n$ such that

$$\nabla h_i(x^*)'y = 0, \quad i = 1, \ldots, m, \qquad \nabla g_j(x^*)'y < 0, \quad \forall \, j \in A(x^*).$$

For the case where there are no equality constraints, CQ2 is known as the *Arrow-Hurwitz-Uzawa constraint qualification*, introduced in [AHU61]. In the more general case where there are equality constraints, it is known as the *Mangasarian-Fromovitz constraint qualification*, introduced in [MaF67].

CQ3: $X = \Re^n$, the functions h_i, $i = 1, \ldots, m$, are affine, and the functions g_j, $j = 1, \ldots, r$, are concave.

CQ4: $X = \Re^n$ and for some integer $\overline{r} < r$, the following superset $\overline{C}$ of the constraint set C,

$$\overline{C} = \big\{ x \mid h_i(x) = 0, \, i = 1, \ldots, m, \, g_j(x) \le 0, \, j = \overline{r} + 1, \ldots, r \big\},$$

is pseudonormal at x^*. Furthermore, there exists a $y \in \Re^n$ such that

$$\nabla h_i(x^*)'y = 0, \quad i = 1, \ldots, m, \qquad \nabla g_j(x^*)'y \le 0, \quad \forall \, j \in A(x^*),$$

$$\nabla g_j(x^*)'y < 0, \quad \forall \, j \in \{1, \ldots, \overline{r}\} \cap A(x^*).$$

Note that since CQ1-CQ3 imply pseudonormality, a fact to be shown in the subsequent Prop. 5.4.1, it can be seen that CQ4 generalizes CQ1-CQ3. The next constraint qualification applies to the case where X is a strict subset of $\Re^n$, and reduces to CQ2 when $X = \Re^n$ and none of the equality constraints is linear.

CQ5:

(a) The equality constraints with index above some $\overline{m} \le m$:

$$h_i(x) = 0, \qquad i = \overline{m} + 1, \ldots, m,$$

are linear.

(b) There does not exist a vector $\lambda = (\lambda_1, \ldots, \lambda_m)$ such that

$$-\sum_{i=1}^{m} \lambda_i \nabla h_i(x^*) \in N_X(x^*) \qquad (5.53)$$

and at least one of the scalars $\lambda_1, \ldots, \lambda_{\overline{m}}$ is nonzero.

(c) The subspace

$$V_L(x^*) = \left\{ y \mid \nabla h_i(x^*)'y = 0, \ i = \overline{m}+1, \ldots, m \right\}$$

has a nonempty intersection with either the interior of $N_X(x^*)^*$, or, in the case where X is convex, with the relative interior of $N_X(x^*)^*$.

(d) There exists a $y \in N_X(x^*)^*$ such that

$$\nabla h_i(x^*)'y = 0, \quad i = 1, \ldots, m, \qquad \nabla g_j(x^*)'y < 0, \quad \forall \, j \in A(x^*).$$

The above constraint qualification has several special cases:

CQ5a:

(a) There does not exist a nonzero vector $\lambda = (\lambda_1, \ldots, \lambda_m)$ such that

$$-\sum_{i=1}^{m} \lambda_i \nabla h_i(x^*) \in N_X(x^*).$$

(b) There exists a $y \in N_X(x^*)^*$ such that

$$\nabla h_i(x^*)'y = 0, \quad i = 1, \ldots, m, \qquad \nabla g_j(x^*)'y < 0, \quad \forall \, j \in A(x^*).$$

CQ5b: There are no inequality constraints, the gradients $\nabla h_i(x^*)$, $i = 1, \ldots, m$, are linearly independent, and the subspace

$$V(x^*) = \left\{ y \mid \nabla h_i(x^*)'y = 0, \ i = 1, \ldots, m \right\}$$

contains a point in the interior of $N_X(x^*)^*$.

CQ5c: X is convex, the functions h_i, $i = 1, \ldots, m$, are affine, and the affine set

$$L = \{x \mid h_i(x) = 0, \, i = 1, \ldots, m\}$$

contains a point in the relative interior of X. Furthermore, the functions g_j are convex and there exists a feasible vector $\overline{x}$ satisfying

$$g_j(\overline{x}) < 0, \qquad \forall \, j \in A(x^*).$$

CQ5a is the special case of CQ5 where all equality constraints are assumed nonlinear. CQ5b is a special case of CQ5, where there are no inequality constraints and no linear equality constraints, based on the fact that if $\nabla h_i(x^*)$, $i = 1, \ldots, m$, are linearly independent and the subspace $V(x^*)$ contains a point in the interior of $N_X(x^*)^*$, then it can be shown that assumption (b) of CQ5 is satisfied. Finally, the convexity assumptions in CQ5c can be used to establish the corresponding assumptions (c) and (d) of CQ5. In particular, if X is convex, we have

$$\{y \mid x^* + y \in X\} \subset T_X(x^*) = N_X(x^*)^*,$$

so the relative interior of $\{y \mid x^* + y \in X\}$ is contained in the relative interior of $N_X(x^*)^*$ [since these two sets have the same affine hull; cf. Exercise 1.30(b)]. It can be seen that this implies that if the affine set L contains a point in the relative interior of X, the subspace $V_L(x^*)$ contains a point in the relative interior of $N_X(x^*)^*$. When there are no equality constraints, CQ5c is a classical constraint qualification, introduced by Slater [Sla50] and known as the *Slater condition* [see also the discussion following the Nonlinear Farkas' Lemma (Prop. 3.5.4)].

Note that in the case where X is convex, condition (c) of CQ5 applies even when the interior of $N_X(x^*)^*$ is empty. Actually, the proof of the subsequent proposition shows that convexity of X is not a strict requirement. It is possible to replace the convexity assumption on X with the assumption that $N_X(x^*)^*$ is convex and contains the set $\{y \mid x^* + y \in X \cap S\}$, for some open sphere S that is centered at x^*.

The following constraint qualification is simpler than the preceding ones, and it is designed so that it immediately implies pseudonormality.

CQ6: The set

$$W = \big\{(\lambda, \mu) \mid \lambda_1, \ldots, \lambda_m, \mu_1, \ldots, \mu_r \text{ satisfy conditions (i) and (ii)}$$
$$\text{of the definition of pseudonormality}\big\}$$
$$(5.54)$$

consists of just the origin $(0, 0)$.

It can be shown that the set W of Eq. (5.54) is the recession cone of the set of R-multipliers, provided that the set of R-multipliers is a nonempty closed convex set (so that we can talk about its recession cone). To see this, note that for any R-multiplier (λ^*, μ^*), and any $(\lambda, \mu) \in W$, we have for all $\alpha \geq 0$,

$$-\left(\nabla f(x^*) + \sum_{i=1}^{m}(\lambda_i^* + \alpha\lambda_i)\nabla h_i(x^*) + \sum_{j=1}^{r}(\mu_j^* + \alpha\mu_j)\nabla g_j(x^*)\right) \in N_X(x^*),$$

since $N_X(x^*)$ is a cone. Furthermore, we have $\mu_j^* + \alpha\mu_j \geq 0$ for all j and $\mu_j^* + \alpha\mu_j = 0$ for all $j \notin A(x^*)$. Thus, $(\lambda^* + \alpha\lambda, \mu^* + \alpha\mu)$ is an R-multiplier for all $\alpha \geq 0$, and (λ, μ) is a direction of recession of the set of R-multipliers. Conversely, if (λ, μ) is a direction of recession, then for all R-multipliers (λ^*, μ^*) and all $\alpha > 0$, we have

$$-\left(\frac{1}{\alpha}\nabla f(x^*) + \sum_{i=1}^{m}\left(\frac{1}{\alpha}\lambda_i^* + \lambda_i\right)\nabla h_i(x^*)\right.$$
$$\left. + \sum_{j=1}^{r}\left(\frac{1}{\alpha}\mu_j^* + \mu_j\right)\nabla g_j(x^*)\right) \in N_X(x^*),$$

as well as $(1/\alpha)\mu_j^* + \mu_j \geq 0$ for all j, and $(1/\alpha)\mu_j^* + \mu_j = 0$ for all $j \notin A(x^*)$. Taking the limit as $\alpha \to \infty$ and using the closedness of $N_X(x^*)$, we see that $(\lambda, \mu) \in W$.

Since compactness of a closed convex set is equivalent to its recession cone consisting of just the origin [cf. Prop. 1.5.1(c)], it follows that if the set of R-multipliers is nonempty, convex, and compact, then CQ6 holds. In view of Prop. 5.2.1, the reverse is also true, provided that the set of R-multipliers is guaranteed to be convex, which is true in particular if $N_X(x^*)$ is convex. Thus, *if $N_X(x^*)$ is convex, CQ6 is equivalent to the set of R-multipliers being nonempty and compact.* It can also be shown that if X is regular at x^*, then CQ6 is equivalent to CQ5a (see Exercise 5.8).

Clearly CQ6 implies pseudonormality, since the vectors in W are not required to satisfy condition (iii) of the definition of pseudonormality. However, CQ3, CQ4, and CQ5 do not preclude unboundedness of the set of Lagrange multipliers and hence do not imply CQ6.

Proposition 5.4.1: A feasible point x^* of problem (5.6)-(5.7) is pseudonormal if any one of the constraint qualifications CQ1-CQ6 is satisfied.

Proof: We will not consider CQ2 since it is a special case of CQ5. It is also evident that CQ6 implies pseudonormality. Thus we will prove the result for the cases CQ1, CQ3, CQ4, and CQ5 in that order. In all cases, the method of proof is by contradiction, i.e., we assume that there are scalars $\lambda_1, \ldots, \lambda_m$, and $\mu_1, \ldots, \mu_r$, which satisfy conditions (i)-(iii) of the definition of pseudonormality. We then assume that each of the constraint qualifications CQ1, CQ3, CQ4, and CQ5 is in turn also satisfied, and in each case we arrive at a contradiction.

CQ1: Since $X = \Re^n$, implying that $N_X(x^*) = \{0\}$, and we also have $\mu_j = 0$ for all $j \notin A(x^*)$ by condition (ii), we can write condition (i) as

$$\sum_{i=1}^{m} \lambda_i \nabla h_i(x^*) + \sum_{j \in A(x^*)} \mu_j \nabla g_j(x^*) = 0.$$

Linear independence of $\nabla h_i(x^*)$, $i = 1, \ldots, m$, and $\nabla g_j(x^*)$, $j \in A(x^*)$, implies that $\lambda_i = 0$ for all i and $\mu_j = 0$ for all $j \in A(x^*)$. This, together with the condition $\mu_j = 0$ for all $j \notin A(x^*)$, contradicts condition (iii).

CQ3: By the linearity of h_i and the concavity of g_j, we have for all $x \in \Re^n$,

$$h_i(x) = h_i(x^*) + \nabla h_i(x^*)'(x - x^*), \qquad i = 1, \ldots, m,$$

$$g_j(x) \le g_j(x^*) + \nabla g_j(x^*)'(x - x^*), \qquad j = 1, \ldots, r.$$

By multiplying these two relations with λ_i and μ_j, and by adding over i and j, respectively, we obtain

$$\sum_{i=1}^{m} \lambda_i h_i(x) + \sum_{j=1}^{r} \mu_j g_j(x) \le \sum_{i=1}^{m} \lambda_i h_i(x^*) + \sum_{j=1}^{r} \mu_j g_j(x^*)$$

$$+ \left(\sum_{i=1}^{m} \lambda_i \nabla h_i(x^*) + \sum_{j=1}^{r} \mu_j \nabla g_j(x^*) \right)' (x - x^*)$$

$$= 0,$$

(5.55)

where the last equality holds because we have $\lambda_i h_i(x^*) = 0$ for all i and $\mu_j g_j(x^*) = 0$ for all j [by condition (ii)], and

$$\sum_{i=1}^{m} \lambda_i \nabla h_i(x^*) + \sum_{j=1}^{r} \mu_j \nabla g_j(x^*) = 0$$

[by condition (i)]. On the other hand, by condition (iii), there is an x satisfying $\sum_{i=1}^{m} \lambda_i h_i(x) + \sum_{j=1}^{r} \mu_j g_j(x) > 0$, which contradicts Eq. (5.55).

CQ4: It is not possible that $\mu_j = 0$ for all $j \in \{1, \ldots, \bar{r}\}$, since if this were so, the pseudonormality assumption for $\overline{C}$ would be violated. Thus we have $\mu_j > 0$ for some $j \in \{1, \ldots, \bar{r}\} \cap A(x^*)$. It follows that for the vector

y appearing in the statement of CQ4, we have $\sum_{j=1}^{\overline{r}} \mu_j \nabla g_j(x^*)'y < 0$, so that

$$\sum_{i=1}^{m} \lambda_i \nabla h_i(x^*)'y + \sum_{j=1}^{r} \mu_j \nabla g_j(x^*)'y < 0.$$

This contradicts the equation

$$\sum_{i=1}^{m} \lambda_i \nabla h_i(x^*) + \sum_{j=1}^{r} \mu_j \nabla g_j(x^*) = 0,$$

[cf. condition (i)].

CQ5: We first show by contradiction that at least one of $\lambda_1, \ldots, \lambda_{\overline{m}}$ and μ_j, $j \in A(x^*)$ must be nonzero. If this is not so, then by using a translation argument we may assume that x^* is the origin, and the linear constraints have the form $a_i'x = 0$, $i = \overline{m} + 1, \ldots, m$. Using condition (i) we have

$$- \sum_{i=\overline{m}+1}^{m} \lambda_i a_i \in N_X(x^*). \tag{5.56}$$

Consider first the case where X is not necessarily convex and there is an interior point $\overline{y}$ of $N_X(x^*)^*$ that satisfies $a_i'\overline{y} = 0$ for all $i = \overline{m}+1, \ldots, m$. Let S be an open sphere centered at the origin such that $\overline{y} + d \in N_X(x^*)^*$ for all $d \in S$. We have from Eq. (5.56),

$$\sum_{i=\overline{m}+1}^{m} \lambda_i a_i'd \geq 0, \qquad \forall\, d \in S,$$

from which we obtain $\sum_{i=\overline{m}+1}^{m} \lambda_i a_i = 0$. This contradicts condition (iii), which requires that for some $x \in S \cap X$ we have $\sum_{i=\overline{m}+1}^{m} \lambda_i a_i'x > 0$.

Consider now the alternative case where X is convex and there is a relative interior point $\overline{y}$ of $N_X(x^*)^*$ that satisfies $a_i'\overline{y} = 0$ for all $i = \overline{m} + 1, \ldots, m$. Then, we have

$$\sum_{i=\overline{m}+1}^{m} \lambda_i a_i'\overline{y} = 0,$$

while from Eq. (5.56), we have

$$\sum_{i=\overline{m}+1}^{m} \lambda_i a_i'y \geq 0, \qquad \forall\, y \in N_X(x^*)^*.$$

The convexity of X implies that $X - \{x^*\} \subset T(x^*) = N_X(x^*)^*$ and that $N_X(x^*)^*$ is convex. Since the linear function $\sum_{i=\overline{m}+1}^{m} \lambda_i a_i'y$ attains a minimum over $N_X(x^*)^*$ at the relative interior point $\overline{y}$, it follows from Prop.

1.4.2 that this linear function is constant over $N_X(x^*)^*$. Thus, we have $\sum_{i=\overline{m}+1}^{m} \lambda_i a_i' y = 0$ for all $y \in N_X(x^*)^*$, and hence [since $X - \{x^*\} \subset N_X(x^*)^*$ and $\lambda_i a_i' x^* = 0$ for all i]

$$\sum_{i=\overline{m}+1}^{m} \lambda_i a_i' x = 0, \qquad \forall\, x \in X.$$

This contradicts condition (iii), which requires that for some $x \in X$ we have $\sum_{i=\overline{m}+1}^{m} \lambda_i a_i' x > 0$. We have thus completed the proof that at least one of $\lambda_1, \ldots, \lambda_{\overline{m}}$ and μ_j, $j \in A(x^*)$ must be nonzero.

Next we show by contradiction that we cannot have $\mu_j = 0$ for all j. If this were so, by condition (i) there would exist a nonzero vector $\lambda = (\lambda_1, \ldots, \lambda_m)$ such that

$$-\sum_{i=1}^{m} \lambda_i \nabla h_i(x^*) \in N_X(x^*). \tag{5.57}$$

By what we have proved above, the multipliers $\lambda_1, \ldots, \lambda_{\overline{m}}$ of the nonlinear constraints cannot be all zero, so Eq. (5.57) contradicts assumption (b) of CQ5.

Hence we must have $\mu_j > 0$ for at least one j, and since $\mu_j \geq 0$ for all j with $\mu_j = 0$ for $j \notin A(x^*)$, we obtain

$$\sum_{i=1}^{m} \lambda_i \nabla h_i(x^*)' y + \sum_{j=1}^{r} \mu_j \nabla g_j(x^*)' y < 0,$$

for the vector y of $N_X(x^*)^*$ that appears in assumption (d) of CQ5. Thus,

$$-\left(\sum_{i=1}^{m} \lambda_i \nabla h_i(x^*) + \sum_{j=1}^{r} \mu_j \nabla g_j(x^*) \right) \notin \left(N_X(x^*)^* \right)^*.$$

Since $N_X(x^*) \subset \left(N_X(x^*)^* \right)^*$, this contradicts condition (i).   **Q.E.D.**

Note that the constraint qualifications CQ5 and CQ6 guarantee pseudonormality, as per the preceding proposition, but do not guarantee that the constraint set admits Lagrange multipliers at a point x^*, unless X is regular at x^*. As an illustration, in Example 5.3.1, it can be verified that x^* satisfies CQ5 and CQ6, and is therefore pseudonormal. However, as we have seen, in this example, the constraint set does not admit Lagrange multipliers, although there do exist R-multipliers for every smooth cost function f, consistently with the pseudonormality of x^*.

As an example of application of Prop. 5.4.1, let us show a strengthened version of Farkas' Lemma.

Proposition 5.4.2: (Enhanced Farkas' Lemma) Let $a_1, \ldots, a_r$ and c be given vectors in $\Re^n$, and assume that $c \neq 0$. We have

$$c'y \leq 0, \qquad \text{for all } y \text{ such that } a_j'y \leq 0, \; \forall \, j = 1, \ldots, r,$$

if and only if there exist nonnegative scalars $\mu_1, \ldots, \mu_r$ and a vector $\overline{y} \in \Re^n$ such that

$$c = \mu_1 a_1 + \cdots + \mu_r a_r,$$

$c'\overline{y} > 0$, $a_j'\overline{y} > 0$ for all j with $\mu_j > 0$, and $a_j'\overline{y} \leq 0$ for all j with $\mu_j = 0$.

Proof: If $c = \sum_{j=1}^{r} \mu_j a_j$ for some scalars $\mu_j \geq 0$, then for every y satisfying $a_j'y \leq 0$ for all j, we have $c'y = \sum_{j=1}^{r} \mu_j a_j'y \leq 0$.

Conversely, if c satisfies $c'y \leq 0$ for all y such that $a_j'y \leq 0$ for all j, then $y^* = 0$ minimizes $-c'y$ subject to $a_j'y \leq 0$, $j = 1, \ldots, r$. Since the constraint qualification CQ3 holds for this problem, by Prop. 5.4.1, there is a nonempty set of Lagrange multipliers M, which has the form

$$M = \{\mu \geq 0 \mid c = \mu_1 a_1 + \cdots + \mu_r a_r\}.$$

Lemma 5.3.1 applies with $N = \{0\}$, so we have

$$-\|\mu^*\|^2 \leq -c'y + \tfrac{1}{2} \sum_{j=1}^{r} \left((a_j'y)^+\right)^2, \qquad \forall \, y \in \Re^n,$$

where μ^* is the Lagrange multiplier vector of minimum norm. Minimization of the right-hand side above over $y \in \Re^n$ is a quadratic program, which is bounded below, so it follows from the results of Section 2.3 that there is a vector $\overline{y} \in \Re^n$ that minimizes this right-hand side. Applying Lemma 5.3.1 again, we have

$$-c'\overline{y} = -\|\mu^*\|^2, \qquad (a_j'\overline{y})^+ = \mu_j^*, \quad j = 1, \ldots, r.$$

Since by assumption $c \neq 0$, we have $\mu^* \neq 0$ and the result follows. **Q.E.D.**

The classical version of Farkas' Lemma does not include the assertion on the existence of a $\overline{y}$ that satisfies $c'\overline{y} > 0$ while violating precisely those inequalities that correspond to positive multipliers. Another way to state this assertion is that given a nonzero vector c in $\text{cone}(\{a_1, \ldots, a_r\})$, there is a nonempty index set $J \subset \{1, \ldots, r\}$ such that:

(1) The vector c can be represented as a positive combination of the vectors a_j, $j \in J$.

(2) There is a hyperplane that passes through the origin, and contains the vectors a_j, $j \in J$ in one of its open halfspaces and the vectors a_j, $j \notin J$ in the complementary closed halfspace.

The above two conditions capture the essence of an informative Lagrange multiplier in the case where there is no abstract constraint set ($X = \Re^n$); see also Fig. 5.1.3.

It is interesting to note that Farkas' Lemma can be proved in many different ways, and it can in turn be used to prove the existence of Lagrange multipliers for problems with linear constraints. We have followed here the reverse route: starting from the enhanced Fritz John conditions, we showed the existence of informative Lagrange multipliers for problems with linear constraints, and from there, we showed the enhanced version of the Farkas' Lemma given above. Note that the CV condition (iv) of Prop. 5.2.1 is critical for this line of development.

Proposition 5.4.1 does not apply to the important class of problems where the constraint functions h_i and g_j are affine, and the set X is a polyhedral strict subset of $\Re^n$. Unfortunately, in this case pseudonormality may be violated, as will be shown by an example later. However, it turns out that there always exists a Lagrange multiplier vector for this class of problems. We can show this from first principles, by using Farkas' Lemma, but we can also show it by applying Prop. 5.4.1 to a problem involving an "extended" representation of the problem constraints. In particular, we introduce an auxiliary and equivalent problem where the linear constraints embodied in X are added to the other linear constraints corresponding to h_i and g_j, so that there is no abstract set constraint. For the auxiliary problem, CQ3 is satisfied and Prop. 5.4.1 guarantees the existence of Lagrange multipliers. We can then show that the multipliers corresponding to h_i and g_j form a Lagrange multiplier vector for the original problem. We defer to Section 5.6 a fuller and more general discussion of this approach.

Finally, let us mention the connection between pseudonormality and the classical notion of quasiregularity. When $X = \Re^n$, a local minimum is called quasiregular if

$$T(x^*) = \left\{ y \mid \nabla h_i(x^*)'y = 0,\ \forall\ i,\ \nabla g_j(x^*)'y \leq 0,\ \forall\ j \text{ with } g_j(x^*) = 0 \right\}.$$

It can be shown, using a line a proof of Hestenes [Hes75] (also reproduced in Bertsekas [Ber99a]) that quasiregularity is implied by a property called *quasinormality*, which is in turn implied by pseudonormality. Unfortunately, however, in the case where $X \neq \Re^n$, there is no useful extension of the notion of quasiregularity that can serve as a unifying constraint qualification (see the discussion and the references given at the end of the chapter).

5.5   EXACT PENALTY FUNCTIONS

In this section, we relate the problem

$$\begin{aligned} \text{minimize} \quad & f(x) \\ \text{subject to} \quad & x \in C, \end{aligned}$$

$$(5.58)$$

where

$$C = X \cap \left\{x \mid h_1(x) = 0, \ldots, h_m(x) = 0\right\} \cap \left\{x \mid g_1(x) \leq 0, \ldots, g_r(x) \leq 0\right\},$$
$$\tag{5.59}$$

with another problem that involves minimization over X of the cost function

$$F_c(x) = f(x) + c\left(\sum_{i=1}^{m} |h_i(x)| + \sum_{j=1}^{r} g_j^+(x)\right),$$

where c is a positive scalar, and as earlier, we use the notation

$$g_j^+(x) = \max\{0, g_j(x)\}.$$

Here the equality and inequality constraints are eliminated, and instead the cost is augmented with a term that penalizes the violation of these constraints. The severity of the penalty is controlled by c, which is called the *penalty parameter*.

Let us say that the constraint set C *admits an exact penalty* at the feasible point x^* if for every smooth function f for which x^* is a strict local minimum of f over C, there is a scalar $c > 0$ such that x^* is also a local minimum of the function F_c over X. In the absence of additional assumptions, it is essential for our analysis to require that x^* be a strict local minimum in the definition of admittance of an exact penalty. This restriction may not be important in analytical studies, since we can replace a cost function $f(x)$ with the cost function $f(x) + \|x - x^*\|^2$ without affecting the problem's Lagrange multipliers. On the other hand, if we allow functions f involving multiple local minima, it is hard to relate constraint qualifications such as the ones of the preceding section, the admittance of an exact penalty, and the admittance of Lagrange multipliers. This is illustrated in the following example.

Example 5.5.1:

Consider the 2-dimensional constraint set specified by

$$h_1(x) = \frac{x_2}{x_1^2 + 1} = 0, \qquad x \in X = \Re^2.$$

The feasible points are of the form $x = (x_1, 0)$ with $x_1 \in \Re$, and at each of them the gradient $\nabla h_1(x)$ is nonzero, so x satisfies the LICQ (CQ1 holds). If $f(x) = x_2$, every feasible point is a local minimum, yet for any $c > 0$, we have

$$\inf_{x \in \Re^2}\left\{x_2 + c\frac{|x_2|}{x_1^2 + 1}\right\} = -\infty$$

(take $x_1 = x_2$ as $x_2 \to -\infty$). Thus, the penalty function is not exact for any $c > 0$. It follows that the LICQ would not imply the admittance of an exact penalty if we were to change the definition of the latter to allow cost functions with nonstrict local minima.

We will show that pseudonormality of x^* implies that the constraint set admits an exact penalty, which in turn, together with regularity of X at x^*, implies that the constraint set admits Lagrange multipliers. We first use the constraint qualification CQ5 to obtain a necessary condition for a local minimum of the exact penalty function.

Proposition 5.5.1: Let x^* be a local minimum of

$$F_c(x) = f(x) + c\left(\sum_{i=1}^{m}|h_i(x)| + \sum_{j=1}^{r}g_j^+(x)\right)$$

over X. Then there exist $\lambda_1^*,\ldots,\lambda_m^*$ and $\mu_1^*,\ldots,\mu_r^*$ such that

$$-\left(\nabla f(x^*) + c\left(\sum_{i=1}^{m}\lambda_i^*\nabla h_i(x^*) + \sum_{j=1}^{r}\mu_j^*\nabla g_j(x^*)\right)\right) \in N_X(x^*),$$

$$\lambda_i^* = 1 \quad \text{if } h_i(x^*) > 0, \qquad \lambda_i^* = -1 \quad \text{if } h_i(x^*) < 0,$$

$$\lambda_i^* \in [-1,1] \quad \text{if } h_i(x^*) = 0,$$

$$\mu_j^* = 1 \quad \text{if } g_j(x^*) > 0, \qquad \mu_j^* = 0 \quad \text{if } g_j(x^*) < 0,$$

$$\mu_j^* \in [0,1] \quad \text{if } g_j(x^*) = 0.$$

Proof: The problem of minimizing $F_c(x)$ over $x \in X$ can be converted to the problem

$$\text{minimize} \quad f(x) + c\left(\sum_{i=1}^{m}w_i + \sum_{j=1}^{r}v_j\right)$$

$$\text{subject to} \quad x \in X, \ h_i(x) \leq w_i, \ -h_i(x) \leq w_i, \ i = 1,\ldots,m,$$

$$g_j(x) \leq v_j, \ 0 \leq v_j, \ j = 1,\ldots,r,$$

which involves the auxiliary variables w_i and v_j. It can be seen that the constraint qualification CQ5 is satisfied at the local minimum of this problem that corresponds to x^*. Thus, by Prop. 5.4.1, this local minimum is pseudonormal, and hence there exist multipliers satisfying the enhanced Fritz John conditions (Prop. 5.2.1) with $\mu_0^* = 1$. With straightforward calculation, these conditions yield scalars $\lambda_1^*, \ldots, \lambda_m^*$, and $\mu_1^*, \ldots, \mu_r^*$, satisfying the desired conditions.   **Q.E.D.**

Proposition 5.5.2: If x^* is a pseudonormal vector of problem (5.58)-(5.59), the constraint set admits an exact penalty at x^*.

Proof: Assume the contrary, i.e., that there exists a smooth f such that x^* is a strict local minimum of f over the constraint set C, while x^* is not a local minimum over $x \in X$ of the function

$$F_k(x) = f(x) + k \left(\sum_{i=1}^m |h_i(x)| + \sum_{j=1}^r g_j^+(x) \right),$$

for all $k = 1, 2, \ldots$. Let $\epsilon > 0$ be such that

$$f(x^*) < f(x), \qquad \forall\, x \in C \text{ with } x \neq x^* \text{ and } \|x - x^*\| \leq \epsilon. \qquad (5.60)$$

Suppose that x^k minimizes $F_k(x)$ over the (compact) set of all $x \in X$ satisfying $\|x - x^*\| \leq \epsilon$. Then, since x^* is not a local minimum of F_k over X, we must have that $x^k \neq x^*$, and that x^k is infeasible for problem (5.59), i.e.,

$$\sum_{i=1}^m |h_i(x^k)| + \sum_{j=1}^r g_j^+(x^k) > 0. \qquad (5.61)$$

We have

$$F_k(x^k) = f(x^k) + k \left(\sum_{i=1}^m |h_i(x^k)| + \sum_{j=1}^r g_j^+(x^k) \right) \leq f(x^*), \qquad (5.62)$$

so it follows that $h_i(x^k) \to 0$ for all i and $g_j^+(x^k) \to 0$ for all j. The sequence $\{x^k\}$ is bounded and if $\bar{x}$ is any of its limit points, we have that $\bar{x}$ is feasible. From Eqs. (5.60) and (5.62) it then follows that $\bar{x} = x^*$. Thus $\{x^k\}$ converges to x^* and we have $\|x^k - x^*\| < \epsilon$ for all sufficiently large k. This implies the following necessary condition for optimality of x^k (cf. Prop. 5.5.1):

$$-\left(\frac{1}{k} \nabla f(x^k) + \sum_{i=1}^m \lambda_i^k \nabla h_i(x^k) + \sum_{j=1}^r \mu_j^k \nabla g_j(x^k) \right) \in N_X(x^k), \qquad (5.63)$$

where
$$\lambda_i^k = 1 \quad \text{if } h_i(x^k) > 0, \qquad \lambda_i^k = -1 \quad \text{if } h_i(x^k) < 0,$$

$$\lambda_i^k \in [-1, 1] \qquad \text{if } h_i(x^k) = 0,$$

$$\mu_j^k = 1 \quad \text{if } g_j(x^k) > 0, \qquad \mu_j^k = 0 \quad \text{if } g_j(x^k) < 0,$$

$$\mu_j^k \in [0, 1] \qquad \text{if } g_j(x^k) = 0.$$

In view of Eq. (5.61), we can find a subsequence $\{(\lambda^k, \mu^k)\}_{\mathcal{K}}$ such that $|\lambda_i^k| = 1$ and $h_i(x^k) \neq 0$ for some equality constraint index i and all $k \in \mathcal{K}$, or $\mu_j^k = 1$ and $g_j(x^k) > 0$ for some inequality constraint index j and all $k \in \mathcal{K}$. Let (λ, μ) be a limit point of this subsequence. We then have $(\lambda, \mu) \neq (0,0)$, $\mu \geq 0$. By taking the limit in Eq. (5.63), using also the definition of the normal cone, we obtain

$$-\left(\sum_{i=1}^m \lambda_i \nabla h_i(x^*) + \sum_{j=1}^r \mu_j \nabla g_j(x^*) \right) \in N_X(x^*). \qquad (5.64)$$

Finally, for all $k \in \mathcal{K}$, we have $\lambda_i^k h_i(x^k) \geq 0$ for all i, and $\mu_j^k g_j(x^k) \geq 0$ for all j. Thus, for all $k \in \mathcal{K}$, we have $\lambda_i h_i(x^k) \geq 0$ for all i, $\mu_j g_j(x^k) \geq 0$ for all j. Since by construction of the subsequence $\{\lambda^k, \mu^k\}_{\mathcal{K}}$, we have for some i and all $k \in \mathcal{K}$, $|\lambda_i^k| = 1$ and $h_i(x^k) \neq 0$, or for some j and all $k \in \mathcal{K}$, $\mu_j^k = 1$ and $g_j(x^k) > 0$, it follows that for all $k \in \mathcal{K}$,

$$\sum_{i=1}^m \lambda_i h_i(x^k) + \sum_{j=1}^r \mu_j g_j(x^k) > 0. \qquad (5.65)$$

Thus, Eqs. (5.64) and (5.65) violate the hypothesis that x^* is pseudonormal. **Q.E.D.**

The following example shows that the converse of Prop. 5.5.2 does not hold. In particular, the admittance of an exact penalty function at a point x^* does not imply pseudonormality.

Example 5.5.2:

Here we provide a case where $X = \Re^n$, and the admittance of an exact penalty function does not imply quasiregularity. Since as mentioned at the end of Section 5.4, pseudonormality implies quasiregularity, it follows that admittance of an exact penalty function does not imply pseudonormality. Let

$$C = \left\{ x \in \Re^2 \mid g_1(x) \leq 0, \ g_2(x) \leq 0, \ g_3(x) \leq 0 \right\},$$

where

$$g_1(x) = -(x_1 + 1)^2 - (x_2)^2 + 1,$$

$$g_2(x) = x_1^2 + (x_2 + 1)^2 - 1,$$

$$g_3(x) = -x_2,$$

(see Fig. 5.5.1). The only feasible solution is $x^* = (0,0)$ and the constraint gradients are given by

$$\nabla g_1(x^*) = (-2,0), \qquad \nabla g_2(x^*) = (0,2), \qquad \nabla g_3(x^*) = (0,-1).$$

At $x^* = (0,0)$, the cone of first order feasible variations $V(x^*)$ is equal to the nonnegative portion of the x_1 axis and strictly contains $T(x^*)$, which is equal to $\{(0,0)\}$. Therefore x^* is not a quasiregular point.

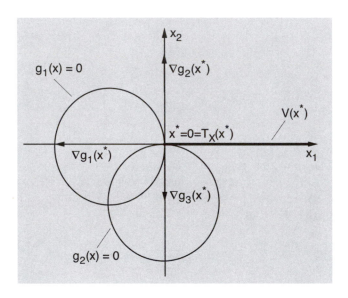

Figure 5.5.1. Constraints of Example 5.5.2. The only feasible point is $x^* = (0,0)$. The tangent cone $T(x^*)$ and the cone of first order feasible variations $V(x^*)$ are also illustrated in the figure.

However, it can be seen that the directional derivative of the function $P(x) = \sum_{j=1}^{3} g_j^+(x)$ at x^* is positive in all directions. This implies that for any smooth function f with a local minimum at x^*, we can choose a sufficiently large penalty parameter c, so that x^* is a local minimum of the function $F_c(x)$. Therefore, the constraint set admits an exact penalty function at x^*.

The following proposition establishes the connection between admittance of an exact penalty and admittance of Lagrange multipliers. Regularity of X is an important condition for this connection.

Proposition 5.5.3: Let x^* be a feasible vector of problem (5.58)-(5.59), and let X be regular at x^*. If the constraint set admits an exact penalty at x^*, it admits Lagrange multipliers at x^*.

Proof: Suppose that a given smooth function $f(x)$ has a local minimum at x^*. Then the function $f(x) + \|x - x^*\|^2$ has a strict local minimum at x^*. Since C admits an exact penalty at x^*, there exist λ_i^* and μ_j^* satisfying the conditions of Prop. 5.5.1. (The term $\|x - x^*\|^2$ in the cost function is inconsequential, since its gradient at x^* is 0.) In view of the regularity of X at x^*, the λ_i^* and μ_j^* are Lagrange multipliers.   **Q.E.D.**

Note that because Prop. 5.5.1 does not require regularity of X, the proof of Prop. 5.5.3 can be used to establish that *admittance of an exact penalty implies the admittance of R-multipliers*, as defined in Section 5.3. On the other hand, Example 5.3.1 shows that the regularity assumption on X in Prop. 5.5.3 cannot be dispensed with. Indeed, in that example, x^* is pseudonormal, the constraint set admits an exact penalty at x^* (consistently with Prop. 5.5.2), but it does not admit Lagrange multipliers.

The relations shown thus far are summarized in Fig. 5.5.2, which illustrates the unifying role of pseudonormality. In this figure, unless indicated otherwise, the implications cannot be established in the opposite direction without additional assumptions (based on the preceding examples and the discussion).

5.6   USING THE EXTENDED REPRESENTATION

In practice, the set X can often be described in terms of smooth equality and inequality constraints:

$$X = \big\{ x \mid h_i(x) = 0, \ i = m + 1, \ldots, \overline{m}, \ g_j(x) \le 0, \ j = r + 1, \ldots, \overline{r} \big\}.$$

Then the constraint set C can alternatively be described without an abstract set constraint, in terms of all of the constraint functions

$$h_i(x) = 0, \quad i = 1, \ldots, \overline{m}, \qquad g_j(x) \le 0, \quad j = 1, \ldots, \overline{r}.$$

We call this the *extended representation* of C, to contrast it with the representation (5.59), which we call the *original representation*. Issues relating to exact penalty functions and Lagrange multipliers can be investigated for the extended representation, and results can be carried over to the original representation by using the following proposition.

Proposition 5.6.1:

  (a) If the constraint set admits Lagrange multipliers in the extended representation, it admits Lagrange multipliers in the original representation.

  (b) If the constraint set admits an exact penalty in the extended representation, it admits an exact penalty in the original representation.

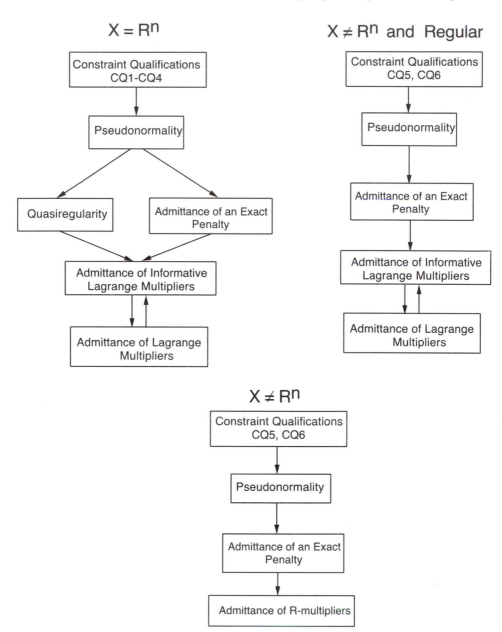

Figure 5.5.2. Relations between various conditions, which when satisfied at a local minimum x^*, guarantee the admittance of an exact penalty and corresponding multipliers. In the case where X is regular, the tangent and normal cones are convex. Hence, by Prop. 5.3.1(a), the admittance of Lagrange multipliers implies the admittance of an informative Lagrange multiplier, while by Prop. 5.5.1, pseudonormality implies the admittance of an exact penalty.

Proof: (a) The hypothesis implies that for every smooth cost function f for which x^* is a local minimum, there exist scalars $\lambda_1^*, \ldots, \lambda_{\overline{m}}^*$, and $\mu_1^*, \ldots, \mu_{\overline{r}}^*$ satisfying

$$\nabla f(x^*) + \sum_{i=1}^{\overline{m}} \lambda_i^* \nabla h_i(x^*) + \sum_{j=1}^{\overline{r}} \mu_j^* \nabla g_j(x^*) = 0, \tag{5.66}$$

$$\mu_j^* \geq 0, \qquad \forall\, j = 0, 1, \ldots, \overline{r},$$

$$\mu_j^* = 0, \qquad \forall\, j \notin \overline{A}(x^*),$$

where

$$\overline{A}(x^*) = \{ j \mid g_j(x^*) = 0, \; j = 1, \ldots, \overline{r} \}.$$

It can be shown that for all $y \in T_X(x^*)$, we have $\nabla h_i(x^*)'y = 0$ for all $i = m+1, \ldots, \overline{m}$, and $\nabla g_j(x^*)'y \leq 0$ for all $j = r+1, \ldots, \overline{r}$ with $j \in \overline{A}(x^*)$ [this is given as Exercise 4.27(b)]. Hence Eq. (5.66) implies that

$$\left(\nabla f(x^*) + \sum_{i=1}^{m} \lambda_i^* \nabla h_i(x^*) + \sum_{j=1}^{r} \mu_j^* \nabla g_j(x^*) \right)' y \geq 0, \qquad \forall\, y \in T_X(x^*),$$

and it follows that $\lambda_1^*, \ldots, \lambda_m^*, \mu_1^*, \ldots, \mu_r^*$ are Lagrange multipliers for the original representation.

(b) Consider the exact penalty function for the extended representation:

$$\overline{F}_c(x) = f(x) + c \left(\sum_{i=1}^{\overline{m}} |h_i(x)| + \sum_{j=1}^{\overline{r}} g_j^+(x) \right).$$

We have $F_c(x) = \overline{F}_c(x)$ for all $x \in X$. Hence if x^* is an unconstrained local minimum of $\overline{F}_c(x)$, it is also a local minimum of $F_c(x)$ over $x \in X$. Thus, for a given $c > 0$, if x^* is both a strict local minimum of f over C and an unconstrained local minimum of $\overline{F}_c(x)$, it is also a local minimum of $F_c(x)$ over $x \in X$. **Q.E.D.**

Proposition 5.6.1 can be used to assert the existence of Lagrange multipliers in some cases where the constraint set is not pseudonormal. An important such case is when all the constraints are linear and X is a polyhedral set. Here, the constraint set need not satisfy pseudonormality (see the following example). However, by Prop. 5.4.1, it satisfies pseudonormality in the extended representation, so in view of Prop. 5.6.1, it admits Lagrange multipliers and an exact penalty at any feasible point in the original representation.

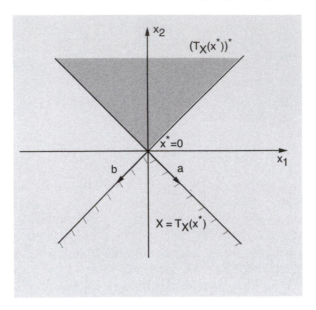

Figure 5.6.1. Constraints of Example 5.6.1. The only feasible point is $x^* = (0,0)$. The tangent cone $T_X(x^*)$ and its polar $T_X(x^*)^*$ are shown in the figure.

Example 5.6.1:

Let the constraint set be

$$C = \{x \in X \mid a'x \leq 0, \; b'x \leq 0\},$$

where $a = (1,-1)$, $b = (-1,-1)$, and

$$X = \{x \in \Re^2 \mid a'x \geq 0, \; b'x \geq 0\};$$

see Fig. 5.6.1. Thus the constraints are linear and X is polyhedral.

The only feasible point is $x^* = (0,0)$. By choosing $\mu = (1,1)$, we get

$$-(\mu_1 a + \mu_2 b) = -(a+b) \in T_X(x^*)^*,$$

while in every neighborhood N of x^* there is an $x \in X \cap N$ such that $a'x > 0$ and $b'x > 0$ simultaneously. Hence x^* is not pseudonormal. This constraint set, however, admits Lagrange multipliers at $x^* = (0,0)$ with respect to its extended representation (cf. Prop. 5.6.1), and hence it admits Lagrange multipliers at $x^* = (0,0)$ with respect to the original representation.

Note that part (a) of Prop. 5.6.1 does not guarantee the existence of informative Lagrange multipliers in the original representation, and indeed in the following example, there exists an informative Lagrange multiplier in the extended representation, but there exists none in the original representation. For this to happen, of course, the tangent cone $T_X(x^*)$ must be nonconvex, for otherwise Prop. 5.3.1(a) applies.

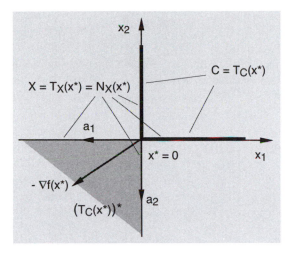

Figure 5.6.2. Constraints and relevant cones for the problem in Example 5.6.2.

Example 5.6.2:

Let the constraint set be represented in extended form without an abstract set constraint as

$$C = \big\{ x \in \Re^2 \mid a_1' x \le 0,\ a_2' x \le 0,\ (a_1' x)(a_2' x) = 0 \big\},$$

where $a_1 = (-1, 0)$ and $a_2 = (0, -1)$. Consider the vector $x^* = (0, 0)$. It can be verified that the constraint set admits Lagrange multipliers at x^* in the extended representation, so it also admits informative Lagrange multipliers, as shown by Prop. 5.3.1(a). Indeed, for any f for which x^* is a local minimum, we must have $-\nabla f(x^*) \in T_C(x^*)^*$ (see Fig. 5.6.2). Thus, $-\nabla f(x^*)$ must lie in the nonpositive orthant and can be represented as a nonnegative combination of the two constraint gradients a_1 and a_2 [the third constraint, $(a_1' x)(a_2' x) = 0$, has gradient equal to 0 at x^*].

  Now let the same constraint set be specified by the two linear constraints $a_1' x \le 0$ and $a_2' x \le 0$ together with the abstract constraint set

$$X = \big\{ x \mid (a_1' x)(a_2' x) = 0 \big\} = \{ x \mid x_1 = 0 \text{ or } x_2 = 0 \}.$$

Here $T_X(x^*) = X$ and $T_X(x^*)^* = \{0\}$. The normal cone $N_X(x^*)$ is also equal to X, so that $N_X(x^*) \ne T_X(x^*)^*$ and X is not regular at x^*. Furthermore, $T_X(x^*)$ is not convex, so Prop. 5.3.1(a) cannot be used to guarantee the admittance of an informative Lagrange multiplier. Again, for any f for which x^* is a local minimum, we must have $-\nabla f(x^*) \in T_C(x^*)^*$. The candidate multipliers are determined from the requirement that

$$-\left(\nabla f(x^*) + \sum_{j=1}^{2} \mu_j a_j \right) \in T_X(x^*)^* = \{0\},$$

which uniquely determines μ_1 and μ_2. If $\nabla f(x^*)$ lies in the interior of the positive orthant, we must have $\mu_1 > 0$ and $\mu_2 > 0$. However, there exists no $x \in X$ that violates both constraints $a_1'x \leq 0$ and $a_2'x \leq 0$, so the multipliers do not qualify as informative. Thus, the constraint set does not admit informative Lagrange multipliers in the original representation.

5.7 EXTENSIONS UNDER CONVEXITY ASSUMPTIONS

In this section, we extend the theory of the preceding sections to the case where some or all of the functions f and g_j are nondifferentiable but are instead assumed convex. We thus consider the problem

$$\text{minimize} \quad f(x)$$
$$\text{subject to} \quad x \in C, \tag{5.67}$$

where

$$C = X \cap \{x \mid h_1(x) = 0, \ldots, h_m(x) = 0\} \cap \{x \mid g_1(x) \leq 0, \ldots, g_r(x) \leq 0\}, \tag{5.68}$$

X is a subset of $\Re^n$, and f and g_j are real-valued functions defined on $\Re^n$. We introduce the following assumption:

Assumption 5.7.1: (Smoothness or Convexity) The set X is nonempty and closed, the functions h_i are smooth, and each of the functions f and g_j is either smooth or is real-valued and convex over $\Re^n$.

The theory of the preceding sections can be generalized under the above assumption, once we substitute the gradients of the convex but nondifferentiable functions with subgradients. To see this, we recall the necessary optimality condition given in Section 4.7 for the problem of minimizing a sum $F_1(x) + F_2(x)$ over X, where F_1 is convex and F_2 is smooth: if x^* is a local minimum and the tangent cone $T_X(x^*)$ is convex, then

$$-\nabla F_2(x^*) \in \partial F_1(x^*) + T_X(x^*)^*. \tag{5.69}$$

For a smooth convex function F, we will use the notation

$$\partial F(x) = \{\nabla F(x)\}$$

even if F is not convex, so the necessary condition (5.69) can be written as

$$0 \in \partial F_1(x^*) + \partial F_2(x^*) + T_X(x^*)^*. \tag{5.70}$$

By a nearly verbatim repetition of the proof of Prop. 5.2.1, while using the preceding necessary condition in place of $-\nabla F^k(x^k) \in T_X(x^k)^*$, together with the closedness of $N_X(x^*)$, we obtain the following extension.

Proposition 5.7.1: Let x^* be a local minimum of problem (5.67)-(5.68). Let Assumption 5.7.1 (Smoothness or Convexity) hold and assume that the tangent cone $T_X(x^*)$ is convex. Then there exist scalars $\mu_0^*, \lambda_1^*, \ldots, \lambda_m^*, \mu_1^*, \ldots, \mu_r^*$, satisfying the following conditions:

  (i) $0 \in \mu_0^* \partial f(x^*) + \sum_{i=1}^m \lambda_i^* \partial h_i(x^*) + \sum_{j=1}^r \mu_j^* \partial g_j(x^*) + N_X(x^*)$.

 (ii) $\mu_j^* \geq 0$ for all $j = 0, 1, \ldots, r$.

(iii) $\mu_0^*, \lambda_1^*, \ldots, \lambda_m^*, \mu_1^*, \ldots, \mu_r^*$ are not all equal to 0.

(iv) If the index set $I \cup J$ is nonempty where

$$I = \{i \mid \lambda_i^* \neq 0\}, \qquad J = \{j \neq 0 \mid \mu_j^* > 0\},$$

there exists a sequence $\{x^k\} \subset X$ that converges to x^* and is such that for all k,

$$f(x^k) < f(x^*),$$

$$\lambda_i^* h_i(x^k) > 0, \quad \forall\, i \in I, \qquad \mu_j^* g_j(x^k) > 0, \quad \forall\, j \in J,$$

$$|h_i(x^k)| = o\big(w(x^k)\big), \quad \forall\, i \notin I, \qquad g_j^+(x^k) = o\big(w(x^k)\big), \quad \forall\, j \notin J,$$

where

$$w(x) = \min\left\{ \min_{i \in I} |h_i(x)|, \min_{j \in J} g_j^+(x) \right\}.$$

The natural definition of a Lagrange multiplier is as follows.

Definition 5.7.1: Consider problem (5.58) under Assumption 5.7.1 (Smoothness or Convexity), and let x^* be a local minimum. A pair (λ^*, μ^*) is called a *Lagrange multiplier vector corresponding to f and x^** if

$$0 \in \partial f(x^*) + \sum_{i=1}^m \lambda_i^* \partial h_i(x^*) + \sum_{j=1}^r \mu_j^* \partial g_j(x^*) + T_X(x^*)^*, \tag{5.71}$$

$$\mu^* \geq 0, \qquad \mu^{*\prime} g(x^*) = 0. \tag{5.72}$$

If we can guarantee that the multiplier μ_0^* in Prop. 5.7.1 is nonzero, and the tangent cone $T_X(x^*)$ is convex, then there exists a Lagrange multiplier vector corresponding to x^*, which also satisfies the extra condition (iv) of the proposition.

The Convex Case

We now consider the problem

$$
\begin{aligned}
&\text{minimize} \quad f(x) \\
&\text{subject to} \quad x \in X, \qquad g_j(x) \leq 0, \quad j = 1, \ldots, r,
\end{aligned}
\tag{5.73}
$$

under some convexity assumptions. The extension of the analysis to the case where there are affine equality constraints is straightforward: we replace each equality constraint with two affine inequality constraints.†

Assumption 5.7.2: (Closedness and Convexity) The set X is convex, and the functions f and g_j, viewed as functions from X to $\Re$, are closed and convex.

Under the preceding assumption, we can show the following version of the enhanced Fritz John conditions. Because the assumption does not require that f and g_j be convex over $\Re^n$ as in Prop. 5.7.1, the line of proof of Props. 5.2.1 and 5.7.1 breaks down. We use a different line of proof, which does not rely on gradients and subgradients, and is based instead on the minimax theory of Section 2.6.

Proposition 5.7.2: Consider problem (5.73) under Assumption 5.7.2 (Closedness and Convexity), and let x^* be a global minimum. Then there exist a scalar μ_0^* and a vector $\mu^* = (\mu_1^*, \ldots, \mu_r^*)$, satisfying the following conditions:

 (i) $\mu_0^* f(x^*) = \min_{x \in X} \{ \mu_0^* f(x) + \mu^{*\prime} g(x) \}$.

 (ii) $\mu_j^* \geq 0$ for all $j = 0, 1, \ldots, r$.

 (iii) $\mu_0^*, \mu_1^*, \ldots, \mu_r^*$ are not all equal to 0.

---

† Note that the domain of the real-valued functions f and g_j is $\Re^n$. What is meant in Assumption 5.7.2 is that the functions $\tilde{f} : X \mapsto \Re$ and $\tilde{g}_j : X \mapsto \Re$, defined by $\tilde{f}(x) = f(x)$ and $\tilde{g}_j(x) = g_j(x)$ for all $x \in X$, are closed and convex.

(iv) If the index set $J = \{j \neq 0 \mid \mu_j^* > 0\}$ is nonempty, there exists a sequence $\{x^k\} \subset X$ that converges to x^* and is such that

$$\lim_{k \to \infty} f(x^k) = f(x^*), \qquad \limsup_{k \to \infty} g(x^k) \leq 0,$$

and for all k,

$$f(x^k) < f(x^*),$$

$$g_j(x^k) > 0, \quad \forall\, j \in J, \qquad g_j^+(x^k) = o\left(\min_{j \in J} g_j^+(x^k)\right), \quad \forall\, j \notin J.$$

Proof: The line of proof is similar to the one used for the enhanced Fritz John conditions of Section 5.2. However, we introduce a quadratic perturbation/penalty indirectly, through a minimax framework, thereby bypassing the need to use gradients, subgradients, and necessary optimality conditions.

For positive integers k and m, we consider the function

$$L_{k,m}(x, \xi) = f(x) + \frac{1}{k^3}\|x - x^*\|^2 + \xi' g(x) - \frac{\|\xi\|^2}{2m}.$$

We note that, for fixed $\xi \geq 0$, $L_{k,m}(x, \xi)$, viewed as a function from X to $\Re$, is closed and convex, because of the Closedness and Convexity Assumption. Furthermore, for fixed x, $L_{k,m}(x, \xi)$ is negative definite quadratic in ξ. For each k, we consider the set

$$X^k = X \cap \{x \mid \|x - x^*\| \leq k\}.$$

Since f and g_j are closed and convex when restricted to X, they are closed, convex, and coercive when restricted to X^k. Hence, we can use the Saddle Point Theorem (Prop. 2.6.9) to assert that $L_{k,m}$ has a saddle point over $x \in X^k$ and $\xi \geq 0$. This saddle point is denoted by $(x^{k,m}, \xi^{k,m})$.

The infimum of $L_{k,m}(x, \xi^{k,m})$ over $x \in X^k$ is attained at $x^{k,m}$, implying that

$$f(x^{k,m}) + \frac{1}{k^3}\|x^{k,m} - x^*\|^2 + \xi^{k,m\prime} g(x^{k,m})$$

$$= \inf_{x \in X^k} \left\{ f(x) + \frac{1}{k^3}\|x - x^*\|^2 + \xi^{k,m\prime} g(x) \right\}$$

$$\leq \inf_{x \in X^k,\, g(x) \leq 0} \left\{ f(x) + \frac{1}{k^3}\|x - x^*\|^2 + \xi^{k,m\prime} g(x) \right\} \qquad (5.74)$$

$$\leq \inf_{x \in X^k,\, g(x) \leq 0} \left\{ f(x) + \frac{1}{k^3}\|x - x^*\|^2 \right\}$$

$$= f(x^*).$$

Hence, we have

$$L_{k,m}(x^{k,m}, \xi^{k,m}) = f(x^{k,m}) + \frac{1}{k^3} \|x^{k,m} - x^*\|^2 + \xi^{k,m\prime} g(x^{k,m}) - \frac{1}{2m} \|\xi^{k,m}\|^2$$

$$\leq f(x^{k,m}) + \frac{1}{k^3} \|x^{k,m} - x^*\|^2 + \xi^{k,m\prime} g(x^{k,m})$$

$$\leq f(x^*).$$

(5.75)

Since $L_{k,m}$ is quadratic in ξ, the supremum of $L_{k,m}(x^{k,m}, \xi)$ over $\xi \geq 0$ is attained at

$$\xi_j^{k,m} = m g_j^+(x^{k,m}), \qquad j = 1, \ldots, r. \tag{5.76}$$

This implies that

$$L_{k,m}(x^{k,m}, \xi^{k,m}) = f(x^{k,m}) + \frac{1}{k^3} \|x^{k,m} - x^*\|^2 + \frac{m}{2} \|g^+(x^{k,m})\|^2$$

$$\geq f(x^{k,m}) + \frac{1}{k^3} \|x^{k,m} - x^*\|^2 \tag{5.77}$$

$$\geq f(x^{k,m}).$$

From Eqs. (5.75) and (5.77), we see that the sequence $\{x^{k,m}\}$ belongs to the set $\{x \in X^k \mid f(x) \leq f(x^*)\}$, which is compact. Hence, $\{x^{k,m}\}$ has a limit point (as $m \to \infty$), denoted by $\bar{x}^k$, which belongs to $\{x \in X^k \mid f(x) \leq f(x^*)\}$. By passing to a subsequence if necessary, we can assume without loss of generality that $\{x^{k,m}\}$ converges to $\bar{x}^k$ as $m \to \infty$. For each k, the sequence $\{f(x^{k,m})\}$ is bounded from below by $\inf_{x \in X^k} f(x)$, which is finite by Weierstrass' Theorem since f is closed and coercive when restricted to X^k. Also, for each k, $L_{k,m}(x^{k,m}, \xi^{k,m})$ is bounded from above by $f(x^*)$ [cf. Eq. (5.75)], so the equality in Eq. (5.77) implies that

$$\limsup_{m \to \infty} g_j(x^{k,m}) \leq 0, \qquad \forall\, j = 1, \ldots, r.$$

Therefore, by using the lower semicontinuity of g_j, we obtain $g(\bar{x}^k) \leq 0$, implying that $\bar{x}^k$ is a feasible point of problem (5.73), so that $f(\bar{x}^k) \geq f(x^*)$. Using Eqs. (5.75) and (5.77) together with the lower semicontinuity of f, we also have

$$f(\bar{x}^k) \leq \liminf_{m \to \infty} f(x^{k,m}) \leq \limsup_{m \to \infty} f(x^{k,m}) \leq f(x^*),$$

thereby showing that for each k,

$$\lim_{m \to \infty} f(x^{k,m}) = f(x^*).$$

Together with Eqs. (5.75) and (5.77), this also implies that for each k,

$$\lim_{m \to \infty} x^{k,m} = x^*.$$

Combining the preceding relations with Eqs. (5.75) and (5.77), for each k, we obtain

$$\lim_{m \to \infty} \left(f(x^{k,m}) - f(x^*) + \xi^{k,m'} g(x^{k,m}) \right) = 0. \tag{5.78}$$

Denote

$$\delta^{k,m} = \sqrt{1 + \sum_{j=1}^{r} (\xi_j^{k,m})^2},$$

$$\mu_0^{k,m} = \frac{1}{\delta^{k,m}}, \qquad \mu_j^{k,m} = \frac{\xi_j^{k,m}}{\delta^{k,m}}, \qquad j = 1, \dots, r. \tag{5.79}$$

Since $\delta^{k,m}$ is bounded from below, by dividing Eq. (5.78) by $\delta^{k,m}$, we obtain

$$\lim_{m \to \infty} \left(\mu_0^{k,m} f(x^{k,m}) - \mu_0^{k,m} f(x^*) + \sum_{j=1}^{r} \mu_j^{k,m} g_j(x^{k,m}) \right) = 0.$$

We now use the preceding relation to fix, for each k, an integer m_k such that

$$\left| \mu_0^{k,m_k} f(x^{k,m_k}) - \mu_0^{k,m_k} f(x^*) + \sum_{j=1}^{r} \mu_j^{k,m_k} g_j(x^{k,m_k}) \right| \le \frac{1}{k}, \tag{5.80}$$

and

$$\|x^{k,m_k} - x^*\| \le \frac{1}{k}, \quad |f(x^{k,m_k}) - f(x^*)| \le \frac{1}{k}, \quad \|g^+(x^{k,m_k})\| \le \frac{1}{k}. \tag{5.81}$$

Dividing both sides of the first relation in Eq. (5.74) by δ^{k,m_k}, and using the definition of X^k, we get

$$\mu_0^{k,m_k} f(x^{k,m_k}) + \frac{1}{k^3 \delta^{k,m_k}} \|x^{k,m_k} - x^*\|^2 + \sum_{j=1}^{r} \mu_j^{k,m_k} g_j(x^{k,m_k})$$

$$\le \mu_0^{k,m_k} f(x) + \sum_{j=1}^{r} \mu_j^{k,m_k} g_j(x) + \frac{1}{k \delta^{k,m_k}}, \qquad \forall\, x \in X^k.$$

Since the sequence $\{(\mu_0^{k,m_k}, \mu_1^{k,m_k}, \dots, \mu_r^{k,m_k})\}$ is bounded, it has a limit point, denoted by $(\mu_0^*, \mu_1^*, \dots, \mu_r^*)$. Taking the limit along the relevant subsequence in the preceding relation, and using Eq. (5.80), we obtain

$$\mu_0^* f(x^*) \le \mu_0^* f(x) + \mu^{*'} g(x), \qquad \forall\, x \in X.$$

This implies that

$$\mu_0^* f(x^*) \leq \inf_{x \in X} \left\{ \mu_0^* f(x) + \mu^{*\prime} g(x) \right\}$$

$$\leq \inf_{x \in X, \, g(x) \leq 0} \left\{ \mu_0^* f(x) + \mu^{*\prime} g(x) \right\}$$

$$\leq \inf_{x \in X, \, g(x) \leq 0} \mu_0^* f(x)$$

$$= \mu_0^* f(x^*).$$

Thus we have

$$\mu_0^* f(x^*) = \inf_{x \in X} \left\{ \mu_0^* f(x) + \mu^{*\prime} g(x) \right\},$$

so that $\mu_0^*, \mu_1^*, \ldots, \mu_r^*$ satisfy conditions (i), (ii), and (iii) of the proposition.

Assume now that the index set $J = \{ j \neq 0 \mid \mu_j^* > 0 \}$ is nonempty. Then, for sufficiently large k, we have $\xi_j^{k,m_k} > 0$ and hence $g_j(x^{k,m_k}) > 0$ for all $j \in J$. Thus, for each k, we can choose the index m_k so that we have $x^{k,m_k} \neq x^*$, in addition to Eq. (5.81). Dividing both sides of Eq. (5.76) by δ^{k,m_k}, and using Eq. (5.79) and the fact that $\mu_j^{k,m_k} \to \mu_j^*$, along the relevant subsequence, we obtain

$$\lim_{k \to \infty} \frac{m_k g_j^+(x^{k,m_k})}{\delta^{k,m_k}} = \mu_j^*, \qquad j = 1, \ldots, r.$$

From Eqs. (5.75) and (5.77), we also have

$$f(x^{k,m_k}) < f(x^*),$$

while from Eq. (5.81), we have $f(x^{k,m_k}) \to f(x^*)$, $g^+(x^{k,m_k}) \to 0$, and $x^{k,m_k} \to x^*$. It follows that the sequence $\{x^{k,m_k}\}$ satisfies condition (iv) of the proposition as well, concluding the proof.   **Q.E.D.**

The proof of the preceding proposition can be explained in terms of the construction shown in Fig. 5.7.1. Consider the function $L_{k,m}$ introduced in the proof,

$$L_{k,m}(x, \xi) = f(x) + \frac{1}{k^3} \|x - x^*\|^2 + \xi' g(x) - \frac{\|\xi\|^2}{2m}.$$

Note that the term $(1/k^3)\|x - x^*\|^2$ ensures that x^* is a strict local minimum of the function $f(x) + (1/k^3)\|x - x^*\|^2$. To simplify the following discussion, let us assume that f is strictly convex, so that this term can be omitted from the definition of $L_{k,m}$.

For any nonnegative vector $u \in \Re^r$, let $p^k(u)$ denote the optimal value of the problem

$$\begin{aligned}
& \text{minimize} && f(x) \\
& \text{subject to} && g(x) \leq u, \\
& && x \in X^k = X \cap \left\{ x \mid \|x - x^*\| \leq k \right\}.
\end{aligned} \tag{5.82}$$

For each k and m, the saddle point of the function $L_{k,m}(x, \xi)$, denoted by $(x^{k,m}, \xi^{k,m})$, can be characterized in terms of $p^k(u)$ as follows. The maximization of $L_{k,m}(x, \xi)$ over $\xi \geq 0$ for any fixed $x \in X^k$ yields

$$\xi_j = mg_j^+(x), \qquad j = 1, \ldots, r, \tag{5.83}$$

so that we have

$$
\begin{aligned}
L_{k,m}(x^{k,m}, \xi^{k,m}) &= \inf_{x \in X^k} \sup_{\xi \geq 0} \left\{ f(x) + \xi' g(x) - \frac{\|\xi\|^2}{2m} \right\} \\
&= \inf_{x \in X^k} \left\{ f(x) + \frac{m}{2} \|g^+(x)\|^2 \right\}.
\end{aligned}
$$

This minimization can also be written as

$$
\begin{aligned}
L_{k,m}(x^{k,m}, \xi^{k,m}) &= \inf_{x \in X^k} \inf_{u \in \Re^r, \ g(x) \leq u} \left\{ f(x) + \frac{m}{2} \|u^+\|^2 \right\} \\
&= \inf_{u \in \Re^r} \inf_{x \in X^k, \ g(x) \leq u} \left\{ f(x) + \frac{m}{2} \|u^+\|^2 \right\} \tag{5.84} \\
&= \inf_{u \in \Re^r} \left\{ p^k(u) + \frac{m}{2} \|u^+\|^2 \right\}.
\end{aligned}
$$

The vector $u^{k,m} = g(x^{k,m})$ attains the infimum in the preceding relation. This minimization can be visualized geometrically as in Fig. 5.7.1. The point of contact of the functions $p^k(u)$ and $L_{k,m}(x^{k,m}, \xi^{k,m}) - m/2\|u^+\|^2$ corresponds to the vector $u^{k,m}$ that attains the infimum in Eq. (5.84).

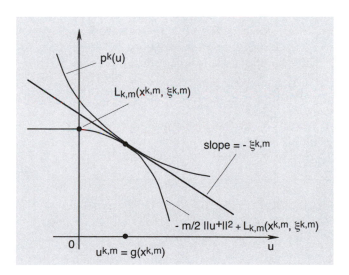

Figure 5.7.1. Illustration of the saddle point of the function $L_{k,m}(x, \xi)$ over $x \in X^k$ and $\xi \geq 0$ in terms of the function $p^k(u)$, which is the optimal value of problem (5.82) as a function of u.

We can also interpret $\xi^{k,m}$ in terms of the function p^k. In particular, the infimum of $L_{k,m}(x, \xi^{k,m})$ over $x \in X^k$ is attained at $x^{k,m}$, implying that

$$f(x^{k,m}) + \xi^{k,m'} g(x^{k,m}) = \inf_{x \in X^k} \{ f(x) + \xi^{k,m'} g(x) \}$$

$$= \inf_{u \in \Re^r} \{ p^k(u) + \xi^{k,m'} u \}.$$

Replacing $g(x^{k,m})$ by $u^{k,m}$ in the preceding relation, and using the fact that $x^{k,m}$ is feasible for problem (5.82) with $u = u^{k,m}$, we obtain

$$p^k(u^{k,m}) \leq f(x^{k,m}) = \inf_{u \in \Re^r} \{ p^k(u) + \xi^{k,m'}(u - u^{k,m}) \}.$$

Thus, we see that

$$p^k(u^{k,m}) \leq p^k(u) + \xi^{k,m'}(u - u^{k,m}), \qquad \forall \, u \in \Re^r,$$

which, by the definition of the subgradient of a function, implies that

$$-\xi^{k,m} \in \partial p^k(u^{k,m})$$

(cf. Fig. 5.7.1). It can be seen from this interpretation that the limit of $L_{k,m}(x^{k,m}, \xi^{k,m})$ as $m \to \infty$ is equal to $p^k(0)$, which is equal to $f(x^*)$ for each k. The limit of the normalized sequence

$$\left\{ \frac{(1, \xi^{k,m})}{\sqrt{1 + \|\xi^{k,m}\|^2}} \right\}$$

as $k \to \infty$ and $m \to \infty$ yields the Fritz John multipliers that satisfy conditions (i)-(iii) of Prop. 5.7.2, and the sequence $\{x^{k,m}\}$ is used to construct the sequence that satisfies condition (iv) of the proposition.

Lagrange Multipliers and Pseudonormality

Under the Closedness and Convexity Assumption, the definition of a Lagrange multiplier is adapted as follows, so that it does not involve gradients or subgradients.

Definition 5.7.2: Consider problem (5.73) under Assumption 5.7.2 (Closedness and Convexity), and let x^* be a global minimum. A vector $\mu^* \geq 0$ is called a *Lagrange multiplier vector corresponding to f and x^** if

$$f(x^*) = \min_{x \in X} \{ f(x) + \mu^{*'} g(x) \}, \qquad \mu^{*'} g(x^*) = 0. \qquad (5.85)$$

Note that given the feasibility of x^*, which implies $g(x^*) \leq 0$, and the nonnegativity of μ^*, the condition $\mu^{*\prime}g(x^*) = 0$ of Eq. (5.85) is equivalent to the complementary slackness condition $\mu_j^* = 0$ for $j \in A(x^*)$ [cf. Eq. (5.72)]. Furthermore, when f and g_j are real-valued and convex over $\Re^n$, the condition that x^* minimizes $f(x) + \mu^{*\prime}g(x)$ over a convex set X is equivalent to the condition

$$0 \in \partial f(x^*) + \sum_{j=1}^{r} \mu_j^* \partial g_j(x^*) + T_X(x^*)^*$$

[cf. Eq. (5.71)].

Similarly, the definition of pseudonormality is adapted as follows.

Definition 5.7.3: Consider problem (5.73), and assume that the set X is convex, and the functions g_j, viewed as functions from X to $\Re$, are convex. The constraint set of the problem is said to be *pseudo-normal* if one cannot find a vector $\mu \geq 0$ and a vector $\tilde{x} \in X$ such that:

   (i) $0 = \inf_{x \in X} \mu'g(x)$.

   (ii) $\mu'g(\tilde{x}) > 0$.

To provide a geometric interpretation of pseudonormality in the convex case, let us introduce the set

$$G = \big\{ g(x) \mid x \in X \big\}$$

and consider hyperplanes that support this set and pass through 0. As Fig. 5.7.2 illustrates, *pseudonormality means that there is no hyperplane with a normal $\mu \geq 0$ that properly separates the sets $\{0\}$ and G, and contains G in its positive halfspace.*

Consider now a global minimum x^*, and multipliers μ_0^* and μ^* satisfying the enhanced Fritz John conditions of Prop. 5.7.2. If the constraint set is pseudonormal, we cannot have $\mu_0^* = 0$, since otherwise, $\mu^* \neq 0$ and the vectors of the sequence $\{x^k\}$ would satisfy $\mu^{*\prime}g(x^k) > 0$, thereby violating condition (ii) of pseudonormality. It follows that μ^*/μ_0^* is a Lagrange multiplier vector, which also satisfies the extra CV condition (iv) of Prop. 5.7.2.

The analysis of Section 5.4 is easily extended to show that the constraint set is pseudonormal under either one of the following two criteria:

(a) *Linearity criterion:* $X = \Re^n$ and the functions g_j are affine.

(b) *Slater criterion:* X is convex, the functions g_j are convex, and there exists a feasible vector $\bar{x}$ such that

$$g_j(\bar{x}) < 0, \qquad j = 1, \ldots, r.$$

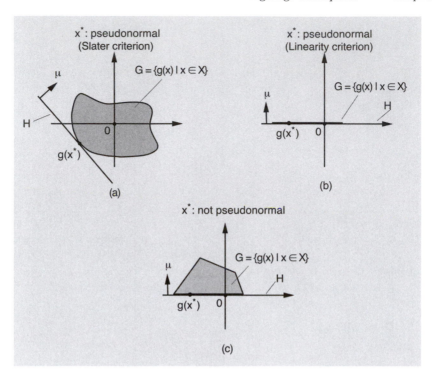

Figure 5.7.2. Geometric interpretation of pseudonormality. Consider the set

$$G = \big\{ g(x) \mid x \in X \big\}$$

and hyperplanes that support this set. For feasibility, G should intersect the nonpositive orthant $\{z \mid z \leq 0\}$. The first condition $[0 = \inf_{x \in X} \mu' g(x)]$ in the definition of pseudonormality means that there is a hyperplane with normal $\mu \geq 0$, which passes through 0, supports G, and contains G in its positive halfspace [note that, as illustrated in figure (a), this cannot happen if G intersects the interior of the nonpositive orthant; cf. the Slater criterion]. The second condition means that H does not fully contain G [cf. figures (b) and (c)]. If the Slater criterion holds, the first condition cannot be satisfied. If the linearity criterion holds, the set G is an affine set and the second condition cannot be satisfied (this depends critically on $X = \Re^n$ rather than X being a general polyhedron).

Thus, under either one of these criteria, and assuming the Convexity and Closedness Assumption, by Prop. 5.7.2, a Lagrange multiplier vector, satisfying the extra CV condition (iv) of that proposition, is guaranteed to exist.

Figure 5.7.2 illustrates why the linearity and Slater criteria guarantee pseudonormality. In particular, the Slater criterion is equivalent to the statement that the set G intersects the interior of the nonpositive orthant. Clearly, if this is so, there cannot exist a hyperplane with a normal $\mu \geq 0$

that simultaneously supports G and passes through 0. Similarly, if $X = \Re^n$ and g_j are affine, the set G is affine, and is fully contained in H. Thus the linearity and Slater criteria imply pseudonormality.

If X is a polyhedron (rather than $X = \Re^n$) and the functions g_j are affine, the constraint set need not be pseudonormal. The reason is that there may exist a hyperplane with a normal $\mu \geq 0$ that properly separates the sets $\{0\}$ and G, and contains G in its positive halfspace, as Fig. 5.7.2(c) illustrates (compare also with the discussion in Section 5.6). However, we can prove existence of a Lagrange multiplier, by combining the linearity criterion with the extended representation of the problem as in Section 5.6. Furthermore, we can extend the Slater criterion to the case where there are additional linear equality constraints. Then, for existence of a Lagrange multiplier, we should require that there exists a feasible point in the relative interior of X, in addition to the condition $g_j(\bar{x}) < 0$ for all j (see the constraint qualification CQ5c in Section 5.4; also the analysis of the next chapter).

5.8 NOTES, SOURCES, AND EXERCISES

Lagrange multipliers were originally introduced for problems with equality constraints. Research on inequality constrained problems started considerably later, and highlighted the major role of convexity in optimization. Important early works are those of Karush [Kar39] (an unpublished MS thesis), John [Joh48], and Kuhn and Tucker [KuT51]. The survey by Kuhn [Kuh76] gives a historical view of the development of the subject.

There has been much work in the 60s and early 70s on constraint qualifications. Important examples are those of Arrow, Hurwicz, and Uzawa [AHU61], Abadie [Aba67], Mangasarian and Fromovitz [MaF67], and Guignard [Gui69]. For textbook treatments, see Mangasarian [Man69] and Hestenes [Hes75]. There has been much subsequent work on the subject, some of which addresses nondifferentiable problems, e.g., Gould and Tolle [GoT71], [GoT72], Bazaraa, Goode, and Shetty [BGS72], Clarke [Cla83], Demjanov and Vasilév [DeV85], Mordukhovich [Mor88], and Rockafellar [Roc93].

In the case where there are just equality and inequality constraints ($X = \Re^n$), the classical lines of analysis (which are either based on quasiregularity, or show existence of Lagrange multipliers using a distinct proof for each constraint qualification) are cumbersome and long-winded. Our approach, based on the unifying power of pseudonormality, results in substantially shorter proofs, and is more general because it applies to the case where $X \neq \Re^n$ as well.

Most of the works mentioned above consider problems with equality and inequality constraints only. Abstract set constraints (in addition to equality and inequality constraints) have been considered along two different lines:

(a) For convex programs (convex f, g_j, and X, and affine h_i), and in the context of the geometric multiplier theory to be developed in Chapter 6. Here the abstract set constraint does not cause significant complications, because for convex X, the tangent cone is conveniently defined in terms of feasible directions, and nonsmooth analysis issues of nonregularity do not arise.

(b) For the nonconvex setting of this chapter, where the abstract set constraint causes significant difficulties because the classical approach that is based on quasiregularity is not fully satisfactory.

The difficulties in case (b) above may be addressed using classical concepts in the case where the abstract set constraint X is convex (the textbook by Bertsekas [Ber99a] gives the corresponding analysis). However, for a general set X, a resort to nonsmooth analysis ideas appears to be necessary. In particular, the concepts of the normal cone $N_X(x^*)$ (Mordukhovich [Mor76]) and regularity $[N_X(x^*) = T_X(x^*)^*]$ are essential for delineating the class of problems for which a satisfactory Lagrange multiplier theory exists. Rockafellar [Roc93], following the work of Clarke [Cla83] and other researchers, used the Lagrange multiplier definition given in Section 5.3 (what we have called R-multiplier), but he did not develop or use the main ideas of this chapter, i.e., the enhanced Fritz John conditions, informative Lagrange multipliers, and pseudonormality. Instead he assumed the constraint qualification CQ6, which, as discussed in Section 5.4, is restrictive because, when X is regular, it implies that the set of Lagrange multipliers is not only nonempty but also compact.

The starting point for the line of analysis of this chapter may be traced to the penalty-based proof of the classical Fritz John conditions given by McShane [McS73]. Hestenes [Hes75] observed that McShane's proof can be used to strengthen the CS condition to assert the existence, within any neighborhood B of x^*, of an $x \in B \cap X$ such that

$$\lambda_i^* h_i(x) > 0, \quad \forall\, i \in I, \qquad g_j(x) > 0, \quad \forall\, j \in J,$$

which is slightly weaker than CV as defined here [there is no requirement that x, simultaneously with violation of the constraints with nonzero multipliers, satisfies $f(x) < f(x^*)$ and Eq. (5.13)]. Furthermore, Hestenes showed the existence of Lagrange multipliers under the constraint qualifications CQ1-CQ3 by using his enhanced Fritz John conditions.

McShane and Hestenes considered only the case where $X = \Re^n$. The case where X is a closed convex set was considered in Bertsekas [Ber99a], where a special case of CQ5 was also proved. The extension to the case where X is a general closed set and the strengthened version of condition (iv) were given in Bertsekas and Ozdaglar [BeO00a], [BeO00b]. In particular, these works first gave the enhanced Fritz John conditions of Prop. 5.2.1, introduced the notion of an informative Lagrange multiplier and the notion of pseudonormality, and established the constraint qualification results of Section 5.4 (including the introduction of CQ5) and the exact penalty results of Section 5.5.

A notion that is related to pseudonormality, called *quasinormality* and implied by pseudonormality, is given by Hestenes [Hes75] (for the case where $X = \Re^n$) and Bertsekas [Ber99a] (for the case where X is a closed convex set). It is argued in Bertsekas and Ozdaglar [BeO00a] that pseudonormality is better suited as a unifying vehicle for Lagrange multiplier theory. In particular, some of the relations shown in Fig. 5.5.2 do not hold if pseudonormality is replaced by quasinormality. Furthermore, quasinormality does not admit an intuitive geometric interpretation.

There is an interesting relation between pseudonormality and an extension of the notion of quasiregularity, proposed by Gould and Tolle [GoT72], and Guignard [Gui69], to the more general setting where X may be a strict subset of $\Re^n$. In this extension, x^* is said to be quasiregular if

$$T_C(x^*) = V(x^*) \cap T_X(x^*), \tag{5.86}$$

where $V(x^*)$ is the cone of first order feasible variations

$$V(x^*) = \left\{ y \mid \nabla h_i(x^*)'y = 0, \ \forall \ i, \ \nabla g_j(x^*)'y \leq 0, \ \forall \ j \text{ with } g_j(x^*) = 0 \right\}.$$

It is shown by Ozdaglar [Ozd03], and Ozdaglar and Bertsekas [OzB03] that under a regularity assumption on X, quasiregularity is implied by pseudonormality. It is also shown in these references that contrary to the case where $X = \Re^n$, quasiregularity is not sufficient to guarantee the existence of a Lagrange multiplier. Thus the importance of quasiregularity, the classical pathway to Lagrange multipliers when $X = \Re^n$, diminishes when $X \neq \Re^n$. By contrast, pseudonormality provides satisfactory unification of the theory. What is happening here is that the constraint set admits Lagrange multipliers at x^* if and only if

$$T_C(x^*)^* = T_X(x^*)^* + V(x^*)^*; \tag{5.87}$$

this is a classical result derived in various forms by Gould and Tolle [GoT72], Guignard [Gui69], and Rockafellar [Roc93]. When $X = \Re^n$, this condition reduces to $T_C(x^*)^* = V(x^*)^*$, and is implied by quasiregularity $[T_C(x^*) = V(x^*)]$. However, when $X \neq \Re^n$ the natural definition of quasiregularity, given by Eq. (5.86), does not imply the Lagrange multiplier condition (5.87), unless substantial additional assumptions are imposed.

The study of exact penalty functions was initiated by Zangwill [Zan67], who treated the case of the constraint qualification CQ2. The case of CQ1 was treated by Pietrzykowski [Pie69]. Han and Mangasarian [HaM79] provide a survey of research on the subject. Exact penalty functions have traditionally been viewed as a computational device and they have not been integrated earlier within the theory of constraint qualifications in the manner described here.

Exact penalty functions are related to the notion of *calmness*, introduced and suggested as a constraint qualification by Clarke [Cla76], [Cla83]. However, there are some important differences between the notions of calmness and admittance of an exact penalty. In particular, calmness is a property of the problem (5.58) and depends on the cost function f, while admittance of an exact penalty is a property of the constraint set and is independent of the cost function. More importantly for the purposes of this work, calmness is not useful as a unifying theoretical vehicle because it does not relate well with other major constraint qualifications. For example, it can be shown that CQ1, the most common constraint qualification, does not imply calmness of problem (5.58), and reversely problem calmness does not imply CQ1.

The enhanced Fritz John conditions of Section 5.7 for problems simultaneously involving convex, possibly nonsmooth functions, and the associated notions of pseudonormality are new, and have been developed by Bertsekas, Ozdaglar, and Tseng [BOT02]. Further analysis of enhanced Fritz John conditions for the case of convex problems, but in the absence of an optimal solution x^*, will be given in Section 6.6.

Nonsmoothness in optimization arises in a variety of ways, most importantly in the context of duality (see Chapter 6). Within this latter context, nonsmoothness is accompanied by convexity, so the framework of Section 5.7 applies. It is possible to derive optimality conditions, including Lagrange multiplier theorems, for problems involving nonsmooth and nonconvex functions (see e.g., Clarke [Cla83], Rockafellar and Wets [RoW98]), but we have not pursued this subject here: it requires the full machinery of nonsmooth analysis, and the conditions obtained are not as convenient as the ones for the smooth case. It should also be noted that nonsmooth/nonconvex problems are not nearly as interesting or important as their convex counterparts. Not only they do not arise as dual problems, but also they can often be converted to smooth problems. For example, the minimax problem

$$\text{minimize} \quad F(x)$$
$$\text{subject to} \quad h_i(x) = 0, \quad i = 1, \ldots, m,$$

where

$$F(x) = \max\{f_1(x), \ldots, f_p(x)\},$$

and the functions f_j and h_i are smooth, can be converted to the smooth problem

$$\text{minimize} \quad z$$
$$\text{subject to} \quad f_j(x) \leq z, \quad j = 1, \ldots, p, \quad h_i(x) = 0, \quad i = 1, \ldots, m.$$

Similar transformations are possible in many practical nonsmooth contexts where duality is not involved.

EXERCISES

5.1 (Second Order Sufficiency Conditions for Equality-Constrained Problems)

Define the Lagrangian function $L(x, \lambda)$ to be

$$L(x, \lambda) = f(x) + \lambda' h(x).$$

Assume that f and h are twice continuously differentiable, and let $x^* \in \Re^n$ and $\lambda^* \in \Re^m$ satisfy

$$\nabla_x L(x^*, \lambda^*) = 0, \qquad \nabla_\lambda L(x^*, \lambda^*) - 0,$$

$$y' \nabla_{xx}^2 L(x^*, \lambda^*) y > 0, \qquad \forall \, y \neq 0 \text{ with } \nabla h(x^*)' y = 0.$$

Show that x^* is a strict local minimum of f subject to $h(x) = 0$.

5.2 (Second Order Sufficiency Conditions for Inequality-Constrained Problems)

Define the Lagrangian function $L(x, \lambda, \mu)$ to be

$$L(x, \lambda, \mu) = f(x) + \lambda' h(x) + \mu' g(x).$$

Assume that f, h, and g are twice continuously differentiable, and let $x^* \in \Re^n$, $\lambda^* \in \Re^m$, and $\mu^* \in \Re^r$ satisfy

$$\nabla_x L(x^*, \lambda^*, \mu^*) = 0, \qquad h(x^*) = 0, \qquad g(x^*) \leq 0,$$

$$\mu_j^* > 0, \;\; \forall \, j \in A(x^*), \qquad \mu_j^* = 0, \;\; \forall \, j \notin A(x^*),$$

$$y' \nabla_{xx}^2 L(x^*, \lambda^*, \mu^*) y > 0,$$

for all $y \neq 0$ such that

$$\nabla h_i(x^*)' y = 0, \;\; \forall \, i = 1, \ldots, m, \qquad \nabla g_j(x^*)' y = 0, \;\; \forall \, j \in A(x^*).$$

Show that x^* is a strict local minimum of f subject to $h(x) = 0$, $g(x) \leq 0$.

5.3 (Sensitivity Under Second Order Conditions)

Let x^* and (λ^*, μ^*) be a local minimum and Lagrange multiplier, respectively, of the problem

$$\text{minimize} \quad f(x)$$
$$\text{subject to} \quad h_1(x) = 0, \ldots, h_m(x) = 0, \tag{5.88}$$
$$g_1(x) \leq 0, \ldots, g_r(x) \leq 0,$$

satisfying the second order sufficiency conditions of Exercise 5.2. Assume that the gradients $\nabla h_i(x^*)$, $i = 1, \ldots, m$, $\nabla g_j(x^*)$, $j \in A(x^*)$, are linearly independent. Consider the family of problems

$$\text{minimize} \quad f(x)$$
$$\text{subject to} \quad h(x) = u, \qquad g(x) \leq v, \tag{5.89}$$

parameterized by the vectors $u \in \Re^m$ and $v \in \Re^r$. Then there exists an open sphere S centered at $(u, v) = (0, 0)$ such that for every $(u, v) \in S$ there is an $x(u, v) \in \Re^n$ and $\lambda(u, v) \in \Re^m$, $\mu(u, v) \in \Re^r$, which are a local minimum and associated Lagrange multiplier vectors of problem (5.89). Furthermore, $x(\cdot, \cdot)$, $\lambda(\cdot, \cdot)$, and $\mu(\cdot, \cdot)$ are continuously differentiable in S and we have $x(0, 0) = x^*$, $\lambda(0, 0) = \lambda^*$, $\mu(0, 0) = \mu^*$. In addition, for all $(u, v) \in S$, there holds

$$\nabla_u p(u, v) = -\lambda(u, v),$$
$$\nabla_v p(u, v) = -\mu(u, v),$$

where $p(u, v)$ is the optimal cost parameterized by (u, v),

$$p(u, v) = f\big(x(u, v)\big).$$

5.4 (General Sufficiency Condition)

Consider the problem

$$\text{minimize} \quad f(x)$$
$$\text{subject to} \quad x \in X, \qquad g_j(x) \leq 0, \quad j = 1, \ldots, r,$$

where f and g_j are real valued functions on $\Re^n$, and X is a subset of $\Re^n$. Let x^* be a feasible point, which together with a vector $\mu^* = (\mu_1^*, \ldots, \mu_r^*)$, satisfies

$$\mu_j^* \geq 0, \qquad j = 1, \ldots, r,$$
$$\mu_j^* = 0, \qquad \forall\, j \notin A(x^*),$$

and minimizes the Lagrangian function $L(x, \mu^*)$ over $x \in X$:

$$x^* \in \arg\min_{x \in X} L(x, \mu^*).$$

Show that x^* is a global minimum of the problem.

5.5

The purpose of this exercise is to work out an alternative proof of Lemma 5.3.1, assuming that $N = \{0\}$ [which corresponds to the case where there is no abstract set constraint $(X = \Re^n)$]. Let $a_0, \ldots, a_r$ be given vectors in $\Re^n$. Suppose that the set

$$M = \left\{ \mu \geq 0 \;\middle|\; a_0 + \sum_{j=1}^{r} \mu_j a_j = 0 \right\}$$

is nonempty, and let μ^* be the vector of minimum norm in M. For any $\gamma > 0$, consider the function

$$L_\gamma(d, \mu) = \left(a_0 + \sum_{j=1}^{r} \mu_j a_j \right)' d + \frac{\gamma}{2} \|d\|^2 - \frac{1}{2}\|\mu\|^2.$$

(a) Show that

$$-\frac{1}{2}\|\mu^*\|^2 = \sup_{\mu \geq 0} \; \inf_{d \in \Re^n} \; L_0(d, \mu)$$

$$\leq \inf_{d \in \Re^n} \; \sup_{\mu \geq 0} L_0(d, \mu) \tag{5.90}$$

$$= \inf_{d \in \Re^n} \left\{ a_0' d + \frac{1}{2} \sum_{j=1}^{r} \left((a_j' d)^+ \right)^2 \right\}.$$

(b) Use the lower bound of part (a) and the theory of Section 2.3 on the existence of solutions of quadratic programs to conclude that the infimum in the right-hand side above is attained for some $d^* \in \Re^n$.

(c) Show that for every $\gamma > 0$, L_γ has a saddle point (d^γ, μ^γ) such that

$$\mu_j^\gamma = (a_j' d^\gamma)^+, \qquad j = 1, \ldots, r.$$

Furthermore,

$$L_\gamma(d^\gamma, \mu^\gamma) = -\frac{\left\| a_0 + \sum_{j=1}^{r} \mu_j^\gamma a_j \right\|^2}{2\gamma} - \frac{1}{2}\|\mu^\gamma\|^2 \geq -\frac{1}{2}\|\mu^*\|^2.$$

(d) Use part (c) to show that $\|\mu^\gamma\| \leq \|\mu^*\|$, and use the minimum norm property of μ^* to conclude that as $\gamma \to 0$, we have $\mu^\gamma \to \mu^*$ and $L_\gamma(d^\gamma, \mu^\gamma) \to -(1/2)\|\mu^*\|^2$.

(e) Use part (d) and Eq. (5.90) to show that (d^*, μ^*) is a saddle point of L_0, and that

$$a_0' d^* = -\|\mu^*\|^2, \qquad (a_j' d^*)^+ = \mu_j^*, \qquad j = 1, \ldots, r.$$

5.6 (Strict Complementarity)

Consider the problem

$$\text{minimize} \quad f(x)$$
$$\text{subject to} \quad g_j(x) \le 0, \qquad j = 1, \ldots, r,$$

where $f : \Re^n \mapsto \Re$ and $g_j : \Re^n \mapsto \Re$ are smooth functions. A Lagrange multiplier $\{\mu_1^*, \ldots, \mu_r^*\}$, corresponding to a local minimum x^*, is said to satisfy *strict complementarity* if for all j such that $g_j(x^*) = 0$, we have $\mu_j^* > 0$. Show that a Lagrange multiplier that satisfies strict complementarity need not be informative, and conversely, a Lagrange multiplier that is informative need not satisfy strict complementarity.

5.7

Consider the problem

$$\text{minimize} \quad \sum_{i=1}^{n} f_i(x_i)$$
$$\text{subject to} \quad x \in S, \qquad x_i \in X_i, \quad i = 1, \ldots, n,$$

where $f_i : \Re \mapsto \Re$ are smooth functions, X_i are closed intervals of real numbers of $\Re^n$, and S is a subspace of $\Re^n$. Let x^* be a local minimum. Introduce artificial optimization variables $z_1, \ldots, z_n$ and the linear constraints $x_i = z_i$, $i = 1, \ldots, n$, while replacing the constraint $x \in S$ with $z \in S$, so that the problem becomes

$$\text{minimize} \quad \sum_{i=1}^{n} f_i(x_i)$$
$$\text{subject to} \quad z \in S, \qquad x_i \in X_i, \qquad x_i = z_i, \quad i = 1, \ldots, n.$$

Show that there exists a Lagrange multiplier vector $\lambda^* = (\lambda_1^*, \ldots, \lambda_n^*)$ such that $\lambda^* \in S^\perp$ and

$$\left(\nabla f_i(x_i^*) + \lambda_i^* \right)(x_i - x_i^*) \ge 0, \qquad \forall \ x_i \in X_i, \ i = 1, \ldots, n.$$

5.8

Show that if X is regular at x^* the constraint qualifications CQ5a and CQ6 are equivalent.

5.9 (Minimax Problems)

Derive Lagrange multiplier-like optimality conditions for the minimax problem

$$\text{minimize} \quad \max\{f_1(x), \ldots, f_p(x)\}$$
$$\text{subject to} \quad x \in X,$$

where X is a closed set, and the functions f_i are smooth. *Hint:* Convert the problem to the smooth problem

$$\text{minimize} \quad z$$
$$\text{subject to} \quad x \in X, \quad f_i(x) \leq z, \quad i = 1, \ldots, p,$$

and show that CQ5 holds.

5.10 (Exact Penalty Functions)

Consider the problem

$$\text{minimize} \quad f(x)$$
$$\text{subject to} \quad x \in C, \tag{5.91}$$

where

$$C = X \cap \{x \mid h_1(x) = 0, \ldots, h_m(x) = 0\} \cap \{x \mid g_1(x) \leq 0, \ldots, g_r(x) \leq 0\},$$

and assume that f, h_i, and g_j are smooth functions. Let F_c be the exact penalty function, i.e.,

$$F_c(x) = f(x) + c\left(\sum_{i=1}^{m} |h_i(x)| + \sum_{j=1}^{r} g_j^+(x)\right),$$

where c is a positive scalar.

(a) Suppose that x^* is a local minimum of problem (5.91), and that for some given $c > 0$, x^* is also a local minimum of F_c over X. Show that there exists an R-multiplier vector (λ^*, μ^*) for problem (5.91) such that

$$|\lambda_i^*| \leq c, \quad i = 1, \ldots, m, \qquad \mu_j^* \in [0, c], \quad j = 1, \ldots, r. \tag{5.92}$$

(b) Derive conditions that guarantee that if x^* is a local minimum of problem (5.91) and (λ^*, μ^*) is a corresponding Lagrange multiplier vector, then x^* is also a local minimum of F_c over X for all c such that Eq. (5.92) holds.

5.11 (Extended Representations)

This exercise generalizes Prop. 5.6.1 by including an additional set constraint in the extended representation. Assume that the set constraint can be described as

$$X = \big\{x \in \overline{X} \mid h_i(x) = 0, \ i = m+1, \ldots, \overline{m}, \ g_j(x) \leq 0, \ j = r+1, \ldots, \overline{r}\big\},$$

so that C is represented alternatively as

$$C = \overline{X} \cap \big\{x \mid h_1(x) = 0, \ldots, h_{\overline{m}}(x) = 0\big\} \cap \big\{x \mid g_1(x) \leq 0, \ldots, g_{\overline{r}}(x) \leq 0\big\}.$$

We call this the extended representation of C. Assuming that $\overline{X}$ is closed and that all the functions h_i and g_j are smooth, show the following:

(a) If the constraint set admits Lagrange multipliers in the extended representation, it admits Lagrange multipliers in the original representation.

(b) If the constraint set admits an exact penalty in the extended representation, it admits an exact penalty in the original representation.

6

Lagrangian Duality

Contents

6.1. Geometric Multipliers . . . . p. 346
6.2. Duality Theory . . . . p. 355
6.3. Linear and Quadratic Programming Duality . . . . . p. 362
6.4. Existence of Geometric Multipliers . . . . p. 367
    6.4.1. Convex Cost – Linear Constraints . . . . p. 368
    6.4.2. Convex Cost – Convex Constraints . . . . p. 371
6.5. Strong Duality and the Primal Function . . . . p. 374
    6.5.1. Duality Gap and the Primal Function . . . . . . p. 374
    6.5.2. Conditions for No Duality Gap . . . . p. 377
    6.5.3. Subgradients of the Primal Function . . . . p. 382
    6.5.4. Sensitivity Analysis . . . . p. 383
6.6. Fritz John Conditions when there is no Optimal Solution p. 384
    6.6.1. Enhanced Fritz John Conditions . . . . p. 390
    6.6.2. Informative Geometric Multipliers . . . . p. 406
6.7. Notes, Sources, and Exercises . . . . p. 413

In this chapter, we take a geometric approach towards Lagrange multipliers, with a view towards duality. Because of its geometric character, duality theory admits insightful visualization, through the use of hyperplanes, and their convex set support and separation properties. The min common and max crossing problems, discussed in Section 2.6 are principal examples of the geometric constructions that underlie the theory. Using such constructions, we will interpret multipliers in terms of hyperplanes and we will view the dual problem as a special case of the max crossing problem.

We will first introduce a certain type of multipliers, called *geometric*, in Section 6.1, and we will highlight how under convexity assumptions, they are related to the Lagrange multipliers of the preceding chapter. In Section 6.2, we will define the *dual function* and the associated dual optimization problem, and we will relate its optimal solutions to geometric and Lagrange multipliers. In Section 6.3, we will consider the special cases of linear and quadratic programs with finite optimal value, and we will show that the sets of Lagrange multipliers, geometric multipliers, and dual optimal solutions coincide. In Section 6.4, we will delineate other conditions that guarantee that the sets of geometric multipliers and dual optimal solutions are nonempty and coincide, and also coincide with the set of Lagrange multipliers. In Section 6.5, we relate the duality theory of this chapter with the min common/max crossing duality of Section 2.5 and the minimax theory of Section 2.6, and we derive conditions under which there is no duality gap (the optimal values of the original problem and the dual problem are equal). In the process, we introduce the *primal function*, which captures many of the important characteristics of the problem structure. Finally, in Section 6.6, we will derive enhanced Fritz John conditions, similar to the ones of Chapter 5, but modified to conform to the geometric spirit of the present chapter.

6.1 GEOMETRIC MULTIPLIERS

We consider the problem

$$\text{minimize} \quad f(x)$$
$$\text{subject to} \quad x \in X, \quad h(x) = 0, \quad g(x) \leq 0, \tag{P}$$

where $f : \Re^n \mapsto \Re$, $h_i : \Re^n \mapsto \Re$, $g_j : \Re^n \mapsto \Re$ are given functions, and X is a nonempty subset of $\Re^n$, and we use the notation

$$h(x) = \big(h_1(x), \ldots, h_m(x)\big), \qquad g(x) = \big(g_1(x), \ldots, g_r(x)\big).$$

We refer to this as the *primal problem* and we denote by f^* its optimal value:

$$f^* = \inf_{\substack{x \in X \\ h_i(x)=0,\, i=1,\ldots,m \\ g_j(x)\leq 0,\, j=1,\ldots,r}} f(x).$$

Note that, throughout this chapter, the functions f, h_i, and g_j are defined on all of $\Re^n$ and are real-valued. On occasion, however, we will make assumptions about properties of these functions, such as convexity, that hold over only some subset of $\Re^n$.

For many of the results of this chapter, we will assume that the cost function of the primal problem (P) is bounded from below, i.e., $-\infty < f^*$. We will also mostly assume that the primal problem is feasible, in which case by our convention regarding minimization over the empty set, we write $f^* < \infty$.

We want to define a notion of multiplier vector that is not tied to a specific local or global minimum, and does not assume differentiability or even continuity of the cost and constraint functions. To motivate the definition, we take as a starting point the Lagrange multiplier theory of Chapter 5, and consider the special case where X is closed and convex, h_i are affine, and f and g_j are convex and smooth. If we know that there is a minimizer x^*, the theory of Chapter 5 asserts that under any one of the constraint qualifications that imply pseudonormality, there exist vectors $\lambda^* = (\lambda_1^*, \ldots, \lambda_m^*)$ and $\mu^* = (\mu_1^*, \ldots, \mu_r^*)$ that satisfy the conditions

$$\nabla_x L(x^*, \lambda^*, \mu^*)'(x - x^*) \geq 0, \qquad \forall\, x \in X, \tag{6.1}$$

$$\mu^* \geq 0, \qquad \mu^{*'} g(x^*) = 0, \tag{6.2}$$

where $L : \Re^{n+m+r} \mapsto \Re$ is the Lagrangian function

$$L(x, \lambda, \mu) = f(x) + \lambda' h(x) + \mu' g(x).$$

In view of the convexity of X and the function $L(\cdot, \lambda^*, \mu^*)$, Eq. (6.1) implies that x^* minimizes $L(\cdot, \lambda^*, \mu^*)$ over X (cf. Prop. 4.7.2). By using also the optimality of x^*, and the conditions $h(x^*) = 0$ and $\mu^{*'} g(x^*) = 0$ [cf. Eq. (6.2)], we obtain

$$\begin{aligned}
f^* &= f(x^*) \\
&= f(x^*) + \lambda^{*'} h(x^*) + \mu^{*'} g(x^*) \\
&= L(x^*, \lambda^*, \mu^*) \\
&= \min_{x \in X} L(x, \lambda^*, \mu^*).
\end{aligned}$$

This condition does not depend on the differentiability of the cost and constraint functions, and is the basis for the following alternative multiplier definition:

Definition 6.1.1: A vector $(\lambda^*, \mu^*) = (\lambda_1^*, \ldots, \lambda_m^*, \mu_1^*, \ldots, \mu_r^*)$ is said to be a *geometric multiplier vector* (or simply *a geometric multiplier*) for the primal problem (P) if $\mu^* \geq 0$ and

$$f^* = \inf_{x \in X} L(x, \lambda^*, \mu^*).$$

To visualize the definition of a geometric multiplier and other concepts related to duality, we consider hyperplanes in the space of constraint-cost pairs $\big(h(x), g(x), f(x)\big)$ (viewed as vectors in $\Re^{m+r+1}$). A hyperplane H of this type is specified by a linear equation involving a nonzero normal vector (λ, μ, μ_0), where $\lambda \in \Re^m$, $\mu \in \Re^r$, $\mu_0 \in \Re$, and by a scalar c:

$$H = \big\{(v, u, w) \mid v \in \Re^m,\ u \in \Re^r,\ w \in \Re,\ \mu_0 w + \lambda' v + \mu' u = c\big\}.$$

Any vector $(\overline{v}, \overline{u}, \overline{w})$ that belongs to H specifies c as

$$c = \mu_0 \overline{w} + \lambda' \overline{v} + \mu' \overline{u}.$$

Thus, the hyperplane with normal (λ, μ, μ_0) that passes through a vector $(\overline{v}, \overline{u}, \overline{w})$ is the set of vectors (v, u, w) that satisfy the equation

$$\mu_0 w + \lambda' v + \mu' u = \mu_0 \overline{w} + \lambda' \overline{v} + \mu' \overline{u}.$$

This hyperplane defines two closed halfspaces: the positive closed halfspace

$$\big\{(v, u, w) \mid \mu_0 w + \lambda' v + \mu' u \geq \mu_0 \overline{w} + \lambda' \overline{v} + \mu' \overline{u}\big\}$$

and the negative closed halfspace

$$\big\{(v, u, w) \mid \mu_0 w + \lambda' v + \mu' u \leq \mu_0 \overline{w} + \lambda' \overline{v} + \mu' \overline{u}\big\}.$$

As in Section 2.5, a hyperplane with normal (λ, μ, μ_0) where $\mu_0 \neq 0$ is referred to as *nonvertical*. By dividing the normal vector of such a hyperplane by μ_0, we can restrict attention to the case where $\mu_0 = 1$.

We will now use the above definitions to interpret geometric multipliers as nonvertical hyperplanes with a certain orientation to the set of all constraint-cost pairs as x ranges over X, i.e., the subset of $\Re^{m+r+1}$ given by

$$S = \Big\{\big(h(x), g(x), f(x)\big) \mid x \in X\Big\}. \tag{6.3}$$

We have the following lemma, which is graphically illustrated in Fig. 6.1.1.

Visualization Lemma: Assume that $-\infty < f^* < \infty$.

(a) The hyperplane with normal $(\lambda, \mu, 1)$ that passes through a vector $\big(h(x), g(x), f(x)\big)$ intercepts the vertical axis $\big\{(0, 0, w) \mid w \in \Re\big\}$ at the level $L(x, \lambda, \mu)$.

(b) Among all hyperplanes with a given normal $(\lambda, \mu, 1)$ that contain in their positive closed halfspace the set S of Eq. (6.3), the highest attained level of interception of the vertical axis is $\inf_{x \in X} L(x, \lambda, \mu)$.

(c) The vector (λ^*, μ^*) is a geometric multiplier if and only if $\mu^* \geq 0$ and among all hyperplanes with normal $(\lambda^*, \mu^*, 1)$ that contain in their positive closed halfspace the set S, the highest attained level of interception of the vertical axis is f^*.

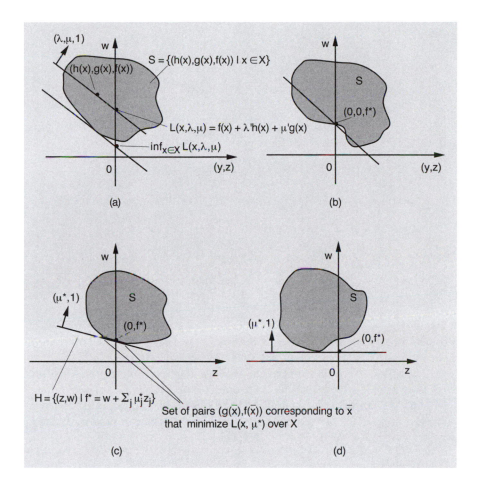

Figure 6.1.1. (a) Geometrical interpretation of the Lagrangian value $L(x, \lambda, \mu)$ as the level of interception of the vertical axis with the hyperplane with normal $(\lambda, \mu, 1)$ that passes through $\big(h(x), g(x), f(x)\big)$.

(b) A case where there is no geometric multiplier. Here, there is no hyperplane that passes through the point $(0, 0, f^*)$ and contains S in its positive closed half-space.

(c), (d) Illustration of a geometric multiplier vector μ^* in the cases $\mu^* \neq 0$ and $\mu^* = 0$, respectively, assuming no equality constraints. It defines a hyperplane H that has normal $(\mu^*, 1)$, passes through $(0, f^*)$, and contains S in its positive closed halfspace. Note that the common points of H and S (if any) are the pairs $(g(\bar{x}), f(\bar{x}))$ corresponding to points $\bar{x}$ that minimize the Lagrangian $L(x, \mu^*)$ over $x \in X$. Some of these $\bar{x}$ may be infeasible as in (c). Note that in (c), the point $(0, f^*)$ belongs to S and corresponds to optimal primal solutions for which the constraint is active, while in (d), the point $(0, f^*)$ does not belong to S and corresponds to an optimal primal solution for which the constraint is inactive.

Proof: (a) As discussed earlier, the hyperplane with normal $(\lambda, \mu, 1)$ that passes through $(h(x), g(x), f(x))$ is the set of vectors (v, u, w) satisfying

$$w + \lambda'v + \mu'u = f(x) + \lambda'h(x) + \mu'g(x) = L(x, \lambda, \mu).$$

It is seen that the only vector on the vertical axis that satisfies this equation is $(0, 0, L(x, \lambda, \mu))$.

(b) The hyperplane with normal $(\lambda, \mu, 1)$ that intercepts the vertical axis at level c is the set of vectors (v, u, w) that satisfy the equation

$$w + \lambda'v + \mu'u = c.$$

This hyperplane contains S in its positive closed halfspace if and only if

$$L(x, \lambda, \mu) = f(x) + \lambda'h(x) + \mu'g(x) \geq c, \qquad \forall\, x \in X.$$

Therefore the maximum point of interception is $\inf_{x \in X} L(x, \lambda, \mu)$.

(c) Follows from the definition of a geometric multiplier and part (b). **Q.E.D.**

The hyperplane with normal $(\lambda, \mu, 1)$ that attains the highest level of interception of the vertical axis, as in part (b) of the Visualization Lemma, is seen to support the set S in the sense of Section 2.4. Figure 6.1.2 gives some examples where there exist one or more geometric multipliers. Figure 6.1.3 shows cases where there is no geometric multiplier.

The Visualization Lemma also shows that a geometric multiplier can be interpreted as an optimal solution of a max crossing problem. Indeed, consider the subset M of $\Re^{m+r+1}$ given by

$$M = \{(v, u, w) \mid x \in X,\ h(x) = v,\ g(x) \leq u,\ f(x) \leq w\}$$
$$= \{(v, u, w) \mid \text{for some } (\overline{v}, \overline{u}, \overline{w}) \in S,\ \overline{v} = v,\ \overline{u} \leq u,\ \overline{w} \leq w\}.$$

Then it can be seen that *the optimal value of the corresponding min common problem is equal to f^*. Furthermore, μ^* is a geometric multiplier if and only if the min common and max crossing values are equal, and μ^* solves the max crossing problem.*

Primal Optimality Condition

Suppose that a geometric multiplier (λ^*, μ^*) is known. Then all optimal solutions x^* can be obtained by minimizing the Lagrangian $L(x, \lambda^*, \mu^*)$ over $x \in X$, as indicated in Fig. 6.1.1(c) and shown in the following proposition. However, there may be vectors that minimize $L(x, \lambda^*, \mu^*)$ over $x \in X$ but are not optimal solutions because they do not satisfy the constraints $h(x) = 0$ and $g(x) \leq 0$, or the complementary slackness condition $\mu^{*'}g(x) = 0$ [cf. Figs. 6.1.1(c) and 6.1.2(a)].

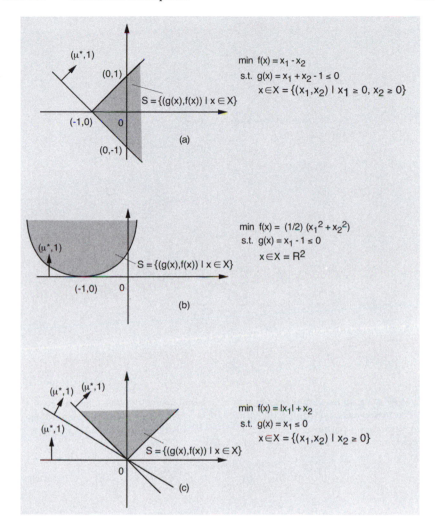

Figure 6.1.2. Examples where a geometric multiplier exists. In (a), there is a unique geometric multiplier, $\mu^* = 1$. In (b), there is a unique geometric multiplier, $\mu^* = 0$. In (c), the set of geometric multipliers is the interval $[0, 1]$.

Proposition 6.1.1: Let (λ^*, μ^*) be a geometric multiplier. Then x^* is a global minimum of the primal problem if and only if x^* is feasible and

$$L(x^*, \lambda^*, \mu^*) = \min_{x \in X} L(x, \lambda^*, \mu^*), \qquad \mu^{*\prime} g(x^*) = 0. \qquad (6.4)$$

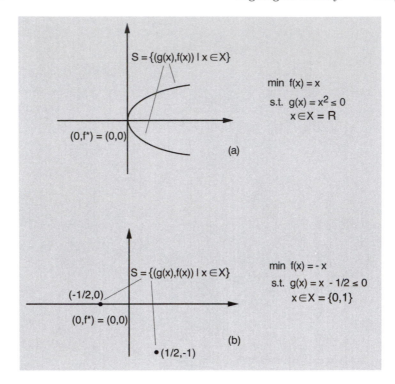

Figure 6.1.3. Examples where a geometric multiplier does not exist. In (a), the only hyperplane that supports S at $(0, f^*)$ is vertical. In (b), we have an integer programming problem, where the set X is discrete. Here, there is no hyperplane that supports S and passes through $(0, f^*)$.

Proof: If x^* is a global minimum, then x^* is feasible and furthermore,

$$
\begin{aligned}
f^* &= f(x^*) \\
&\geq f(x^*) + \lambda^{*'}h(x^*) + \mu^{*'}g(x^*) \\
&= L(x^*, \lambda^*, \mu^*) \\
&\geq \inf_{x \in X} L(x, \lambda^*, \mu^*),
\end{aligned}
\tag{6.5}
$$

where the first inequality follows from the feasibility of x^* [$h(x^*) = 0$ and $g(x^*) \leq 0$] and the definition of a geometric multiplier [$\mu^* \geq 0$, which together with $g(x^*) \leq 0$, implies that $\mu^{*'}g(x^*) \leq 0$]. Using again the definition of a geometric multiplier, we have $f^* = \inf_{x \in X} L(x, \lambda^*, \mu^*)$, so that equality holds throughout in Eq. (6.5), implying Eq. (6.4).

Conversely, if x^* is feasible and Eq. (6.4) holds, we have, using also the definition of a geometric multiplier,

$$
f(x^*) = L(x^*, \lambda^*, \mu^*) = \min_{x \in X} L(x, \lambda^*, \mu^*) = f^*,
$$

so x^* is a global minimum.   **Q.E.D.**

As Prop. 6.1.1 indicates, a geometric multiplier can be very useful for finding an optimal solution of the primal problem. We will see in the next section, that geometric multipliers, when they exist, can be found by solving a certain dual problem. However, as shown by the examples of Fig. 6.1.3, a geometric multiplier is by no means guaranteed to exist. In fact when X is a finite set as in discrete optimization problems, typically there is no geometric multiplier. Still, however, even in this case, we will see that the dual problem is useful and forms the basis for important discrete optimization methods.

Relation Between Geometric and Lagrange Multipliers

As indicated by the preceding discussion, there is a strong relation between geometric and Lagrange multipliers for problems with a convex structure. We introduce some types of problems for which Lagrange multipliers were defined in Chapter 5.

Definition 6.1.2: We say that the primal problem (P) is:

(a) *Semiconvex* if each of the functions f and g_j is either convex or smooth, the functions h_i are smooth, and X is a closed set.

(b) *Convex* if each of the functions f and g_j is convex, the functions h_i are affine, and X is a closed convex set.

Let us restate for convenience the definition of Lagrange multiplier, introduced in Section 5.7, so that it applies to semiconvex problems.

Definition 6.1.3: Given an optimal solution x^* of a primal problem (P), which is semiconvex, we say that (λ^*, μ^*) is a Lagrange multiplier associated with x^* if

$$0 \in \partial f(x^*) + \sum_{i=1}^{m} \lambda_i^* \partial h_i(x^*) + \sum_{j=1}^{r} \mu_j^* \partial g_j(x^*) + T_X(x^*)^*, \qquad (6.6)$$

$$\mu^* \geq 0, \qquad \mu^{*\prime} g(x^*) = 0, \qquad (6.7)$$

where for a function $F : \Re^n \mapsto \Re$, we denote by $\partial F(x)$ the subdifferential of F at x, if F is convex, and the singleton set $\{\nabla F(x)\}$, if F is smooth.

The following proposition shows how geometric multipliers are related to Lagrange multipliers.

Proposition 6.1.2: Assume that the primal problem (P) has at least one optimal solution x^*.

(a) If the problem is semiconvex and the tangent cone $T_X(x^*)$ is convex, then every geometric multiplier is a Lagrange multiplier associated with x^*.

(b) If the problem is convex, the set of Lagrange multipliers associated with x^* and the set of geometric multipliers coincide.

Proof: (a) Let (λ^*, μ^*) be a geometric multiplier. Then, $\mu^* \geq 0$, and furthermore, by Prop. 6.1.1, we have $\mu^{*\prime}g(x^*) = 0$, and x^* minimizes $L(\cdot, \lambda^*, \mu^*)$ over X. Hence, by the necessary condition of Prop. 4.7.3, we obtain

$$0 \in \partial f(x^*) + \sum_{i=1}^{m} \lambda_i^* \partial h_i(x^*) + \sum_{j=1}^{r} \mu_j^* \partial g_j(x^*) + T_X(x^*)^*.$$

Thus, all the conditions of Definition 6.1.3 are fulfilled and (λ^*, μ^*) is a Lagrange multiplier.

(b) Let (λ^*, μ^*) be a Lagrange multiplier. Then Definition 6.1.3 and the convexity assumption imply that x^* minimizes $L(\cdot, \lambda^*, \mu^*)$ over X. Hence, using also the conditions $h(x^*) = 0$ and $\mu^{*\prime}g(x^*) = 0$, we have

$$
\begin{aligned}
f^* &= f(x^*) \\
&= f(x^*) + \lambda^{*\prime}h(x^*) + \mu^{*\prime}g(x^*) \\
&= L(x^*, \lambda^*, \mu^*) \\
&= \min_{x \in X} L(x, \lambda^*, \mu^*).
\end{aligned}
$$

It follows that (λ^*, μ^*) is a geometric multiplier. **Q.E.D.**

Note an implication of Prop. 6.1.2(b): for a convex problem that has multiple optimal solutions, *all the optimal solutions have the same set of associated Lagrange multipliers*, namely the set of geometric multipliers. Despite the connections, however, there is still considerable difference between the notions of geometric and Lagrange multipliers, even for convex problems. In particular, there may exist geometric multipliers, but no optimal solution and hence no Lagrange multipliers. As an example, consider the one-dimensional convex problem of minimizing e^x subject to the single inequality constraint $x \leq 0$; it has the optimal value $f^* = 0$ and the geometric multiplier $\mu^* = 0$, but it has no optimal solution.

Furthermore, when the problem is semiconvex, there may exist Lagrange multipliers but no geometric multipliers. As an example, consider the one-dimensional problem of minimizing the concave function $-x^2$ subject to the single equality constraint $x = 0$; it has the optimal value $f^* = 0$ and the Lagrange multiplier $\lambda^* = 0$, but it has no geometric multiplier because for all $\lambda \in \Re$, we have $\inf_{x \in \Re}\{-x^2 + \lambda x\} = -\infty < f^*$.

6.2 DUALITY THEORY

We consider the *dual function q* defined for $(\lambda, \mu) \in \Re^{m+r}$ by

$$q(\lambda, \mu) = \inf_{x \in X} L(x, \lambda, \mu).$$

This definition is illustrated in Fig. 6.2.1, where $q(\lambda, \mu)$ is interpreted as the highest point of interception with the vertical axis over all hyperplanes with normal $(\lambda, \mu, 1)$, which contain the set

$$S = \Big\{ \big(h(x), g(x), f(x)\big) \mid x \in X \Big\}$$

in their positive closed halfspace [compare also with the Visualization Lemma and Fig. 6.1.1(a)]. The *dual problem* is

$$\begin{aligned} \text{maximize} \quad & q(\lambda, \mu) \\ \text{subject to} \quad & \lambda \in \Re^m, \ \mu \in \Re^r, \ \mu \geq 0, \end{aligned} \tag{D}$$

and corresponds to finding the maximum point of interception, over all hyperplanes with normal $(\lambda, \mu, 1)$ where $\mu \geq 0$. Thus, *the dual problem is equivalent to the max crossing problem that corresponds to the subset M of* $\Re^{m+r+1}$ *given by*

$$\begin{aligned} M &= \big\{(v, u, w) \mid x \in X, \ h(x) = v, \ g(x) \leq u, \ f(x) \leq w\big\} \\ &= \big\{(v, u, w) \mid \text{for some } (\overline{v}, \overline{u}, \overline{w}) \in S, \ \overline{v} = v, \ \overline{u} \leq u, \ \overline{w} \leq w\big\}. \end{aligned}$$

The dual value $q(\lambda, \mu)$ may be equal to $-\infty$ for some (λ, μ), so the dual problem embodies in effect the additional constraint

$$(\lambda, \mu) \in Q,$$

where Q is the set

$$Q = \big\{(\lambda, \mu) \mid q(\lambda, \mu) > -\infty\big\}.$$

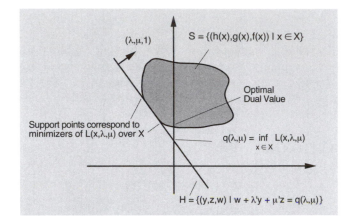

Figure 6.2.1. Geometrical interpretation of the dual function.

Note that it may happen that $q(\lambda, \mu) = -\infty$ for all $(\lambda, \mu) \in Q$ with $\mu \geq 0$. Then, the dual optimal value

$$q^* = \sup_{\lambda \in \Re^m, \, \mu \geq 0} q(\lambda, \mu)$$

is equal to $-\infty$, and all (λ, μ) with $\mu \geq 0$ are optimal solutions of the dual problem (situations where this can happen will be discussed later).

Regardless of the structure of the cost and constraints of the primal problem, the dual problem has nice properties, as shown by the following proposition, which is a special case of the corresponding min common/max crossing result (cf. Prop. 2.5.2).

Proposition 6.2.1: The dual function $q : \Re^{m+r} \mapsto [-\infty, \infty)$ is concave and upper semicontinuous over $\Re^{m+r}$.

Proof: This follows by viewing q as the infimum over $x \in X$ of the collection of the affine functions $\{L(x, \cdot, \cdot) \mid x \in X\}$ [cf. Prop. 1.2.4(c)]. **Q.E.D.**

Another important property is that the optimal dual value is always an underestimate of the optimal primal value, as is evident from Fig. 6.2.1. This is consistent with the view of the primal and dual problems as min common and max crossing problems, respectively.

Proposition 6.2.2: (Weak Duality Theorem) We have

$$q^* \leq f^*.$$

Proof: This is a special case of Prop. 2.5.3, but for convenience we repeat the proof argument. For all $\lambda \in \Re^m$, $\mu \geq 0$, and $x \in X$ with $h(x) = 0$ and $g(x) \leq 0$, we have

$$q(\lambda, \mu) = \inf_{z \in X} L(z, \lambda, \mu) \leq L(x, \lambda, \mu) = f(x) + \lambda' h(x) + \mu' g(x) \leq f(x),$$

so

$$q^* = \sup_{\lambda \in \Re^m, \mu \geq 0} q(\lambda, \mu) \leq \inf_{\substack{x \in X \\ h(x) = 0, g(x) \leq 0}} f(x) = f^*.$$

Q.E.D.

If $q^* = f^*$ we say that *there is no duality gap* or that *strong duality holds*, and if $q^* < f^*$ we say that *there is a duality gap*. Note that if there exists a geometric multiplier μ^*, the Weak Duality Theorem ($q^* \leq f^*$) and the definition of a geometric multiplier [$f^* = q(\lambda^*, \mu^*) \leq q^*$] imply that there is no duality gap. However, the converse is not true. In particular, it is possible that no geometric multiplier exists even though there is no duality gap [cf. Fig. 6.1.3(a)]; in this case the dual problem does not have an optimal solution, as implied by the following proposition.

Proposition 6.2.3:

   (a) If there is no duality gap, the set of geometric multipliers is equal to the set of optimal dual solutions.

   (b) If there is a duality gap, the set of geometric multipliers is empty.

Proof: By definition, a vector (λ^*, μ^*) with $\mu^* \geq 0$ is a geometric multiplier if and only if $f^* = q(\lambda^*, \mu^*) \leq q^*$, which by the Weak Duality Theorem, holds if and only if there is no duality gap and (λ^*, μ^*) is a dual optimal solution.   **Q.E.D.**

The preceding results are illustrated in Figs. 6.2.2 and 6.2.3 for the problems of Figs. 6.1.2 and 6.1.3, respectively.

Duality theory is most useful when there is no duality gap, since then the dual problem can be solved to yield q^* (which is also equal to the optimal primal value f^*) and to help obtain an optimal primal solution (via the optimality conditions of Prop. 6.1.1). To guarantee that there is no duality gap and that a geometric multiplier exists, it is typically necessary to impose various types of convexity conditions on the cost and the constraints of the primal problem; we will develop in Section 6.4 some conditions of this type. However, even with a duality gap, the dual problem can be useful, as indicated by the following example.

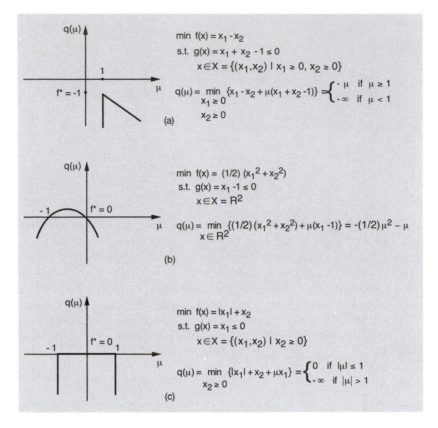

Figure 6.2.2. The duals of the problems of Fig. 6.1.2. In all these problems, there is no duality gap and the set of geometric multipliers is equal to the set of dual optimal solutions (cf. Prop. 6.2.3). In (a), there is a unique dual optimal solution $\mu^* = 1$. In (b), there is a unique dual optimal solution $\mu^* = 0$; at the corresponding primal optimal solution ($x^* = 0$) the constraint is inactive. In (c), the set of dual optimal solutions is $\{\mu^* \mid 0 \leq \mu^* \leq 1\}$.

Example 6.2.1: (Integer Programming, and Branch-and-Bound)

Many important practical optimization problems of the form

$$\text{minimize} \quad f(x)$$
$$\text{subject to} \quad x \in X, \qquad g_j(x) \leq 0, \quad j = 1, \ldots, r,$$

have a finite constraint set X. An example is *integer programming*, where the components of x must be integers from a bounded range (usually 0 or 1). An important special case is the linear 0-1 integer programming problem

$$\text{minimize} \quad c'x$$
$$\text{subject to} \quad Ax \leq b, \qquad x_i = 0 \text{ or } 1, \quad i = 1, \ldots, n.$$

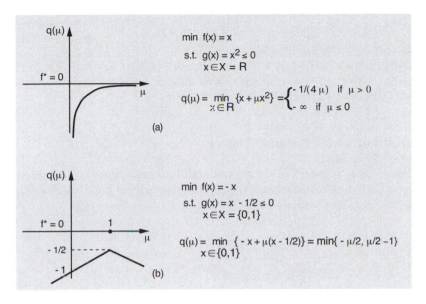

Figure 6.2.3. The duals of the problems of Fig. 6.1.3. In these problems, there is no geometric multiplier. In (a), there is no duality gap and the dual problem has no optimal solution (cf. Prop. 6.2.2). In (b), there is a duality gap ($f^* - q^* = 1/2$).

A principal approach for solving such problems is the *branch-and-bound method*. This method relies on obtaining lower bounds to the optimal cost of restricted problems of the form

$$\text{minimize} \quad f(x)$$
$$\text{subject to} \quad x \in \tilde{X}, \qquad g_j(x) \le 0, \quad j = 1, \ldots, r,$$

where $\tilde{X}$ is a subset of X; for example in the 0-1 integer case where X specifies that all x_i should be 0 or 1, $\tilde{X}$ may be the set of all 0-1 vectors x such that one or more components x_i are restricted to satisfy $x_i = 0$ for all $x \in \tilde{X}$ or $x_i = 1$ for all $x \in \tilde{X}$. These lower bounds can often be obtained by finding a dual-feasible (possibly dual-optimal) solution μ of this problem and the corresponding dual value

$$q(\mu) = \inf_{x \in \tilde{X}} \left\{ f(x) + \sum_{j=1}^{r} \mu_j g_j(x) \right\},$$

which by the Weak Duality Theorem, is a lower bound to the optimal value of the restricted problem $\min_{x \in \tilde{X}, \, g(x) \le 0} f(x)$.

One is interested in finding as tight lower bounds as possible, so the usual approach is to start with some dual feasible solution and iteratively improve it by using some algorithm. A major difficulty here is that the dual function $q(\mu)$ is typically nondifferentiable, so the methods developed so far cannot be used. We will discuss special methods for optimization of nondifferentiable cost functions in Chapter 8. Note that in many problems of interest that have favorable structure, the value $q(\mu)$ can be calculated easily, and we will see in Chapter 8 that other quantities, such as subgradients, that are needed for application of nondifferentiable optimization methods are also easily obtained together with $q(\mu)$.

Connection with Minimax Theory

As discussed in Section 2.6, the primal and dual problems can be viewed in terms of the minimax problem involving the Lagrangian function $L(x, \lambda, \mu)$. In particular, since we have

$$\sup_{\lambda \in \Re^m, \, \mu \in \Re^r, \, \mu \geq 0} \{ f(x) + \lambda'h(x) + \mu'g(x) \} = \begin{cases} f(x) & \text{if } h(x) = 0, \ g(x) \leq 0, \\ \infty & \text{otherwise,} \end{cases}$$

the primal problem (P) is equivalent to

$$\text{minimize} \quad \sup_{\lambda \in \Re^m, \, \mu \in \Re^r, \, \mu \geq 0} L(x, \lambda, \mu)$$

$$\text{subject to} \quad x \in X,$$

while the dual problem (D) is, by definition,

$$\text{maximize} \quad \inf_{x \in X} L(x, \lambda, \mu)$$

$$\text{subject to} \quad \lambda \in \Re^m, \ \mu \in \Re^r, \ \mu \geq 0.$$

Based on this view, we can obtain powerful characterizations of primal and dual optimal solution pairs, given in the following two propositions. Note, however, that these characterizations are useful only if there is no duality gap, since otherwise there is no geometric multiplier [cf. Prop. 6.1.3(b)], even if the dual problem has an optimal solution.

Proposition 6.2.4: (Lagrangian Saddle Point Theorem) The vectors (x^*, λ^*, μ^*) form an optimal solution-geometric multiplier pair if and only if (x^*, λ^*, μ^*) is a saddle point of the Lagrangian in the sense that $x^* \in X$, $\mu^* \geq 0$, and

$$L(x^*, \lambda, \mu) \leq L(x^*, \lambda^*, \mu^*) \leq L(x, \lambda^*, \mu^*), \quad \forall \, x \in X, \ \lambda \in \Re^m, \ \mu \geq 0. \tag{6.8}$$

Proof: By Prop. 2.6.1, (x^*, λ^*, μ^*) is a saddle point of L if and only if the following three conditions hold:

$$\inf_{x \in X} \sup_{\lambda \in \Re^m, \, \mu \geq 0} L(x, \lambda, \mu) = \sup_{\lambda \in \Re^m, \, \mu \geq 0} \inf_{x \in X} L(x, \lambda, \mu), \tag{6.9}$$

$$\sup_{\lambda \in \Re^m, \, \mu \geq 0} L(x^*, \lambda, \mu) = \min_{x \in X} \left\{ \sup_{\lambda \in \Re^m, \, \mu \geq 0} L(x, \lambda, \mu) \right\}, \qquad (6.10)$$

$$\inf_{x \in X} L(x, \lambda^*, \mu^*) = \max_{\lambda \in \Re^m, \, \mu \geq 0} \left\{ \inf_{x \in X} L(x, \lambda, \mu) \right\}. \qquad (6.11)$$

We have

$$\inf_{x \in X} L(x, \lambda, \mu) = q(\lambda, \mu)$$

and

$$\sup_{\lambda \in \Re^m, \, \mu \geq 0} L(x, \lambda, \mu) = \begin{cases} f(x) & \text{if } h(x) = 0, \ g(x) \leq 0, \\ \infty & \text{otherwise.} \end{cases}$$

Thus the minimax equality (6.9) is equivalent to $q^* = f^*$, the condition (6.10) is equivalent to x^* being an optimal solution, while the condition (6.11) is equivalent to (λ^*, μ^*) being a dual optimal solution, and hence also a geometric multiplier since $q^* = f^*$.   **Q.E.D.**

Proposition 6.2.5: (Necessary and Sufficient Optimality Conditions) The vectors (x^*, λ^*, μ^*) form an optimal solution-geometric multiplier pair if and only if the following four conditions hold:

$$x^* \in X, \ h(x^*) = 0, \ g(x^*) \leq 0, \qquad \text{(Primal Feasibility),}$$

$$\mu^* \geq 0, \qquad \text{(Dual Feasibility),}$$

$$L(x^*, \lambda^*, \mu^*) = \min_{x \in X} L(x, \lambda^*, \mu^*), \qquad \text{(Lagrangian Optimality),}$$

$$\mu^{*\prime} g(x^*) = 0, \qquad \text{(Complementary Slackness).}$$

Proof: If (x^*, λ^*, μ^*) is an optimal solution-geometric multiplier pair, then x^* is primal feasible and (λ^*, μ^*) is dual feasible. The Lagrangian optimality and complementary slackness conditions follow from Prop. 6.1.1.

Conversely, using the four conditions, we obtain

$$f^* \leq f(x^*) = L(x^*, \lambda^*, \mu^*) = \min_{x \in X} L(x, \lambda^*, \mu^*) = q(\lambda^*, \mu^*) \leq q^*.$$

Using the Weak Duality Theorem (Prop. 6.2.2), we see that equality holds throughout in the preceding relation. It follows that x^* is primal optimal and (λ^*, μ^*) is dual optimal, while there is no duality gap.   **Q.E.D.**

An alternative proof of the preceding proposition is to verify that its four conditions hold if and only if x^* minimizes $L(x, \lambda^*, \mu^*)$ over $x \in X$, and (λ^*, μ^*) maximizes $L(x^*, \lambda, \mu)$ over $\lambda \in \Re^m$, $\mu \geq 0$. The latter statement is in turn true if and only if (x^*, λ^*, μ^*) is a saddle point of L over $x \in X$ and $\lambda \in \Re^m$, $\mu \geq 0$, or equivalently, by the preceding Lagrangian Saddle Point Theorem, if and only if (x^*, λ^*, μ^*) is an optimal solution-geometric multiplier pair.

The Case of an Infeasible or Unbounded Primal Problem

Let us now consider what happens when f^* is not finite. Suppose that the primal problem is *unbounded*, i.e.,

$$f^* = -\infty.$$

Then, the Weak Duality Theorem still applies and we have $q(\lambda, \mu) = -\infty$ for all $\lambda \in \Re^m$ and $\mu \geq 0$. As a result the dual problem is infeasible.

Suppose now that X is nonempty but the primal problem is infeasible, i.e.,

$$f^* = \infty.$$

The dual function $q(\lambda, \mu) = \inf_{x \in X} L(x, \lambda, \mu)$ satisfies $q(\lambda, \mu) < \infty$ for all (λ, μ) (since X has been assumed nonempty), but it is possible that $q^* = \infty$, in which case the dual problem is unbounded. It is also possible, however, that $q^* < \infty$ or even that $q^* = -\infty$, as the examples of Fig. 6.2.4 show. For linear and convex quadratic programs, it will be shown in the next section that if $-\infty < f^* < \infty$, then there is no duality gap. However, even for linear programs, it is possible that both the primal and the dual problems are infeasible, i.e., $f^* = \infty$ and $q^* = -\infty$ (see Exercise 6.1).

6.3 LINEAR AND QUADRATIC PROGRAMMING DUALITY

We will now derive the dual problems for linear and for convex quadratic programs. These problems are special because they are guaranteed to possess optimal solutions as long as their optimal values are finite (cf. Prop. 2.3.4). We will take advantage of this property in our analysis.

The Dual of a Linear Program

Consider the linear program

$$\begin{aligned} &\text{minimize} \quad c'x \\ &\text{subject to} \ \ x \geq 0, \qquad e_i'x = d_i, \quad i = 1, \ldots, m, \end{aligned} \tag{LP}$$

where c and e_i are vectors in $\Re^n$, and d_i are scalars. We consider the dual function

$$q(\lambda) = \inf_{x \geq 0} \left\{ \sum_{j=1}^{n} \left(c_j - \sum_{i=1}^{m} \lambda_i e_{ij} \right) x_j + \sum_{i=1}^{m} \lambda_i d_i \right\}, \tag{6.12}$$

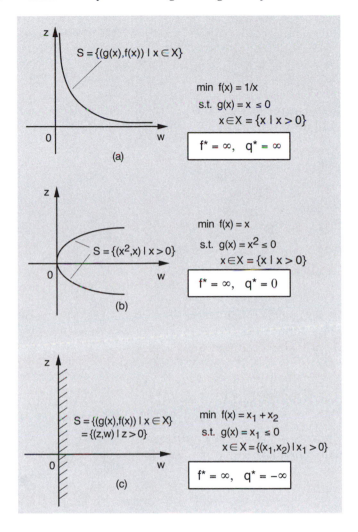

Figure 6.2.4. Examples where X is nonempty but the primal problem is infeasible. In (a), we have $f^* = q^* = \infty$. In (b), we have $f^* = \infty$ and $-\infty < q^* < \infty$. In (c), we have $f^* = \infty$ and $q^* = -\infty$.

where e_{ij} is the jth component of the vector e_i. We see that if

$$c_j - \sum_{i=1}^{m} \lambda_i e_{ij} \geq 0, \qquad \forall\, j = 1, \dots, n,$$

the infimum above is attained for $x = 0$, and we have

$$q(\lambda) = \sum_{i=1}^{m} \lambda_i d_i.$$

On the other hand, if

$$c_j - \sum_{i=1}^{m} \lambda_i e_{ij} < 0, \qquad \text{for some } j,$$

we can make the expression in braces in the definition of the dual function [cf. Eq. (6.12)] arbitrarily small by taking x_j sufficiently large, so that we have

$$q(\lambda) = -\infty.$$

Thus, the dual problem is

$$\begin{aligned} \text{maximize} \quad & \sum_{i=1}^{m} \lambda_i d_i \\ \text{subject to} \quad & \sum_{i=1}^{m} \lambda_i e_{ij} \leq c_j, \quad j = 1, \dots, n. \end{aligned} \qquad \text{(DLP)}$$

By the optimality conditions of Prop. 6.2.5, (x^*, λ^*) is a primal and dual optimal solution pair if and only if x^* is primal feasible, and the Lagrangian optimality condition

$$x^* \in \arg\min_{x \geq 0} \left\{ \left(c - \sum_{i=1}^{m} \lambda_i^* e_i \right)' x + \sum_{i=1}^{m} \lambda_i^* d_i \right\}$$

holds. The above condition is equivalent to $x^* \geq 0$, $c \geq \sum_{i=1}^{m} \lambda_i^* e_i$ (otherwise the minimum above would not be achieved at x^*), and the following two relations

$$x_j^* > 0 \quad \Rightarrow \quad \sum_{i=1}^{m} \lambda_i^* e_{ij} = c_j, \quad j = 1, \dots, n, \qquad (6.13)$$

$$\sum_{i=1}^{m} \lambda_i^* e_{ij} < c_j \quad \Rightarrow \quad x_j^* = 0, \quad j = 1, \dots, n, \qquad (6.14)$$

known as the *complementary slackness conditions for linear programming*.

Let us see now what happens when we dualize the dual problem (DLP). We first convert this program into the equivalent minimization problem

$$\begin{aligned} \text{minimize} \quad & \sum_{i=1}^{m} (-d_i) \lambda_i \\ \text{subject to} \quad & \sum_{i=1}^{m} \lambda_i e_{ij} \leq c_j, \quad j = 1, \dots, n. \end{aligned}$$

Assigning a geometric multiplier x_j to the jth inequality constraint, the dual function of this problem is given by

$$p(x) = \min_{\lambda \in \Re^m} \left\{ \sum_{i=1}^{m} \left(\sum_{j=1}^{n} e_{ij} x_j - d_i \right) \lambda_i - \sum_{j=1}^{n} c_j x_j \right\}$$

$$= \begin{cases} -c'x & \text{if } e_i'x = d_i, \quad i = 1, \ldots, m, \\ -\infty & \text{otherwise.} \end{cases}$$

The corresponding dual problem is

$$\text{maximize} \quad p(x)$$
$$\text{subject to} \quad x \geq 0,$$

or equivalently

$$\text{minimize} \quad c'x$$
$$\text{subject to} \quad x \geq 0, \qquad e_i'x = d_i, \quad i = 1, \ldots, m,$$

which is identical to the primal problem (LP). We have thus shown that the duality is symmetric, i.e., *the dual of the dual linear program (DLP) is the primal problem (LP).*

The pair of primal and dual linear programs (LP) and (DLP) can also be written compactly in terms of the $m \times n$ matrix E having rows $e_1', \ldots, e_m'$, and the vector d having components $d_1, \ldots, d_m$:

$$\min_{Ex=d,\ x\geq 0} c'x \qquad \Longleftrightarrow \qquad \max_{E'\lambda \leq c} d'\lambda.$$

Other linear programming duality relations that can be verified by the reader include the following:

$$\min_{A'x \geq b} c'x \qquad \Longleftrightarrow \qquad \max_{A\mu=c,\ \mu \geq 0} b'\mu,$$

$$\min_{A'x \geq b,\ x \geq 0} c'x \qquad \Longleftrightarrow \qquad \max_{A\mu \leq c,\ \mu \geq 0} b'\mu;$$

see Exercise 6.4.

The Dual of a Quadratic Program

Consider the quadratic programming problem

$$\text{minimize} \quad \tfrac{1}{2}x'Qx + c'x$$
$$\text{subject to} \quad Ax \leq b, \qquad \text{(QP)}$$

where Q is a symmetric positive definite $n \times n$ matrix, A is an $r \times n$ matrix, b is a vector in $\Re^r$, and c is a vector in $\Re^n$. The dual function is

$$q(\mu) = \inf_{x \in \Re^n} \left\{ \tfrac{1}{2}x'Qx + c'x + \mu'(Ax - b) \right\}.$$

The infimum is attained for $x = -Q^{-1}(c + A'\mu)$, and a straightforward calculation after substituting this expression in the preceding relation, yields

$$q(\mu) = -\tfrac{1}{2}\mu'AQ^{-1}A'\mu - \mu'(b + AQ^{-1}c) - \tfrac{1}{2}c'Q^{-1}c.$$

The dual problem, after dropping the constant $\tfrac{1}{2}c'Q^{-1}c$ and changing the minus sign to convert the maximization to a minimization, can be written as

$$\begin{aligned} \text{minimize} \quad & \tfrac{1}{2}\mu'P\mu + t'\mu \\ \text{subject to} \quad & \mu \geq 0, \end{aligned} \qquad \text{(DQP)}$$

where

$$P = AQ^{-1}A', \qquad t = b + AQ^{-1}c.$$

If μ^* is any dual optimal solution, then the optimal solution of the primal problem is

$$x^* = -Q^{-1}(c + A'\mu^*).$$

Note that the dual problem is also a quadratic program, but it has simpler constraints than the primal. Furthermore, if the row dimension r of A is smaller than its column dimension n, the dual problem is defined on a space of smaller dimension than the primal, and this can be algorithmically significant.

    Let us now relax the assumption that Q is positive definite, and assume instead that Q is positive semidefinite. We know from Prop. 2.3.4 that a feasible positive semidefinite quadratic program (including the special case of a linear program) possesses an optimal solution if and only if its optimal value is finite. We also know that for any optimal solution x^* there exist Lagrange multipliers associated with x^*, since the constraints are linear (cf. the constraint qualification CQ3 in Section 5.4). Furthermore, based on Prop. 6.1.2(b), the set of Lagrange multipliers associated with every optimal solution, and the set of geometric multipliers coincide. We thus obtain the following.

Proposition 6.3.1: Consider the quadratic program

$$\begin{aligned} \text{minimize} \quad & c'x + \tfrac{1}{2}x'Qx \\ \text{subject to} \quad & a_j'x \leq b_j, \quad j = 1, \ldots, r, \end{aligned}$$

where Q is a symmetric positive semidefinite $n \times n$ matrix, $c, a_1, \ldots, a_r$ are vectors in $\Re^n$, and $b_1, \ldots, b_r$ are scalars, and assume that its optimal value is finite. Then there exist at least one optimal primal solution and at least one geometric multiplier.

The preceding proposition can be extended to the case where there are linear equality constraints $e_i'x = d_i$, by representing every such constraint as two inequalities $e_i'x \leq d_i$ and $-e_i'x \leq -d_i$. It can also be extended to the case where there is an additional abstract set constraint $x \in X$, where X is a polyhedral set; see Exercise 6.3.

6.4 EXISTENCE OF GEOMETRIC MULTIPLIERS

In this section we will develop conditions that guarantee the existence of a geometric multiplier. These conditions parallel some of the constraint qualifications of Section 5.4, but do not require that the primal problem has at least one optimal solution. It is possible to use the theory of the min common/max crossing framework (e.g., the theorems proved in Sections 2.5 and 3.5) to analyze the existence of geometric multipliers. However, all the necessary machinery has already been captured in the nonlinear version of Farkas' Lemma of Section 3.5 (Prop. 3.5.4). It will suffice for our purposes to use this lemma, which we repeat for the purpose of easy reference.

Proposition 6.4.1: (Nonlinear Farkas' Lemma) Let C be a nonempty convex subset of $\Re^n$, and let $f : C \mapsto \Re$ and $g_j : C \mapsto \Re$, $j = 1, \ldots, r$, be convex functions. Consider the set F given by

$$F = \{x \in C \mid g(x) \leq 0\},$$

where $g(x) = (g_1(x), \ldots, g_r(x))$, and assume that

$$f(x) \geq 0, \qquad \forall\, x \in F. \tag{6.15}$$

Consider the subset Q^* of $\Re^r$ given by

$$Q^* = \{\mu \mid \mu \geq 0,\ f(x) + \mu'g(x) \geq 0,\ \forall\, x \in C\}.$$

(a) Q^* is nonempty and compact if and only if there exists a vector $\overline{x} \in C$ such that

$$g_j(\overline{x}) < 0, \qquad \forall\, j = 1, \ldots, r.$$

(b) Q^* is nonempty if the functions g_j, $j = 1, \ldots, r$, are affine, and F contains a relative interior point of C.

We now use the preceding Nonlinear Farkas' Lemma to assert the existence of geometric multipliers under some specific assumptions.

6.4.1   Convex Cost – Linear Constraints

We first consider the linearly constrained problem

$$\begin{array}{ll} \text{minimize} & f(x) \\ \text{subject to} & x \in X, \quad e_i'x - d_i = 0, \quad i = 1, \ldots, m, \\ & a_j'x - b_j \leq 0, \quad j = 1, \ldots, r, \end{array} \qquad (6.16)$$

where X is the intersection of a polyhedral set with some other convex set. We will show that problem (6.16) possesses geometric multipliers under a convexity assumption on f and an assumption that involves a relative interior condition. This latter assumption is not needed when X itself is a polyhedral set.

Assumption 6.4.1: (Convexity and Linear Constraints) The optimal value f^* of problem (6.16) is finite, and the following hold:

  (1) The set X is the intersection of a polyhedral set and a convex set C.

  (2) The cost function $f : \Re^n \mapsto \Re$ is convex over C.

  (3) There exists a feasible solution of the problem that belongs to the relative interior of C.

The following is the main result of this section.

Proposition 6.4.2: (Strong Duality Theorem - Linear Constraints) Let Assumption 6.4.1 hold for problem (6.16). Then there is no duality gap and there exists at least one geometric multiplier.

Proof: The proof is based on the Nonlinear Farkas' Lemma [Prop. 6.4.1(b)]. Without loss of generality, we assume that there are no equality constraints, so we are dealing with the problem

$$\begin{array}{ll} \text{minimize} & f(x) \\ \text{subject to} & x \in X, \quad a_j'x - b_j \leq 0, \quad j = 1, \ldots, r, \end{array}$$

(each equality constraint can be converted into two inequality constraints, as discussed earlier). Also without loss of generality, we assume that $f^* = 0$ [otherwise, we replace $f(x)$ by $f(x) - f^*$].

Let $X = P \cap C$, where P is a polyhedral set that is expressed in terms of linear inequalities as

$$P = \{x \mid a_j' x - b_j \le 0, \; j = r+1, \ldots, p\}$$

for some integer $p > r$. By applying the Nonlinear Farkas' Lemma (Prop. 6.4.1) with F being the set

$$\{x \in C \mid a_j' x - b_j \le 0, \; j = 1, \ldots, p\},$$

we see that there exist nonnegative $\mu_1^*, \ldots, \mu_p^*$ such that

$$f(x) + \sum_{j=1}^{p} \mu_j^*(a_j' x - b_j) \ge 0, \qquad \forall \; x \in C.$$

Since for $x \in P$, we have $\mu_j^*(a_j' x - b_j) \le 0$ for all $j = r+1, \ldots, p$, the above equation yields

$$f(x) + \sum_{j=1}^{r} \mu_j^*(a_j' x - b_j) > 0, \qquad \forall \; x \in P \cap C = X,$$

from which we obtain

$$\inf_{x \in X} L(x, \mu^*) = q(\mu^*) \ge 0 = f^*.$$

From the Weak Duality Theorem (Prop. 6.2.2) we have $q(\mu^*) \le q^* \le f^*$, so it follows that $q^* = f^*$ and that μ^* is a geometric multiplier.   **Q.E.D.**

We note that convexity of f over X is not enough for Prop. 6.4.2 to hold; it is essential that f be convex over the entire set C, as the following example shows.

Example 6.4.1:

Consider the two-dimensional problem

$$\text{minimize} \quad f(x)$$
$$\text{subject to} \quad x_1 \le 0, \qquad x \in X = \{x \mid x \ge 0\},$$

where
$$f(x) = e^{-\sqrt{x_1 x_2}}, \qquad \forall \; x \in X,$$

and $f(x)$ is arbitrarily defined for $x \notin X$. Here it can be verified that f is convex over X (its Hessian matrix is positive definite in the interior of X, so it is convex over $\text{int}(X)$, and since it is also continuous, it is convex over X).

Since for feasibility, we must have $x_1 = 0$, we see that $f^* = 1$. On the other hand, for all $\mu \geq 0$ we have

$$q(\mu) = \inf_{x \geq 0} \left\{ e^{-\sqrt{x_1 x_2}} + \mu x_1 \right\} = 0,$$

since the expression in braces is nonnegative for $x \geq 0$ and can approach zero by taking $x_1 \to 0$ and $x_1 x_2 \to \infty$. It follows that $q^* = 0$, so there is a duality gap, $f^* - q^* = 1$.

The difficulty here is that if we let $C = X$, then there is no feasible solution that belongs to the relative interior of C. On the other hand, if we let $C = \Re^2$, then $f(x)$ cannot be defined as a convex function over C. In either case, Assumption 6.4.1 is violated.

Finally, let us use the Strong Duality Theorem of this section to show a result (due to Xin Chen – private communication), which we stated in Section 5.3 in connection with sensitivity of nonconvex problems.

Example 6.4.2: (Existence of the Steepest Descent Direction)

Consider the problem

$$\text{minimize} \quad a_0' d + \tfrac{1}{2} \sum_{j=1}^{r} \left((a_j' d)^+ \right)^2 \tag{6.17}$$
$$\text{subject to} \quad d \in N^*,$$

given in Lemma 5.3.1, where $a_0, a_1, \ldots, a_r$ are given vectors and N is a nonempty closed convex cone. We assume that there exists a vector $\bar{\mu} \geq 0$ such that

$$-\left(a_0 + \sum_{j=1}^{r} \bar{\mu}_j a_j \right) \in \text{ri}(N), \tag{6.18}$$

where $\text{ri}(N)$ denotes the relative interior of N, and we will use a duality argument to show that problem (6.17) has at least one optimal solution. This solution was interpreted as the steepest descent direction in the sensitivity context of Section 5.3.

Consider the problem

$$\text{minimize} \quad \tfrac{1}{2} \|\mu\|^2$$
$$\text{subject to} \quad y = -\left(a_0 + \sum_{j=1}^{r} \mu_j a_j \right) \tag{6.19}$$
$$y \in N, \qquad \mu \geq 0.$$

Let us view the set $\big\{ (y, \mu) \mid y \in N, \, \mu \geq 0 \big\}$ as the abstract set constraint of this problem, and view the equation

$$-\left(y + a_0 + \sum_{j=1}^{r} \mu_j a_j \right) = 0$$

as a linear equality constraint. The abstract set constraint $\{(y, \mu) \mid y \in N, \mu \geq 0\}$ can be written as the intersection of the polyhedral set $\{(y, \mu) \mid \mu \geq 0\}$ and the set $C = \{(y, \mu) \mid y \in N\}$. Since $\mathrm{ri}(C) = \{(y, \mu) \mid y \in \mathrm{ri}(N)\}$, based on the relative interior assumption (6.18), we can apply Prop. 6.4.2, and assert the existence of a geometric multiplier.

Denoting by d the multiplier that corresponds to the equality constraint, we see that the dual function is given by

$$q(d) = \inf_{y \in N, \, \mu \geq 0} \left\{ \tfrac{1}{2}\|\mu\|^2 - \left(y + a_0 + \textstyle\sum_{j=1}^r \mu_j a_j \right)' d \right\}$$

$$= \inf_{y \in N} (-y'd) + \inf_{\mu \geq 0} \left\{ \tfrac{1}{2}\|\mu\|^2 - \left(a_0 + \textstyle\sum_{j=1}^r \mu_j a_j \right)' d \right\}.$$

The minimization over $y \in N$ of $-y'd$ yields a value of $-\infty$ if $d \notin N^*$ and a value of 0 otherwise. For each j, the minimum over $\mu_j \geq 0$ is attained at $\mu_j = (a_j'd)^+$, so after a straightforward calculation, we obtain

$$q(d) = \begin{cases} -a_0'd - \tfrac{1}{2} \sum_{j=1}^r \left((a_j'd)^+ \right)^2 & \text{if } d \in N^*, \\ -\infty & \text{otherwise.} \end{cases}$$

Thus the dual problem of maximizing $q(d)$ over $d \in \Re^r$ is equivalent to problem (6.17), and the set of geometric multipliers of problem (6.19), which is nonempty as discussed earlier, coincides with the set of optimal solutions of problem (6.17).

6.4.2   Convex Cost – Convex Constraints

We now consider the nonlinearly constrained problem

$$\begin{aligned} \text{minimize} \quad & f(x) \\ \text{subject to} \quad & x \in X, \quad g_j(x) \leq 0, \; j = 1, \ldots, r, \end{aligned} \qquad (6.20)$$

under the following assumption.

Assumption 6.4.2: (Slater Condition) The optimal value f^* of problem (6.20) is finite, the set X is convex, and the functions $f : \Re^n \mapsto \Re$ and $g_j : \Re^n \mapsto \Re$ are convex over X. Furthermore, there exists a vector $\overline{x} \in X$ such that $g_j(\overline{x}) < 0$ for all $j = 1, \ldots, r$.

We have the following proposition.

Proposition 6.4.3: (Strong Duality Theorem - Nonlinear Constraints) Let Assumption 6.4.2 hold for problem (6.20). Then there is no duality gap, and the set of geometric multipliers is nonempty and compact.

Proof: Similar to the proof of Prop. 6.4.2, the result follows by applying the Nonlinear Farkas' Lemma [Prop. 6.4.1(a)].   **Q.E.D.**

As can be inferred from part (a) of the Nonlinear Farkas' Lemma, a converse to the above proposition can be shown, namely that if f^* is finite, X is convex, the functions $f : \Re^n \mapsto \Re$ and $g_j : \Re^n \mapsto \Re$ are convex over X, and the set of geometric multipliers is nonempty and compact, then the Slater condition must hold. We leave this as an exercise for the reader.

Here is an example where g is nonlinear, and there is a duality gap, even though $X = \Re^n$, and f and g are convex over $\Re^n$.

Example 6.4.3:

Consider the two-dimensional problem

$$\text{minimize}\quad e^{x_2}$$

$$\text{subject to}\ \ \|x\| - x_1 \leq 0,\qquad x \in X = \Re^2.$$

It can be seen that a vector (x_1, x_2) is feasible if and only if

$$x_1 \geq 0,\qquad x_2 = 0.$$

Furthermore, all feasible points attain the optimal value, which is $f^* = 1$. Consider now the dual function

$$q(\mu) = \inf_{x \in \Re^2} \left\{ e^{x_2} + \mu \left(\sqrt{x_1^2 + x_2^2} - x_1 \right) \right\}.$$

We claim that $q(\mu) = 0$ for all $\mu \geq 0$. Indeed, we clearly have $q(\mu) \geq 0$ for all $\mu \geq 0$. To prove the reverse relation, we note that if we restrict x to vary so that $x_1 = x_2^4$, we have

$$\sqrt{x_1^2 + x_2^2} - x_1 = \frac{x_2^2}{\sqrt{x_1^2 + x_2^2} + x_1} = \frac{x_2^2}{\sqrt{x_2^8 + x_2^2} + x_2^4}.$$

Thus,

$$\sqrt{x_1^2 + x_2^2} - x_1 \to 0,\qquad \text{as } x_2 \to -\infty \text{ and } x_1 = x_2^4,$$

and it follows that

$$q(\mu) \leq \inf_{x_2 < 0,\, x_1 = x_2^4} \left\{ e^{x_2} + \mu \left(\sqrt{x_1^2 + x_2^2} - x_1 \right) \right\} = 0,\qquad \forall\, \mu \in \Re.$$

showing that $q(\mu) = 0$ for all $\mu \geq 0$. Thus,

$$q^* = \sup_{\mu \geq 0} q(\mu) = 0,$$

and there is a duality gap, $f^* - q^* = 1$. The difficulty here is that g is nonlinear and there is no $\overline{x} \in X$ such that $g(\overline{x}) < 0$, so the Slater condition (Assumption 6.4.2) is violated.

We finally consider a generalization of problems (6.16) and (6.20), where there are linear equality and inequality constraints, as well as convex inequality constraints:

minimize $f(x)$

subject to $x \in X, \quad g_j(x) \leq 0, \quad j = 1, \ldots, \bar{r},$

$$e_i' x - d_i = 0, \ i = 1, \ldots, m, \ a_j' x - b_j \leq 0, \ j = \bar{r} + 1, \ldots, r.$$
$$(6.21)$$

To cover this case, we suitably extend Assumptions 6.4.1 and 6.4.2.

Assumption 6.4.3: (Linear and Nonlinear Constraints) The optimal value f^* of problem (6.21) is finite, and the following hold:

(1) The set X is the intersection of a polyhedral set and a convex set C.

(2) The functions $f : \Re^n \mapsto \Re$ and $g_j : \Re^n \mapsto \Re$ are convex over C.

(3) There exists a feasible vector $\bar{x}$ such that $g_j(\bar{x}) < 0$ for all $j = 1, \ldots, \bar{r}$.

(4) There exists a vector that satisfies the linear constraints [but not necessarily the constraints $g_j(x) \leq 0, \ j = 1, \ldots, \bar{r}$], and belongs to X and to the relative interior of C.

Proposition 6.4.4: (Strong Duality Theorem - For Linear and Nonlinear Constraints) Let Assumption 6.4.3 hold for problem (6.21). Then there is no duality gap and there exists at least one geometric multiplier.

Proof: Using Prop. 6.4.3, we argue that there exist $\mu_j^* \geq 0, \ j = 1, \ldots, \bar{r}$, such that

$$f^* = \inf_{\substack{x \in X, \, a_j' x - b_j \leq 0, \, j = \bar{r}+1, \ldots, r \\ e_i' x - d_i = 0, \, i = 1, \ldots, m}} \left\{ f(x) + \sum_{j=1}^{\bar{r}} \mu_j^* g_j(x) \right\}.$$

Then we apply Prop. 6.4.2 to the minimization problem in the right-hand side of the above equation to show that there exist $\lambda_i^*, \ i = 1, \ldots, m$, and $\mu_j^* \geq 0, \ j = \bar{r} + 1, \ldots, r$, such that

$$f^* = \inf_{x \in X} \left\{ f(x) + \sum_{i=1}^{m} \lambda_i^* (e_i' x - d_i) + \sum_{j=\bar{r}+1}^{r} \mu_j^* (a_j' x - b_j) + \sum_{j=1}^{\bar{r}} \mu_j^* g_j(x) \right\}.$$

Thus, the vector $(\lambda_1^*, \ldots, \lambda_m^*, \mu_1^*, \ldots, \mu_r^*)$ is a geometric multiplier. **Q.E.D.**

6.5 STRONG DUALITY AND THE PRIMAL FUNCTION

The primal and dual problems can be viewed in terms of a minimax problem involving the Lagrangian function $L(x, \lambda, \mu)$. We discussed this viewpoint in Section 2.6 and also in connection with the Lagrangian Saddle Point Theorem (Prop. 6.2.4). Furthermore, we showed in Section 2.6 that the properties of the function

$$p(v, u) = \inf_{x \in X} \sup_{\lambda \in \Re^m, \, \mu \in \Re^r, \, \mu \geq 0} \left\{ L(x, \lambda, \mu) - \lambda' v - \mu' u \right\}$$

around $(v, u) = (0, 0)$ are critical in analyzing the existence of dual optimal solutions and the presence of a duality gap.

The function p is known as the *primal function* of the constrained optimization problem (P). Note that we have

$$\sup_{\lambda \in \Re^m, \, \mu \in \Re^r, \, \mu \geq 0} \left\{ L(x, \lambda, \mu) - \lambda' v - \mu' u \right\}$$

$$= \sup_{\lambda \in \Re^m, \, \mu \in \Re^r, \, \mu \geq 0} \left\{ f(x) + \lambda' \big(h(x) - v \big) + \mu' \big(g(x) - u \big) \right\}$$

$$= \begin{cases} f(x) & \text{if } h(x) = v, \, g(x) \leq u, \\ \infty & \text{otherwise,} \end{cases}$$

so the primal function has the form

$$p(v, u) = \inf_{x \in X, \, h(x) = v, \, g(x) \leq u} f(x).$$

Thus $p(v, u)$ can be interpreted as the optimal value of a modified version of the primal problem (P), where the right hand sides of the equality and inequality constraints have been perturbed by v and u, respectively.

In this section, we will discuss several issues relating to the primal function, including the presence of a duality gap.

6.5.1 Duality Gap and the Primal Function

We will first use the minimax theory of Section 2.6 to derive conditions under which there is no duality gap. For simplicity, we assume that there are just inequality constraints, but as usual, the theory readily extends to the case where there are equality constraints as well. We thus consider the problem

$$\text{minimize} \quad f(x)$$

$$\text{subject to} \quad x \in X, \quad g(x) \leq 0,$$

where $f : \Re^n \mapsto \Re$ is a function, X is a nonempty subset of $\Re^n$,

$$g(x) = \big(g_1(x), \ldots, g_r(x)\big),$$

and $g_j : \Re^n \mapsto \Re$, $j = 1, \ldots, r$, are some functions.

The primal function then takes the form

$$p(u) = \inf_{x \in X, \, g(x) \le u} f(x).$$

We view p as a function mapping $\Re^r$ into $[-\infty, \infty]$. The properties of p and the connection with the issue of existence of a duality gap have been discussed in Section 2.6. We have the following two propositions.

Proposition 6.5.1: Assume that the set X is convex, and f and g_j are convex over X. Then p is convex.

Proof: Apply Lemma 2.6.1 (or Prop. 2.3.5, on which Lemma 2.6.1 is based). **Q.E.D.**

Proposition 6.5.2: Assume that $f^* < \infty$, the set X is convex, and f and g_j are convex over X. Then, there is no duality gap if and only if p is lower semicontinuous at $u = 0$.

Proof: Apply Prop. 2.6.2. **Q.E.D.**

Let us illustrate how p fails to be lower semicontinuous at $u = 0$ for the two examples of the preceding section where there is a duality gap. In both cases, there is a strong qualitative change in the minimization defining $p(u)$ as u increases from 0 to positive values.

Example 6.5.1:

Consider the two-dimensional problem of Example 6.4.1:

$$\text{minimize} \quad e^{-\sqrt{x_1 x_2}}$$
$$\text{subject to} \quad x_1 \le 0, \qquad x \in X = \{x \mid x \ge 0\}.$$

The primal function is

$$p(u) = \inf_{x \ge 0, \, x_1 \le u} e^{-\sqrt{x_1 x_2}}.$$

As Fig. 6.5.1 shows, when $u = 0$, the feasible set of the minimization in the definition of the primal function is just the nonnegative vertical half-axis, where the cost function $e^{-\sqrt{x_1 x_2}}$ is identically equal to 1, so $p(0) = 1$. When $u > 0$, the feasible set changes qualitatively. The feasible set now includes vectors (x_1, x_2) with $x_1 = u$ and $x_2 \to \infty$ for which the cost function tends to 0, so $p(u) = 0$ for $u > 0$.

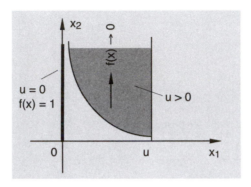

Figure 6.5.1. Illustration of the minimization in the definition of the primal function

$$p(u) = \inf_{x \geq 0,\, x_1 \leq u} e^{-\sqrt{x_1 x_2}}$$

of Example 6.5.1. The figure shows the intersection of a typical level set of the cost function and the constraint set for the cases where $u = 0$ and $u > 0$. In the case where $u > 0$, the cost function recedes asymptotically to 0 along the direction of recession $(0, 1)$.

Example 6.5.2:

Consider the two-dimensional problem of Example 6.4.3:

$$\text{minimize} \quad e^{x_2}$$
$$\text{subject to} \quad \|x\| - x_1 \leq 0, \qquad x \in X = \Re^2.$$

The primal function is

$$p(u) = \inf_{\|x\| - x_1 \leq u} e^{x_2}.$$

Figure 6.5.2 shows the feasible set of the above minimization for $u = 0$ and for $u > 0$. When $u = 0$, the feasible set $\{x \mid \|x\| - x_1 \leq 0\}$ consists of just the horizontal nonnegative axis, the cost function e^{x_2} is constant and equal to 1 on that set, and we have $p(0) = 1$. On the other hand, when $u > 0$, the feasible set is $\{x \mid \|x\| - x_1 \leq u\}$, the infimum of the cost function e^{x_2} is not attained, and we have $p(u) = 0$.

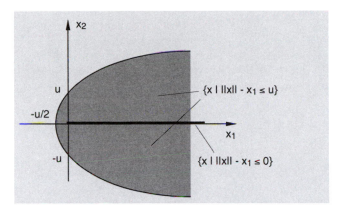

Figure 6.5.2. The feasible set $\{x \mid \|x\| - x_1 \leq u\}$ of the minimization in the definition of the primal function

$$p(u) = \inf_{\|x\| - x_1 \leq u} e^{x_2}$$

of Example 6.5.2 for the cases where $u = 0$ and $u > 0$.

6.5.2   Conditions for No Duality Gap

We now discuss conditions under which the primal function p is lower semicontinuous, in which case, by Prop. 6.5.2, there is no duality gap. The following propositions, are applications of the results of Section 2.3 on the preservation of closedness under partial minimization (cf. Props. 2.3.6-2.3.9); see also the corresponding results in Section 2.6 on the validity of the minimax equality (cf. Props. 2.6.4-2.6.7). These results show that under convexity assumptions, there is no duality gap under standard conditions for the attainment of the infimum in the definition of p (cf. Section 2.3).

Proposition 6.5.3: Assume that X is a convex set, and that for each $\mu \geq 0$, the function $t_\mu : \Re^n \mapsto (-\infty, \infty]$ defined by

$$t_\mu(x) = \begin{cases} L(x, \mu) & \text{if } x \in X, \\ \infty & \text{otherwise,} \end{cases}$$

is closed and convex. Assume further that $f^* < \infty$ and that the set of common directions of recession of all the functions t_μ, $\mu \geq 0$, consists of the zero vector only. Then, there is no duality gap, and the primal problem has a nonempty and compact optimal solution set. Furthermore, the primal function p is closed, proper, and convex.

Proof: We consider the minimax problem associated with the Lagrangian function L, and we apply Prop. 2.6.4, and the proposition on the preservation of closedness under partial minimization, on which Prop. 2.6.4 relies (Prop. 2.3.6). The directions of recession assumption of the present proposition is equivalent to the function t given by

$$t(x) = \sup_{\mu \geq 0} t_\mu(x)$$

$$= \begin{cases} \sup_{\mu \geq 0} L(x,\mu) & \text{if } x \in X, \\ \infty & \text{otherwise,} \end{cases}$$

$$= \begin{cases} f(x) & \text{if } x \in X, \ g(x) \leq 0, \\ \infty & \text{otherwise,} \end{cases}$$

having compact level sets. It can thus be seen that our assumptions are equivalent to the assumptions of Prop. 2.6.4, and from the conclusions of that proposition, it follows that there is no duality gap, and that the primal problem has a nonempty and compact optimal solution set. The convexity and closedness properties of p follow from Prop. 2.3.6.   **Q.E.D.**

Note that if X is compact, and f and g_j are convex functions, the condition on directions of recession of the above proposition is satisfied. The same is true if instead of assuming that X is compact, we assume that X is closed, and that the primal optimal solution set is nonempty and compact (which is equivalent to X, f, and g_j having no common nonzero direction of recession). Note also that while these conditions imply that there is no duality gap, as per the above proposition, they do not guarantee the existence of a geometric multiplier/optimal dual solution [see the example of Fig. 6.2.3(a)].

We will now establish that there is no duality gap under convexity assumptions together with some assumption involving directions of recession, which guarantees that the primal problem has an optimal solution. The analysis involves the use of recession cones and lineality spaces, and is patterned after the existence results of Section 2.3 and the minimax results of Section 2.6. The notation of Section 2.3 will be used as follows:

R_X and L_X denote the recession cone and the lineality space of X.

R_f and L_f denote the recession cone and the constancy space of f.

R_{g_j} and L_{g_j} denote the recession cone and the constancy space of g_j.

Proposition 6.5.4: Assume that the set X is closed and convex , and that the functions f and g_j are convex. Assume further that $f^* < \infty$ and that

$$R_X \cap R_f \cap \left(\cap_{j=1}^r R_{g_j}\right) = L_X \cap L_f \cap \left(\cap_{j=1}^r L_{g_j}\right).$$

Then, there is no duality gap, and the primal problem has a nonempty optimal solution set. Furthermore, the primal function p is closed, proper, and convex.

Proof: Similar to the proof of Prop. 6.5.3, apply Prop. 2.6.5 and Prop. 2.3.7. **Q.E.D.**

By using a similar proof, it can be shown that the result of the preceding proposition also holds under a slightly weaker assumption: instead of assuming that f and g_j are convex, we may assume that if we restrict f and g_j to have domain X, they are closed, proper, and convex, i.e., that the functions $\tilde{f}$ and $\tilde{g}_j$, defined by

$$\tilde{f}(x) = \begin{cases} f(x) & \text{if } x \in X, \\ \infty & \text{otherwise,} \end{cases} \qquad \tilde{g}_j(x) = \begin{cases} g_j(x) & \text{if } x \in X, \\ \infty & \text{otherwise,} \end{cases}$$

are closed, proper, and convex.

If the set X is polyhedral, the assumptions of Prop. 6.5.4 can be weakened, as shown in the following proposition.

Proposition 6.5.5: Assume that the set X is polyhedral, and that the functions f and g_j are convex. Assume further that $f^* < \infty$ and that

$$R_X \cap R_f \cap \left(\cap_{j=1}^r R_{g_j}\right) \subset L_f \cap \left(\cap_{j=1}^r L_{g_j}\right).$$

Then, there is no duality gap, and the primal problem has a nonempty optimal solution set. Furthermore, the primal function p is closed, proper, and convex.

Proof: Similar to the proof of Prop. 6.5.3, apply Prop. 2.6.6 and Prop. 2.3.8. **Q.E.D.**

Here is an example where the recession cone assumptions of the preceding two propositions are violated, and there is a duality gap, even though the set of primal optimal solutions is nonempty.

Example 6.5.3:

Consider the two-dimensional problem

$$\text{minimize} \quad e^{x_2}$$

$$\text{subject to} \quad \|x\| - x_1 \leq 0, \qquad x \in X = \Re^2.$$

As discussed in Example 6.4.3, the set of optimal solutions of this problem coincides with the set of feasible points,

$$\{x \mid x_1 \geq 0, \ x_2 = 0\},$$

but there is a duality gap. Here, the condition on common directions of recession of Prop. 6.5.4 is violated, while the other conditions hold. In particular, we have

$$R_X = \Re^2, \quad R_f = \{y \mid y_1 \in \Re, \ y_2 \leq 0\}, \quad R_{g_1} = \{y \mid y_1 \geq 0, \ y_2 = 0\},$$

so that

$$R_X \cap R_f \cap R_{g_1} = \{y \mid y_1 \geq 0, \ y_2 = 0\}.$$

On the other hand, we have $L_X \cap L_f \cap L_{g_1} = \{0\}$, since $L_{g_1} = \{0\}$, so

$$R_X \cap R_f \cap R_{g_1} \neq L_X \cap L_f \cap L_{g_1},$$

and the assumption of Prop. 6.5.4 is violated. Furthermore, $R_X \cap R_f \cap R_{g_1}$ is not contained in $L_f \cap L_{g_1}$, so the assumption of Prop. 6.5.5 is violated.

We next consider the case where f and g_j are convex quadratic functions.

Proposition 6.5.6: Assume that $X = \Re^n$, and that the functions f and g_j, $j = 1, \ldots, r$, are quadratic of the form

$$f(x) = x'Qx + a'x + b,$$

$$g_j(x) = x'Q_jx + a_j'x + b_j \leq 0, \quad \forall \ j = 1, \ldots, r,$$

where Q and Q_j, $j = 1, \ldots, r$, are symmetric positive semidefinite $n \times n$ matrices, a and a_j, $j = 1, \ldots, r$, are vectors in $\Re^n$, and b and b_j, $j = 1, \ldots, r$, are scalars. Assume further that $-\infty < f^* < \infty$. Then, there is no duality gap, and the primal problem has a nonempty optimal solution set. Furthermore, the primal function p is closed, proper, and convex.

Proof: Similar to the proof of Prop. 6.5.3, apply Prop. 2.6.7 or Prop. 2.3.9. **Q.E.D.**

Note that under the assumptions of the preceding proposition, a geometric multiplier may not exist, as illustrated by the example of Fig. 6.2.3(a). The following proposition provides a generalization to the case where X is given in terms of convex quadratic inequalities (rather than $X = \Re^n$).

Proposition 6.5.7: Assume that X has the form

$$X = \{x \mid x'Q_l x + a_l' x + b_l \leq 0, \quad \forall \, l = 1, \ldots, \bar{r}\},$$

where Q_l, $l = 1, \ldots, \bar{r}$, are symmetric positive semidefinite $n \times n$ matrices, a_l, $l = 1, \ldots, \bar{r}$, are vectors in $\Re^n$, and b_l, $l = 1, \ldots, \bar{r}$, are scalars. Let f and g_j, $j = 1, \ldots, r$, be as in Prop. 6.5.6, and assume that $-\infty < f^* < \infty$. Then, there is no duality gap, and the primal problem has a nonempty optimal solution set. Furthermore, the primal function p is closed, proper, and convex.

Proof: Let us denote

$$g_{r+l}(x) = x'Q_l x + a_l' x + b_l \leq 0, \quad \forall \, l = 1, \ldots, \bar{r},$$

and view the problem as minimizing f subject to $g_j(x) \leq 0$, $j = 1, \ldots, r+\bar{r}$. Then, by Prop. 6.5.6, there is no duality gap, so that

$$f^* = \sup_{\mu \geq 0, \nu \geq 0} \inf_x \left\{ f(x) + \sum_{j=1}^r \mu_j g_j(x) + \sum_{j=r+1}^{r+\bar{r}} \nu_j g_j(x) \right\},$$

where $\mu \in \Re^r$, $\nu \in \Re^{\bar{r}}$, $\mu \geq 0$, and $\nu \geq 0$. We have

$$\inf_x \left\{ f(x) + \sum_{j=1}^r \mu_j g_j(x) + \sum_{j=r+1}^{r+\bar{r}} \nu_j g_j(x) \right\} \leq \inf_{x \in X} \left\{ f(x) + \sum_{j=1}^r \mu_j g_j(x) \right\}$$
$$= q(\mu).$$

and therefore,

$$f^* = \sup_{\mu \geq 0, \nu \geq 0} \inf_x \left\{ f(x) + \sum_{j=1}^r \mu_j g_j(x) + \sum_{j=r+1}^{r+\bar{r}} \nu_j g_j(x) \right\} \leq \sup_{\mu \geq 0} q(\mu) = q^*.$$

This and the weak duality relation $q^* \leq f^*$ imply that $q^* = f^*$.   **Q.E.D.**

The preceding proposition can be generalized to the case where f and g_j are bidirectionally flat functions (see Exercise 2.7), since the key result on preservation of closedness under partial minimization (Prop. 2.3.9) holds for such functions as well.

6.5.3   Subgradients of the Primal Function

If the primal function p is convex and $p(0)$ is finite, the subdifferential $\partial p(0)$ is closely related to the set of geometric multipliers. To show this, we first note that for any $\mu \geq 0$, we have

$$
\begin{aligned}
q(\mu) &= \inf_{x \in X} \left\{ f(x) + \mu' g(x) \right\} \\
&= \inf_{u \in \Re^r} \inf_{x \in X,\, g(x) \leq u} \left\{ f(x) + \mu' g(x) \right\} \\
&= \inf_{u \in \Re^r} \left\{ p(u) + \mu' u \right\}.
\end{aligned}
\tag{6.22}
$$

Thus, if μ is a geometric multiplier, we have $q(\mu) = f^* = p(0)$, and from Eq. (6.22) we see that

$$
p(0) \leq p(u) + \mu' u, \qquad \forall\, u \in \Re^r.
\tag{6.23}
$$

This implies that p is proper and that $-\mu$ is a subgradient of $p(u)$ at 0. Conversely, if p is proper and the above equation holds for some μ, then since $p(u)$ is monotonically nonincreasing with respect to the components of u, we have $\mu \geq 0$ [otherwise the right-hand side of Eq. (6.23) would be unbounded below]. Furthermore, from Eqs. (6.23) and (6.22), it follows that

$$
f^* = p(0) \leq \inf_{u \in \Re^r} \left\{ p(u) + \mu' u \right\} = q(\mu),
$$

which, in view of the Weak Duality Theorem, implies that μ is a geometric multiplier. We thus obtain the following proposition:

Proposition 6.5.8:   Assume that p is convex and $p(0)$ is finite. Then:

(a) The vector μ is a geometric multiplier if and only if p is proper and $-\mu$ is a subgradient of p at $u = 0$.

(b) If the origin lies in the relative interior of the effective domain of p, then there exists a geometric multiplier.

(c) If the origin lies in the interior of the effective domain of p, the set of geometric multipliers is nonempty and compact.

Proof: Part (a) follows from the analysis preceding the proposition. Parts (b) and (c) follow from part (a) and Prop. 4.4.2.   **Q.E.D.**

More generally, by replacing the constraint $g(x) \leq 0$ with a constraint $g(x) \leq u$, where u is such that $p(u)$ is finite, we see that under the assumptions of Prop. 6.5.8, the subdifferential of p at u is equal to the set of geometric multipliers of the problem $\min_{x \in X,\, g(x) \leq u} f(x)$. This subdifferential is in turn nonempty if u is in the relative interior of $\mathrm{dom}(p)$, and is also compact if u is in the interior of $\mathrm{dom}(p)$.

6.5.4   Sensitivity Analysis

The preceding analysis also points to a sensitivity interpretation of geometric multipliers. In particular, if p is convex and differentiable at 0,
then $-\nabla p(0)$ is equal to the unique geometric multiplier. If μ_j^* is the jth
component of this multiplier, we have

$$\mu_j^* = -\frac{\partial p(0)}{\partial u_j},$$

so μ_j^* (being nonnegative) gives the rate of improvement of p (the optimal
cost) as the constraint $g(x) \leq 0$ is violated.

   For another sensitivity view, let μ^* be a geometric multiplier, and
consider a vector u_j^γ of the form

$$u_j^\gamma = (0, \ldots, 0, \gamma, 0, \ldots, 0)$$

where γ is a scalar in the jth position. Then from Eq. (6.23), which is just
the subgradient inequality at $u = 0$, we obtain

$$\lim_{\gamma \downarrow 0} \frac{p(0) - p(u_j^\gamma)}{\gamma} \leq \mu_j^* \leq \lim_{\gamma \uparrow 0} \frac{p(0) - p(u_j^\gamma)}{\gamma},$$

or

$$\lim_{\gamma \uparrow 0} \frac{p(u_j^\gamma) - p(0)}{\gamma} \leq -\mu_j^* \leq \lim_{\gamma \downarrow 0} \frac{p(u_j^\gamma) - p(0)}{\gamma}.$$

Thus $-\mu_j^*$ lies between the left and the right slope of p in the direction of
the jth axis starting at $u = 0$.

   Finally, assume that p is convex and finite in a neighborhood of 0.
Then the subdifferential $\partial p(0)$ is nonempty and compact (cf. Prop. 4.4.2),
and the directional derivative $p'(0; y)$ is a real-valued convex function of y
satisfying

$$p'(0; y) = \max_{g \in \partial p(0)} y'g$$

(cf. Prop. 4.2.2). Consider the direction of steepest descent of p at 0, i.e.,
the vector $\bar{y}$ that minimizes $p'(0; y)$ subject to $\|y\| \leq 1$. We have

$$p'(0; \bar{y}) = \min_{\|y\| \leq 1} p'(0; y) = \min_{\|y\| \leq 1} \max_{g \in \partial p(0)} y'g = \max_{g \in \partial p(0)} \min_{\|y\| \leq 1} y'g,$$

where the last equality follows from the Saddle Point Theorem of Section
2.6 (Prop. 2.6.9). The minimum over $\|y\| \leq 1$ in the right-hand side above
is attained for $y = -g/\|g\|$ (assuming $g \neq 0$), and if $0 \notin \partial p(0)$, it can be
seen that $\bar{y} = -g^*/\|g^*\|$, where g^* is the subgradient of minimum norm in
$\partial p(0)$. Furthermore, we have

$$p'(0; \bar{y}) = -\|g^*\|.$$

Equivalently, if μ^* is the geometric multiplier of minimum norm and $\mu^* \neq 0$, the direction of steepest descent of p at 0 is

$$\overline{y} = \frac{\mu^*}{\|\mu^*\|},$$

while the rate of steepest descent (per unit norm of constraint violation) is $\|\mu^*\|$. This result is consistent with the theory of informative Lagrange multipliers of Section 5.2, where the Lagrange multiplier of minimum norm was interpreted as providing the rate of improvement of the cost function per unit constraint violation along the direction of steepest descent.

6.6 FRITZ JOHN CONDITIONS WHEN THERE IS NO OPTIMAL SOLUTION

In this section, we consider the problem

$$\begin{aligned} \text{minimize} \quad & f(x) \\ \text{subject to} \quad & x \in X, \ g(x) \leq 0, \end{aligned} \tag{6.24}$$

where $g(x) = \big(g_1(x), \ldots, g_r(x)\big)$. We denote by f^* the optimal value of the problem. Under various convexity assumptions, we derive optimality conditions similar to the Fritz John conditions of Chapter 5. For simplicity, we assume no equality constraints. Our analysis extends to the case where we have linear equality constraints, by replacing each equality constraint by two linear inequality constraints.

We have already derived in Section 5.7 the Fritz John conditions in the case where there exists an optimal solution x^*, and furthermore X is convex, and f and g_j are convex (over $\Re^n$). These conditions were shown in their enhanced form, which includes the CV condition and relates to the notion of pseudonormality (see also the discussion in Section 6.1).

In this section, consistent with the spirit of the present chapter, we do not associate the Fritz John multipliers with any particular optimal solution, and in fact we allow the problem to have no optimal solution at all. We develop several results under a variety of assumptions. The following, a classical result, requires no assumptions other than convexity.

Proposition 6.6.1: (Fritz John Conditions) Consider problem (6.24), and assume that the set X is convex, the functions f and g_j are convex over X, and $f^* < \infty$. Then there exist a scalar μ_0^* and a vector $\mu^* = (\mu_1^*, \ldots, \mu_r^*)$ satisfying the following conditions:

  (i) $\mu_0^* f^* = \inf_{x \in X}\big\{\mu_0^* f(x) + \mu^{*\prime} g(x)\big\}$.

  (ii) $\mu_j^* \geq 0$ for all $j = 0, 1, \ldots, r$.

  (iii) $\mu_0^*, \mu_1^*, \ldots, \mu_r^*$ are not all equal to 0.

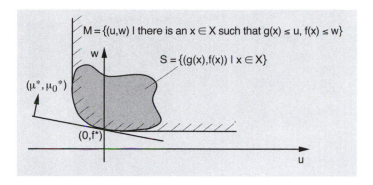

Figure 6.6.1. Illustration of the set

$$S = \Big\{ \big(g(x), f(x)\big) \mid x \in X \Big\}$$

and the set

$$M = \Big\{ (u, w) \mid \text{ there exists } x \in X \text{ such that } g(x) \leq u, \ f(x) \leq w \Big\}$$

used in the proof of Prop. 6.6.1. The idea of the proof is to show that M is convex and that $(0, f^*)$ is not an interior point of M. A hyperplane passing through $(0, f^*)$ and supporting M is used to define the Fritz John multipliers.

Proof: If $f^* = -\infty$, the given conditions hold with $\mu_0^* = 1$ and $\mu^* = 0$. We may thus assume that f^* is finite. Consider the subset of $\Re^{r+1}$ given by

$$M = \big\{ (u_1, \ldots, u_r, w) \mid \text{ there exists } x \in X \text{ such that }$$
$$g_j(x) \leq u_j, \ j = 1, \ldots, r, \ f(x) \leq w \big\}$$

(cf. Fig. 6.6.1). We first show that M is convex. To this end, we consider vectors $(u, w) \in M$ and $(\tilde{u}, \tilde{w}) \in M$, and we show that their convex combinations lie in M. The definition of M implies that for some $x \in X$ and $\tilde{x} \in X$, we have

$$f(x) \leq w, \qquad g_j(x) \leq u_j, \qquad j = 1, \ldots, r,$$
$$f(\tilde{x}) \leq \tilde{w}, \qquad g_j(\tilde{x}) \leq \tilde{u}_j, \qquad j = 1, \ldots, r.$$

For any $\alpha \in [0, 1]$, we multiply these relations with α and $1-\alpha$, respectively, and add them. By using the convexity of f and g_j, we obtain

$$f\big(\alpha x + (1 - \alpha)\tilde{x}\big) \leq \alpha f(x) + (1 - \alpha)f(\tilde{x}) \leq \alpha w + (1 - \alpha)\tilde{w},$$
$$g_j\big(\alpha x + (1-\alpha)\tilde{x}\big) \leq \alpha g_j(x) + (1-\alpha)g_j(\tilde{x}) \leq \alpha u_j + (1-\alpha)\tilde{u}_j, \quad j = 1, \ldots, r.$$

In view of the convexity of X, we have $\alpha x + (1 - \alpha)\tilde{x} \in X$, so these equations imply that the convex combination of (u, w) and $(\tilde{u}, \tilde{w})$, i.e., $\big(\alpha u + (1 - \alpha)\tilde{u}, \alpha w + (1 - \alpha)\tilde{w}\big)$, belongs to M. This proves the convexity of M.

We next observe that $(0, f^*)$ is not an interior point of M; otherwise, for some $\epsilon > 0$, the point $(0, f^* - \epsilon)$ would belong to M, contradicting the definition of f^* as the optimal value. Therefore, there exists a hyperplane passing through $(0, f^*)$ and containing M in one of its closed halfspaces, i.e., there exists a vector $(\mu^*, \mu_0^*) \neq (0, 0)$ such that

$$\mu_0^* f^* \leq \mu_0^* w + \mu^{*\prime} u, \qquad \forall\, (u, w) \in M. \tag{6.25}$$

This equation implies that

$$\mu_0^* \geq 0, \qquad \mu_j^* \geq 0, \qquad \forall\, j = 1, \ldots, r,$$

since for each $(u, w) \in M$, we have that $(u, w + \gamma) \in M$ and $(u_1, \ldots, u_j + \gamma, \ldots, u_r, w) \in M$ for all $\gamma > 0$ and j.

Finally, since for all $x \in X$, we have $\bigl(g(x), f(x)\bigr) \in M$, Eq. (6.25) implies that

$$\mu_0^* f^* \leq \mu_0^* f(x) + \mu^{*\prime} g(x), \qquad \forall\, x \in X.$$

Taking the infimum over all $x \in X$, it follows that

$$
\begin{aligned}
\mu_0^* f^* &\leq \inf_{x \in X} \bigl\{\mu_0^* f(x) + \mu^{*\prime} g(x)\bigr\} \\
&\leq \inf_{x \in X,\ g(x) \leq 0} \bigl\{\mu_0^* f(x) + \mu^{*\prime} g(x)\bigr\} \\
&\leq \inf_{x \in X,\ g(x) \leq 0} \mu_0^* f(x) \\
&= \mu_0^* f^*.
\end{aligned}
$$

Hence, equality holds throughout above, proving the desired result. **Q.E.D.**

If the multiplier μ_0^* in the preceding proposition can be proved to be positive, then μ^*/μ_0^* is a geometric multiplier for problem (6.24). This can be used to show the existence of a geometric multiplier in the case where the Slater condition holds (cf. Assumption 6.4.2). Indeed, if there exists a vector $\bar{x} \in X$ such that $g(\bar{x}) < 0$, the multiplier μ_0^* cannot be 0, since if it were, then according to the proposition, we would have

$$0 = \inf_{x \in X} \mu^{*\prime} g(x)$$

for some vector $\mu^* \geq 0$ with $\mu^* \neq 0$, while for this vector, we would also have $\mu^{*\prime} g(\bar{x}) < 0$, a contradiction.

On the other hand, the preceding proposition is insufficient to show the existence of a geometric multiplier in the case of linear constraints (cf. Assumption 6.4.3). The following proposition strengthens the Fritz John conditions for this case, so that they suffice for the proof of the corresponding existence result.

Proposition 6.6.2: (Fritz John Conditions for Linear Constraints) Consider problem (6.24), and assume that the set X is convex, the function f is convex over X, the functions g_j are affine, and $f^* < \infty$. Then there exist a scalar μ_0^* and a vector $\mu^* = (\mu_1^*, \ldots, \mu_r^*)$, satisfying the following conditions:

(i) $\mu_0^* f^* = \inf_{x \in X} \{\mu_0^* f(x) + \mu^{*\prime} g(x)\}$.

(ii) $\mu_j^* \geq 0$ for all $j = 0, 1, \ldots, r$.

(iii) $\mu_0^*, \mu_1^*, \ldots, \mu_r^*$ are not all equal to 0.

(iv) If the index set $J = \{j \neq 0 \mid \mu_j^* > 0\}$ is nonempty, there exists a vector $\tilde{x} \in X$ such that

$$f(\tilde{x}) < f^*, \qquad \mu^{*\prime} g(\tilde{x}) > 0.$$

Proof: In the case where $\inf_{x \in X} f(x) = f^*$, we may choose $\mu_0^* = 1$, and $\mu^* = 0$, and conditions (i)-(iv) are automatically satisfied. We will thus assume that $\inf_{x \subset X} f(x) < f^*$, which also implies that f^* is finite.

Let the affine constraint function be represented as

$$g(x) = Ax - b.$$

Consider the nonempty convex sets

$$C_1 = \{(x, w) \mid \text{there is a vector } x \in X \text{ such that } f(x) < w\},$$

$$C_2 = \{(x, f^*) \mid Ax - b \leq 0\}.$$

Note that C_1 and C_2 are disjoint. The reason is that if $(x, f^*) \in C_1 \cap C_2$, then we must have $x \in X$, $Ax - b \leq 0$, and $f(x) < f^*$, contradicting the fact that f^* is the optimal value of the problem.

Since C_2 is polyhedral, by the Polyhedral Proper Separation Theorem (cf. Prop. 3.5.1), there exists a hyperplane that separates C_1 and C_2 and does not contain C_1, i.e., a vector (ξ, μ_0^*) such that

$$\mu_0^* f^* + \xi' z \leq \mu_0^* w + \xi' x, \quad \forall\, x \in X, w, z \text{ with } f(x) < w, Az - b \leq 0, \quad (6.26)$$

$$\inf_{(x,w) \in C_1} \{\mu_0^* w + \xi' x\} < \sup_{(x,w) \in C_1} \{\mu_0^* w + \xi' x\}.$$

These relations imply that

$$\mu_0^* f^* + \sup_{Az - b \leq 0} \xi' z \leq \inf_{(x,w) \in C_1} \{\mu_0^* w + \xi' x\} < \sup_{(x,w) \in C_1} \{\mu_0^* w + \xi' x\}, \quad (6.27)$$

and that $\mu_0^* \geq 0$ [since w can be taken arbitrarily large in Eq. (6.26)].
Consider the linear program in Eq. (6.27):

$$\text{maximize } \xi'z$$
$$\text{subject to } Az - b \leq 0.$$

By Eq. (6.27), this program is bounded and therefore, by Prop. 2.3.4, it
has an optimal solution, which we denote by z^*. The dual of this program
is

$$\text{minimize } b'\mu$$
$$\text{subject to } \xi = A'\mu, \quad \mu \geq 0,$$

(see Section 6.3). By linear programming duality (cf. Prop. 6.3.1), it follows
that this problem has a dual optimal solution $\mu^* \geq 0$ satisfying

$$\sup_{Az-b\leq 0} \xi'z = \xi'z^* = \mu^{*'}b, \qquad \xi = A'\mu^*. \tag{6.28}$$

Note that μ_0^* and μ^* satisfy the nonnegativity condition (ii). Furthermore,
we cannot have both $\mu_0^* = 0$ and $\mu^* = 0$, since then by Eq. (6.28), we would
also have $\xi = 0$, and Eq. (6.27) would be violated. Thus, μ_0^* and μ^* also
satisfy condition (iii) of the proposition.
From Eq. (6.27), we have

$$\mu_0^* f^* + \sup_{Az-b\leq 0} \xi'z \leq \mu_0^* w + \xi'x, \qquad \forall\, x \in X \text{ with } f(x) < w,$$

which together with Eq. (6.28), implies that

$$\mu_0^* f^* + \mu^{*'}b \leq \mu_0^* w + \mu^{*'}Ax, \qquad \forall\, x \in X \text{ with } f(x) < w,$$

or

$$\mu_0^* f^* \leq \inf_{x\in X,\, f(x)<w} \left\{\mu_0^* w + \mu^{*'}(Ax - b)\right\}. \tag{6.29}$$

Similarly, from Eqs. (6.27) and (6.28), we have

$$\mu_0^* f^* < \sup_{x\in X,\, f(x)<w} \left\{\mu_0^* w + \mu^{*'}(Ax - b)\right\}. \tag{6.30}$$

Using Eq. (6.29), we obtain

$$\mu_0^* f^* \leq \inf_{x\in X} \left\{\mu_0^* f(x) + \mu^{*'}(Ax - b)\right\}$$
$$\leq \inf_{x\in X,\, Ax-b\leq 0} \left\{\mu_0^* f(x) + \mu^{*'}(Ax - b)\right\}$$
$$\leq \inf_{x\in X,\, Ax-b\leq 0} \mu_0^* f(x)$$
$$= \mu_0^* f^*.$$

Hence, equality holds throughout above, which proves condition (i) of the proposition.

We will finally show that the vector μ^* also satisfies condition (iv). To this end, we consider separately the cases where $\mu_0^* > 0$ and $\mu_0^* = 0$.

If $\mu_0^* > 0$, let $\tilde{x} \in X$ be such that $f(\tilde{x}) < f^*$ [based on our earlier assumption that $\inf_{x \in X} f(x) < f^*$]. Then condition (i) yields

$$\mu_0^* f^* \le \mu_0^* f(\tilde{x}) + \mu^{*\prime}(A\tilde{x} - b),$$

implying that $0 < \mu_0^*(f^* - f(\tilde{x})) \le \mu^{*\prime}(A\tilde{x} - b)$, and showing condition (iv).

If $\mu_0^* = 0$, condition (i) together with Eq. (6.30) yields

$$0 = \inf_{x \in X} \mu^{*\prime}(Ax - b) < \sup_{x \in X} \mu^{*\prime}(Ax - b). \tag{6.31}$$

The above relation implies the existence of a vector $\hat{x} \in X$ such that $\mu^{*\prime}(A\hat{x} - b) > 0$. Let $\overline{x} \in X$ be such that $f(\overline{x}) < f^*$, and consider a vector of the form

$$\tilde{x} = \alpha\hat{x} + (1 - \alpha)\overline{x},$$

where $\alpha \in (0, 1)$. Note that $\tilde{x} \in X$ for all $\alpha \in (0, 1)$, since X is convex. From Eq. (6.31), we have $\mu^{*\prime}(A\overline{x} - b) \ge 0$, which combined with the inequality $\mu^{*\prime}(A\hat{x} - b) > 0$, implies that

$$\mu^{*\prime}(A\tilde{x} - b) = \alpha\mu^{*\prime}(A\hat{x} - b) + (1 - \alpha)\mu^{*\prime}(A\overline{x} - b) > 0, \quad \forall \alpha \in (0, 1). \tag{6.32}$$

Furthermore, since f is convex, we have

$$f(\tilde{x}) \le \alpha f(\hat{x}) + (1 - \alpha)f(\overline{x}) = f^* + (f(\overline{x}) - f^*) + \alpha(f(\hat{x}) - f(\overline{x})), \quad \forall \alpha \in (0, 1).$$

Thus, for α small enough so that $\alpha(f(\hat{x}) - f(\overline{x})) < f^* - f(\overline{x})$, we have $f(\tilde{x}) < f^*$ as well as $\mu^{*\prime}(A\tilde{x} - b) > 0$ [cf. Eq. (6.32)].   **Q.E.D.**

Let us now recall the definition of pseudonormality, given in Section 5.7.

Definition 6.6.1: Consider problem (6.24), and assume that the set X is convex, and the functions g_j, viewed as functions from X to $\Re$, are convex. The constraint set of the problem is said to be *pseudonormal* if one cannot find a vector $\mu \ge 0$ and a vector $\tilde{x} \in X$ such that:

(i) $0 = \inf_{x \in X} \mu' g(x)$.

(ii) $\mu' g(\tilde{x}) > 0$.

Proposition 6.6.2 shows that if the problem has a convex cost function f, affine constraint functions g_j, and a pseudonormal constraint set, there exists a geometric multiplier satisfying the special condition (iv) of Prop. 6.6.2. As discussed in Section 5.7, the constraint set is pseudonormal if $X = \Re^n$ and g_j, $j = 1,\ldots,r$, are affine.

Consider now the question of pseudonormality and existence of geometric multipliers in the case of Assumption 6.4.1, where $X \neq \Re^n$, but instead X is the intersection of a polyhedral set and a convex set C, and there exists a feasible solution that belongs to the relative interior of C. Then, as discussed in Section 5.7, the constraint set need not be pseudonormal. However, it is pseudonormal in the extended representation (i.e., when the linear inequalities that represent the polyhedral part are lumped with the remaining linear inequality constraints and we have $X = C$.) To see this, let $g_j(x) = a_j'x - b_j$, and assume that there is a vector μ that satisfies condition (i) of the pseudonormality definition, i.e.,

$$0 \leq \sum_{j=1}^{r} \mu_j(a_j'x - b_j), \qquad \forall\, x \in C.$$

Since there exists some $\bar{x} \in \mathrm{ri}(C)$ such that $a_j'\bar{x} - b_j \leq 0$, the preceding relation implies that

$$\sum_{j=1}^{r} \mu_j(a_j'\bar{x} - b_j) = 0,$$

thus showing, by Prop. 1.4.2, that

$$\sum_{j=1}^{r} \mu_j(a_j'x - b_j) = 0, \qquad \forall\, x \in C.$$

This, however, contradicts condition (ii) of pseudonormality. Thus, the constraint set is pseudonormal in the extended representation, and from Prop. 6.6.2, it follows that there exists a geometric multiplier. From this, as in the proof of Prop. 6.4.2, it follows that there exists a geometric multiplier in the original representation as well (see also Exercise 6.2).

The preceding analysis shows that the Fritz John conditions of Props. 6.6.1 and 6.6.2 are sufficiently powerful to provide alternative proofs of the results on the existence of geometric multiplier (cf. Props. 6.4.2-6.4.4 in Section 6.4, where a line of proof based on the Nonlinear Farkas' Lemma was used). Note, however, that in the case of linear constraints, both lines of proof ultimately rely on the fundamental Polyhedral Proper Separation Theorem (Prop. 3.5.1).

6.6.1   Enhanced Fritz John Conditions

The Fritz John conditions of the preceding two propositions are weaker than the ones that we have encountered so far in that they do not include conditions analogous to the CV condition, which formed the basis for the notion of pseudonormality and the developments of Chapter 5. A natural form of this condition would assert, assuming that $\mu^* \neq 0$, the existence of a sequence $\{x^k\} \subset X$ such that

$$\lim_{k \to \infty} f(x^k) = f^*, \qquad \limsup_{k \to \infty} g(x^k) \leq 0, \qquad (6.33)$$

and for all k,

$$f(x^k) < f^*, \qquad (6.34)$$

$$g_j(x^k) > 0, \quad \forall\, j \in J, \qquad g_j^+(x^k) = o\left(\min_{j \in J} g_j^+(x^k)\right), \quad \forall\, j \notin J, \quad (6.35)$$

where $J = \{j \neq 0 \mid \mu_j^* > 0\}$. Unfortunately, such a condition does not hold in the absence of additional assumptions, as can be seen from the following example.

Example 6.6.1

Consider the one-dimensional problem

$$\text{minimize} \ \ f(x)$$

$$\text{subject to} \ \ g(x) = x \leq 0, \ x \in X = \{x \mid x \geq 0\},$$

where

$$f(x) = \begin{cases} -1 & \text{if } x > 0, \\ 0 & \text{if } x = 0, \\ 1 & \text{if } x < 0. \end{cases}$$

Then f is convex over X and the assumptions of Props. 6.6.1 and 6.6.2 are satisfied. Indeed the Fritz John multipliers that satisfy conditions (i)-(iii) of Prop. 6.6.1 or conditions (i)-(iv) of Prop. 6.6.2 must have the form $\mu_0^* = 0$ and $\mu^* > 0$ (cf. Fig. 6.6.2). However, here we have $f^* = 0$, and for all x with $g(x) > 0$, we have $x > 0$ and $f(x) = -1$. Thus, there is no sequence $\{x^k\} \subset X$ satisfying the CV conditions (6.33)- (6.35).

The following proposition imposes stronger assumptions in order to derive an enhanced set of Fritz John conditions. In particular, we assume the following:

Assumption 6.6.1: (Closedness and Convexity) The set X is convex, and the functions f and g_j, viewed as functions from X to $\Re$, are closed and convex. Furthermore we have $f^* < \infty$.

The proof of the following proposition bears similarity with the proof of Prop. 5.7.2, but is more complicated because an optimal primal solution may not exist.

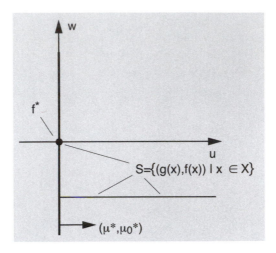

Figure 6.6.2. Illustration of the set

$$S = \left\{ \big(g(x), f(x)\big) \mid x \in X \right\}$$

in Example 6.6.1. Even though $\mu^* > 0$, there is no sequence $\{x^k\} \subset X$ such that $g(x^k) > 0$ for all k, and $f(x^k) \to f^*$.

Proposition 6.6.3: (Enhanced Fritz John Conditions) Consider problem (6.24) under Assumption 6.6.1 (Closedness and Convexity). Then there exist a scalar μ_0^* and a vector $\mu^* = (\mu_1^*, \ldots, \mu_r^*)$, satisfying the following conditions:

(i) $\mu_0^* f^* = \inf_{x \in X} \{ \mu_0^* f(x) + \mu^{*\prime} g(x) \}$.

(ii) $\mu_j^* \geq 0$ for all $j = 0, 1, \ldots, r$.

(iii) $\mu_0^*, \mu_1^*, \ldots, \mu_r^*$ are not all equal to 0.

(iv) If the index set $J = \{ j \neq 0 \mid \mu_j^* > 0 \}$ is nonempty, there exists a sequence $\{x^k\} \subset X$ such that

$$\lim_{k \to \infty} f(x^k) = f^*, \qquad \limsup_{k \to \infty} g(x^k) \leq 0,$$

and for all k,

$$f(x^k) < f^*,$$

$$g_j(x^k) > 0, \quad \forall\, j \in J, \qquad g_j^+(x^k) = o\left(\min_{j \in J} g_j^+(x^k) \right), \quad \forall\, j \notin J.$$

Proof: If $f(x) \geq f^*$ for all $x \in X$, then we set $\mu_0^* = 1$ and $\mu^* = 0$, and we are done. We will thus assume that there exists some $\overline{x} \in X$ such that $f(\overline{x}) < f^*$. In this case, f^* is finite. Consider the problem

$$\begin{aligned}\text{minimize} \quad & f(x) \\ \text{subject to} \quad & g(x) \leq 0, \quad x \in X^k,\end{aligned} \tag{6.36}$$

where

$$X^k = X \cap \left\{ x \mid \|x\| \leq \beta k \right\}, \qquad k = 1, 2, \ldots,$$

where β is a scalar that is large enough so that for all k, the constraint set $\{x \in X^k \mid g(x) \leq 0\}$ is nonempty. By the Closedness and Convexity Assumption, f and g_j are closed and convex when restricted to X, so they are closed, convex, and coercive when restricted to X^k. Hence, problem (6.36) has an optimal solution, which we denote by $\overline{x}^k$. Since this is a more constrained problem than the original, we have $f^* \leq f(\overline{x}^k)$ and $f(\overline{x}^k) \downarrow f^*$ as $k \to \infty$. Let $\gamma^k = f(\overline{x}^k) - f^*$. Note that if $\gamma^k = 0$ for some k, then $\overline{x}^k$ is an optimal solution for problem (6.24), and the result follows by the enhanced Fritz John conditions for convex problems with an optimal solution (cf. Prop. 5.7.2). Therefore, we assume that $\gamma^k > 0$ for all k.

For positive integers k and positive scalars m, we consider the function

$$L_{k,m}(x, \xi) = f(x) + \frac{(\gamma^k)^2}{4k^2}\|x - \overline{x}^k\|^2 + \xi' g(x) - \frac{\|\xi\|^2}{2m}.$$

We note that $L_{k,m}(x, \xi)$, viewed as a function from X^k to $\Re$, for fixed $\xi \geq 0$, is closed, convex, and coercive, in view of the Closedness and Convexity Assumption. Furthermore, $L_{k,m}(x, \xi)$ is negative definite quadratic in ξ, for fixed x. Hence, we can use the Saddle Point Theorem (Prop. 2.6.9) to assert that $L_{k,m}$ has a saddle point over $x \in X^k$ and $\xi \geq 0$, which we denote by $(x^{k,m}, \xi^{k,m})$.

We now derive several properties of the saddle points $(x^{k,m}, \xi^{k,m})$, which set the stage for the main argument. The first of these properties is

$$f(x^{k,m}) \leq L_{k,m}(x^{k,m}, \xi^{k,m}) \leq f(\overline{x}^k),$$

which is shown in the next paragraph.

The infimum of $L_{k,m}(x, \xi^{k,m})$ over $x \in X^k$ is attained at $x^{k,m}$, implying that

$$\begin{aligned}
f(x^{k,m}) &+ \frac{(\gamma^k)^2}{4k^2}\|x^{k,m} - \overline{x}^k\|^2 + \xi^{k,m'} g(x^{k,m}) \\
&= \inf_{x \in X^k} \left\{ f(x) + \frac{(\gamma^k)^2}{4k^2}\|x - \overline{x}^k\|^2 + \xi^{k,m'} g(x) \right\} \\
&\leq \inf_{x \in X^k, \, g(x) \leq 0} \left\{ f(x) + \frac{(\gamma^k)^2}{4k^2}\|x - \overline{x}^k\|^2 + \xi^{k,m'} g(x) \right\} \qquad (6.37) \\
&\leq \inf_{x \in X^k, \, g(x) \leq 0} \left\{ f(x) + \frac{(\gamma^k)^2}{4k^2}\|x - \overline{x}^k\|^2 \right\} \\
&= f(\overline{x}^k).
\end{aligned}$$

Hence, we have

$$
\begin{aligned}
L_{k,m}(x^{k,m}, \xi^{k,m}) &= f(x^{k,m}) + \frac{(\gamma^k)^2}{4k^2}\|x^{k,m} - \overline{x}^k\|^2 + \xi^{k,m\prime}g(x^{k,m}) \\
&\quad - \frac{1}{2m}\|\xi^{k,m}\|^2 \\
&\leq f(x^{k,m}) + \frac{(\gamma^k)^2}{4k^2}\|x^{k,m} - \overline{x}^k\|^2 + \xi^{k,m\prime}g(x^{k,m}) \\
&\leq f(\overline{x}^k).
\end{aligned}
\tag{6.38}
$$

Since $L_{k,m}$ is quadratic in ξ, the supremum of $L_{k,m}(x^{k,m}, \xi)$ over $\xi \geq 0$ is attained at

$$
\xi_j^{k,m} = mg_j^+(x^{k,m}), \qquad j = 1, \ldots, r. \tag{6.39}
$$

This implies that

$$
\begin{aligned}
L_{k,m}(x^{k,m}, \xi^{k,m}) &= f(x^{k,m}) + \frac{(\gamma^k)^2}{4k^2}\|x^{k,m} - \overline{x}^k\|^2 + \frac{m}{2}\|g^+(x^{k,m})\|^2 \\
&\geq f(x^{k,m}).
\end{aligned}
\tag{6.40}
$$

We next show another property of the saddle points $(x^{k,m}, \xi^{k,m})$, namely that for each k, we have

$$
\lim_{m \to \infty} f(x^{k,m}) = f(\overline{x}^k) = f^* + \gamma^k. \tag{6.41}
$$

For a fixed k and any sequence of integers m that tends to ∞, consider the corresponding sequence $\{x^{k,m}\}$. From Eqs. (6.38) and (6.40), we see that $\{x^{k,m}\}$ belongs to the set $\{x \in X^k \mid f(x) \leq f(\overline{x}^k)\}$, which is compact, since f is closed. Hence, $\{x^{k,m}\}$ has a limit point, denoted by $\hat{x}^k$, which belongs to $\{x \in X^k \mid f(x) \leq f(\overline{x}^k)\}$. By passing to a subsequence if necessary, we can assume without loss of generality that $\{x^{k,m}\}$ converges to $\hat{x}^k$. We claim that $\hat{x}^k$ is feasible for problem (6.36), i.e., $\hat{x}^k \in X^k$ and $g(\hat{x}^k) \leq 0$. Indeed, the sequence $\{f(x^{k,m})\}$ is bounded from below by $\inf_{x \in X^k} f(x)$, which is finite by Weierstrass' Theorem since f is closed and coercive when restricted to X^k. Also, for each k, $L_{k,m}(x^{k,m}, \xi^{k,m})$ is bounded from above by $f(\overline{x}^k)$ [cf. Eq. (6.38)], so Eq. (6.40) implies that

$$
\limsup_{m \to \infty} g_j(x^{k,m}) \leq 0, \qquad \forall\, j = 1, \ldots, r.
$$

Therefore, by using the closedness of g_j, we obtain $g(\hat{x}^k) \leq 0$, implying that $\hat{x}^k$ is a feasible point of problem (6.36). Thus, $f(\hat{x}^k) \geq f(\overline{x}^k)$. Using Eqs. (6.38) and (6.40) together with the closedness of f, we also have

$$
f(\hat{x}^k) \leq \liminf_{m \to \infty} f(x^{k,m}) \leq \limsup_{m \to \infty} f(x^{k,m}) \leq f(\overline{x}^k),
$$

thereby showing Eq. (6.41).

The next step in the proof is given in the following lemma:

Lemma 6.6.1: For all sufficiently large k, and for all scalars $m \leq 1/\sqrt{\gamma^k}$, we have

$$f(x^{k,m}) \leq f^* - \frac{\gamma^k}{2}. \tag{6.42}$$

Furthermore, there exists a scalar $m_k \geq 1/\sqrt{\gamma^k}$ such that

$$f(x^{k,m_k}) = f^* - \frac{\gamma^k}{2}. \tag{6.43}$$

Proof: Let $\gamma = f^* - f(\overline{x})$, where $\overline{x}$ was defined earlier as the vector in X such that $f(\overline{x}) < f^*$. For sufficiently large k, we have $\overline{x} \in X^k$ and $\gamma^k < \gamma$. Consider the vector

$$z^k = \left(1 - \frac{2\gamma^k}{\gamma^k + \gamma}\right)\overline{x}^k + \frac{2\gamma^k}{\gamma^k + \gamma}\overline{x},$$

which belongs to X^k for sufficiently large k [by the convexity of X^k and the fact that $2\gamma^k/(\gamma^k + \gamma) < 1$]. By the convexity of f, we have

$$
\begin{aligned}
f(z^k) &\leq \left(1 - \frac{2\gamma^k}{\gamma^k + \gamma}\right) f(\overline{x}^k) + \frac{2\gamma^k}{\gamma^k + \gamma} f(\overline{x}) \\
&= \left(1 - \frac{2\gamma^k}{\gamma^k + \gamma}\right)(f^* + \gamma^k) + \frac{2\gamma^k}{\gamma^k + \gamma}(f^* - \gamma) \\
&= f^* - \gamma^k.
\end{aligned}
\tag{6.44}
$$

Similarly, by the convexity of g_j, we have

$$g_j(z^k) \leq \left(1 - \frac{2\gamma^k}{\gamma^k + \gamma}\right) g_j(\overline{x}^k) + \frac{2\gamma^k}{\gamma^k + \gamma} g_j(\overline{x}) \leq \frac{2\gamma^k}{\gamma^k + \gamma} g_j(\overline{x}). \tag{6.45}$$

Using Eq. (6.40), we obtain

$$
\begin{aligned}
f(x^{k,m}) &\leq L_{k,m}(x^{k,m}, \xi^{k,m}) \\
&= \inf_{x \in X^k} \sup_{\xi \geq 0} L_{k,m}(x, \xi) \\
&= \inf_{x \in X^k} \left\{ f(x) + \frac{(\gamma^k)^2}{4k^2}\|x - \overline{x}^k\|^2 + \frac{m}{2}\|g^+(x)\|^2 \right\} \\
&\leq f(x) + (\beta\gamma^k)^2 + \frac{m}{2}\|g^+(x)\|^2, \quad \forall\, x \in X^k,
\end{aligned}
$$

where in the last inequality we also use the definition of X^k so that $\|x - \bar{x}^k\| \leq 2\beta k$ for all $x \in X^k$. Substituting $x = z^k$ in the preceding relation, and using Eqs. (6.44) and (6.45), we see that for large k,

$$f(x^{k,m}) \leq f^* - \gamma^k + (\beta\gamma^k)^2 + \frac{2m(\gamma^k)^2}{(\gamma^k + \gamma)^2}\|g^+(\bar{x})\|^2.$$

Since $\gamma^k \to 0$, this implies that for sufficiently large k and for all scalars $m \leq 1/\sqrt{\gamma^k}$, we have

$$f(x^{k,m}) \leq f^* - \frac{\gamma^k}{2},$$

i.e., Eq. (6.42) holds.

We next show that there exists a scalar $m_k \geq 1/\sqrt{\gamma^k}$ such that Eq. (6.43) holds. In the process, we show that, for fixed k, $L_{k,m}(x^{k,m}, \xi^{k,m})$ changes continuously with m, i.e, for all $\overline{m} > 0$, we have $L_{k,m}(x^{k,m}, \xi^{k,m}) \to L_{k,\overline{m}}(x^{k,\overline{m}}, \xi^{k,\overline{m}})$ as $m \to \overline{m}$. [By this we mean, for every sequence $\{m^t\}$ that converges to $\overline{m}$, the corresponding sequence $L_{k,m^t}(x^{k,m^t}, \xi^{k,m^t})$ converges to $L_{k,\overline{m}}(x^{k,\overline{m}}, \xi^{k,\overline{m}})$.] Denote

$$\overline{f}(x) = f(x) + \frac{(\gamma^k)^2}{4k^2}\|x - \bar{x}^k\|^2.$$

Note that we have

$$L_{k,m}(x^{k,m}, \xi^{k,m}) = \overline{f}(x^{k,m}) + \frac{m}{2}\|g^+(x^{k,m})\|^2 = \inf_{x \in X^k}\left\{\overline{f}(x) + \frac{m}{2}\|g^+(x)\|^2\right\},$$

so that for all $m \geq \overline{m}$, we obtain

$$L_{k,\overline{m}}(x^{k,\overline{m}}, \xi^{k,\overline{m}}) = \overline{f}(x^{k,\overline{m}}) + \frac{\overline{m}}{2}\|g^+(x^{k,\overline{m}})\|^2$$
$$\leq \overline{f}(x^{k,m}) + \frac{\overline{m}}{2}\|g^+(x^{k,m})\|^2$$
$$\leq \overline{f}(x^{k,m}) + \frac{m}{2}\|g^+(x^{k,m})\|^2$$
$$\leq \overline{f}(x^{k,\overline{m}}) + \frac{m}{2}\|g^+(x^{k,\overline{m}})\|^2.$$

It follows that $L_{k,m}(x^{k,m}, \xi^{k,m}) \to L_{k,\overline{m}}(x^{k,\overline{m}}, \xi^{k,\overline{m}})$ as $m \downarrow \overline{m}$. Similarly, we have for all $m \leq \overline{m}$,

$$\overline{f}(x^{k,\overline{m}}) + \frac{m}{2}\|g^+(x^{k,\overline{m}})\|^2 \leq \overline{f}(x^{k,\overline{m}}) + \frac{\overline{m}}{2}\|g^+(x^{k,\overline{m}})\|^2$$
$$\leq \overline{f}(x^{k,m}) + \frac{\overline{m}}{2}\|g^+(x^{k,m})\|^2$$
$$= \overline{f}(x^{k,m}) + \frac{m}{2}\|g^+(x^{k,m})\|^2 + \frac{\overline{m} - m}{2}\|g^+(x^{k,m})\|^2$$
$$\leq \overline{f}(x^{k,\overline{m}}) + \frac{m}{2}\|g^+(x^{k,\overline{m}})\|^2 + \frac{\overline{m} - m}{2}\|g^+(x^{k,m})\|^2.$$

For each k, $g_j(x^{k,m})$ is bounded from below by $\inf_{x \in X^k} g_j(x)$, which is finite by Weierstrass' Theorem since g_j is closed and coercive when restricted to X^k. Therefore, we have from the preceding relation that $L_{k,m}(x^{k,m}, \xi^{k,m}) \to L_{k,\overline{m}}(x^{k,\overline{m}}, \xi^{k,\overline{m}})$ as $m \uparrow \overline{m}$, which shows that $L_{k,m}(x^{k,m}, \xi^{k,m})$ changes continuously with m.

Next, we show that, for fixed k, $x^{k,m} \to x^{k,\overline{m}}$ as $m \to \overline{m}$. Since, for each k, $x^{k,m}$ belongs to the compact set $\{x \in X^k \mid f(x) \le f(\overline{x}^k)\}$, it has a limit point as $m \to \overline{m}$. Let $\hat{x}$ be a limit point of $x^{k,m}$. Using the continuity of $L_{k,m}(x^{k,m}, \xi^{k,m})$ in m, and the closedness of f and g_j, we obtain

$$
\begin{aligned}
L_{k,\overline{m}}(x^{k,\overline{m}}, \xi^{k,\overline{m}}) &= \lim_{m \to \overline{m}} L_{k,m}(x^{k,m}, \xi^{k,m}) \\
&= \lim_{m \to \overline{m}} \left\{ \overline{f}(x^{k,m}) + \frac{m}{2} \|g^+(x^{k,m})\|^2 \right\} \\
&\ge \overline{f}(\hat{x}) + \frac{\overline{m}}{2} \|g^+(\hat{x})\|^2 \\
&\ge \inf_{x \in X^k} \left\{ \overline{f}(x) + \frac{\overline{m}}{2} \|g^+(x)\|^2 \right\} \\
&= L_{k,\overline{m}}(x^{k,\overline{m}}, \xi^{k,\overline{m}}).
\end{aligned}
$$

This shows that $\hat{x}$ attains the infimum of $\overline{f}(x) + \frac{\overline{m}}{2} \|g^+(x)\|^2$ over $x \in X^k$. Since this function is strictly convex, it has a unique optimal solution, showing that $\hat{x} = x^{k,\overline{m}}$.

Finally, we show that $f(x^{k,m}) \to f(x^{k,\overline{m}})$ as $m \to \overline{m}$. Assume that $f(x^{k,\overline{m}}) < \limsup_{m \to \overline{m}} f(x^{k,m})$. Using the continuity of $L_{k,m}(x^{k,m}, \xi^{k,m})$ in m and the fact that $x^{k,m} \to x^{k,\overline{m}}$ as $m \to \overline{m}$, we have

$$
\begin{aligned}
\overline{f}(x^{k,\overline{m}}) + \liminf_{m \to \overline{m}} \|g^+(x^{k,m})\|^2 &< \limsup_{m \to \overline{m}} L_{k,m}(x^{k,m}, \xi^{k,m}) \\
&= L_{k,\overline{m}}(x^{k,\overline{m}}, \xi^{k,\overline{m}}) \\
&= \overline{f}(x^{k,\overline{m}}) + \|g^+(x^{k,\overline{m}})\|^2.
\end{aligned}
$$

But this contradicts the lower semicontinuity of g_j, hence showing that $f(x^{k,\overline{m}}) \ge \limsup_{m \to \overline{m}} f(x^{k,m})$, which together with the lower semicontinuity of f yields the desired result. Thus $f(x^{k,m})$ is continuous in m.

From Eqs. (6.41), (6.42), and the continuity of $f(x^{k,m})$ in m, we see that there exists some scalar $m_k \ge 1/\sqrt{\gamma^k}$ such that Eq. (6.43) holds. **Q.E.D.**

We are now ready to construct Fritz John multipliers with the desired properties. By combining Eqs. (6.43), (6.38), and (6.40) (for $m = m_k$), together with the facts that $f(\overline{x}^k) \to f^*$ and $\gamma^k \to 0$ as $k \to \infty$, we obtain

$$
\lim_{k \to \infty} \left(f(x^{k,m_k}) - f^* + \frac{(\gamma^k)^2}{4k^2} \|x^{k,m_k} - \overline{x}^k\|^2 + \xi^{k,m'_k} g(x^{k,m_k}) \right) = 0.
$$

$$
(6.46)
$$

Denote

$$\delta^k = \sqrt{1 + \sum_{j=1}^{r}(\xi_j^{k,m_k})^2},$$

$$\mu_0^k = \frac{1}{\delta^k}, \qquad \mu_j^k = \frac{\xi_j^{k,m_k}}{\delta^k}, \qquad j = 1, \dots, r. \tag{6.47}$$

Since δ^k is bounded from below, Eq. (6.46) yields

$$\lim_{k \to \infty} \left(\mu_0^k f(x^{k,m_k}) - \mu_0^k f^* + \frac{(\gamma^k)^2}{4k^2\delta^k} \|x^{k,m_k} - \bar{x}^k\|^2 + \sum_{j=1}^{r} \mu_j^k g_j(x^{k,m_k}) \right) = 0.$$

$$\tag{6.48}$$

Substituting $m = m_k$ in the first relation of Eq. (6.37) and dividing by δ^k, we obtain

$$\mu_0^k f(x^{k,m_k}) + \frac{(\gamma^k)^2}{4k^2\delta^k} \|x^{k,m_k} - \bar{x}^k\|^2 + \sum_{j=1}^{r} \mu_j^k g_j(x^{k,m_k})$$

$$\leq \mu_0^k f(x) + \sum_{j=1}^{r} \mu_j^k g_j(x) + \frac{(\beta\gamma^k)^2}{\delta^k}, \qquad \forall\, x \in X^k,$$

where we also use the fact that $\|x - \bar{x}^k\| \leq 2\beta k$ for all $x \in X^k$ (cf. the definition of X^k). Since the sequence $\{\mu_0^k, \mu_1^k, \dots, \mu_r^k\}$ is bounded, it has a limit point, denoted by $\{\mu_0^*, \mu_1^*, \dots, \mu_r^*\}$, which satisfies conditions (ii) and (iii) of the proposition. Without loss of generality, we will assume that the entire sequence $\{\mu_0^k, \mu_1^k, \dots, \mu_r^k\}$ converges to $\{\mu_0^*, \mu_1^*, \dots, \mu_r^*\}$. Taking the limit as $k \to \infty$ in the preceding relation together with Eq. (6.48) yields

$$\mu_0^* f^* \leq \mu_0^* f(x) + \mu^{*\prime} g(x), \qquad \forall\, x \in X,$$

which implies that

$$\mu_0^* f^* \leq \inf_{x \in X} \left\{ \mu_0^* f(x) + \mu^{*\prime} g(x) \right\}$$

$$\leq \inf_{x \in X,\, g(x) \leq 0} \left\{ \mu_0^* f(x) + \mu^{*\prime} g(x) \right\}$$

$$\leq \inf_{x \in X,\, g(x) \leq 0} \mu_0^* f(x)$$

$$= \mu_0^* f^*.$$

Thus we have

$$\mu_0^* f^* = \inf_{x \in X} \left\{ \mu_0^* f(x) + \mu^{*\prime} g(x) \right\},$$

so that $\mu_0^*, \mu_1^*, \dots, \mu_r^*$ satisfy condition (i) of the proposition.

Finally, we use the relation

$$\xi_j^{k,m_k} = m_k g_j^+(x^{k,m_k}), \qquad j = 1, \ldots, r,$$

[cf. Eq. (6.39)] to establish condition (iv). Dividing by δ^k, and using Eq. (6.47) and the fact that $\mu_j^k \to \mu_j^*$, as $k \to \infty$, we obtain

$$\mu_j^* = \lim_{k \to \infty} \frac{m_k g_j^+(x^{k,m_k})}{\delta^k}, \qquad j = 1, \ldots, r.$$

Note that the existence of $\bar{x} \in X$ such that $f(\bar{x}) < f^*$, together with condition (i), imply that $\mu_1^*, \ldots, \mu_r^*$ cannot all be equal to 0, i.e. $J \neq \emptyset$. From Eqs. (6.38) and (6.40) with $m = m_k$, we have

$$\frac{m_k}{2} \|g^+(x^{k,m_k})\|^2 \leq f(\bar{x}^k).$$

Since $f(\bar{x}^k) \to f^*$ and $m_k \geq 1/\sqrt{\gamma^k} \to \infty$, this yields $g^+(x^{k,m_k}) \to 0$. Since, we also have from Eq. (6.43) that

$$f(x^{k,m_k}) < f^*, \qquad \lim_{k \to \infty} f(x^{k,m_k}) = f^*,$$

it follows that the sequence $\{x^{k,m_k}\}$ satisfies condition (iv) of the proposition, thereby completing the proof.   **Q.E.D.**

The Fritz John multipliers of Props. 6.6.1-6.6.3 define a hyperplane with normal (μ^*, μ_0^*) that supports the set of constraint-cost pairs (i.e., the set M of Fig. 6.6.1) at $(0, f^*)$. On the other hand, it is possible to construct a hyperplane that supports the set M at the point $(0, q^*)$, where q^* is the optimal dual value

$$q^* = \sup_{\mu \geq 0} q(\mu) = \sup_{\mu \geq 0} \inf_{x \in X} \{f(x) + \mu' g(x)\},$$

while asserting the existence of a sequence that satisfies a condition analogous to the CV condition. This is the subject of the next proposition.

Proposition 6.6.4: (Enhanced Dual Fritz John Conditions)
Consider problem (6.24) under Assumption 6.6.1 (Closedness and Convexity), and assume that $-\infty < q^*$. Then there exists a scalar μ_0^* and a vector $\mu^* = (\mu_1^*, \ldots, \mu_r^*)$, satisfying the following conditions:

(i) $\mu_0^* q^* = \inf_{x \in X} \{\mu_0^* f(x) + \mu^{*\prime} g(x)\}$.

(ii) $\mu_j^* \geq 0$ for all $j = 0, 1, \ldots, r$.

(iii) $\mu_0^*, \mu_1^*, \ldots, \mu_r^*$ are not all equal to 0.

(iv) If the index set $J = \{j \neq 0 \mid \mu_j^* > 0\}$ is nonempty, there exists a sequence $\{x^k\} \subset X$ such that

$$\lim_{k \to \infty} f(x^k) = q^*, \qquad \limsup_{k \to \infty} g(x^k) \leq 0,$$

and for all k,

$$f(x^k) < q^*,$$

$$g_j(x^k) > 0, \quad \forall \, j \in J, \qquad g_j^+(x^k) = o\left(\min_{j \in J} g_j(x^k)\right), \quad \forall \, j \notin J.$$

Proof: We first show the following lemma.

Lemma 6.6.2: Consider problem (6.24), and assume that the set X is convex, the functions f and g_j are convex over X, and $f^* < \infty$, where f^* is the optimal value. For each $\delta > 0$, let

$$f^\delta = \inf_{\substack{x \in X \\ g_j(x) \leq \delta, \, j = 1, \dots, r}} f(x). \qquad (6.49)$$

Then the dual optimal value q^* satisfies $f^\delta \leq q^*$ for all $\delta > 0$ and

$$q^* = \lim_{\delta \downarrow 0} f^\delta.$$

Proof: We first note that either $\lim_{\delta \downarrow 0} f^\delta$ exists and is a scalar, or else $\lim_{\delta \downarrow 0} f^\delta = -\infty$, since f^δ is monotonically nondecreasing as $\delta \downarrow 0$, and $f^\delta \leq f^*$ for all $\delta > 0$. Since $f^* < \infty$, there exists some $\bar{x} \in X$ such that $g(\bar{x}) \leq 0$. Thus, for each $\delta > 0$ such that $f^\delta > -\infty$, the Slater condition is satisfied for problem (6.49), and by Prop. 6.4.3, there exists a nonnegative geometric multiplier μ^δ such that

$$\begin{aligned}
f^\delta &= \inf_{x \in X} \left\{ f(x) + {\mu^\delta}' g(x) - \delta \sum_{j=1}^r \mu_j^\delta \right\} \\
&\leq \inf_{x \in X} \left\{ f(x) + {\mu^\delta}' g(x) \right\} \\
&= q(\mu^\delta) \\
&\leq q^*.
\end{aligned}$$

For each $\delta > 0$ such that $f^\delta = -\infty$, we also have $f^\delta \leq q^*$, so that

$$f^\delta \leq q^*, \qquad \forall \, \delta > 0.$$

By taking the limit as $\delta \downarrow 0$, we obtain

$$\lim_{\delta \downarrow 0} f^\delta \leq q^*.$$

To show the reverse inequality, we consider two cases: (1) $f^\delta > -\infty$ for all $\delta > 0$ that are sufficiently small, and (2) $f^\delta = -\infty$ for all $\delta > 0$. In case (1), for each $\delta > 0$ with $f^\delta > -\infty$, choose $x^\delta \in X$ such that $g_j(x^\delta) \leq \delta$ for all j and $f(x^\delta) \leq f^\delta + \delta$. Then, for any $\mu \geq 0$,

$$q(\mu) = \inf_{x \in X} \left\{ f(x) + \mu'g(x) \right\} \leq f(x^\delta) + \mu'g(x^\delta) \leq f^\delta + \delta + \delta \sum_{j=1}^{r} \mu_j.$$

Taking the limit as $\delta \downarrow 0$, we obtain

$$q(\mu) \leq \lim_{\delta \downarrow 0} f^\delta,$$

so that $q^* \leq \lim_{\delta \downarrow 0} f^\delta$. In case (2), choose $x^\delta \in X$ such that $g_j(x^\delta) \leq \delta$ for all j and $f(x^\delta) \leq -1/\delta$. Then, similarly, for any $\mu \geq 0$, we have

$$q(\mu) \leq f(x^\delta) + \mu'g(x^\delta) \leq -\frac{1}{\delta} + \delta \sum_{j=1}^{r} \mu_j,$$

so by taking $\delta \downarrow 0$, we obtain $q(\mu) = -\infty$ for all $\mu \geq 0$, and hence also $q^* = -\infty = \lim_{\delta \downarrow 0} f^\delta$. **Q.E.D.**

We now return to the proof of the proposition. Since by assumption, we have $-\infty < q^*$ and $f^* < \infty$, it follows from the weak duality relation $q^* \leq f^*$ that both q^* and f^* are finite. For $k = 1, 2, \ldots$, consider the problem

$$\text{minimize} \quad f(x)$$
$$\text{subject to} \quad x \in X, \ g_j(x) \leq \frac{1}{k^4}, \quad j = 1, \ldots, r.$$

By the preceding lemma, for each k, the optimal value of this problem is less than or equal to q^*. For each k, let $\tilde{x}^k \in X$ be a vector that satisfies

$$f(\tilde{x}^k) \leq q^* + \frac{1}{k^2}, \qquad g_j(\tilde{x}^k) \leq \frac{1}{k^4}, \quad j = 1, \ldots, r.$$

Consider also the problem

$$\text{minimize} \quad f(x)$$
$$\text{subject to} \quad g_j(x) \leq \frac{1}{k^2}, \quad j = 1, \ldots, r, \tag{6.50}$$
$$x \in \tilde{X}^k = X \cap \left\{ x \ \Big| \ \|x\| \leq k \left(\max_{1 \leq i \leq k} \|\tilde{x}^i\| + 1 \right) \right\}.$$

Since f and g_j are closed and convex when restricted to X, they are closed, convex, and coercive when restricted to $\tilde{X}^k$. Hence, problem (6.50) has an optimal solution, which we denote by $\overline{x}^k$. Note that since $\tilde{x}^k$ belongs to the feasible set of this problem, we have

$$f(\overline{x}^k) \le f(\tilde{x}^k) \le q^* + \frac{1}{k^2}. \tag{6.51}$$

For each k, we consider the function

$$L_k(x, \xi) = f(x) + \xi' g(x) - \frac{\|\xi\|^2}{2k},$$

and the set

$$X^k = \tilde{X}^k \cap \{x \mid g_j(x) \le k, \, j = 1, \ldots, r\}. \tag{6.52}$$

We note that $L_k(x, \xi)$, for fixed $\xi \ge 0$, is closed, convex, and coercive in x, when restricted to X^k, and negative definite quadratic in ξ for fixed x. Hence, using the Saddle Point Theorem (Prop. 2.6.9), we can assert that L_k has a saddle point over $x \in X^k$ and $\xi \ge 0$, denoted by (x^k, ξ^k).

Since L_k is quadratic in ξ, the supremum of $L_k(x^k, \xi)$ over $\xi \ge 0$ is attained at

$$\xi_j^k = k g_j^+(x^k), \qquad j = 1, \ldots, r. \tag{6.53}$$

Similarly, the infimum of $L_k(x, \xi^k)$ over $x \in X^k$ is attained at x^k, implying that

$$
\begin{aligned}
f(x^k) + \xi^{k'} g(x^k) &= \inf_{x \in X^k} \left\{ f(x) + \xi^{k'} g(x) \right\} \\
&= \inf_{x \in X^k} \left\{ f(x) + k g^+(x^k)' g(x) \right\} \\
&\le \inf_{x \in X^k, \, g_j(x) \le \frac{1}{k^4}, j=1,\ldots,r} \left\{ f(x) + k \sum_{j=1}^{r} g_j^+(x^k)' g_j(x) \right\} \\
&\le \inf_{x \in X^k, \, g_j(x) \le \frac{1}{k^4}, j=1,\ldots,r} \left\{ f(x) + \frac{r}{k^2} \right\} \\
&= f(\overline{x}^k) + \frac{r}{k^2} \\
&\le q^* + \frac{r+1}{k^2},
\end{aligned}
$$

$$\tag{6.54}$$

where the second inequality follows using the fact $g_j^+(x^k) \le k$, $j = 1, \ldots, r$ [cf. Eq. (6.52)], and the third inequality follows from Eq. (6.51).

Since q^* is finite, we may select a nonnegative sequence $\{\zeta^k\}$ such that

$$q(\zeta^k) \to q^*, \qquad \frac{\|\zeta^k\|^2}{2k} \to 0. \tag{6.55}$$

[For example, we can take ζ^k to be any maximizer of $q(\zeta)$ subject to $\zeta \geq 0$ and $\|\zeta\| \leq k^{1/3}$.] Then, we have for all k,

$$
\begin{aligned}
L_k(x^k, \xi^k) &= \sup_{\xi \geq 0} \inf_{x \in X^k} L_k(x, \xi) \\
&\geq \sup_{\xi \geq 0} \inf_{x \in X} L_k(x, \xi) \\
&= \sup_{\xi \geq 0} \left\{ \inf_{x \in X} \{ f(x) + \xi' g(x) \} - \frac{\|\xi\|^2}{2k} \right\} \\
&= \sup_{\xi \geq 0} \left\{ q(\xi) - \frac{\|\xi\|^2}{2k} \right\} \\
&\geq q(\zeta^k) - \frac{\|\zeta^k\|^2}{2k}.
\end{aligned}
\tag{6.56}
$$

Combining Eqs. (6.56) and (6.54), we obtain

$$
\begin{aligned}
q(\zeta^k) - \frac{\|\zeta^k\|^2}{2k} &\leq L_k(x^k, \xi^k) \\
&= f(x^k) + \xi^{k'} g(x^k) - \frac{\|\xi^k\|^2}{2k} \\
&\leq f(x^k) + \xi^{k'} g(x^k) \\
&\leq q^* + \frac{r+1}{k^2}.
\end{aligned}
\tag{6.57}
$$

Taking the limit in the preceding relation, and using Eq. (6.55), we obtain

$$
\lim_{k \to \infty} \left\{ f(x^k) - q^* + \xi^{k'} g(x^k) \right\} = 0.
\tag{6.58}
$$

Denote

$$
\delta^k = \sqrt{1 + \sum_{j=1}^{r} (\xi_j^k)^2},
$$

$$
\mu_0^k = \frac{1}{\delta^k}, \qquad \mu_j^k = \frac{\xi_j^k}{\delta^k}, \quad j = 1, \ldots, r.
\tag{6.59}
$$

Since δ^k is bounded from below, Eq. (6.58) yields

$$
\lim_{k \to \infty} \left\{ \mu_0^k \big(f(x^k) - q^* \big) + \sum_{j=1}^{r} \mu_j^k g_j(x^k) \right\} = 0.
\tag{6.60}
$$

Dividing both sides of the first relation in Eq. (6.54) by δ^k, we get

$$
\mu_0^k f(x^k) + \sum_{j=1}^{r} \mu_j^k g_j(x^k) \leq \mu_0^k f(x) + \sum_{j=1}^{r} \mu_j^k g_j(x), \qquad \forall \, x \in X^k.
$$

Since the sequence $\{\mu_0^k, \mu_1^k, \ldots, \mu_r^k\}$ is bounded, it has a limit point, denoted by $\{\mu_0^*, \mu_1^*, \ldots, \mu_r^*\}$. This limit point satisfies conditions (ii) and (iii) of the proposition. Without loss of generality, we assume that the entire sequence converges. Taking the limit as $k \to \infty$ in the preceding relation and using Eq. (6.60) yields

$$\mu_0^* q^* \leq \inf_{x \in X} \left\{ \mu_0^* f(x) + \sum_{j=1}^r \mu_j^* g_j(x) \right\}.$$

We consider separately the two cases, $\mu_0^* > 0$ and $\mu_0^* = 0$, in the above relation to show that $\mu_0^*, \ldots, \mu_r^*$ satisfy condition (i) of the proposition. Indeed, if $\mu_0^* > 0$, by dividing with μ_0^*, we have

$$q^* \leq \inf_{x \in X} \left\{ f(x) + \sum_{j=1}^r \frac{\mu_j^*}{\mu_0^*} g_j(x) \right\} = q\left(\frac{\mu^*}{\mu_0^*}\right) \leq q^*.$$

Similarly, if $\mu_0^* = 0$, it can be seen that

$$0 = \inf_{x \in X} \mu^{*'} g(x)$$

[since $f^* < \infty$, so that there exists an $x \in X$ such that $g(x) \leq 0$ and $\mu^{*'} g(x) \leq 0$]. Hence, in both cases, we have

$$\mu_0^* q^* = \inf_{x \in X} \left\{ \mu_0^* f(x) + \sum_{j=1}^r \mu_j^* g_j(x) \right\},$$

thus showing condition (i).

Assume that the index set

$$J = \{j \neq 0 \mid \mu_j^* > 0\}$$

is nonempty. Dividing both sides of Eq. (6.53) by δ^k, and using Eq. (6.59) and the fact that $\mu_j^k \to \mu_j^*$, for all $j = 1, \ldots, r$, we obtain

$$\lim_{k \to \infty} \frac{k g_j^+(x^k)}{\delta^k} = \mu_j^*, \qquad j = 1, \ldots, r.$$

This implies that for all sufficiently large k,

$$g_j(x^k) > 0, \qquad \forall \, j \in J,$$

and

$$g_j(x^k) = o\left(\min_{j \in J} g_j(x^k)\right), \qquad \forall \, j \notin J.$$

Note also that from Eq. (6.57) we have

$$k\big(f(x^k) - q^*\big) + \xi^{k\prime} kg(x^k) \leq \frac{r+1}{k}, \qquad \forall\, k = 1, 2, \ldots.$$

Using Eq. (6.53), this yields

$$k\big(f(x^k) - q^*\big) + \sum_{j=1}^{r} (\xi_j^k)^2 \leq \frac{r+1}{k}.$$

Dividing by $(\delta^k)^2$ and taking the limit, we obtain, using also Eq. (6.59),

$$\limsup_{k \to \infty} \frac{k\big(f(x^k) - q^*\big)}{(\delta^k)^2} \leq -\sum_{j=1}^{r} (\mu_j^*)^2, \qquad (6.61)$$

implying that $f(x^k) < q^*$ for all sufficiently large k, since the index set J is nonempty.

We finally show that $f(x^k) \to q^*$ and $\limsup_{k \to \infty} g(x^k) \leq 0$. By Eqs. (6.57) and (6.55), we have

$$\lim_{k \to \infty} \frac{\|\xi^k\|^2}{2k} = 0. \qquad (6.62)$$

By Eq. (6.53), we have

$$\xi^{k\prime} g(x^k) = \frac{1}{k} \|\xi^k\|^2,$$

so using also Eqs. (6.57) and (6.55), we obtain

$$\lim_{k \to \infty} \big(f(x^k) - q^*\big) + \frac{\|\xi^k\|^2}{2k} = 0,$$

which together with Eq. (6.62) shows that $f(x^k) \to q^*$. Moreover, Eqs. (6.62) and (6.53) imply that

$$\lim_{k \to \infty} k \sum_{j=1}^{r} \big(g_j^+(x^k)\big)^2 = 0,$$

showing that $\limsup_{k \to \infty} g(x^k) \leq 0$. Therefore, the sequence $\{x^k\}$ satisfies condition (iv) of the proposition, completing the proof.   **Q.E.D.**

Note that the proof of Prop. 6.6.4 is similar to the proof of Prop. 5.7.2. The idea is to generate saddle points of the function

$$L_k(x, \xi) = f(x) + \xi' g(x) - \frac{\|\xi\|^2}{2k},$$

over $x \in X^k$ [cf. Eq. (6.52)] and $\xi \geq 0$. It can be shown that

$$L_k(x, \xi) = \inf_{u \in \Re^r} \left\{ p^k(u) + \frac{k}{2} \|u^+\|^2 \right\},$$

where $p^k(u)$ is the optimal value of the problem

$$\text{minimize} \quad f(x)$$
$$\text{subject to} \quad g(x) \leq u, \ x \in X^k, \tag{6.63}$$

(see the discussion following the proof of Prop. 5.7.2 in Chapter 5). For each k, the value $L_k(x^k, \xi^k)$ can be visualized geometrically as in Fig. 5.7.1. However, here the rate at which X^k approaches X is chosen high enough so that $L_k(x^k, \xi^k)$ converges to q^* as $k \to \infty$ [cf. Eq. (6.57)], and not to f^*, as in the proof of Prop. 5.7.2 of Chapter 5.

6.6.2   Informative Geometric Multipliers

We will now focus on geometric multipliers and dual optimal solutions, which are special in that they satisfy conditions analogous to the CV condition. Consistent with our analysis in Chapter 5, we refer to such multipliers as being *informative*, since they provide sensitivity information by indicating the constraints to violate in order to effect a cost reduction.

Definition 6.6.2: A vector $\mu^* \geq 0$ is said to be an *informative geometric multiplier* if the following two conditions hold:

(i) $f^* = \inf_{x \in X} \left\{ f(x) + \mu^{*\prime} g(x) \right\}$.

(ii) If the index set $J = \{ j \neq 0 \mid \mu_j^* > 0 \}$ is nonempty, there exists a sequence $\{x^k\} \subset X$ such that

$$\lim_{k \to \infty} f(x^k) = f^*, \qquad \limsup_{k \to \infty} g(x^k) \leq 0,$$

and for all k,

$$f(x^k) < f^*,$$

$$g_j(x^k) > 0, \quad \forall \, j \in J, \qquad g_j^+(x^k) = o\left(\min_{j \in J} g_j(x^k) \right), \quad \forall \, j \notin J.$$

In the next proposition, assuming there exists at least one geometric multiplier, we show the existence of an informative geometric multiplier under very general assumptions.

> **Proposition 6.6.5: (Existence of Informative Geometric Multipliers)** Consider problem (6.24) under Assumption 6.6.1 (Closedness and Convexity). If the set of geometric multipliers is nonempty, then the vector of minimum norm in this set is an informative geometric multiplier.

Proof: If $f^* = -\infty$, the vector $\mu^* = 0$ is the geometric multiplier of minimum norm, and is also informative. We may thus assume that $f^* > -\infty$, in which case f^* is finite, in view of the Closedness and Convexity Assumption. Let $\{\tilde{x}^k\}$ be a sequence of feasible points for problem (6.24) such that $f(\tilde{x}^k) \to f^*$ and

$$f(\tilde{x}^k) \le f^* + \frac{1}{k^2}, \qquad k = 1, 2, \ldots$$

Consider the problem

$$\text{minimize} \quad f(x)$$
$$\text{subject to} \quad g(x) \le 0,$$
$$x \in X^k = X \cap \left\{ x \;\Big|\; \|x\| \le k \left(\max_{1 \le i \le k} \|\tilde{x}^i\| + 1 \right) \right\}.$$

By the Closedness and Convexity Assumption, f and g_j are closed and convex when restricted to X, so they are closed, convex, and coercive when restricted to X^k. Thus the above problem has an optimal solution, which is denoted by $\overline{x}^k$. Note that since $\tilde{x}^k$ belongs to the feasible set of this problem, we have

$$f(\overline{x}^k) \le f(\tilde{x}^k) \le f^* + \frac{1}{k^2}. \tag{6.64}$$

For each k, we consider the function

$$L_k(x, \mu) = f(x) + \mu' g(x) - \frac{1}{2k} \|\mu\|^2,$$

and we note that $L_k(x, \mu)$, for fixed $\mu \ge 0$, is closed, convex, and coercive in x, when restricted to X^k, and negative definite quadratic in μ for fixed x. Hence, using the Saddle Point Theorem (Prop. 2.6.9), we can assert that L_k has a saddle point over $x \in X^k$ and $\mu \ge 0$, denoted by (x^k, μ^k).

Let M be the set of geometric multipliers,

$$M = \left\{ \mu \ge 0 \;\Big|\; f^* = \inf_{x \in X} \{ f(x) + \mu' g(x) \} \right\},$$

and let μ^* be the vector of minimum norm in M. If $\mu^* = 0$, then μ^* is an informative geometric multiplier and we are done, so assume that $\mu^* \neq 0$. For any $\mu \in M$, we have

$$\inf_{x \in X} \{f(x) + \mu'g(x)\} = f^*,$$

so that

$$\inf_{x \in X^k} L_k(x, \mu) \geq \inf_{x \in X} L_k(x, \mu) = f^* - \frac{1}{2k}\|\mu\|^2.$$

Therefore,

$$\begin{aligned}
L_k(x^k, \mu^k) &= \sup_{\mu \geq 0} \inf_{x \in X^k} L_k(x, \mu) \\
&\geq \sup_{\mu \in M} \inf_{x \in X^k} L_k(x, \mu) \\
&\geq \sup_{\mu \in M} \left(f^* - \frac{1}{2k}\|\mu\|^2\right) \\
&= f^* - \frac{1}{2k}\|\mu^*\|^2.
\end{aligned} \tag{6.65}$$

Since (x^k, μ^k) is a saddle point of L_k over $x \in X^k$ and $\mu \geq 0$, the infimum in the left hand side of

$$\inf_{x \in X^k} L_k(x, \mu^k) = \inf_{x \in X^k} \{f(x) + \mu^{k'}g(x)\} - \frac{1}{2k}\|\mu^k\|^2$$

is attained at x^k, implying that

$$\begin{aligned}
f(x^k) + \mu^{k'}g(x^k) &= \inf_{x \in X^k} \{f(x) + \mu^{k'}g(x)\} \\
&\leq \inf_{x \in X^k, \ g(x) \leq 0} \{f(x) + \mu^{k'}g(x)\} \\
&\leq \inf_{x \in X^k, \ g(x) \leq 0} f(x) \\
&= f(\overline{x}^k) \\
&\leq f^* + \frac{1}{k^2},
\end{aligned} \tag{6.66}$$

where the last inequality follows from Eq. (6.64). Combining Eqs. (6.65) and (6.66), we obtain

$$\begin{aligned}
f^* - \frac{1}{2k}\|\mu^*\|^2 &\leq L_k(x^k, \mu^k) \\
&= f(x^k) + \mu^{k'}g(x^k) - \frac{1}{2k}\|\mu^k\|^2 \\
&\leq f^* + \frac{1}{k^2} - \frac{1}{2k}\|\mu^k\|^2.
\end{aligned} \tag{6.67}$$

This relation shows that $\|\mu^k\|^2 \leq \|\mu^*\|^2 + 2/k$, so the sequence $\{\mu^k\}$ is bounded. Let $\bar{\mu}$ be a limit point of $\{\mu^k\}$, and without loss of generality, assume that the entire sequence $\{\mu^k\}$ converges to $\bar{\mu}$. It follows from the preceding relation that

$$\lim_{k \to \infty} \left\{ f(x^k) + \mu^{k'} g(x^k) \right\} = f^*,$$

so taking the limit as $k \to \infty$ in Eq. (6.66) yields

$$f^* \leq \inf_{x \in X} \left\{ f(x) + \bar{\mu}' g(x) \right\} = q(\bar{\mu}) \leq f^*,$$

where the last inequality follows from the weak duality relation. Hence $\bar{\mu}$ belongs to the set M, and since $\|\bar{\mu}\| \leq \|\mu^*\|$ [which follows by taking the limit in Eq. (6.67)], by using the minimum norm property of μ^*, we conclude that any limit point $\bar{\mu}$ of μ^k must be equal to μ^*. Thus $\mu^k \to \mu^*$, and using Eq. (6.67), we obtain

$$\lim_{k \to \infty} k \left(L_k(x^k, \mu^k) - f^* \right) = -\frac{1}{2} \|\mu^*\|^2. \tag{6.68}$$

Since L_k is quadratic in μ, the supremum in $\sup_{\mu \geq 0} L_k(x^k, \mu)$ is attained at

$$\mu_j^k = k g_j(x^k)^+, \qquad j = 1, \ldots, r, \tag{6.69}$$

so that

$$\mu^{k'} g(x^k) = \frac{1}{k} \|\mu^k\|^2$$

and hence

$$L_k(x^k, \mu^k) = \sup_{\mu \geq 0} L_k(x^k, \mu)$$

$$= f(x^k) + \mu^{k'} g(x^k) - \frac{1}{2k} \|\mu^k\|^2$$

$$= f(x^k) + \frac{1}{2k} \|\mu^k\|^2.$$

Combining this relation with Eq. (6.68) and the fact that $\mu^k \to \mu^*$, we obtain

$$\lim_{k \to \infty} k \left(f(x^k) - f^* \right) = -\|\mu^*\|^2,$$

implying that $f(x^k) < q^*$ for all sufficiently large k, since $\mu^* \neq 0$. We have $\mu^k \to \mu^*$, so Eq. (6.69) implies that

$$\lim_{k \to \infty} k g_j^+(x^k) = \mu_j^*, \qquad j = 1, \ldots, r.$$

It follows that the sequence $\{x^k\}$ fulfills condition (ii) of the definition of informative geometric multiplier, thereby completing the proof. **Q.E.D.**

When there is a duality gap, there exists no geometric multiplier, even if there is a dual optimal solution. In this case, we are motivated to investigate the existence of a special dual optimal solution, which satisfies condition (iv) of Prop. 6.6.4. We call such a dual optimal solution *informative*, since it indicates the constraints to violate in order to obtain a cost reduction by an amount that is strictly greater than the size of the duality gap $f^* - q^*$.

Definition 6.6.3: A vector μ^* is said to be an *informative dual optimal solution* if the following two conditions hold:

(i) $q(\mu^*) = q^*$.

(ii) If the index set $J = \{j \neq 0 \mid \mu_j^* > 0\}$ is nonempty, there exists a sequence $\{x^k\} \subset X$ such that

$$\lim_{k \to \infty} f(x^k) = q^*, \qquad \limsup_{k \to \infty} g(x^k) \leq 0,$$

and for all k,

$$f(x^k) < q^*,$$

$$g_j(x^k) > 0, \quad \forall\, j \in J, \qquad g_j^+(x^k) = o\left(\min_{j \in J} g_j(x^k)\right), \quad \forall\, j \notin J.$$

We have the following result, which provides a dual counterpart of the preceding proposition.

Proposition 6.6.6: (Existence of Informative Dual Optimal Solutions) Consider problem (6.24) under Assumption 6.6.1 (Closedness and Convexity), and assume that $-\infty < q^*$. If the set of dual optimal solutions is nonempty, then the vector of minimum norm in this set is an informative dual optimal solution.

Proof: Let μ^* be the dual optimal solution of minimum norm. If $\mu^* = 0$, then μ^* is an informative dual optimal solution and we are done, so assume that $\mu^* \neq 0$. Consider the problem

$$\text{minimize} \quad f(x)$$

$$\text{subject to} \quad x \in X, \quad g_j(x) \leq \frac{1}{k^4}, \; j = 1, \ldots, r.$$

By Lemma 6.6.2, for each k, the optimal value of this problem is less than or equal to q^*. Since q^* is finite (in view of the assumptions $-\infty < q^*$ and $f^* < \infty$, and the weak duality relation $q^* \le f^*$), we may select for each k, a vector $\tilde{x}^k \in X$ that satisfies

$$f(\tilde{x}^k) \le q^* + \frac{1}{k^2}, \qquad g_j(\tilde{x}^k) \le \frac{1}{k^4}, \ j = 1, \ldots, r.$$

Consider also the problem

$$\text{minimize} \quad f(x)$$
$$\text{subject to} \quad g_j(x) \le \frac{1}{k^4}, \ j = 1, \ldots, r,$$
$$x \in \tilde{X}^k = X \cap \left\{ x \ \middle| \ \|x\| \le k \left(\max_{1 \le i \le k} \|\tilde{x}^i\| + 1 \right) \right\}.$$

By the Closedness and Convexity Assumption, f and g_j are closed and convex when restricted to X, so they are closed, convex, and coercive when restricted to $\tilde{X}^k$. Thus, the problem has an optimal solution, which we denote by $\overline{x}^k$. Note that since $\tilde{x}^k$ belongs to the feasible set of this problem, we have

$$f(\overline{x}^k) \le f(\tilde{x}^k) \le q^* + \frac{1}{k^2}. \tag{6.70}$$

For each k, we consider the function

$$L_k(x, \mu) = f(x) + \mu'g(x) - \frac{\|\mu\|^2}{2k},$$

and the set

$$X^k = \tilde{X}^k \cap \left\{ x \mid g_j(x) \le k, \ j = 1, \ldots, r \right\}.$$

We note that $L_k(x, \mu)$, for fixed $\mu \ge 0$, is closed, convex, and coercive in x, when restricted to X^k, and negative definite quadratic in μ for fixed x. Hence, using the Saddle Point Theorem (Prop. 2.6.9), we can assert that L_k has a saddle point over $x \in X^k$ and $\mu \ge 0$, denoted by (x^k, μ^k).

Since L_k is quadratic in μ, the supremum in

$$\sup_{\mu \ge 0} L_k(x^k, \mu)$$

is attained at

$$\mu_j^k = k g_j^+(x^k), \qquad j = 1, \ldots, r. \tag{6.71}$$

Similarly, the infimum in $\inf_{x \in X^k} L_k(x, \mu^k)$ is attained at x^k, implying that

$$
\begin{aligned}
f(x^k) + \mu^{k'} g(x^k) &= \inf_{x \in X^k} \left\{ f(x) + \mu^{k'} g(x) \right\} \\
&= \inf_{x \in X^k} \left\{ f(x) + k g^+(x^k)' g(x) \right\} \\
&\leq \inf_{x \in X^k, \ g_j(x) \leq \frac{1}{k^4}, j=1,\ldots,r} \left\{ f(x) + k \sum_{j=1}^{r} g_j^+(x^k)' g_j(x) \right\} \\
&\leq \inf_{x \in X^k, \ g_j(x) \leq \frac{1}{k^4}, j=1,\ldots,r} \left\{ f(x) + \frac{r}{k^2} \right\} \\
&= f(\overline{x}^k) + \frac{r}{k^2} \\
&\leq q^* + \frac{r+1}{k^2},
\end{aligned}
$$

$$(6.72)$$

where the second inequality holds in view of the fact $x^k \in X^k$, implying that $g_j^+(x^k) \leq k$, $j = 1, \ldots, r$, and the third inequality follows from Eq. (6.70).

We also have

$$
\begin{aligned}
L_k(x^k, \mu^k) &= \sup_{\mu \geq 0} \inf_{x \in X^k} L_k(x, \mu) \\
&\geq \sup_{\mu \geq 0} \inf_{x \in X} L_k(x, \mu) \\
&= \sup_{\mu \geq 0} \left\{ \inf_{x \in X} \left\{ f(x) + \mu' g(x) \right\} - \frac{\|\mu\|^2}{2k} \right\} \\
&= \sup_{\mu \geq 0} \left\{ q(\mu) - \frac{\|\mu\|^2}{2k} \right\} \\
&\geq q(\mu^*) - \frac{\|\mu^*\|^2}{2k} \\
&= q^* - \frac{\|\mu^*\|^2}{2k},
\end{aligned}
$$

$$(6.73)$$

where we recall that μ^* is the dual optimal solution with the minimum norm.

Combining Eqs. (6.73) and (6.72), we obtain

$$
\begin{aligned}
q^* - \frac{1}{2k} \|\mu^*\|^2 &\leq L_k(x^k, \mu^k) \\
&= f(x^k) + \mu^{k'} g(x^k) - \frac{1}{2k} \|\mu^k\|^2 \\
&\leq q^* + \frac{r+1}{k^2} - \frac{1}{2k} \|\mu^k\|^2.
\end{aligned}
$$

$$(6.74)$$

This relation shows that $\|\mu^k\|^2 \leq \|\mu^*\|^2 + 2(r+1)/k$, so the sequence $\{\mu^k\}$ is bounded. Let $\overline{\mu}$ be a limit point of $\{\mu^k\}$. Without loss of generality, we assume that the entire sequence $\{\mu^k\}$ converges to $\bar{\mu}$. We also have from Eq. (6.74) that

$$\lim_{k\to\infty}\left\{f(x^k)+\mu^{k'}g(x^k)\right\}=q^*.$$

Hence, taking the limit as $k\to\infty$ in Eq. (6.72) yields

$$q^*\le\inf_{x\in X}\left\{f(x)+\bar{\mu}'g(x)\right\}=q(\bar{\mu})\le q^*.$$

Hence $\bar{\mu}$ is a dual optimal solution, and since $\|\bar{\mu}\|\le\|\mu^*\|$ [which follows by taking the limit in Eq. (6.74)], by using the minimum norm property of μ^*, we conclude that any limit point $\bar{\mu}$ of μ^k must be equal to μ^*. Thus $\mu^k\to\mu^*$, and using Eq. (6.74), we obtain

$$\lim_{k\to\infty}k\left(L_k(x^k,\mu^k)-q^*\right)=-\frac{1}{2}\|\mu^*\|^2. \tag{6.75}$$

Using Eq. (6.71), it follows that

$$L_k(x^k,\mu^k)=\sup_{\mu\ge0}L_k(x^k,\mu)=f(x^k)+\frac{1}{2k}\|\mu^k\|^2,$$

which combined with Eq. (6.75) yields

$$\lim_{k\to\infty}k\left(f(x^k)-q^*\right)=-\|\mu^*\|^2,$$

implying that $f(x^k)<q^*$ for all sufficiently large k, since $\mu^*\ne0$. Since, $\mu^k\to\mu^*$, Eq. (6.71) also implies that

$$\lim_{k\to\infty}kg_j^+(x^k)=\mu_j^*,\qquad j=1,\dots,r.$$

It follows that the sequence $\{x^k\}$ fulfills condition (ii) of the definition of informative dual optimal solution, thereby completing the proof.   **Q.E.D.**

6.7   NOTES, SOURCES, AND EXERCISES

Duality theory has its origins in the work of von Neuman on zero sum games. Linear programming duality was formulated by Gale, Kuhn, and Tucker [GKT51]. The connection between Lagrange multipliers and saddle points of the Lagrangian function was articulated by Kuhn and Tucker [KuT51], who also focused attention on the role of convexity. The paper by Slater [Sla50] gave the first constraint qualification for (not necessarily differentiable) convex programs. The constraint qualification of Assumption

6.4.3 and the corresponding strong duality result of Prop. 6.4.4 are slightly more general than those found in the literature.

The question of absence of a duality gap has been addressed in the literature primarily through strong duality theorems that assert the existence of geometric multipliers (cf. the theory of Section 6.4). As a result, it has not been sufficiently recognized that absence of a duality gap and existence of geometric multipliers are separate issues, which require different lines of analysis. In particular, the absence of duality gap is intimately connected with questions of preservation of closedness under partial minimization (the closedness of the primal function, cf. Sections 2.3 and 6.5.1), while the existence of geometric multipliers is tied to the existence of nonvertical max crossing hyperplanes in the min common/max crossing framework (the subdifferentiability of the primal function, cf. Sections 2.5 and 6.5.2).

Proposition 6.5.3, which asserts that there is no duality gap if the set of optimal primal solutions is nonempty and compact is given by Rockafellar [Roc70]. Proposition 6.5.6, which deals with the case where the cost function and the constraints are quadratic, is due to Terlaky [Ter85], who proved it by using a different method. Propositions 6.5.4 and 6.5.5 are new. Some other conditions that guarantee the absence of a duality gap are given by Rockafellar [Roc71], and by Auslender and Teboulle ([AuT03], p. 162).

The Fritz John conditions of Section 6.6, except for Prop. 6.6.1, which is classical, are new and have been developed by Bertsekas, Ozdaglar, and Tseng [BOT02]. These conditions provide a fully satisfactory alternative line of analysis of the existence of geometric multipliers under the standard conditions (the Slater and linearity criteria). The results on the existence of informative geometric multipliers and informative dual optimal solutions for convex programming problems are also new, and are due to Bertsekas, Ozdaglar, and Tseng [BOT02].

EXERCISES

6.1

Show that the dual of the (infeasible) linear program

$$\text{minimize} \quad x_1 - x_2$$
$$\text{subject to} \quad x \in X = \{x \mid x_1 \geq 0, \, x_2 \geq 0\}, \quad x_1 + 1 \leq 0, \, 1 - x_1 - x_2 \leq 0$$

is the (infeasible) linear program

$$\text{maximize} \quad \mu_1 + \mu_2$$

$$\text{subject to} \quad \mu_1 \geq 0, \ \mu_2 \geq 0, \ -\mu_1 + \mu_2 - 1 \leq 0, \ \mu_2 + 1 \leq 0.$$

6.2 (Extended Representation)

Consider problem (P) and assume that the set X is described by equality and inequality constraints as

$$X = \left\{ x \mid h_i(x) = 0, \ i = m+1, \ldots, \overline{m}, \ g_j(x) \leq 0, \ j = r+1, \ldots, \overline{r} \right\}.$$

Then the problem can alternatively be described without an abstract set constraint, in terms of all of the constraint functions

$$h_i(x) = 0, \quad i = 1, \ldots, \overline{m}, \qquad g_j(x) \leq 0, \quad j = 1, \ldots, \overline{r}.$$

We call this the *extended representation* of (P). Show if there exists a geometric multiplier for the extended representation, there exists a geometric multiplier for the original problem (P).

6.3 (Quadratic Programming Duality)

This exercise is an extension of Prop. 6.3.1. Consider the quadratic program

$$\text{minimize} \quad c'x + \tfrac{1}{2}x'Qx$$

$$\text{subject to} \quad x \in X, \qquad a_j'x \leq b_j, \quad j = 1, \ldots, r,$$

where X is a polyhedral set, Q is a symmetric positive semidefinite $n \times n$ matrix, $c, a_1, \ldots, a_r$ are vectors in $\Re^n$, and $b_1, \ldots, b_r$ are scalars, and assume that its optimal value is finite. Then there exist at least one optimal solution and at least one geometric multiplier. *Hint*: Use the extended representation of Exercise 6.2.

6.4 (Sensitivity)

Consider the class of problems

$$\text{minimize} \quad f(x)$$

$$\text{subject to} \quad x \in X, \qquad g_j(x) \leq u_j, \quad j = 1, \ldots, r,$$

where $u = (u_1, \ldots, u_r)$ is a vector parameterizing the right-hand side of the constraints. Given two distinct values $\overline{u}$ and $\tilde{u}$ of u, let $\overline{f}$ and $\tilde{f}$ be the corresponding optimal values, and assume that $-\infty < \overline{f} < \infty$ and $-\infty < \tilde{f} < \infty$, and that $\overline{\mu}$ and $\tilde{\mu}$ are corresponding geometric multipliers. Show that

$$\tilde{\mu}'(\tilde{u} - \overline{u}) \leq \overline{f} - \tilde{f} \leq \overline{\mu}'(\tilde{u} - \overline{u}).$$

6.5

Verify the linear programming duality relations

$$\min_{A'x \geq b} c'x \qquad \Longleftrightarrow \qquad \max_{A\mu = c, \ \mu \geq 0} b'\mu,$$

$$\min_{A'x \geq b, \ x \geq 0} c'x \qquad \Longleftrightarrow \qquad \max_{A\mu \leq c, \ \mu \geq 0} b'\mu,$$

show that they are symmetric, and derive the corresponding complementary slackness conditions [cf. Eqs. (6.13) and (6.14)].

6.6 (Duality and Zero Sum Games)

Let A be an $n \times m$ matrix, and let X and Z be the unit simplices in $\Re^n$ and $\Re^m$, respectively:

$$X = \left\{ x \ \Big| \ \sum_{i=1}^{n} x_i = 1, \ x_i \geq 0, \ i = 1, \dots, n \right\},$$

$$Z = \left\{ z \ \Big| \ \sum_{j=1}^{m} z_j = 1, \ z_j \geq 0, \ j = 1, \dots, m \right\}.$$

Show that the minimax equality

$$\max_{z \in Z} \min_{x \in X} x'Az = \min_{x \in X} \max_{z \in Z} x'Az$$

is a special case of linear programming duality. *Hint*: For a fixed z, $\min_{x \in X} x'Az$ is equal to the minimum component of the vector Az, so

$$\max_{z \in Z} \min_{x \in X} x'Az = \max_{z \in Z} \min\big\{ (Az)_1, \dots, (Az)_n \big\} = \max_{\xi e \leq Az, \ z \in Z} \xi, \qquad (6.76)$$

where e is the unit vector in $\Re^n$ (all components are equal to 1). Similarly,

$$\min_{x \in X} \max_{z \in Z} x'Az = \min_{\zeta e \geq A'x, \ x \in X} \zeta. \qquad (6.77)$$

Show that the linear programs in the right-hand sides of Eqs. (6.76) and (6.77) are dual to each other.

6.7 (Goldman-Tucker Complementarity Theorem [GoT56])

Consider the linear programming problem

$$\text{minimize} \quad c'x$$
$$\text{subject to} \quad Ax = b, \quad \cdot \; x \geq 0, \tag{LP}$$

where A is an $m \times n$ matrix, c is a vector in $\Re^n$, and b is a vector in $\Re^m$. Consider also the dual problem

$$\text{maximize} \quad b'\lambda$$
$$\text{subject to} \quad A'\lambda \leq c. \tag{DLP}$$

Assume that the sets of optimal solutions of LP and DLP, denoted X^* and Λ^*, respectively, are nonempty. Show that the index set $\{1, \ldots, n\}$ can be partitioned into two disjoint subsets I and $\bar{I}$ with the following two properties:

(1) For all $x^* \in X^*$ and $\lambda^* \in \Lambda^*$, we have

$$x_i^* = 0, \quad \forall \, i \in \bar{I}, \qquad (A'\lambda^*)_i = c_i, \quad \forall \, i \in I,$$

where x_i^* and $(A'\lambda^*)_i$ are the ith components of x^* and $A'\lambda^*$, respectively.

(2) There exist vectors $x^* \in X^*$ and $\lambda^* \in \Lambda^*$ such that

$$x_i^* > 0, \quad \forall \, i \in I, \qquad x_i^* = 0, \quad \forall \, i \in \bar{I},$$
$$(A'\lambda^*)_i = c_i, \quad \forall \, i \in I, \qquad (A'\lambda^*)_i < c_i, \quad \forall \, i \in \bar{I}.$$

Hint: Apply the Tucker Complementarity Theorem (Exercise 3.32).

6.8

Use duality to show that in three-dimensional space, the (minimum) distance from the origin to a line is equal to the maximum over all (minimum) distances of the origin from planes that contain the line.

6.9

Consider the problem

$$\text{minimize} \quad \sum_{i=0}^{m} f_i(x)$$
$$\text{subject to} \quad x \in X_i, \qquad i = 0, 1, \ldots, m,$$

where $f_i : \Re^n \mapsto \Re$ are convex functions and X_i are bounded polyhedral subsets of $\Re^n$ with nonempty intersection. Show that a dual problem is given by

$$\text{maximize} \quad q_0(\lambda_1 + \cdots + \lambda_m) + \sum_{i=1}^{m} q_i(\lambda_i)$$
$$\text{subject to} \quad \lambda_i \in \Re^n, \qquad i = 1, \ldots, m,$$

where the functions $q_i : \Re^n \mapsto \Re$ are given by

$$q_0(\lambda) = \min_{x \in X_0} \left\{ f_0(x) - \lambda' x \right\},$$

$$q_i(\lambda) = \min_{x \in X_i} \left\{ f_i(x) + \lambda' x \right\}, \qquad i = 1, \ldots, m.$$

Show also that the primal and dual problems have optimal solutions, and that there is no duality gap. *Hint*: Introduce artificial optimization variables $z_1, \ldots, z_m$ and the linear constraints $x = z_i$, $i = 1, \ldots, m$.

6.10

Consider the problem

$$\text{minimize} \quad f(x)$$

$$\text{subject to} \quad x \in X, \quad g_j(x) \leq 0, \ j = 1, \ldots, r,$$

and assume that f^* is finite, X is convex, and the functions $f : \Re^n \mapsto \Re$ and $g_j : \Re^n \mapsto \Re$ are convex over X. Show that if the set of geometric multipliers is nonempty and compact, then the Slater condition holds.

6.11 (Inconsistent Convex Systems of Inequalities)

Let $g_j : \Re^n \mapsto \Re$, $j = 1, \ldots, r$, be convex functions over the nonempty convex subset of $\Re^n$. Show that the system

$$g_j(x) < 0, \qquad j = 1, \ldots, r,$$

has no solution within X if and only if there exists a vector $\mu \in \Re^r$ such that

$$\sum_{j=1}^{r} \mu_j = 1, \qquad \mu \geq 0,$$

$$\mu' g(x) \geq 0, \qquad \forall \, x \in X.$$

Hint: Consider the convex program

$$\text{minimize} \quad y$$

$$\text{subject to} \quad x \in X, \quad y \in \Re, \quad g_j(x) \leq y, \quad j = 1, \ldots, r.$$

6.12

This exercise is a refinement of the Enhanced Farkas' Lemma (Prop. 5.4.2). Let N be a closed cone in $\Re^n$, let $a_1, \ldots, a_r$ be vectors in $\Re^n$, and let c be a vector in $\mathrm{cone}(\{a_1, \ldots, a_r\}) + \mathrm{ri}(N)$ such that $c \notin N$. Show that there is a nonempty index set $J \subset \{1, \ldots, r\}$ such that:

(1) The vector c can be represented as a positive combination of the vectors a_j, $j \in J$, plus a vector in N.

(2) There is a hyperplane that passes through the origin, and contains the vectors a_j, $j \in J$, in one of its open halfspaces and the vectors a_j, $j \notin J$, in the complementary closed halfspace.

Hint: Combine Example 6.4.2 with Lemma 5.3.1.

6.13 (Pareto Optimality)

A decisionmaker wishes to choose a vector $x \in X$, which keeps the values of *two* cost functions $f_1 : \Re^n \mapsto \Re$ and $f_2 : \Re^n \mapsto \Re$ reasonably small. Since a vector x^* minimizing simultaneously both f_1 and f_2 over X need not exist, he/she decides to settle for a *Pareto optimal solution*, i.e., a vector $x^* \in X$ with the property that there does not exist any vector $\bar{x} \in X$ that is strictly better than x^*, in the sense that either

$$f_1(\bar{x}) \le f_1(x^*), \qquad f_2(\bar{x}) < f_2(x^*),$$

or

$$f_1(\bar{x}) < f_1(x^*), \qquad f_2(\bar{x}) \le f_2(x^*).$$

(a) Show that if x^* is a vector in X, and λ_1^* and λ_2^* are two positive scalars such that

$$\lambda_1^* f_1(x^*) + \lambda_2^* f_2(x^*) = \min_{x \in X} \left\{ \lambda_1^* f_1(x) + \lambda_2^* f_2(x) \right\},$$

then x^* is a Pareto optimal solution.

(b) Assume that X is convex and f_1, f_2 are convex over X. Show that if x^* is a Pareto optimal solution, then there exist non-negative scalars λ_1^*, λ_2^*, not both zero, such that

$$\lambda_1^* f_1(x^*) + \lambda_2^* f_2(x^*) = \min_{x \in X} \left\{ \lambda_1^* f_1(x) + \lambda_2^* f_2(x) \right\}.$$

Hint: Consider the set

$$A = \left\{ (z_1, z_2) \mid \text{there exists } x \in X \text{ such that } f_1(x) \le z_1, \ f_2(x) \le z_2 \right\}$$

and show that it is a convex set. Use hyperplane separation arguments.

(c) Generalize the results of (a) and (b) to the case where there are m cost functions rather than two.

6.14 (Polyhedral Programming)

Consider the problem

$$\text{minimize} \quad f(x)$$

$$\text{subject to} \quad x \in X, \qquad g_j(x) \le 0, \quad j = 1, \ldots, r,$$

where X is a polyhedral set, and f and g_j are real-valued polyhedral functions. Assume that the optimal value is finite. Show that the primal function is proper and polyhedral.

7

Conjugate Duality

Contents

7.1. Conjugate Functions . . . . p. 424
7.2. Fenchel Duality Theorems . . . . p. 434
    7.2.1. Connection of Fenchel Duality and Minimax Theory p. 437
    7.2.2. Conic Duality . . . . p. 439
7.3. Exact Penalty Functions . . . . p. 441
7.4. Notes, Sources, and Exercises . . . . p. 446

In this chapter we will develop an alternative form of duality, called *conjugate* or *Fenchel duality*. This duality is based on an important convexity notion, the conjugacy transformation, which associates with any function f, a convex function, called the *conjugate* of f. Under convexity assumptions, the transformation is typically symmetric, in that f can be recovered by taking the conjugate of the conjugate of f. Aside from its use in Fenchel duality, the conjugacy transformation has many applications. For example, in the context of the Lagrangian duality of the preceding chapter, it relates the dual function with the primal function, introduced in Section 6.5.

The basic context for Fenchel duality is the problem

$$\begin{aligned} \text{minimize} \quad & f_1(x) - f_2(x) \\ \text{subject to} \quad & x \in X_1 \cap X_2, \end{aligned} \tag{7.1}$$

where $f_1 : \Re^n \mapsto \Re$ and $f_2 : \Re^n \mapsto \Re$ are given functions on $\Re^n$, and X_1 and X_2 are given subsets of $\Re^n$. While this problem may appear quite different from other constrained optimization problems discussed so far, it may be more conveniently applicable depending on the context. We will see examples where this is so in what follows.

It is important to realize that Fenchel duality is not fundamentally different from the Lagrangian duality of Chapter 6. In fact, we can derive Fenchel duality via Lagrangian duality, by converting problem (7.1) to the following equivalent problem in the variables $y \in \Re^n$ and $z \in \Re^n$

$$\begin{aligned} \text{minimize} \quad & f_1(y) - f_2(z) \\ \text{subject to} \quad & z = y, \qquad y \in X_1, \qquad z \in X_2, \end{aligned} \tag{7.2}$$

and by dualizing the constraint $z = y$. The dual function is

$$\begin{aligned} q(\lambda) &= \inf_{y \in X_1, \, z \in X_2} \big\{ f_1(y) - f_2(z) + (z - y)'\lambda \big\} \\ &= \inf_{z \in X_2} \big\{ z'\lambda - f_2(z) \big\} + \inf_{y \in X_1} \big\{ f_1(y) - y'\lambda \big\} \\ &= g_2(\lambda) - g_1(\lambda), \end{aligned}$$

where the functions g_1 and g_2 are defined by

$$g_1(\lambda) = \sup_{x \in X_1} \big\{ x'\lambda - f_1(x) \big\}, \qquad g_2(\lambda) = \inf_{x \in X_2} \big\{ x'\lambda - f_2(x) \big\}.$$

The dual problem is given by

$$\begin{aligned} \text{maximize} \quad & g_2(\lambda) - g_1(\lambda) \\ \text{subject to} \quad & \lambda \in \Lambda_1 \cap \Lambda_2, \end{aligned} \tag{7.3}$$

where Λ_1 and Λ_2 are the sets

$$\Lambda_1 = \big\{ \lambda \mid g_1(\lambda) < \infty \big\}, \qquad \Lambda_2 = \big\{ \lambda \mid g_2(\lambda) > -\infty \big\}.$$

It can be seen that there is symmetry between the primal and dual problems (7.1) and (7.3).

Note that the sets Λ_1 and Λ_2 need not appear explicitly in the dual problem (7.3) if we are willing to view g_1 and g_2 as extended real-valued functions $g_1 : \Re^n \mapsto (-\infty, \infty]$ and $g_2 : \Re^n \mapsto [-\infty, \infty)$, and write the dual problem as

$$\text{maximize} \quad g_2(\lambda) - g_1(\lambda)$$
$$\text{subject to} \quad \lambda \in \Re^n.$$

Similarly, we may view the functions f_1 and f_2 as extended real-valued functions that take the values ∞ or $-\infty$, respectively, whenever $x \notin X_1$ or $x \notin X_2$, respectively. With this convention, the primal problem is written as

$$\text{minimize} \quad f_1(x) - f_2(x)$$
$$\text{subject to} \quad x \in \Re^n, \tag{7.4}$$

where $f_1 : \Re^n \mapsto (-\infty, \infty]$ and $f_2 : \Re^n \mapsto [-\infty, \infty)$ are given functions, while the definitions of g_1 and g_2 become

$$g_1(\lambda) = \sup_{x \in \Re^n} \left\{ x'\lambda - f_1(x) \right\}, \qquad g_2(\lambda) = \inf_{x \in \Re^n} \left\{ x'\lambda - f_2(x) \right\}. \tag{7.5}$$

By suppressing the constraints $x \in X_1 \cap X_2$ and $\lambda \in \Lambda_1 \cap \Lambda_2$, and replacing f_1, f_2, g_1, g_2 with the corresponding extended real-valued versions, we gain some notational simplification, while maintaining the symmetry between the primal and dual problems. Furthermore, despite the fact that the Lagrangian duality theory of Chapter 6 involves real-valued functions, it can still be applied to problem (7.4) once this problem is written in the form

$$\text{minimize} \quad f_1(y) - f_2(z)$$
$$\text{subject to} \quad z = y, \qquad y \in \text{dom}(f_1), \qquad z \in \text{dom}(-f_2),$$

[cf. Eq. (7.2)]. Thus, it is convenient to formulate Fenchel duality in terms of extended real-valued functions, and this is the convention that we will adopt in this chapter.

The function g_1 of Eq. (7.5) is called the *conjugate convex function* of f_1, while the function g_2 is called the *conjugate concave function* of f_2 (see Fig. 7.0.1 for a geometrical interpretation of these conjugates). Figure 7.0.2 shows some examples of conjugate convex functions. In this figure, all the functions are convex, and it can be verified that the conjugate of the conjugate yields the original function. This is a manifestation of a more general result to be shown in the next section.

The conjugacy operations just described are central to this chapter. We develop some of their theory in the next section. In Section 7.2, we return to Fenchel duality. We derive conditions that guarantee that there

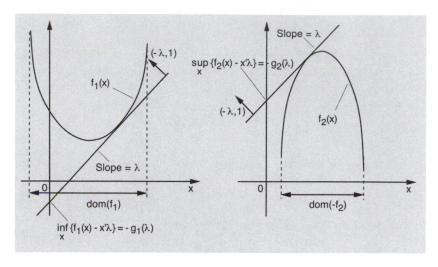

Figure 7.0.1. Visualization of the conjugate convex function

$$g_1(\lambda) = \sup_{x \in \Re^n} \left\{ x'\lambda - f_1(x) \right\}$$

and the conjugate concave function

$$g_2(\lambda) = \inf_{x \in \Re^n} \left\{ x'\lambda - f_2(x) \right\}.$$

The point of interception of the vertical axis with the hyperplane that has normal $(-\lambda, 1)$ and supports the epigraph of f_1 is $\inf_{x \in \Re^n} \{ f_1(x) - x'\lambda \}$, which by definition is equal to $-g_1(\lambda)$. This is illustrated in the figure on the left. The figure on the right gives a similar interpretation of $g_2(\lambda)$.

is no duality gap, essentially by applying the corresponding Lagrangian duality results of Section 6.4 to the equivalent problem (7.2). Finally, in Section 7.3, we discuss exact penalty functions for convex problems, and we visualize their properties by using Fenchel duality.

7.1 CONJUGATE FUNCTIONS

In this section we focus on an extended real-valued function $f : \Re^n \mapsto [-\infty, \infty]$ and we discuss its convex conjugate function. We recall that f is called closed if its epigraph, epi(f), is a closed set. The *convex closure of f*, is the function that has as epigraph the closure of the convex hull of epi(f) (which is also the smallest closed and convex set containing the epigraph of f).

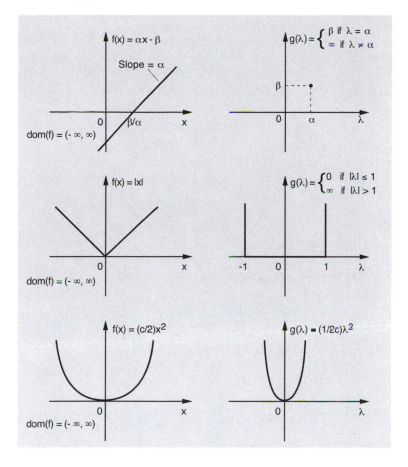

Figure 7.0.2. Some examples of conjugate convex functions. It can also be verified that in each case, the conjugate of the conjugate is the original, i.e., the conjugates of the functions on the right are the corresponding functions on the left (see Section 7.1).

Consider the convex conjugate of f, introduced earlier and illustrated in Figs. 7.0.1 and 7.0.2. This is the function g defined by

$$g(\lambda) = \sup_{x \in \Re^n} \{x'\lambda - f(x)\}, \qquad \lambda \in \Re^n.$$

Note that regardless of the structure of f, by Prop. 1.2.4(c), g is a closed convex function, since it is the pointwise supremum of the collection of affine functions $x'\lambda - f(x)$, $x \in \text{dom}(f)$. Note also that g need not be proper, even if f is. We will show, however, in the following proposition that if f is convex, then g is proper if and only if f is.

The conjugate of the conjugate function can be constructed as shown in Fig. 7.1.1. As this figure suggests and as the following proposition shows, by constructing the conjugate of the conjugate of f, we obtain the convex closure of f (provided the convex closure does not take the value $-\infty$ at any point). In particular, if f is proper, closed, and convex, the conjugate of the conjugate of f is again f. Thus in this case, conjugacy is a symmetric property, i.e., g and f are the conjugates of each other.

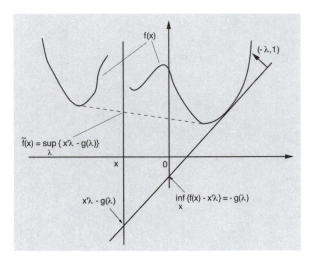

Figure 7.1.1. Visualization of the conjugate of the conjugate

$$\tilde{f}(x) = \sup_{\lambda \in \Re^n} \left\{ \lambda' x - g(\lambda) \right\}$$

of a function f, where g is the conjugate of f,

$$g(\lambda) = \sup_{x \in \Re^n} \left\{ x' \lambda - f(x) \right\}.$$

For each $x \in \Re^n$, we consider the vertical line in $\Re^{n+1}$ that passes through the point $(x, 0)$, and for each λ in the effective domain of g, we consider the interception point of this line with the hyperplane with normal $(-\lambda, 1)$ that supports the graph of f [and therefore passes through the point $(0, -g(\lambda))$]. The vertical level at this point of interception is $\lambda' x - g(\lambda)$, so $\tilde{f}(x)$ is equal to the highest level of interception. As the figure indicates (and Prop. 7.1.1 shows), the conjugate of the conjugate $\tilde{f}$ is the convex closure of f (barring the exceptional case where the convex closure takes the value $-\infty$ at some point, in which case the figure above is not valid).

Proposition 7.1.1: (Conjugacy Theorem) Let $f : \Re^n \mapsto [-\infty, \infty]$ be a function, let g be its convex conjugate, and consider the conjugate of g,

$$\tilde{f}(x) = \sup_{\lambda \in \Re^n} \{\lambda' x - g(\lambda)\}, \qquad x \in \Re^n.$$

(a) We have

$$f(x) \geq \tilde{f}(x), \qquad \forall\, x \in \Re^n.$$

(b) If f is convex, then properness of any one of the functions f, g, and $\tilde{f}$ implies properness of the other two.

(c) If f is closed, proper, and convex, then

$$f(x) = \tilde{f}(x), \qquad \forall\, x \in \Re^n.$$

(d) Let $\hat{f}$ be the convex closure of f. If $\hat{f}$ satisfies $\hat{f}(x) > -\infty$ for all $x \in \Re^n$, then

$$\hat{f}(x) = \tilde{f}(x), \qquad \forall\, x \in \Re^n.$$

Proof: (a) For all x and λ, we have

$$g(\lambda) \geq x'\lambda - f(x),$$

so that

$$f(x) \geq x'\lambda - g(\lambda), \qquad \forall\, x, \lambda \in \Re^n.$$

Hence

$$f(x) \geq \sup_{\lambda \in \Re^n} \{x'\lambda - g(\lambda)\} = \tilde{f}(x), \qquad \forall\, x \in \Re^n.$$

(b) Assume that f is proper, in addition to being convex. Then its epigraph is nonempty and convex, and contains no vertical line. By applying Prop. 2.5.1(a), with C being the set epi(f), it follows that there exists a nonvertical hyperplane with normal $(\lambda, 1)$ that contains epi(f) in its positive halfspace. In particular, this implies the existence of a vector λ and a scalar c such that

$$\lambda' x + f(x) \geq c, \qquad \forall\, x \in \Re^n.$$

We thus obtain

$$g(-\lambda) = \sup_{x \in \Re^n} \{-\lambda' x - f(x)\} \leq -c,$$

so that g is not identically equal to ∞. Also, by the properness of f, there exists a vector $\overline{x}$ such that $f(\overline{x})$ is finite. For every $\lambda \in \Re^n$, we have $g(\lambda) \geq \lambda'\overline{x} - f(\overline{x})$, so $g(\lambda) > -\infty$ for all $\lambda \in \Re^n$. Thus, g is proper.

Conversely, assume that g is proper. The preceding argument shows that properness of g implies properness of its conjugate, $\tilde{f}$, so that $\tilde{f}(x) > -\infty$ for all $x \in \Re^n$. In view of part (a), it follows that $f(x) > -\infty$ for all $x \in \Re^n$. Also, f cannot be identically equal to ∞, since then by its definition, g would be identically equal to $-\infty$. Thus f is proper.

We have thus shown that a convex function is proper if and only if its conjugate is proper, and the result follows in view of the conjugacy relations between f, g, and $\tilde{f}$.

(c) We will apply Prop. 2.5.1, with C being the closed and convex set epi$(\tilde{f})$, which contains no vertical line since f is proper. Let (x, γ) belong to epi$(\tilde{f})$, i.e., $x \in \text{dom}(\tilde{f})$, $\gamma \geq \tilde{f}(x)$, and suppose, to arrive at a contradiction, that (x, γ) does not belong to epi(f). Then by Prop. 2.5.1(b), there exists a nonvertical hyperplane with normal (λ, ζ), where $\zeta \neq 0$, and a scalar c such that

$$\lambda'z + \zeta w < c < \lambda'x + \zeta\gamma, \qquad \forall\, (z, w) \in \text{epi}(f).$$

Since w can be made arbitrarily large, we have $\zeta < 0$, and we can take $\zeta = -1$, so that

$$\lambda'z - w < c < \lambda'x - \gamma, \qquad \forall\, (z, w) \in \text{epi}(f).$$

Since $\gamma \geq \tilde{f}(x)$ and $(z, f(z)) \in \text{epi}(f)$ for all $z \in \text{dom}(f)$, we obtain

$$\lambda'z - f(z) < c < \lambda'x - \tilde{f}(x), \qquad \forall\, z \in \text{dom}(f).$$

Hence

$$\sup_{z \in \Re^n} \left\{ \lambda'z - f(z) \right\} \leq c < \lambda'x - \tilde{f}(x),$$

or

$$g(\lambda) < \lambda'x - \tilde{f}(x),$$

which contradicts the definition $\tilde{f}(x) = \sup_{\lambda \in \Re^n} \{\lambda'x - g(\lambda)\}$. Thus, we have epi$(\tilde{f}) \subset$ epi(f), which implies that $f(x) \leq \tilde{f}(x)$ for all $x \in \Re^n$. This, together with part (a), shows that $\tilde{f}(x) = f(x)$ for all x.

(d) If f is identically equal to ∞, then it can be verified that both $\hat{f}$ and $\check{f}$ are identically equal to ∞, so the result holds. We thus assume that f is not identically equal to ∞, so in view of the assumption $\hat{f}(x) > -\infty$ for all x, we may assume that f is proper. Let $\hat{g}$ be the conjugate of the convex closure $\check{f}$. For any λ, $-g(\lambda)$ and $-\hat{g}(\lambda)$ are the supremum levels of interception of the vertical axis with the nonvertical hyperplanes that contain the sets epi(f) and cl$(\text{conv}(\text{epi}(f)))$, respectively, in one of their closed halfspaces (cf. Fig. 7.0.1). Since the nonvertical hyperplanes that contain these two sets are the same, it follows that $g(\lambda) = \hat{g}(\lambda)$ for all λ. Thus, the conjugate of g is equal to the conjugate of $\hat{g}$, which is equal to

$\hat{f}$ by part (c) [we can use this part because of the properness of f, which together with the assumption $\hat{f}(x) > -\infty$ for all x, implies properness of $\hat{f}$]. **Q.E.D.**

Note that the assumption $\hat{f}(x) > -\infty$ for all x is essential for the validity of part (d) of the preceding proposition, as illustrated by the following example.

Example 7.1.1:

Consider the function $f : \Re \mapsto (-\infty, \infty]$ given by

$$f(x) = \begin{cases} \ln x & \text{if } x > 0, \\ \infty & \text{if } x \le 0. \end{cases}$$

Then, it can be seen that the convex closure of f is given by

$$\hat{f}(x) = \begin{cases} -\infty & \text{if } x \ge 0, \\ \infty & \text{if } x < 0. \end{cases}$$

The conjugate of f is given by

$$g(\lambda) = \infty, \qquad \forall \, \lambda \in \Re^n,$$

and its conjugate takes the value $-\infty$ everywhere, so it is not equal to the convex closure of f.

The exceptional behavior in the above example can be attributed to a subtle difference between the construction of the conjugate of a conjugate function and the construction of the convex closure: while the conjugate functions g and $\tilde{f}$ are defined exclusively in terms of nonvertical hyperplanes, via the construction of Fig. 7.1.1, the epigraph of the convex closure $\hat{f}$ is defined in terms of nonvertical *and* vertical hyperplanes. This difference is inconsequential when there exists at least one nonvertical hyperplane containing the epigraph of f in one of its closed halfspaces [this is equivalent to $\hat{f}(x) > -\infty$ for all $x \in \Re^n$; see Prop. 2.5.1(a)]. The reason is that, in this case, the epigraph of $\hat{f}$ can equivalently be defined by using just nonvertical hyperplanes [this can be seen using Prop. 2.5.1(b)].

Note that as Fig. 7.1.1 suggests, there is a close connection between the construction of the conjugate function and the construction underlying the min common/max crossing duality theory discussed in Section 2.5. In particular, for a given $x \in \Re^n$, we can view $f(x)$ as the minimum common point of the epigraph of f and the (translated) axis $\{(x, w) \mid w \in \Re\}$. Each $\lambda \in \Re^n$ defines a nonvertical supporting hyperplane with normal $(-\lambda, 1)$, whose corresponding crossing point is equal to $x'\lambda - g(x)$ (cf. Fig. 7.1.1). Thus the maximum crossing point is $\sup_\lambda \{x'\lambda - g(x)\}$, and when there is no duality gap, it is equal to the minimum common point $f(x)$. According to the Conjugacy Theorem (Prop. 7.1.1), this is so when $f(x)$ is equal to the value of the convex closure of f at x. This assertion is also the essence of the first min common/max crossing duality theorem of Section 2.5 (Prop. 2.5.4).

There are several interesting types of conjugate pairs. For example, it can be verified that the conjugate of a strictly convex quadratic function is also strictly convex quadratic (cf. Fig. 7.0.2). The following examples discuss some other important cases.

Example 7.1.2: (Support Functions)

Given a nonempty set X, consider the *indicator function* of X, given by

$$\delta_X(x) = \begin{cases} 0 & \text{if } x \in X, \\ \infty & \text{if } x \notin X. \end{cases}$$

The conjugate of δ_X, is given by

$$\sigma_X(\lambda) = \sup_{x \in X} \lambda' x$$

and is called the *support function of X*.

Note that the support function satisfies

$$\sigma_X(\alpha\lambda) = \alpha\sigma_X(\lambda), \qquad \forall\, \alpha > 0,\ \forall\, \lambda \in \Re^n.$$

This is called *positive homogeneity*. Formally, an extended real-valued function is said to be *positively homogeneous* if its epigraph is a cone. Note also that if X is closed and convex, then δ_X is closed and convex, and by the Conjugacy Theorem (cf. Prop. 7.1.1), the conjugate of its support function is the indicator function of δ_X. If X is not closed and convex, it has the same support function as $\mathrm{cl}\big(\mathrm{conv}(X)\big)$ (cf. Prop. 7.1.1).

For an example, consider the case where X is a bounded ellipsoid of the form

$$X = \big\{ x \mid (x - \bar{x})' Q (x - \bar{x}) \leq b \big\},$$

where Q is a symmetric positive definite matrix, $\bar{x}$ is a vector, and b is a positive scalar. The support function is given by

$$\sigma_X(\lambda) = \sup_{(x - \bar{x})' Q (x - \bar{x}) \leq b} \lambda' x,$$

and can be calculated by solving the maximization problem above for each λ. By introducing a Lagrange multiplier μ for the inequality constraint, and by setting the gradient of the corresponding Lagrangian function to 0, we obtain the maximizing point

$$x(\lambda) = \bar{x} + \frac{1}{2\mu} Q^{-1} \lambda.$$

It can be seen that for $\lambda \neq 0$, the inequality constraint is active at $x(\lambda)$, so we can determine μ by solving the equation

$$\big(x(\lambda) - \bar{x}\big)' Q \big(x(\lambda) - \bar{x}\big) = b,$$

which yields

$$\mu = \left(\frac{\lambda'Q^{-1}\lambda}{4b}\right)^{1/2}.$$

By combining the preceding equations, we thus obtain

$$\sigma_X(\lambda) = \lambda'x(\lambda) = \lambda'\left(\bar{x} + \frac{1}{2\mu}Q^{-1}\lambda\right) = \lambda'\left(\bar{x} + \frac{1}{2\left(\lambda'Q^{-1}\lambda/4b\right)^{1/2}}Q^{-1}\lambda\right),$$

and finally

$$\sigma_X(\lambda) = \lambda'\bar{x} + \left(b\,\lambda'Q^{-1}\lambda\right)^{1/2}.$$

Example 7.1.3: (Support Function of a Cone)

Consider a cone C in $\Re^n$ and its support function

$$\sigma_C(\lambda) = \sup_{x \in C} \lambda'x.$$

If $\lambda'x \le 0$ for all $x \in C$, i.e., if λ belongs to the polar cone C^*, we have $\sigma_C(\lambda) = 0$, since 0 is a closure point of C. On the other hand, if $\lambda'x > 0$ for some $x \in C$, we have $\sigma_C(\lambda) = \infty$, since C is a cone and therefore contains αx for all $\alpha > 0$. Thus, we have

$$\sigma_C(\lambda) = \begin{cases} 0 & \text{if } \lambda \in C^*, \\ \infty & \text{if } \lambda \notin C^*, \end{cases}$$

i.e., the support function of C is equal to the indicator function of the polar cone of C. In view of the characterization of the preceding example, we see that the indicator functions of a closed convex cone C and its polar C^* are conjugate to each other. Thus, the Conjugacy Theorem [cf. Prop. 7.1.1(c)], when specialized to the case of the indicator function of a closed convex cone C, yields $C = (C^*)^*$, which is the Polar Cone Theorem [Prop. 3.1.1(b)].

Example 7.1.4: (Support Function of a Polyhedral Set)

Consider a polyhedral set X, and let it have a Minkowski-Weyl representation of the form

$$X = \text{conv}\left(\{v_1, \ldots, v_m\}\right) + \text{cone}\left(\{y_1, \ldots, y_r\}\right)$$

for some vectors $v_1, \ldots, v_m, y_1, \ldots, y_r$ (cf. Prop. 3.2.2). The support function of X takes the form

$$\sigma_X(\lambda) = \sup_{x \in X} \lambda'x$$

$$= \sup_{\substack{\alpha_1,\ldots,\alpha_m,\beta_1,\ldots,\beta_r \ge 0 \\ \sum_{i=1}^{m}\alpha_i = 1}} \left\{\sum_{i=1}^{m}\alpha_i v_i'\lambda + \sum_{j=1}^{r}\beta_j y_j'\lambda\right\}$$

$$= \begin{cases} \max_{i=1,\ldots,m} v_i'\lambda & \text{if } y_j'\lambda \le 0, \ j = 1,\ldots,r, \\ \infty & \text{otherwise.} \end{cases}$$

Thus the support function of a polyhedral set is a polyhedral function that is positively homogeneous. By using a similar argument, or by appealing to the Conjugacy Theorem [Prop. 7.1.1(c)], it can be seen that the conjugate of a positively homogeneous polyhedral function is the support function of some polyhedral set.

Example 7.1.5: (Conjugates of Polyhedral Functions)

We know that a function can be equivalently specified in terms of its epigraph. As a consequence, we will see that the conjugate of a function can be specified in terms of the support function of its epigraph. We will derive the corresponding formula, and we will apply it to the case of a polyhedral function, in order to show that the conjugate of a polyhedral function is polyhedral.

Indeed, the expression for the conjugate of f,

$$g(\lambda) = \sup_{x \in \Re^n} \left\{ x'\lambda - f(x) \right\},$$

can equivalently be written as

$$g(\lambda) = \sup_{(x,w) \in \text{epi}(f)} \left\{ x'\lambda - w \right\}.$$

Since the expression in braces in the right-hand side is the inner product of the vectors (x, w) and $(\lambda, -1)$, the supremum above is the value of the support function of $\text{epi}(f)$ evaluated at $(\lambda, -1)$:

$$g(\lambda) = \sigma_{\text{epi}(f)}(\lambda, -1), \qquad \forall\, \lambda \in \Re^n.$$

Consider now the case where f is a polyhedral function, so that $\text{epi}(f)$ is a polyhedral set. Then it follows from the analysis of Example 7.1.4 that the support function $\sigma_{\text{epi}(f)}$ is a polyhedral function, and it can be seen that the function $\sigma_{\text{epi}(f)}(\cdot, -1)$, viewed as a function of λ, is polyhedral. We thus conclude that the conjugate of a polyhedral function is polyhedral.

Example 7.1.6: (Conjugacy Correspondence of Primal and Dual Functions)

Consider the problem

$$\text{minimize} \quad f(x)$$

$$\text{subject to} \quad x \in X, \qquad g_j(x) \le 0, \quad j = 1, \dots, r,$$

where $f : \Re^n \mapsto \Re$ and $g_j : \Re^n \mapsto \Re$, $j = 1, \dots, r$, are given functions, and X is a given subset of $\Re^n$. The primal function is given by

$$p(u) = \inf_{\substack{x \in X,\, g_j(x) \le u_j \\ j=1,\dots,r}} f(x).$$

Consider the dual function q. For every $\mu \geq 0$, we have

$$q(\mu) = \inf_{x \in X} \left\{ f(x) + \sum_{j=1}^{r} \mu_j g_j(x) \right\}$$

$$= \inf_{\{(u,x) \mid x \in X,\, g_j(x) \leq u_j,\, j=1,\ldots,r\}} \left\{ f(x) + \sum_{j=1}^{r} \mu_j g_j(x) \right\}$$

$$= \inf_{\{(u,x) \mid x \in X,\, g_j(x) \leq u_j,\, j=1,\ldots,r\}} \left\{ f(x) + \sum_{j=1}^{r} \mu_j u_j \right\}$$

$$= \inf_{u \in \Re^r} \inf_{\substack{x \in X,\, g_j(x) \leq u_j \\ j=1,\ldots,r}} \left\{ f(x) + \sum_{j=1}^{r} \mu_j u_j \right\},$$

and finally

$$q(\mu) = \inf_{u \in \Re^r} \left\{ p(u) + \mu' u \right\}, \qquad \forall\, \mu \geq 0,$$

as illustrated in Fig. 7.1.2.

Thus, we have $q(\mu) = -\sup_{u \in \Re^r} \left\{ -\mu' u - p(u) \right\}$ or

$$q(\mu) = -h(-\mu), \qquad \forall\, \mu \geq 0,$$

where h is the conjugate convex function of p:

$$h(\nu) = \sup_{u \in \Re^r} \left\{ \nu' u - p(u) \right\}.$$

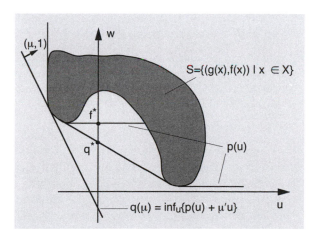

Figure 7.1.2. Illustration of the primal function and its conjugacy relation with the dual function.

7.2 FENCHEL DUALITY THEOREMS

We will now consider a slightly more general version of the Fenchel duality framework, which we discussed in the introduction to this chapter. Consider the problem

$$\text{minimize} \quad f_1(x) - f_2(Qx)$$
$$\text{subject to} \quad x \in \Re^n, \tag{7.6}$$

where Q is an $m \times n$ matrix, and $f_1 : \Re^n \mapsto (-\infty, \infty]$ and $f_2 : \Re^m \mapsto [-\infty, \infty)$ are extended real-valued functions.

As mentioned in the introduction to this chapter, the Lagrangian duality theory, developed in Chapter 6, can be applied to problem (7.6), once this problem is written in terms of the effective domains of f_1 and $-f_2$ as

$$\text{minimize} \quad f_1(y) - f_2(z)$$
$$\text{subject to} \quad z - Qy = 0, \qquad y \in \text{dom}(f_1), \qquad z \in \text{dom}(-f_2). \tag{7.7}$$

By dualizing the equality constraint $z - Qy = 0$, we obtain the dual function

$$
\begin{aligned}
q(\lambda) &= \inf_{y \in \Re^n, \, z \in \Re^m} \left\{ f_1(y) - f_2(z) + (z - Qy)'\lambda \right\} \\
&= \inf_{z \in \Re^m} \left\{ z'\lambda - f_2(z) \right\} + \inf_{y \in \Re^n} \left\{ f_1(y) - y'Q'\lambda \right\} \\
&= g_2(\lambda) - g_1(Q'\lambda),
\end{aligned}
$$

where $g_1 : \Re^n \mapsto (-\infty, \infty]$ and $g_2 : \Re^m \mapsto [-\infty, \infty)$ are given by

$$g_1(\lambda) = \sup_{x \in \Re^n} \left\{ x'\lambda - f_1(x) \right\}, \qquad g_2(\lambda) = \inf_{z \in \Re^m} \left\{ z'\lambda - f_2(z) \right\}.$$

The dual problem is

$$\text{maximize} \quad g_2(\lambda) - g_1(Q'\lambda)$$
$$\text{subject to} \quad \lambda \in \Re^m. \tag{7.8}$$

We can thus apply the theory of Chapter 6 to problem (7.7), which involves the optimization variables (y, z), the linear equality constraint $z - Qy = 0$, and the abstract set constraint $(y, z) \in \text{dom}(f_1) \times \text{dom}(-f_2)$. Using Prop. 6.2.5, we obtain necessary and sufficient conditions for (y^*, z^*) and λ^* to be optimal primal and dual solutions, respectively. When these conditions are translated to the framework of the original problem (7.6), we obtain that (x^*, λ^*) is an optimal primal and dual solution pair, and there is no duality gap, if and only if

$$x^* \in \text{dom}(f_1), \quad Qx^* \in \text{dom}(-f_2), \qquad \text{(primal feasibility)},$$

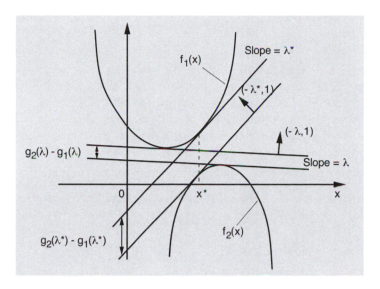

Figure 7.2.1. Illustration of Fenchel's duality theory for the case where Q is the identity matrix. The dual value $g_2(\lambda) - g_1(\lambda)$ becomes maximal when hyperplanes with a common normal $(-\lambda, 1)$ support the graphs of f_1 and f_2 at a common point x^* [cf. the Lagrangian optimality condition (7.9)].

$$Q'\lambda^* \in \mathrm{dom}(g_1), \quad \lambda^* \in \mathrm{dom}(-g_2), \qquad \text{(dual feasibility)},$$

$$x^* \in \arg \max_{x \in \Re^n} \left\{ x'Q'\lambda^* - f_1(x) \right\},$$

$$\text{(Lagrangian optimality)} \quad (7.9)$$

$$Qx^* \in \arg \min_{z \in \Re^m} \left\{ z'\lambda^* - f_2(z) \right\}.$$

The duality between the problems (7.6) and (7.8), and the Lagrangian optimality condition (7.9) are illustrated in Fig. 7.2.1 for the case where Q is the identity matrix.

Note that the first half of the Lagrangian optimality condition (7.9) can be written as

$$x^{*\prime}Q'\lambda^* - f_1(x^*) = \max_{x \in \Re^n} \left\{ x'Q'\lambda^* - f_1(x) \right\} = g_1(Q'\lambda^*),$$

or

$$f_1(x^*) + g_1(Q'\lambda^*) = x^{*\prime}Q'\lambda^*. \qquad (7.10)$$

Similarly, the second half of the Lagrangian optimality condition (7.9) can be written as

$$f_2(Qx^*) + g_2(\lambda^*) = x^{*\prime}Q'\lambda^*. \qquad (7.11)$$

When the functions f_1 and $-f_2$ are proper and convex, conditions (7.10) and (7.11) are equivalent to

$$Q'\lambda^* \in \partial f_1(x^*),$$

and
$$Qx^* \in \partial g_2(\lambda^*),$$

respectively (see Exercise 7.1). It can be seen that there is a remarkable symmetry in the necessary and sufficient conditions for x^* and λ^* to be optimal solutions of the primal and dual problems, respectively.

Under convexity assumptions, we can also apply the theory of Section 6.4 to problem (7.7) (cf. Prop. 6.4.4), to ascertain that there is no duality gap and that there exists a solution to the dual problem.

Proposition 7.2.1: (Primal Fenchel Duality Theorem) Let the functions f_1 and $-f_2$ be proper and convex. Then we have

$$\inf_{x \in \Re^n} \left\{ f_1(x) - f_2(Qx) \right\} = \sup_{\lambda \in \Re^m} \left\{ g_2(\lambda) - g_1(Q'\lambda) \right\}$$

and the supremum in the right-hand side above is attained, if the following conditions hold:

(1) $\mathrm{dom}(f_1)$ is the intersection of a polyhedron P_1 and a convex set C_1, and f_1 can be extended to a real-valued convex function over C_1. [By this we mean that there exists a convex function $\overline{f}_1 : C_1 \mapsto \Re$ such that $f_1(x) = \overline{f}_1(x)$ for all $x \in \mathrm{dom}(f_1)$.]

(2) $\mathrm{dom}(-f_2)$ is the intersection of a polyhedron P_2 and a convex set C_2, and f_2 can be extended to a real-valued concave function over C_2.

(3) The sets $Q \cdot \big(\mathrm{dom}(f_1) \cap \mathrm{ri}(C_1)\big)$ and $\mathrm{dom}(-f_2) \cap \mathrm{ri}(C_2)$ have nonempty intersection.

Proof: Condition (3) guarantees that f^*, the infimum of $f_1(x) - f_2(Qx)$ over $x \in \Re^n$, satisfies $f^* < \infty$ (i.e., the corresponding minimization problem is feasible). If f^* is finite, the result follows from the theory of Section 6.4 (cf. Prop. 6.4.4). If $f^* = -\infty$, the supremum of $g_2(\lambda) - g_1(Q'\lambda)$ is also $-\infty$ (by weak duality), and it is attained by all λ. **Q.E.D.**

We note some special cases of the above proposition. Its assumptions (1)-(3) are satisfied if f_1 and $-f_2$ are proper and convex, and if one of the following two conditions holds:

(a) The sets $Q \cdot \mathrm{ri}\big(\mathrm{dom}(f_1)\big)$ and $\mathrm{ri}\big(\mathrm{dom}(-f_2)\big)$ have nonempty intersection. [Take $C_1 = \mathrm{dom}(f_1)$, $C_2 = \mathrm{dom}(-f_2)$, and $P_1 = \Re^n$, $P_2 = \Re^m$ in Prop. 7.2.1.]

(b) The sets $\mathrm{dom}(f_1)$ and $\mathrm{dom}(-f_2)$ are polyhedral, the sets $Q \cdot \mathrm{dom}(f_1)$ and $\mathrm{dom}(-f_2)$ have nonempty intersection, and f_1 and f_2 can be extended to real-valued convex and concave functions over $\Re^n$ and $\Re^m$, respectively. [Take $C_1 = \Re^n$, $C_2 = \Re^m$, and $P_1 = \mathrm{dom}(f_1)$, $P_2 = \mathrm{dom}(-f_2)$ in Prop. 7.2.1.]

Suppose now that the functions f_1 and $-f_2$ are closed, proper, and convex. Then the Conjugacy Theorem (cf. Prop. 7.1.1) shows that the conjugate convex function of g_1 is f_1 and the conjugate concave function of g_2 is f_2. Under these circumstances, the duality is symmetric; i.e., the dual problem is of the same type as the primal problem, and by dualizing the dual problem, we obtain the primal. Thus, by applying Prop. 7.2.1, with the roles of x and λ reversed, we obtain the following:

Proposition 7.2.2: (Dual Fenchel Duality Theorem) Let the functions f_1 and $-f_2$ be closed, proper, and convex. Then we have

$$\inf_{x \in \Re^n} \big\{ f_1(x) - f_2(Qx) \big\} = \sup_{\lambda \in \Re^m} \big\{ g_2(\lambda) - g_1(Q'\lambda) \big\}$$

and the infimum in the left-hand side above is attained, if the following conditions hold:

(1) $\mathrm{dom}(g_1)$ is the intersection of a polyhedron P_1 and a convex set C_1, and g_1 can be extended to a real-valued convex function over C_1.

(2) $\mathrm{dom}(-g_2)$ is the intersection of a polyhedron P_2 and a convex set C_2, and g_2 can be extended to a real-valued concave function over C_2.

(3) The sets $\mathrm{dom}(g_1) \cap \mathrm{ri}(C_1)$ and $Q' \cdot \big(\mathrm{dom}(-g_2) \cap \mathrm{ri}(C_2)\big)$ have nonempty intersection.

As special cases, the assumptions (1)-(3) of the above proposition are satisfied if one of the following two conditions holds:

(a) The sets $\mathrm{ri}\big(\mathrm{dom}(g_1)\big)$ and $Q' \cdot \mathrm{ri}\big(\mathrm{dom}(-g_2)\big)$ have nonempty intersection. [Take $C_1 = \mathrm{dom}(g_1)$, $C_2 = \mathrm{dom}(-g_2)$, and $P_1 = \Re^n$, $P_2 = \Re^m$ in Prop. 7.2.2.]

(b) The sets $\mathrm{dom}(g_1)$ and $\mathrm{dom}(-g_2)$ are polyhedral, the sets $\mathrm{dom}(g_1)$ and $Q' \cdot \mathrm{dom}(-g_2)$ have nonempty intersection, and g_1 and g_2 can be extended to real-valued convex and concave functions over $\Re^n$ and $\Re^m$, respectively. [Take $C_1 = \Re^n$, $C_2 = \Re^m$, and $P_1 = \mathrm{dom}(g_1)$, $P_2 = \mathrm{dom}(-g_2)$ in Prop. 7.2.2.]

7.2.1   Connection of Fenchel Duality and Minimax Theory

We will now discuss a connection of Fenchel duality with the special minimax theory, developed under polyhedral assumptions in Chapter 3 [see Section 3.5 and Minimax Theorem III (Prop. 3.5.3)]. We will show that the Primal Fenchel Duality Theorem and Minimax Theorem III are essentially equivalent, in the sense that one can be derived from the other. The following discussion assumes that the function $-f_2$ in the Fenchel duality framework is closed, in order to derive the Primal Fenchel Duality Theorem from Minimax Theorem III. This assumption is not needed for the reverse derivation, and can be bypassed to a great extent by using some further analysis (see Exercise 7.6).

Let us return to the problem

$$\text{minimize} \quad f_1(x) - f_2(Qx)$$
$$\text{subject to} \quad x \in \Re^n, \tag{7.12}$$

where Q is an $m \times n$ matrix, and $f_1 : \Re^n \mapsto (-\infty, \infty]$ and $f_2 : \Re^m \mapsto [-\infty, \infty)$ are extended real-valued functions. We assume that the functions f_1 and $-f_2$ are proper and convex. Furthermore, we assume that $-f_2$ is closed. Consider also the function $h : \Re^m \mapsto (-\infty, \infty]$, which is the conjugate of $-f_2$, so that

$$h(z) = \sup_{y \in \Re^m} \left\{ z'y + f_2(y) \right\}.$$

Using the Conjugacy Theorem (cf. Prop. 7.1.1), the conjugate of h, denoted by h^*, satisfies

$$h^*(Qx) = -f_2(Qx) = \sup_{z \in \Re^m} \left\{ z'Qx - h(z) \right\},$$

so that problem (7.12) can be written as

$$\text{minimize} \quad \sup_{z \in \Re^m} \left\{ f_1(x) + z'Qx - h(z) \right\}$$
$$\text{subject to} \quad x \in \Re^n. \tag{7.13}$$

This problem is equivalent to the special minimax problem with linear coupling between x and z, whose theory has been addressed by the Minimax Theorem III of Section 3.5 (cf. Prop. 3.5.3). This theory can be used to analyze the preceding Fenchel duality framework.

If we use the equivalence of problems (7.12) and (7.13), we see that under the assumptions of Prop. 7.2.1, we can apply Minimax Theorem III. We then obtain

$$\inf_{x \in \Re^n} \left\{ f_1(x) - f_2(Qx) \right\} = \inf_{x \in \Re^n} \sup_{z \in \Re^m} \left\{ f_1(x) + z'Qx - h(z) \right\}$$

$$= \sup_{z \in \Re^m} \inf_{x \in \Re^n} \left\{ f_1(x) + z'Qx - h(z) \right\}$$

$$= \sup_{z \in \Re^m} \left\{ \inf_{x \in \Re^n} \left\{ f_1(x) + z'Qx \right\} - \sup_{y \in \Re^m} \left\{ z'y + f_2(y) \right\} \right\}$$

$$= \sup_{z \in \Re^m} \left\{ - \sup_{x \in \Re^n} \left\{ (-z)'Qx - f_1(x) \right\} + \inf_{y \in \Re^m} \left\{ (-z)'y - f_2(y) \right\} \right\}$$

$$= \sup_{\lambda \in \Re^m} \left\{ g_2(\lambda) - g_1(Q'\lambda) \right\},$$

and that the supremum over $\lambda \in \Re^m$ above is attained. This is exactly what Prop. 7.2.1 asserts. We thus conclude that the Primal Fenchel Duality Theorem of Prop. 7.2.1 can be derived from Minimax Theorem III. The preceding analysis can also be used to obtain a reverse derivation, so the two theorems are essentially equivalent.

The preceding derivation of Fenchel duality theory, via Minimax Theorem III, is an alternative to our earlier proof, which is based on the theory of existence of geometric multipliers and the Nonlinear Farkas' Lemma (Section 6.4). Note, however, that both proofs ultimately rely on the ideas embodied in the min common/max crossing framework and the associated analysis.

7.2.2   Conic Duality

Let us now discuss the application of Fenchel duality to a special type of convex programming problem. Consider the problem

$$\text{minimize} \quad f(x)$$
$$\text{subject to} \ \ x \in X \cap C,$$

where X is a convex subset of $\Re^n$, C is a convex cone in $\Re^n$, and $f : X \mapsto \Re$ is convex. We apply Fenchel duality with the definitions

$$f_1(x) = \begin{cases} f(x) & \text{if } x \in X, \\ \infty & \text{if } x \notin X, \end{cases} \qquad f_2(x) = \begin{cases} 0 & \text{if } x \in C, \\ -\infty & \text{if } x \notin C. \end{cases}$$

The corresponding conjugates are given by

$$g_1(\lambda) = \sup_{x \in X}\{\lambda'x - f(x)\}, \qquad g_2(\lambda) = \inf_{x \in C} x'\lambda = \begin{cases} 0 & \text{if } \lambda \in \hat{C}, \\ -\infty & \text{if } \lambda \notin \hat{C}, \end{cases}$$

where $\hat{C}$ is the negative polar cone (sometimes called the *dual cone* of C):

$$\hat{C} = -C^* = \{\lambda \mid x'\lambda \geq 0, \ \forall \ x \in C\}.$$

Thus the equation expressing Fenchel duality can be written as

$$\inf_{x \in X \cap C} f(x) = \sup_{\lambda \in \hat{C}} -g(\lambda),$$

where

$$g(\lambda) = \sup_{x \in X}\{\lambda'x - f(x)\}.$$

Proposition 7.2.1 provides conditions guaranteeing that there is no duality gap and the supremum in the dual problem is attained. For example, it is sufficient that one of the following two conditions holds:

(a) The relative interiors of X and C have nonempty intersection.

(b) f can be extended to a real-valued convex function over $\Re^n$, and X and C are polyhedral.

Similarly, assuming that f is closed and C is closed, Prop. 7.2.2 provides conditions guaranteeing that there is no duality gap and the infimum in the primal problem is attained. For example, it is sufficient that one of the following two conditions holds:

(a) The relative interiors of the sets $\text{dom}(g)$ and $\hat{C}$ have nonempty intersection.

(b) g can be extended to a real-valued convex function over $\Re^n$, and $\text{dom}(g)$ and $\hat{C}$ are polyhedral.

Conic duality has surprisingly many applications, and has been the subject of considerable research. The exercises provide a discussion of a few examples. A special case that has received a lot of attention arises when f is a linear function, i.e.,

$$f(x) = c'x,$$

where c is a vector, and X is an affine set, i.e.,

$$X = b + S,$$

where b is a vector and S is a subspace. Then the primal problem can be written as

$$\begin{aligned} \text{minimize} \quad & c'x \\ \text{subject to} \quad & x - b \in S, \quad x \in C. \end{aligned} \tag{7.14}$$

To derive the dual problem, we note that

$$\begin{aligned} g(\lambda) &= \sup_{x-b \in S} (\lambda - c)'x \\ &= \sup_{y \in S}(\lambda - c)'(y + b) \\ &= \begin{cases} (\lambda - c)'b & \text{if } \lambda - c \in S^\perp, \\ \infty & \text{if } \lambda - c \notin S. \end{cases} \end{aligned}$$

Thus it can be seen that the dual problem, $\max_{\lambda \in \hat{C}} -g(\lambda)$, after converting to minimization, is equivalent to

$$\begin{aligned} \text{minimize} \quad & b'\lambda \\ \text{subject to} \quad & \lambda - c \in S^\perp, \quad \lambda \in \hat{C}. \end{aligned} \tag{7.15}$$

By comparing the primal problem (7.14) and the dual problem (7.15), we see that they have the same form, and that if C is closed, then the dual of the dual yields the primal. Note, however, that there may still be a duality gap, unless one of the conditions given earlier is satisfied [e.g., $X \cap \text{ri}(C) \neq \emptyset$ or C is polyhedral].

7.3 EXACT PENALTY FUNCTIONS

In this section, we return to the exact penalty function theory that we developed in Section 5.5. Under convexity assumptions, we relate a constrained optimization problem with a penalized and less constrained problem. The results that we obtain have similar flavor to those of Section 5.5, but they are more powerful in a number of ways. In particular, there is no assumption that a *strict* local minimum exists, and indeed for part of the theory there is no need for any kind of optimal solution to exist. Furthermore, the threshold that the penalty parameter must exceed in order for the penalty function to be exact is characterized in terms of the geometric multipliers of the problem. In addition, the results admit an insightful visualization via the Fenchel duality theory.

We consider the problem

$$\begin{aligned} &\text{minimize} \quad f(x) \\ &\text{subject to} \quad x \in X, \qquad g_j(x) \le 0, \quad j = 1, \ldots, r, \end{aligned} \tag{7.16}$$

where X is a convex subset of $\Re^n$, and f and g_j are real-valued functions that are convex over X. We denote

$$g(x) = \big(g_1(x), \ldots, g_r(x)\big).$$

We consider penalty functions $P : \Re^r \mapsto \Re$ that are convex and satisfy

$$P(u) = 0, \qquad \forall \, u \le 0, \tag{7.17}$$

$$P(u) > 0, \qquad \text{if } u_j > 0 \text{ for some } j = 1, \ldots, r. \tag{7.18}$$

Such functions impose a penalty only for infeasible points. Examples are the quadratic penalty function,

$$P(u) = \frac{c}{2} \sum_{j=1}^{r} \big(\max\{0, u_j\}\big)^2,$$

and the nondifferentiable exact penalty of Section 5.5,

$$P(u) = c \sum_{j=1}^{r} \max\{0, u_j\},$$

where c is a positive penalty parameter. The conjugate convex function of P is given by

$$Q(\mu) = \sup_{u \in \Re^r} \{u'\mu - P(u)\},$$

and it can be seen that

$$Q(\mu) \ge 0, \qquad \forall \, \mu \in \Re^r,$$

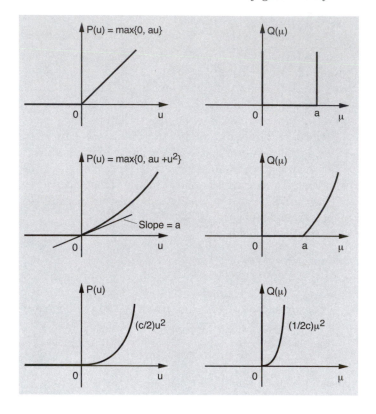

Figure 7.3.1. Illustration of conjugates of various penalty functions.

$$Q(\mu) = \infty, \qquad \text{if } \mu_j < 0 \text{ for some } j = 1, \ldots, r.$$

Some interesting penalty functions P are shown in Fig. 7.3.1, together with their conjugates.

We now consider the "penalized" problem

$$\begin{aligned} \text{minimize} \quad & f(x) + P\big(g(x)\big) \\ \text{subject to} \quad & x \in X, \end{aligned} \qquad (7.19)$$

and the primal function of the original constrained problem, given by

$$p(u) = \inf_{x \in X,\, g(x) \leq u} f(x), \qquad u \in \Re^r.$$

We will assume that the problem has at least one feasible solution, so that $p(0) < \infty$. We will also assume that $p(u) > -\infty$ for all $u \in \Re^r$, so that p is proper (this will be needed for application of the Fenchel Duality Theorem). Exercise 7.4 shows that this assumption is equivalent to the optimal dual value q^* being finite. We have

$$
\begin{aligned}
\inf_{x \in X} \left\{ f(x) + P\big(g(x)\big) \right\} &= \inf_{x \in X} \inf_{u \in \Re^r, \, g(x) \leq u} \left\{ f(x) + P\big(g(x)\big) \right\} \\
&= \inf_{x \in X} \inf_{u \in \Re^r, \, g(x) \leq u} \left\{ f(x) + P(u) \right\} \\
&= \inf_{x \in X, \, u \in \Re^r, \, g(x) \leq u} \left\{ f(x) + P(u) \right\} \qquad (7.20) \\
&= \inf_{u \in \Re^r} \inf_{x \in X, \, g(x) \leq u} \left\{ f(x) + P(u) \right\} \\
&= \inf_{u \in \Re^r} \left\{ p(u) + P(u) \right\}.
\end{aligned}
$$

We can now use the Primal Fenchel Duality Theorem (Prop. 7.2.1) with the identifications $f_1 = p$, $f_2 = -P$, and $Q = I$. We use the conjugacy relation between the primal function p and the dual function q to write

$$
\inf_{u \in \Re^r} \left\{ p(u) + P(u) \right\} = \sup_{\mu \geq 0} \left\{ q(\mu) - Q(\mu) \right\},
$$

so that, in view of Eq. (7.20), we obtain

$$
\inf_{x \in X} \left\{ f(x) + P\big(g(x)\big) \right\} = \sup_{\mu \geq 0} \left\{ q(\mu) - Q(\mu) \right\}; \qquad (7.21)
$$

see Fig. 7.3.2. Note that the conditions for application of the Primal Fenchel Duality Theorem are satisfied since the penalty function P is real-valued, so that the relative interiors of $\operatorname{dom}(p)$ and $\operatorname{dom}(P)$ have nonempty intersection. Furthermore, as part of the conclusions of the Primal Fenchel Duality Theorem, it follows that the supremum over $\mu \geq 0$ in Eq. (7.21) is attained.

It can be seen from Fig. 7.3.2 that in order for the penalized problem (7.19) to have the same optimal value as the original constrained problem (7.16), the conjugate Q must be "flat" along a sufficiently large "area," the size of which is related to the "size" of the dual optimal solutions. In particular, we have the following proposition.

Proposition 7.3.1: Assume that $p(0) < \infty$ and that $p(u) > -\infty$ for all $u \in \Re^r$.

(a) In order for the penalized problem (7.19) and the original constrained problem (7.16) to have equal optimal values, it is necessary and sufficient that there exists a geometric multiplier μ^* such that

$$
u' \mu^* \leq P(u), \qquad \forall \, u \in \Re^r. \qquad (7.22)
$$

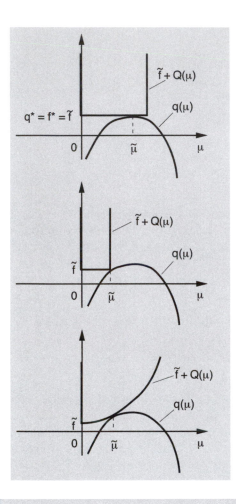

Figure 7.3.2. Illustration of the duality relation (7.21), and the optimal values of the penalized and the dual problem. Here f^* is the optimal value of the original problem, which is assumed to be equal to the optimal dual value q^*, while $\tilde{f}$ is the optimal value of the penalized problem,

$$\tilde{f} = \inf_{x \in X} \big\{ f(x) + P\big(g(x)\big) \big\}.$$

The point of contact of the graphs of the functions $\tilde{f} + Q(\mu)$ and $q(\mu)$ corresponds to the vector $\tilde{\mu}$ that attains the maximum in the relation

$$\tilde{f} = \max_{\mu \geq 0} \big\{ q(\mu) - Q(\mu) \big\}.$$

(b) In order for some optimal solution of the penalized problem (7.19) to be an optimal solution of the constrained problem (7.16), it is necessary that there exists a geometric multiplier μ^* such that

$$u'\mu^* \leq P(u), \qquad \forall\, u \in \Re^r.$$

(c) In order for the penalized problem (7.19) and the constrained problem (7.16) to have the same set of optimal solutions, it is sufficient that there exists a geometric multiplier μ^* such that

$$u'\mu^* < P(u), \qquad \forall\, u \in \Re^r \text{ with } u_j > 0 \text{ for some } j. \qquad (7.23)$$

Proof: (a) Since $P\big(g(x)\big) = 0$ for all feasible solutions of the constrained problem (7.16), we have

$$f^* = \inf_{x \in X,\, g(x) \leq 0} \big\{ f(x) + P\big(g(x)\big) \big\} \geq \inf_{x \in X} \big\{ f(x) + P\big(g(x)\big) \big\}.$$

In view of the Fenchel duality relation (7.21), it follows that for equality to hold in the above relation (i.e., for the optimal values of the penalized problem and the original constrained problem to be equal) it is necessary and sufficient that

$$f^* = q(\mu^*) - Q(\mu^*), \qquad \forall\, \mu^* \in \arg\max_{\mu \geq 0} \big\{ q(\mu) - Q(\mu) \big\}.$$

Since $Q(\mu^*) \geq 0$ for all $\mu^* \geq 0$ and we have $f^* \geq q(\mu^*)$ with equality if and only if μ^* is a geometric multiplier, it follows that the penalized problem and the original constrained problem have equal optimal values if and only if there exists a geometric multiplier μ^* satisfying $Q(\mu^*) = 0$. Using the definition of Q, we see that the relation $Q(\mu^*) = 0$ is equivalent to the desired relation (7.22).

(b) If x^* is an optimal solution of both problems (7.16) and (7.19), then by feasibility of x^*, we have $P\big(g(x^*)\big) = 0$, so these two problems have equal optimal values. The result then follows from part (a).

(c) If x^* is an optimal solution of the constrained problem (7.16), then $P\big(g(x^*)\big) = 0$, so we have

$$f^* = f(x^*) = f(x^*) + P\big(g(x^*)\big) \geq \inf_{x \in X} \big\{ f(x) + P\big(g(x)\big) \big\}.$$

The condition (7.23) implies the condition (7.22), so that by part (a), equality holds throughout in the above relation, showing that x^* is also an optimal solution of the penalized problem (7.19).

Conversely, if x^* is an optimal solution of the penalized problem (7.19), then x^* is either feasible [satisfies $g(x^*) \leq 0$], in which case it is an optimal solution of the constrained problem (7.16) [in view of $P\big(g(x)\big) = 0$ for all feasible vectors x], or it is infeasible in which case $g_j(x^*) > 0$ for some j. In the latter case, by using the given condition (7.23), it follows that there exists an $\epsilon > 0$ such that

$$\mu^{*\prime} g(x^*) + \epsilon < P\big(g(x^*)\big).$$

Let $\tilde{x}$ be a feasible vector such that $f(\tilde{x}) \leq f^* + \epsilon$. Since $P\big(g(\tilde{x})\big) = 0$ and $f^* = \min_{x \in X} \big\{ f(x) + \mu^{*\prime} g(x) \big\}$, we obtain

$$f(\tilde{x}) + P\big(g(\tilde{x})\big) = f(\tilde{x}) \leq f^* + \epsilon \leq f(x^*) + \mu^{*\prime} g(x^*) + \epsilon.$$

By combining the last two equations, we obtain

$$f(\tilde{x}) + P\big(g(\tilde{x})\big) < f(x^*) + P\big(g(x^*)\big),$$

which contradicts the hypothesis that x^* is an optimal solution of the penalized problem (7.19). This completes the proof.   **Q.E.D.**

Note that in the case where the necessary condition (7.22) holds but the sufficient condition (7.23) does not, it is possible that the constrained problem (7.16) has optimal solutions that are not optimal solutions of the penalized problem (7.19), even though the two problems have the same optimal value.

To elaborate on Prop. 7.3.1, consider the penalty function

$$P(u) = c \sum_{j=1}^{r} \max\{0, u_j\},$$

where $c > 0$. The condition $u'\mu^* \leq P(u)$ for all $u \in \Re^r$ [cf. Eq. (7.22)], is equivalent to

$$\mu_j^* \leq c, \qquad \forall\, j = 1, \ldots, r.$$

Similarly, the condition $u'\mu^* < P(u)$ for all $u \in \Re^r$ with $u_j > 0$ for some j [cf. Eq. (7.23)], is equivalent to

$$\mu_j^* < c, \qquad \forall\, j = 1, \ldots, r.$$

7.4 NOTES, SOURCES, AND EXERCISES

The conjugacy transformation was introduced by Fenchel [Fen49] and was further developed in his lecture notes [Fen51]. The Fenchel duality theory is also contained in these notes, together with a lot of other supporting material. The conjugacy transformation was extended to infinite-dimensional spaces by Moreau [Mor62] and by Bronsted [Bro64] (publication from his dissertation written under Fenchel's supervision). A classical version of the conjugacy transformation is the Legendre transformation, which applies to coercive twice continuously differentiable functions with everywhere positive definite Hessian matrix. Support functions were introduced by Minkowski for bounded convex sets. They were also studied for general convex sets by Fenchel [Fen51].

The Fenchel Duality Theorems (Props. 7.2.1 and 7.2.2) are given in [Fen51] for the case where Q is the identity matrix. Refinements that involve a more general Q and some polyhedral conditions were given by Rockafellar [Roc70]. The assumptions of the theorems given here are slightly

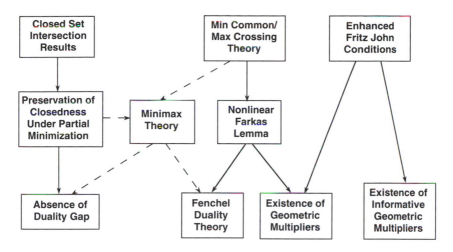

Figure 7.4.1. Lines of analysis for deriving the main results of constrained optimization duality. The broken lines indicate alternative methods of proof using minimax theory.

different than those in [Roc70]. A convex programming problem that fits nicely into the framework of Fenchel duality, but also admits a sharper duality result, is monotropic programming – the minimization of a separable convex function subject to linear equality constraints; see the exercises. The corresponding duality theory was developed by Rockafellar [Roc70], [Roc84] (a textbook treatment is also given in Bertsekas [Ber98]). Conic duality and its application to semidefinite programming and other topics have been discussed extensively. A textbook treatment is given by Ben-Tal and Nemirovski [BeN01] (see also Bonnans and Shapiro [BoS00], and Borwein and Lewis [BoL00]).

The interpretation and analysis of exact penalty functions of Section 7.3, using the Fenchel duality theory, was first given in Bertsekas [Ber75].

With this chapter, we have concluded our discussion of the two fundamental issues of constrained optimization duality: the absence of a duality gap and the existence of a geometric multiplier. As mentioned earlier, these two issues are distinct and require different lines of analysis. In particular, the absence of a duality gap is equivalent to lower semicontinuity of the primal function p at the origin, and can be addressed as a question of preservation of closedness under partial minimization. On the other hand, the existence of a geometric multiplier is equivalent to subdifferentiability of p at the origin, i.e., existence of a nonvertical support hyperplane to the epigraph of p at $(0, f^*)$. Figure 7.4.1 illustrates the corresponding lines of proof that we have used in this book.

Figure 7.4.1 also illustrates the relations between minimax theory and constrained optimization duality. The issue of existence of a duality gap can be addressed as a special case of validity of the minimax equality, and indeed this is the approach that we used in Section 6.5.1. Furthermore, the Primal Fenchel Duality Theorem can essentially be obtained from the special minimax theorem of Section 3.5.3 (cf. Section 7.2.1). Alternatively, the Primal Fenchel Duality Theorem can be derived using the Nonlinear Farkas' Lemma or, equivalently, from the results of Section 6.4 on the existence of a geometric multiplier.

Finally, let us note in Fig. 7.4.1 the two distinct and independent methods by which we have derived the constraint qualifications (Slater condition and linearity of constraint functions) that guarantee the existence of a geometric multiplier. These two methods are based on the Nonlinear Farkas' Lemma of Section 3.5.4, and on the enhanced Fritz John conditions of Section 6.6, respectively. The Fritz John conditions, while requiring a long (yet not unintuitive) proof, also guarantee the existence of an informative geometric multiplier, assuming at least one geometric multiplier exists.

EXERCISES

7.1 (Fenchel's Inequality)

Let $f : \Re^n \mapsto (-\infty, \infty]$ be a closed proper convex function and let g be its convex conjugate.

(a) For any $x \in \Re^n$ and $\lambda \in \Re^n$, we have

$$x'\lambda \leq f(x) + g(\lambda).$$

Furthermore, the following are equivalent:

   (i) $x'\lambda = f(x) + g(\lambda)$.

   (ii) $\lambda \in \partial f(x)$.

   (iii) $x \in \partial g(\lambda)$.

(b) The set of minima of f over $\Re^n$ is $\partial g(0)$.

(c) The set of minima of f over $\Re^n$ is nonempty if $0 \in \mathrm{ri}\big(\mathrm{dom}(g)\big)$, and it is nonempty and compact if and only if $0 \in \mathrm{int}\big(\mathrm{dom}(g)\big)$.

7.2

Let $f : \Re^n \mapsto (-\infty, \infty]$ be a proper convex function, and let g be its conjugate. Show that the lineality space of g is equal to the orthogonal complement of the subspace parallel to $\text{aff}(\text{dom}(f))$.

7.3

Let $f_i : \Re^n \mapsto (-\infty, \infty]$, $i = 1, \dots, m$, be proper convex functions, and let $f = f_1 + \cdots + f_m$. Show that if $\cap_{i=1}^m \text{ri}(\text{dom}(f_i))$ is nonempty, then we have

$$g(\lambda) = \inf_{\substack{\lambda_1 + \cdots + \lambda_m = \lambda \\ \lambda_i \in \Re^n, \, i=1,\dots,m}} \{g_1(\lambda_1) + \cdots + g_m(\lambda_m)\}, \qquad \forall \, \lambda \in \Re^n,$$

where $g, g_1, \dots, g_m$ are the conjugates of $f, f_1, \dots, f_m$, respectively.

7.4 (Finiteness of the Optimal Dual Value)

Consider the problem

$$\text{minimize} \quad f(x)$$
$$\text{subject to} \quad x \in X, \qquad g_j(x) \le 0, \quad j = 1, \dots, r,$$

where X is a convex set, and f and g_j are convex over X. Assume that the problem has at least one feasible solution. Show that the following are equivalent.

  (i) The dual optimal value $q^* = \sup_{\mu \ge 0} q(\mu)$ is finite.

  (ii) The primal function p is proper.

  (iii) The set

$$M = \left\{(u, w) \in \Re^{r+1} \mid \text{there is an } x \in X \text{ such that } g(x) \le u, \, f(x) \le w\right\}$$

does not contain a vertical line.

7.5 (General Perturbations and Min Common/Max Crossing Duality)

Let $F : \Re^{n+m} \mapsto (-\infty, \infty]$ be a proper function, and let $G : \Re^{n+m} \mapsto (-\infty, \infty]$ be its conjugate. Let also p be the function defined by

$$p(u) = \inf_{x \in \Re^n} F(x, u), \qquad u \in \Re^m.$$

Consider the min common/max crossing framework for the set $M = \text{epi}(p)$, and the cost function of the max crossing problem, $q(\lambda) = \inf_{(u,w) \in M} \{w + \lambda' u\}$.

(a) Show that q and the conjugate h of p satisfy

$$h(\lambda) = G(0, \lambda), \qquad q(\lambda) = -G(0, -\lambda), \qquad \forall \lambda \in \Re^m.$$

Show also that these relations generalize Example 7.1.6.

(b) Consider the alternative min common/max crossing framework where

$$M = \big\{(u, w) \mid \text{there is an } x \text{ such that } F(x, u) \leq w\big\}.$$

Show that the optimal values of the corresponding min common and max crossing problems are the same as those corresponding to $M = \text{epi}(p)$.

(c) Show that with $F(x, u) = f_1(x) - f_2(Qx + u)$, the min common/max crossing framework corresponds to the Fenchel duality framework. What are the forms of F that correspond to the minimax and constrained optimization frameworks of Sections 2.6.1 and 6.1?

7.6

Use Minimax Theorem III (Prop. 3.5.3) to derive the following version of the Primal Fenchel Duality Theorem: Let the functions f_1 and $-f_2$ be proper and convex. Then we have

$$\inf_{x \in \Re^n} \big\{f_1(x) - f_2(x)\big\} = \sup_{\lambda \in \Re^m} \big\{g_2(\lambda) - g_1(\lambda)\big\},$$

and the supremum in the right-hand side above is attained, if

$$\text{ri}\big(\text{dom}(f_1)\big) \cap \text{ri}\big(\text{dom}(-f_2)\big) \neq \emptyset.$$

Hint: In view of the results of Exercise 1.35, it is sufficient to show the above equality when f_1 and $-f_2$ are replaced by their closures.

7.7 (Monotropic Programming Duality)

Consider the problem

$$\text{minimize} \quad \sum_{i=1}^n f_i(x_i)$$

$$\text{subject to } x \in S, \qquad x_i \in X_i, \quad i = 1, \dots, n,$$

where $f_i : \Re \mapsto \Re$ are given functions, X_i are intervals of real numbers, and S is a subspace of $\Re^n$. Assume that the problem is feasible and that its optimal value is finite.

(a) Show that a dual problem is

$$\text{minimize} \quad \sum_{i=1}^n g_i(\lambda_i)$$

$$\text{subject to } \lambda \in S^\perp,$$

where the functions $g_i : \Re \mapsto (-\infty, \infty]$ are the conjugate convex functions

$$g_i(\lambda_i) = \sup_{x_i \in X_i} \{\lambda_i x_i - f_i(x_i)\}, \qquad i = 1, \dots, n.$$

(b) Show that the dual problem has an optimal solution and there is no duality gap under one of the following two conditions:

  (1) Each function f_i is convex over X_i and S contains a point in the relative interior of $X_1 \times \cdots \times X_n$.

  (2) The intervals X_i are closed and the functions f_i are convex over the entire real line.

7.8 (Network Optimization and Kirchhoff's Laws)

Consider a linear resistive electric network with node set $\mathcal{N}$ and arc set $\mathcal{A}$. Let v_i be the voltage of node i and let x_{ij} be the current of arc (i, j). Kirchhoff's current law says that for each node i, the total outgoing current is equal to the total incoming current

$$\sum_{\{j | (i,j) \in \mathcal{A}\}} x_{ij} = \sum_{\{j | (j,i) \in \mathcal{A}\}} x_{ji}.$$

Ohm's law says that the current x_{ij} and the voltage drop $v_i - v_j$ along each arc (i, j) are related by

$$v_i - v_j = R_{ij} x_{ij} - t_{ij},$$

where $R_{ij} \geq 0$ is a resistance parameter and t_{ij} is another parameter that is nonzero when there is a voltage source along the arc (i, j) (t_{ij} is positive if the voltage source pushes current in the direction from i to j). Consider the problem

$$
\begin{aligned}
\text{minimize} \quad & \sum_{(i,j) \in \mathcal{A}} \left(\tfrac{1}{2} R_{ij} x_{ij}^2 - t_{ij} x_{ij} \right) \\
\text{subject to} \quad & \sum_{\{j | (i,j) \in \mathcal{A}\}} x_{ij} = \sum_{\{j | (j,i) \in \mathcal{A}\}} x_{ji}, \quad \forall\, i \in \mathcal{N}.
\end{aligned}
\tag{7.24}
$$

Show that a set of variables $\{x_{ij} \mid (i, j) \in \mathcal{A}\}$ and $\{v_i \mid i \in \mathcal{N}\}$ are an optimal solution-Lagrange multiplier pair for this problem if and only if they satisfy Kirchhoff's current law and Ohm's law.

7.9 (Symmetry of Duality)

Consider the primal function

$$p(u) = \inf_{x \in X, \, g(x) \leq u} f(x)$$

of the problem

$$\text{minimize} \quad f(x)$$
$$\text{subject to} \quad x \in X, \qquad g_j(x) \leq 0, \quad j = 1, \ldots, r. \tag{7.25}$$

Consider also the problem

$$\text{minimize} \quad p(u)$$
$$\text{subject to} \quad u \in P, \qquad u \leq 0, \tag{7.26}$$

where P is the effective domain of p,

$$P = \{u \mid \text{there exists } x \in X \text{ with } g(x) \leq u\}.$$

Assume that $-\infty < p(0) < \infty$.

(a) Show that problems (7.25) and (7.26) have equal optimal values, and the same sets of geometric multipliers.

(b) Consider the dual functions of problems (7.25) and (7.26) and show that they are equal on the positive orthant, i.e., for all $\mu \geq 0$,

$$q(\mu) = \inf_{x \in X} \left\{ f(x) + \sum_{j=1}^{r} \mu_j g_j(x) \right\} = \inf_{u \in P} \{p(u) + \mu'u\}.$$

(c) Assume that p is a closed and convex function. Show that u^* is an optimal solution of problem (7.26) if and only if $-u^*$ is a geometric multiplier for the dual problem

$$\text{maximize} \quad q(\mu)$$
$$\text{subject to} \quad \mu \geq 0,$$

in the sense that

$$q^* = \sup_{\mu \geq 0} \{q(\mu) - \mu'u^*\}.$$

7.10 (Second Order Cone Programming)

Consider the problem

$$\text{minimize} \quad c'x$$
$$\text{subject to} \quad \|A_j x + b_j\| \leq e_j'x + d_j, \quad j = 1, \ldots, r,$$

where $x \in \Re^n$, and $c, A_j, b_j, e_j,$ and d_j are given, and have appropriate dimension. Assume that the problem is feasible. Consider the equivalent problem

$$\text{minimize} \quad c'x$$
$$\text{subject to} \quad \|u_j\| \leq t_j, \quad u_j = A_j x + b_j, \quad t_j = e_j'x + d_j, \quad j = 1, \ldots, r, \tag{7.27}$$

where u_j and t_j are auxiliary optimization variables.

(a) Show that problem (7.27) has cone constraints of the type described in Section 7.2.2.

(b) Use the conic duality theory of Section 7.2.2 to show that a dual problem is given by

$$\text{minimize} \quad \sum_{j=1}^{r}(b_j' z_j + d_j w_j)$$

$$\text{subject to} \quad \sum_{j=1}^{r}(A_j' z_j + e_j w_j) = c, \qquad \|z_j\| \le w_j, \quad j = 1, \dots, r. \tag{7.28}$$

Furthermore, show that there is no duality gap if either there exists a feasible solution of problem (7.27) or a feasible solution of problem (7.28) satisfying strictly all the corresponding inequality constraints.

7.11 (Quadratically Constrained Quadratic Problems [LVB98])

Consider the quadratically constrained quadratic problem

$$\text{minimize} \quad x' P_0 x + 2 q_0' x + r_0$$

$$\text{subject to} \quad x' P_i x + 2 q_i' x + r_i \le 0, \quad i = 1, \dots, p,$$

where $P_0, P_1, \dots, P_p$ are symmetric positive definite matrices. Show that the problem can be converted to one of the type described in Exercise 7.10, and derive the corresponding dual problem. *Hint*: Consider the equivalent problem

$$\text{minimize} \quad \|P_0^{1/2} x + P_0^{-1/2} q_0\|$$

$$\text{subject to} \quad \|P_i^{1/2} x + P_i^{-1/2} q_i\| \le (r_i - q_i' P_i^{-1} q_i)^{1/2}, \quad i = 1, \dots, p.$$

7.12 (Minimizing the Sum or the Maximum of Norms [LVB98])

Consider the problems

$$\text{minimize} \quad \sum_{i=1}^{p} \|F_i x + g_i\|$$

$$\text{subject to} \quad x \in \Re^n, \tag{7.29}$$

and

$$\text{minimize} \quad \max_{i=1,\dots,p} \|F_i x + g_i\|$$

$$\text{subject to} \quad x \in \Re^n,$$

where F_i and g_i are given matrices and vectors, respectively. Convert these problems to second-order cone programming problems (cf. Exercise 7.10) and derive the corresponding dual problems.

7.13 (Complex l_1 and l_∞ Approximation [LVB98])

Consider the complex l_1 approximation problem

$$\text{minimize} \quad \|Ax - b\|_1$$
$$\text{subject to} \quad x \in C^n,$$

where C^n is the set of n-dimensional vectors whose components are complex numbers. Show that it is a special case of problem (7.29) and derive the corresponding dual problem. Repeat for the complex l_∞ approximation problem

$$\text{minimize} \quad \|Ax - b\|_\infty$$
$$\text{subject to} \quad x \in C^n.$$

7.14

Consider the case where in Prop. 7.3.1 the function P has the form

$$P(z) = \sum_{j=1}^{r} P_j(z_j),$$

where $P_j : \Re \mapsto \Re$ are convex real-valued functions satisfying

$$P_j(z_j) = 0, \quad \forall\, z_j \leq 0, \qquad P_j(z_j) > 0, \quad \forall\, z_j > 0.$$

Show that the conditions (7.22) and (7.23) of Prop. 7.3.1 are equivalent to

$$\mu_j^* \leq \lim_{z_j \downarrow 0} \frac{P_j(z_j)}{z_j}, \qquad \forall\, j = 1, \ldots, r,$$

and

$$\mu_j^* < \lim_{z_j \downarrow 0} \frac{P_j(z_j)}{z_j}, \qquad \forall\, j = 1, \ldots, r,$$

respectively.

7.15 [Ber99b]

For a vector $y \in \Re^n$, let $d(y)$ be the optimal value of the projection problem

$$\text{minimize} \quad \|y - x\|$$
$$\text{subject to} \quad x \in X, \qquad g_j(x) \leq 0, \qquad j = 1, \ldots, r,$$

where X is a convex subset of $\Re^n$, and the functions $g_j : \Re^n \mapsto \Re$ are convex over X. Assuming that the problem is feasible, show that there exists a constant c such that

$$d(y) \leq c \left\| \left(g(y)\right)^+ \right\|, \qquad \forall\, y \in X,$$

if and only if the projection problem has a geometric multiplier $\mu^*(y)$ such that the set $\{\mu^*(y) \mid y \in X\}$ is bounded.

8

Dual Computational Methods

Contents

8.1. Dual Derivatives and Subgradients . . . . p. 457
8.2. Subgradient Methods . . . . p. 460
    8.2.1. Analysis of Subgradient Methods . . . . p. 470
    8.2.2. Subgradient Methods with Randomization . . . . p. 488
8.3. Cutting Plane Methods . . . . p. 504
8.4. Ascent Methods . . . . p. 509
8.5. Notes, Sources, and Exercises . . . . p. 512

In this chapter we consider dual methods, i.e., computational methods for solving dual problems. In particular, we focus on the primal problem

$$\text{minimize} \quad f(x)$$
$$\text{subject to} \quad x \in X, \qquad g_j(x) \le 0, \quad j = 1, \ldots, r, \tag{P}$$

and its dual

$$\text{maximize} \quad q(\mu)$$
$$\text{subject to} \quad \mu \ge 0, \tag{D}$$

where $f : \Re^n \mapsto \Re$, $g_j : \Re^n \mapsto \Re$ are given functions, X is a subset of $\Re^n$, and

$$q(\mu) = \inf_{x \in X} L(x, \mu)$$

is the dual function, while

$$L(x, \mu) = f(x) + \mu' g(x)$$

is the Lagrangian function. Our algorithms and analysis have straightforward extensions for the case where there are additional equality constraints in the primal problem.

Here are some of the potential incentives for solving the dual problem in place of the primal:

(a) The dual is a concave problem (concave cost, convex constraint set). By contrast, the primal need not be convex.

(b) The dual may have smaller dimension and/or simpler constraints than the primal.

(c) If there is no duality gap and the dual is solved exactly to yield a geometric multiplier μ^*, all optimal primal solutions can be obtained by minimizing the Lagrangian $L(x, \mu^*)$ over $x \in X$ [however, there may be additional minimizers of $L(x, \mu^*)$ that are primal-infeasible]. Furthermore, if the dual is solved approximately to yield an approximate geometric multiplier μ, and x_μ minimizes $L(x, \mu)$ over $x \in X$, then it can be seen that x_μ also solves the problem

$$\text{minimize} \quad f(x)$$
$$\text{subject to} \quad x \in X, \qquad g_j(x) \le g_j(x_\mu), \quad j = 1, \ldots, r.$$

Thus if the constraint violations $g_j(x_\mu)$ are not too large, x_μ may be an acceptable practical solution.

(d) Even if there is a duality gap, for every $\mu \ge 0$, the dual value $q(\mu)$ is a lower bound to the optimal primal value (the weak duality theorem). This lower bound may be useful in the context of discrete optimization and branch-and-bound procedures.

We should also consider some of the difficulties in solving the dual problem. The most important ones are the following:

(a) To evaluate the dual function at any μ requires minimization of the Lagrangian $L(x,\mu)$ over $x \in X$. In effect, this restricts the utility of dual methods to problems where this minimization can either be done in closed form or else is relatively simple.

(b) In many types of problems, the dual function is nondifferentiable, in which case we need special algorithms that can cope with nondifferentiability.

(c) Even if we find an optimal dual solution μ^*, it may be difficult to obtain a primal feasible vector x from the minimization of $L(x,\mu^*)$ over $x \in X$ as required by the primal-dual optimality conditions of Section 6.2, since this minimization can also yield primal-infeasible vectors.

Naturally, the differentiability properties of dual functions are an important determinant of the type of dual method that is appropriate for a given problem. We consequently develop these properties first in Section 8.1. In Sections 8.2 and 8.3, we consider methods for nondifferentiable optimization.

8.1 DUAL DERIVATIVES AND SUBGRADIENTS

For a given $\mu \in \Re^r$, suppose that x_μ minimizes the Lagrangian $L(x,\mu)$ over $x \in X$,

$$L(x_\mu,\mu) = \min_{x \in X} L(x,\mu) = \min_{x \in X}\{f(x) + \mu'g(x)\}.$$

An important fact for our purposes is that $g(x_\mu)$ *is a subgradient of the dual function q at μ*, i.e.,

$$q(\overline{\mu}) \le q(\mu) + (\overline{\mu} - \mu)'g(x_\mu), \qquad \forall\, \overline{\mu} \in \Re^r. \tag{8.1}$$

To see this, we use the definition of q and x_μ to write for all $\overline{\mu} \in \Re^r$,

$$\begin{aligned}
q(\overline{\mu}) &= \inf_{x \in X}\{f(x) + \overline{\mu}'g(x)\} \\
&\le f(x_\mu) + \overline{\mu}'g(x_\mu) \\
&= f(x_\mu) + \mu'g(x_\mu) + (\overline{\mu} - \mu)'g(x_\mu) \\
&= q(\mu) + (\overline{\mu} - \mu)'g(x_\mu).
\end{aligned}$$

Note that this calculation is valid for all $\mu \in \Re^r$ for which there is a minimizing vector x_μ, regardless of whether $\mu \ge 0$.

What is particularly important here is that we need to compute x_μ anyway in order to evaluate the dual function at μ, so *a subgradient $g(x_\mu)$ is obtained essentially at no cost.* The dual methods to be discussed solve the dual problem by computing the dual function value and a subgradient at a sequence of vectors $\{\mu^k\}$. It is not necessary to compute the set of *all* subgradients at μ^k in these methods; a single subgradient is sufficient.

Despite the fact that the following methodology does not require the full set of subgradients at a point, it is still useful to have characterizations of this set. For example it is important to derive conditions under which q is differentiable. We know from our preceding discussion that if q is differentiable at μ, there can be at most one value of $g(x_\mu)$ corresponding to vectors $x_\mu \in X$ minimizing $L(x, \mu)$. This suggests that q is everywhere differentiable (as well real-valued and concave) if for all μ, $L(x, \mu)$ is minimized at a unique $x_\mu \in X$. Indeed this is so, as shown in the following proposition that relies on Danskin's Theorem (Prop. 4.5.1).

Proposition 8.1.1: Let X be a nonempty and compact, and let f and g be continuous over X. Assume also that for every $\mu \in \Re^r$, $L(x, \mu)$ is minimized over $x \in X$ at a unique point x_μ. Then, q is everywhere continuously differentiable and

$$\nabla q(\mu) = g(x_\mu), \qquad \forall\, \mu \in \Re^r.$$

Proof: To assert the uniqueness of the subgradient of q at μ, apply Danskin's Theorem with the identifications $x \sim z$, $X \sim Z$, $\mu \sim x$, and $-L(x, \mu) \sim \phi(x, z)$. The assumptions of this theorem are satisfied because X is compact, while $L(x, \mu)$ is continuous as a function of x and concave (in fact linear) as a function of μ. The continuity of the dual gradient ∇q follows from Prop. 4.1.2. **Q.E.D.**

Note that if the constraint functions g_j are affine, X is convex and compact, and f is *strictly* convex, then the assumptions of Prop. 8.1.1 are satisfied and the dual function q is differentiable. We will focus on this and other related cases in the next section, where we discuss methods for differentiable dual functions.

In the case where X is a discrete set, as for example in integer programming, the continuity and compactness assumptions of Prop. 8.1.1 are satisfied, but there typically exist some μ for which $L(x, \mu)$ has multiple minima, leading to nondifferentiabilities. In fact, it can be shown under some reasonable conditions that *if there exists a duality gap, the dual function is nondifferentiable at every dual optimal solution*; see Exercise 8.1 and Fig. 8.1.1. Thus, nondifferentiabilities tend to arise at the most interesting

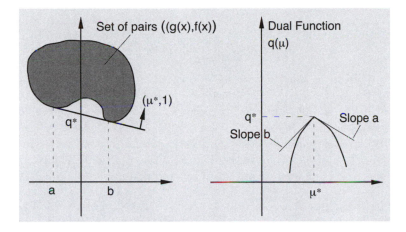

Figure 8.1.1. Illustration of the nondifferentiability of the dual function when there is a duality gap.

points and cannot be ignored in dual methods.

An important special case of a nondifferentiable dual function is when q is polyhedral, i.e., it has the form

$$q(\mu) = \min_{i \in I}\{a_i'\mu + b_i\}, \qquad (8.2)$$

where I is a finite index set, and a_i and b_i are given vectors in $\Re^r$ and scalars, respectively. This case arises, for example, when dealing with a discrete problem where X is a finite set. The set of all subgradients of q at μ is then the convex hull of the vectors a_i for which the minimum is attained in Eq. (8.2), as shown in the following proposition.

Proposition 8.1.2: Let the dual function q be polyhedral of the form (8.2), and for every $\mu \in \Re^r$, let I_μ be the set of indices attaining the minimum in Eq. (8.2),

$$I_\mu = \{i \in I \mid a_i'\mu + b_i = q(\mu)\}.$$

The set of all subgradients of q at μ is given by

$$\partial q(\mu) = \left\{ g \;\middle|\; g = \sum_{i \in I_\mu} \xi_i a_i, \; \xi_i \geq 0, \; \sum_{i \in I_\mu} \xi_i = 1 \right\}.$$

Proof: We apply Danskin's Theorem (Prop. 4.5.1) to the negative of the dual function:

$$-q(\mu) = \max_{i \in I} \{-a_i'\mu - b_i\}.$$

Since I is finite, it may be viewed as a compact set, which is embedded in some Euclidean space. We consider $-a_i'\mu - b_i$ to be a function mapping $\Re^r \times I$ to $\Re$. This function is continuous over $\Re^r \times I$, and is convex and differentiable in μ for each fixed $i \in I$. Thus Danskin's Theorem [part (b)] applies, and provides a characterization of the subdifferential of $\max_{i \in I}\{-a_i'\mu - b_i\}$ as the convex hull of the set $\{-a_i \mid i \in I_\mu\}$. This in turn yields the desired characterization of the subdifferential of q.   **Q.E.D.**

Even though a subgradient may not be a direction of ascent at points μ where $q(\mu)$ is nondifferentiable, it still maintains an important property of the gradient: *it makes an angle less than 90 degrees with all ascent directions at* μ, i.e., all the vectors $\alpha(\overline{\mu} - \mu)$ such that $\alpha > 0$ and $q(\overline{\mu}) > q(\mu)$. In particular, *a small move from* μ *along any subgradient at* μ *decreases the distance to any maximizer* μ^* *of* q. This property follows from Eq. (8.1) and is illustrated in Fig. 8.1.2. It will form the basis for a number of dual methods that use subgradients.

8.2 SUBGRADIENT METHODS

If the dual function is differentiable, some of the classical computational methods, such as steepest descent and Newton's method (in case where second derivatives exist as well), may be used. In this chapter, we focus on the case of a nondifferentiable dual function, where the computational methods rely more strongly on the convexity properties of the dual problem. Note that the dual function is typically nondifferentiable for problems where the constraint set X is discrete and there is a duality gap (cf. Fig. 8.1.1).

We consider the dual problem

$$\text{maximize} \quad q(\mu)$$
$$\text{subject to} \quad \mu \in M,$$

where

$$q(\mu) = \inf_{x \in X} L(x, \mu) = \inf_{x \in X} \{f(x) + \mu'g(x)\}$$

and the constraint set M is given by

$$M = \{\mu \mid \mu \geq 0, \, q(\mu) > -\infty\}. \tag{8.3}$$

We assume that for every $\mu \in M$, some vector x_μ that minimizes $L(x, \mu)$ over $x \in X$ can be calculated, yielding a subgradient $g(x_\mu)$ of q at μ [cf. Eq.

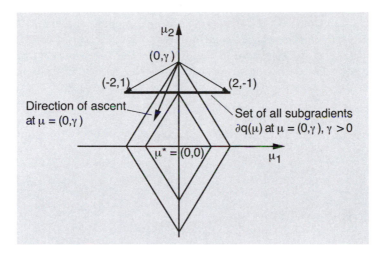

Figure 8.1.2. Illustration of the set of all subgradients $\partial q(\mu)$ at $\mu = (0, 1)$ of the function

$$q(\mu) = -2|\mu_1| - |\mu_2|.$$

By expressing this function as

$$q(\mu) = \min\{-2\mu_1 - \mu_2, 2\mu_1 - \mu_2, -2\mu_1 + \mu_2, 2\mu_1 + \mu_2\},$$

we see that $\partial q(\mu)$ is the convex hull of the gradients of the functions attaining the above minimum (cf. Prop. 8.1.2). Thus, for $\mu = (0, \gamma)$, $\gamma > 0$, the minimum is attained by the first two functions having gradients $(-2, -1)$ and $(2, -1)$. As a result, $\partial q(\mu)$ is as shown in the figure. Note that a subgradient at μ makes an angle less than 90 degrees with every ascent direction at μ. As a result, a small move along a subgradient decreases the distance from the maximizing point $\mu^* = (0, 0)$.

(8.1)]. We also assume that the set M of Eq. (8.3) is closed. This will be true in particular if X is compact, and f and g_j are continuous functions, in which case q is real-valued, so that $M = \{\mu \mid \mu \geq 0\}$.

Subgradient methods are the simplest and among the most popular methods for dual optimization. They generate a sequence of dual feasible points, using a single subgradient at each iteration. The simplest type of subgradient method is given by

$$\mu^{k+1} = P_M(\mu^k + s^k g^k),$$

where g^k denotes the subgradient $g(x_{\mu^k})$, $P_M(\cdot)$ denotes projection on the closed convex set M, and s^k is a positive scalar stepsize. Note that the new iterate may not improve the dual cost for all values of the stepsize; i.e., for some k, we may have

$$q\big(P_M(\mu^k + sg^k)\big) < q(\mu^k), \qquad \forall\, s > 0;$$

see Fig. 8.2.1. However, if the stepsize is small enough, the distance of the current iterate to the optimal solution set is reduced (this was illustrated in Fig. 8.1.2 and is also shown in Fig. 8.2.2 for the case where a projection is involved). The following proposition provides a formal proof and also provides an estimate for the range of appropriate stepsizes.

Proposition 8.2.1: If μ^k is not optimal, then for every dual optimal solution μ^*, we have

$$\|\mu^{k+1} - \mu^*\| < \|\mu^k - \mu^*\|,$$

for all stepsizes s^k such that

$$0 < s^k < \frac{2\big(q(\mu^*) - q(\mu^k)\big)}{\|g^k\|^2}. \tag{8.4}$$

Proof: We have

$$\|\mu^k + s^k g^k - \mu^*\|^2 = \|\mu^k - \mu^*\|^2 - 2s^k(\mu^* - \mu^k)'g^k + (s^k)^2\|g^k\|^2,$$

and by using the subgradient inequality,

$$(\mu^* - \mu^k)'g^k \geq q(\mu^*) - q(\mu^k),$$

we obtain

$$\|\mu^k + s^k g^k - \mu^*\|^2 \leq \|\mu^k - \mu^*\|^2 - 2s^k\big(q(\mu^*) - q(\mu^k)\big) + (s^k)^2\|g^k\|^2.$$

We can be seen that for the range of stepsizes of Eq. (8.4), the sum of the last two terms in the above relation is negative, so that

$$\|\mu^k + s^k g^k - \mu^*\| < \|\mu^k - \mu^*\|.$$

We now observe that since $\mu^* \in M$ and the projection operation is nonexpansive, we have

$$\big\|P_M(\mu^k + s^k g^k) - \mu^*\big\| \leq \|\mu^k + s^k g^k - \mu^*\|.$$

By combining the last two inequalities, the result follows. **Q.E.D.**

The above proposition suggests the stepsize rule

$$s^k = \frac{q^* - q(\mu^k)}{\|g^k\|^2},$$

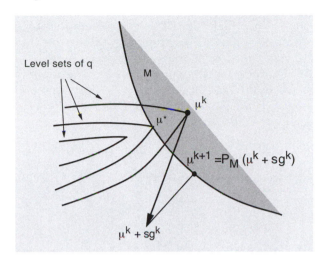

Figure 8.2.1. Illustration of how the iterate $P_M(\mu^k + sg^k)$ may not improve the dual function with a particular choice of subgradient g^k, regardless of the value of the stepsize s.

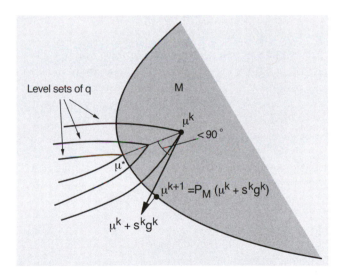

Figure 8.2.2. Illustration of how, given a nonoptimal μ^k, the distance to any optimal solution μ^* is reduced using a subgradient iteration with a sufficiently small stepsize. The crucial fact, which follows from the definition of a subgradient, is that the angle between the subgradient g^k and the vector $\mu^* - \mu^k$ is less than 90 degrees. As a result, if s^k is small enough, the vector $\mu^k + s^k g^k$ is closer to μ^* than μ^k. Through the projection on M, $P_M(\mu^k + s^k g^k)$ gets even closer to μ^*.

where q^* is the optimal dual value. This rule selects the stepsize to be in the middle of the range (8.4). Unfortunately, however, this requires that we know the dual optimal value q^*, which is rare. In practice, one must use some simpler scheme for selecting a stepsize. The simplest possibility is to select s^k to be the same for all k, i.e., $s^k \equiv s$ for some $s > 0$. Then, if the subgradients g^k are bounded ($\|g^k\| \le C$ for some constant C and all k), the preceding proof shows that

$$\|\mu^{k+1} - \mu^*\|^2 \le \|\mu^k - \mu^*\|^2 - 2s\big(q^* - q(\mu^k)\big) + s^2 C^2,$$

and implies that the distance to μ^* decreases ($\|\mu^{k+1} - \mu^*\| < \|\mu^k - \mu^*\|$) if

$$0 < s < \frac{2\big(q^* - q(\mu^k)\big)}{C^2}$$

or equivalently, if μ^k belongs to the level set

$$\left\{ \mu \;\middle|\; q(\mu) < q^* - \frac{sC^2}{2} \right\}.$$

In fact, with a little further analysis based on this argument, we will show in Section 8.2.1 (see Prop. 8.2.2) that the method, at least asymptotically, reaches this level set, i.e.

$$\limsup_{k \to \infty} q(\mu^k) \ge q^* - \frac{sC^2}{2}$$

(see Fig. 8.2.3). Thus, if s is taken to be small enough, the convergence properties of the method are satisfactory. Since a small stepsize may result in slow initial progress of the method, it is common to use a variant of this approach whereby we start with moderate stepsize values s^k, which are progressively reduced up to a small positive value s.

Other possibilities for stepsize choice include a diminishing stepsize, whereby $s^k \to 0$, and schemes that replace the unknown optimal value q^* in Eq. (8.4) with an estimate. We will develop the convergence properties of subgradient methods with such stepsize rules in the context of a more general type of method, which we now introduce.

Incremental Subgradient Methods

We will now discuss an interesting variant of the subgradient method, called *incremental*, which applies to a dual function of the additive form

$$q(\mu) = \sum_{i=1}^{m} q_i(\mu), \tag{8.5}$$

where q_i are concave over M. This structure arises often when separable large scale problems are addressed using dual methods. Here is an example:

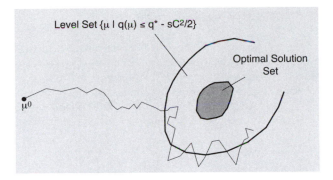

Figure 8.2.3. Illustration of a principal convergence property of the subgradient method with a constant stepsize s, and assuming a bound C on the subgradient norms $\|g^k\|$. When the current iterate is outside the level set

$$\left\{ \mu \;\middle|\; q(\mu) < q^* - \frac{sC^2}{2} \right\},$$

the distance to any optimal solution is reduced at the next iteration. As a result the method gets arbitrarily close to (or inside) this level set.

Example 8.2.1: (Discrete Separable Problems)

Consider a discrete optimization problem of the form

$$\text{minimize} \quad \sum_{i=1}^{m} c_i' x_i$$

$$\text{subject to} \quad x_i \in X_i, \quad i = 1, \ldots, m, \qquad \sum_{i=1}^{m} A_i x_i \le b. \tag{8.6}$$

Here c_i are given vectors in $\Re^p$, X_i is a given finite subset of $\Re^p$, A_i are given $n \times p$ matrices, and b is a given vector in $\Re^n$. By dualizing the coupling constraint $\sum_{i=1}^{m} A_i x_i \le b$, we obtain a dual problem of the form (8.5), where

$$q_i(\mu) = \min_{x_i \in X_i} (c_i + A_i'\mu)' x_i - \beta_i'\mu, \qquad i = 1, \ldots, m, \tag{8.7}$$

and β_i are vectors in $\Re^n$ such that $\beta_1 + \cdots + \beta_m = b$. In this case, the set M is given by $M = \{\mu \in \Re^n \mid \mu \ge 0\}$. Solving dual problems of the type above, possibly within an integer programming context of branch-and-bound, is one of the most important and challenging algorithmic areas of optimization.

Here is another example where the additive structure (8.5) arises in a somewhat different context.

Example 8.2.2: (Stochastic Programming)

A frequently encountered type of large scale optimization problem is *stochastic programming*, which can also be viewed as a two-stage stochastic optimal control problem. Here, a vector $x \in \Re^n$ is first selected from a constraint set X, and following this, a random event occurs that has m possible outcomes numbered $1, \ldots, m$. Another vector $y_i \subset \Re^m$ is then selected with knowledge of the outcome i that occurred, subject to constraints of the form

$$Ax + B_i y_i = b_i, \quad y_i \geq 0, \qquad i = 1, \ldots, m,$$

where the matrices A and B_i, and the vectors b_i are given. The problem is then to minimize the expected cost

$$f_0(x) + \sum_{i=1}^{m} \pi_i d_i' y_i,$$

where $f_0(x)$ is a cost for the choice of x, $d_i' y_i$ is the cost associated with the occurrence of i and the attendant choice of y_i, and π_i is the corresponding probability. Thus the problem can be written as

$$\text{minimize} \quad f_0(x) + \sum_{i=1}^{m} \min_{B_i y_i = b_i - Ax,\, y_i \geq 0} \pi_i d_i' y_i$$

$$\text{subject to } x \in X.$$

Alternatively, by using linear programming duality to replace the minimization over y_i above by its dual, we can write the problem as

$$\text{minimize} \quad f_0(x) + \sum_{i=1}^{m} \max_{B_i' \lambda_i \leq \pi_i d_i} (b_i - Ax)' \lambda_i$$

$$\text{subject to } x \in X.$$

If we introduce the convex functions

$$f_i(x) = \max_{B_i' \lambda_i \leq \pi_i d_i} (b_i - Ax)' \lambda_i, \qquad i = 1, \ldots, m,$$

we can write the problem in the form

$$\text{minimize} \quad \sum_{i=0}^{m} f_i(x)$$

$$\text{subject to } x \in X.$$

Thus the cost function of the problem is additive of the form (8.5).

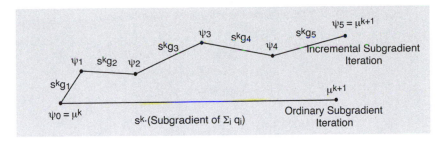

Figure 8.2.4. Illustration of the kth iteration of an incremental subgradient method (here $m = 5$, and we ignore the projection operation, assuming that $M = \Re^n$). The method produces

$$\mu^{k+1} = \mu^k + s^k \sum_{i=1}^{m} g_i$$

after m incremental steps of size $s^k g_i$, where g_i is a subgradient of q_i at ψ_{i-1}. By contrast, the ordinary subgradient method produces $\mu^{k+1} = \mu^k + s^k g$, where g is a subgradient of $q = \sum_{i=1}^{m} q_i$ at μ^k. When there is no projection (i.e., $M = \Re^n$), the difference between the two methods is that the subgradient of q_i, $i = 1, \ldots, m$, is evaluated at ψ_{i-1} in the incremental method versus at ψ_0 in the ordinary method.

The idea of the incremental method is to sequentially take steps along the subgradients of the component functions q_i, with intermediate adjustment of μ after processing each component function. Thus, an iteration is viewed as a cycle of m subiterations. If μ^k is the vector obtained after k cycles, the vector μ^{k+1} obtained after one more cycle is

$$\mu^{k+1} = \psi_m,$$

where ψ_m is obtained after the m steps

$$\psi_i = P_M(\psi_{i-1} + s^k g_i), \qquad i = 1, \ldots, m, \tag{8.8}$$

with g_i being a subgradient of q_i at ψ_{i-1} (see Fig. 8.2.4). The cycle starts with

$$\psi_0 = \mu^k.$$

The motivation for the incremental subgradient method is faster convergence. In particular, we hope that far from the solution, a single cycle of the incremental subgradient method will be as effective as several (as many as m) iterations of the ordinary subgradient method. This type of behavior is illustrated in the following example.

Example 8.2.3: (Incremental Method Behavior)

Assume that μ is a scalar, and that the problem has the form

$$\text{maximize} \quad q(\mu) = \sum_{i=1}^{2(M+N)} q_i(\mu)$$

$$\text{subject to} \quad \mu \in \Re,$$

where the component functions q_i are given by

$$q_i(\mu) = \begin{cases} -|\mu - 1| & \text{for } i = 1, \ldots, M, \\ -|\mu + 1| & \text{for } i = M + 1, \ldots, 2M, \\ -|\mu| & \text{for } i = 2M + 1, \ldots, 2(M + N), \end{cases}$$

and, to illustrate most prominently the effects of incrementalism, we assume that the integers M and N are large and comparable, in the sense that $M/N = O(1)$. For simplicity here, we assume that μ is unconstrained; a similar example can be constructed when there are constraints. The maximum of q is attained for $\mu^* = 0$.

    Consider the incremental subgradient method with a constant stepsize $s \in (0, 1)$. Then for μ outside the interval $[-1, 1]$ that contains the maxima of the component functions q_i, the subgradient of q_i is

$$g_i = \begin{cases} 1 & \text{if } \mu < -1, \\ -1 & \text{if } \mu > 1. \end{cases}$$

Thus each of the steps

$$\psi_i = \psi_{i-1} + sg_i$$

of Eq. (8.8) makes progress towards the maximum $\mu^* = 0$ when $\mu \notin [-1, 1]$, and in fact we have $|\psi_i| = |\psi_{i-1}| - s$.

    However, once the method enters the interval $[-1, 1]$, it exhibits oscillatory behavior. In particular, when a function q_i of the form $-|\mu - 1|$ (or $-|\mu + 1|$) is processed, the method takes a step of size s towards 1 (or -1, respectively). Thus the method generically oscillates, and the asymptotic size of the oscillation within a cycle is roughly proportional to s.

    It is important to note that the size of the oscillation also depends substantially on the *order* in which the functions q_i are processed within a cycle. The maximum oscillation size occurs when N functions $-|\mu|$ are processed, followed by the M functions $-|\mu + 1|$, followed by N functions $-|\mu|$, and followed by the M functions $-|\mu - 1|$. Then it can be seen that the size of the oscillation is of order Ms. The minimum oscillation size occurs when the processing of the M functions $-|\mu + 1|$ is interleaved with the processing of the M functions $-|\mu - 1|$. Then within the interval $[-1, 1]$, the steps corresponding to the functions $-|\mu + 1|$ are essentially canceled by the steps corresponding to the functions $-|\mu - 1|$, and asymptotically the size of the oscillation is a small integer multiple of s (depending on the details of when the component functions $-|\mu|$ are processed).

The preceding example illustrates several characteristics of the incremental method, which tend to manifest themselves in some generality:

(a) When far from the solution, the method can make much faster progress than the nonincremental subgradient method, particularly if the number of component functions m is large. The rate of progress also depends on the stepsize.

(b) When close to the solution, the method oscillates and the size of the oscillation (from start to end of a cycle) is proportional to the stepsize. Thus there is a tradeoff between rapid initial convergence (large stepsize) and size of asymptotic oscillation (small stepsize). With a diminishing stepsize the method is capable of attaining convergence (no asymptotic oscillation).

(c) The size of the oscillation depends also on the order in which the component functions q_i are processed within a cycle.

Order Randomization

To address the potentially detrimental effect of an unfavorable order of processing the component functions, we will introduce *randomization*. In particular, rather than pick each q_i exactly once in each cycle, we will choose the function q_i at random (according to a uniform distribution). We will analyze the potential gains from this approach shortly, but let us first illustrate this idea for the function of Example 8.2.3. In particular, consider a method that at each step, selects independently and with equal probability $1/2(M + N)$ an index i from the range $[1, 2(M + N)]$, and executes the iteration

$$\overline{\mu} = \mu + sg_i,$$

where s is a small constant stepsize s.

If the starting point is an integer multiple of s, then all the points generated by the algorithm will be integer multiples of s. As a result, the algorithm can be modeled by a Markov chain of the random walk type. The stationary probability distribution of the chain can be easily calculated, and one can show that asymptotically, the expected value of μ will be the optimal $\mu^* = 0$, and that its variance will be within a small integer multiple of s, which *does not depend on the number of components* $m = 2(M + N)$ (for large M and N, it essentially depends on the probabilities with which the various components are selected, i.e., the ratio M/N). This is in contrast with the worst-case deterministic processing order, for which, as discussed above, the size of the oscillation is proportional to m (as well as s). Thus, the effect on the oscillation size of a poor order of processing of the components q_i within a cycle is mitigated by randomization.

8.2.1 Analysis of Subgradient Methods

We will now discuss the convergence of subgradient methods. We will provide an analysis of incremental methods, which as a special case, when $m = 1$, applies to the standard (nonincremental) versions as well. We start with methods that use a deterministic fixed order for selecting the component functions q_i for iteration, and we analyze the convergence for three different types of stepsize rules:

(a) A constant stepsize.

(b) A diminishing stepsize.

(c) A dynamically chosen stepsize based on the exact optimal value (cf. Prop. 8.2.1) or a suitable estimate.

We will subsequently, in the next subsection, consider randomized variants of incremental methods with the above stepsize rules. We will analyze the convergence (with probability one) of these methods. For this analysis, we assume that the reader has some knowledge of the concepts and tools of stochastic convergence, and the Supermartingale Convergence Theorem in particular. We will aim to make an important point, namely that *the use of randomization enhances the performance of the method, and mitigates the effects of a poor order for processing the components*, as illustrated in the preceding example.

Rather than focus on the case of a dual problem, we will instead consider a *general minimization problem involving a convex nondifferentiable cost function* (optimization algorithms are usually analyzed in the context of a minimization). This problem is

$$\text{minimize} \quad f(x) = \sum_{i=1}^{m} f_i(x)$$

$$\text{subject to} \quad x \in X,$$

where $f_i : \Re^n \to \Re$ are convex functions, and X is a nonempty, closed, and convex subset of $\Re^n$.

The (nonincremental) subgradient method is given by

$$x_{k+1} = P_X \left(x_k - \alpha_k \sum_{i=1}^{m} d_{i,k} \right),$$

where $d_{i,k}$ is a subgradient of f_i at x_k, α_k is a positive stepsize, and $P_X(\cdot)$ denotes projection on the set X.

In the incremental subgradient method, an iteration is viewed as a cycle of m subiterations. If x_k is the vector obtained after k cycles, the vector x_{k+1} obtained after one more cycle is

$$x_{k+1} = \psi_{m,k}, \tag{8.9}$$

where $\psi_{m,k}$ is obtained after the m steps

$$\psi_{i,k} = P_X\left(\psi_{i-1,k} - \alpha_k g_{i,k}\right), \qquad g_{i,k} \in \partial f_i(\psi_{i-1,k}), \qquad i = 1, \ldots, m, \tag{8.10}$$

starting with

$$\psi_{0,k} = x_k, \tag{8.11}$$

where $\partial f_i(\psi_{i-1,k})$ denotes the subdifferential (set of all subgradients) of f_i at the point $\psi_{i-1,k}$. The updates described by Eq. (8.10) are referred to as the *subiterations* of the kth cycle.

We use the notation

$$f^* = \inf_{x \in X} f(x), \quad X^* = \left\{x \in X \mid f(x) = f^*\right\}, \quad d(x, X^*) = \inf_{x^* \in X^*} \|x - x^*\|,$$

where $\| \cdot \|$ denotes the standard Euclidean norm. *For all the convergence results in this subsection we assume the following*:

Assumption 8.2.1: (Subgradient Boundedness) We have

$$C_i \geq \sup_{k \geq 0}\{\|g\| \mid g \in \partial f_i(x_k) \cup \partial f_i(\psi_{i-1,k})\}, \qquad i = 1, \ldots, m,$$

for some scalars $C_1, \ldots, C_m$.

We note that Assumption 8.2.1 is satisfied if each f_i is real-valued and polyhedral (i.e., f_i is the pointwise maximum of a finite number of affine functions). In particular, Assumption 8.2.1 holds for the dual of an integer programming problem [cf. Eqs. (8.6) and (8.7)], where for each i and all x the set of subgradients $\partial f_i(x)$ is the convex hull of a finite number of points. More generally, since each component f_i is real-valued and convex over the entire space $\Re^n$, the subdifferential $\partial f_i(x)$ is nonempty and compact for all x and i (cf. Prop. 4.2.1). If the set X is compact or the sequences $\{\psi_{i,k}\}$ are bounded, then Assumption 8.2.1 is satisfied since the set $\cup_{x \in B}\partial f_i(x)$ is bounded for any bounded set B [Prop. 4.2.3(a)].

The following is a key lemma that will be used repeatedly in the subsequent convergence analysis.

Lemma 8.2.1: Let $\{x_k\}$ be the sequence generated by the incremental method (8.9)-(8.11). Then for all $y \in X$ and $k \geq 0$, we have

$$\|x_{k+1} - y\|^2 \leq \|x_k - y\|^2 - 2\alpha_k\left(f(x_k) - f(y)\right) + \alpha_k^2 C^2,$$

where $C = \sum_{i=1}^m C_i$ and C_i is as in Assumption 8.2.1.

Proof: Using the nonexpansion property of the projection, the subgradient boundedness (cf. Assumption 8.2.1), and the subgradient inequality for each component function f_i, we obtain for all $y \in X$,

$$
\begin{aligned}
\|\psi_{i,k} - y\|^2 = \left\| P_X \left(\psi_{i-1,k} - \alpha_k g_{i,k} \right) - y \right\|^2 \\
\leq \|\psi_{i-1,k} - \alpha_k g_{i,k} - y\|^2 \\
\leq \|\psi_{i-1,k} - y\|^2 - 2\alpha_k g'_{i,k}(\psi_{i-1,k} - y) + \alpha_k^2 C_i^2 \\
\leq \|\psi_{i-1,k} - y\|^2 - 2\alpha_k \big(f_i(\psi_{i-1,k}) - f_i(y) \big) + \alpha_k^2 C_i^2, \qquad \forall\, i, k.
\end{aligned}
$$

Adding these inequalities over $i = 1, \ldots, m$, we have for all $y \in X$ and k,

$$
\begin{aligned}
\|x_{k+1} - y\|^2 &\leq \|x_k - y\|^2 - 2\alpha_k \sum_{i=1}^{m} \big(f_i(\psi_{i-1,k}) - f_i(y) \big) + \alpha_k^2 \sum_{i=1}^{m} C_i^2 \\
&= \|x_k - y\|^2 - 2\alpha_k \left(f(x_k) - f(y) + \sum_{i=1}^{m} \big(f_i(\psi_{i-1,k}) - f_i(x_k) \big) \right) \\
&\quad + \alpha_k^2 \sum_{i=1}^{m} C_i^2.
\end{aligned}
$$

By strengthening the above inequality, we have for all $y \in X$ and k

$$
\begin{aligned}
\|x_{k+1} - y\|^2 &\leq \|x_k - y\|^2 - 2\alpha_k \big(f(x_k) - f(y) \big) \\
&\quad + 2\alpha_k \sum_{i=1}^{m} C_i \|\psi_{i-1,k} - x_k\| + \alpha_k^2 \sum_{i=1}^{m} C_i^2 \\
&\leq \|x_k - y\|^2 - 2\alpha_k \big(f(x_k) - f(y) \big) \\
&\quad + \alpha_k^2 \left(2 \sum_{i=2}^{m} C_i \left(\sum_{j=1}^{i-1} C_j \right) + \sum_{i=1}^{m} C_i^2 \right) \\
&= \|x_k - y\|^2 - 2\alpha_k \big(f(x_k) - f(y) \big) + \alpha_k^2 \left(\sum_{i=1}^{m} C_i \right)^2 \\
&= \|x_k - y\|^2 - 2\alpha_k \big(f(x_k) - f(y) \big) + \alpha_k^2 C^2,
\end{aligned}
$$

where in the first inequality we use the relation

$$
f_i(x_k) - f_i(\psi_{i-1,k}) \leq \|\tilde{g}_{i,k}\| \cdot \|\psi_{i-1,k} - x_k\| \leq C_i \|\psi_{i-1,k} - x_k\|
$$

with $\tilde{g}_{i,k} \in \partial f_i(x_k)$, and in the second inequality we use the relation

$$
\|\psi_{i,k} - x_k\| \leq \alpha_k \sum_{j=1}^{i} C_j, \qquad i = 1, \ldots, m, \quad k \geq 0,
$$

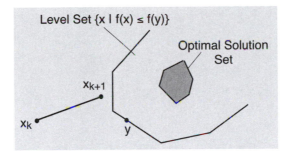

Figure 8.2.5. Illustration of the improvement property of the subgradient method, established in Lemma 8.2.1. For any y, if

$$\alpha_k < \frac{2\big(f(x_k) - f(y)\big)}{C^2},$$

the iterate x_{k+1} is closer to y than x_k.

which follows from Eqs. (8.9)-(8.11) and Assumption 8.2.1.   **Q.E.D.**

Among other things, Lemma 8.2.1 guarantees that given the current iterate x_k and some other point $y \in X$ with lower cost than x_k, the next iterate x_{k+1} will be closer to y than x_k, provided the stepsize α_k is sufficiently small [less than $2\big(f(x_k) - f(y)\big)/C^2$]. This fact, illustrated in Fig. 8.2.5, is used repeatedly, with various choices for y, in the following analysis.

Constant Stepsize Rule

We first consider the case of a constant stepsize rule.

Proposition 8.2.2: Let $\{x_k\}$ be the sequence generated by the incremental method (8.9)-(8.11), with the stepsize α_k fixed at some positive constant α.

(a) If $f^* = -\infty$, then

$$\liminf_{k \to \infty} f(x_k) = f^*.$$

(b) If $f^* > -\infty$, then

$$\liminf_{k \to \infty} f(x_k) \leq f^* + \frac{\alpha C^2}{2},$$

where $C = \sum_{i=1}^{m} C_i$.

Proof: We prove (a) and (b) simultaneously. If the result does not hold, there must exist an $\epsilon > 0$ such that

$$\liminf_{k \to \infty} f(x_k) > f^* + \frac{\alpha C^2}{2} + 2\epsilon.$$

Let $\hat{y} \in X$ be such that

$$\liminf_{k \to \infty} f(x_k) \geq f(\hat{y}) + \frac{\alpha C^2}{2} + 2\epsilon,$$

and let k_0 be large enough so that for all $k \geq k_0$ we have

$$f(x_k) \geq \liminf_{k \to \infty} f(x_k) - \epsilon.$$

By adding the preceding two relations, we obtain for all $k \geq k_0$,

$$f(x_k) - f(\hat{y}) \geq \frac{\alpha C^2}{2} + \epsilon.$$

Using Lemma 8.2.1 for the case where $y = \hat{y}$ together with the above relation, we obtain for all $k \geq k_0$,

$$\|x_{k+1} - \hat{y}\|^2 \leq \|x_k - \hat{y}\|^2 - 2\alpha\epsilon.$$

Thus we have

$$
\begin{aligned}
\|x_{k+1} - \hat{y}\|^2 &\leq \|x_k - \hat{y}\|^2 - 2\alpha\epsilon \\
&\leq \|x_{k-1} - \hat{y}\|^2 - 4\alpha\epsilon \\
&\leq \cdots \\
&\leq \|x_{k_0} - \hat{y}\|^2 - 2(k + 1 - k_0)\alpha\epsilon,
\end{aligned}
$$

which cannot hold for k sufficiently large – a contradiction.   **Q.E.D.**

The preceding proposition involves only the iterates at the end of cycles. However, by shifting the starting index in each cycle and repeating the preceding proof, we see that

$$\liminf_{k \to \infty} f(\psi_{i,k}) \leq f^* + \frac{\alpha C^2}{2}, \qquad \forall\, i = 1, \ldots, m.$$

The next proposition gives an estimate of the number K of cycles needed to guarantee a level of optimality up to the threshold tolerance $\alpha C^2/2$ given in the preceding proposition.

Proposition 8.2.3: Assume that X^* is nonempty. Let $\{x_k\}$ be the sequence generated by the incremental method (8.9)-(8.11) with the stepsize α_k fixed to some positive constant α. Then for any positive scalar ϵ, we have

$$\min_{0 \le k \le K} f(x_k) \le f^* + \frac{\alpha C^2 + \epsilon}{2}, \qquad (8.12)$$

where K is given by

$$K = \left\lfloor \frac{\left(d(x_0, X^*)\right)^2}{\alpha \epsilon} \right\rfloor.$$

Proof: Assume, to arrive at a contradiction, that Eq. (8.12) does not hold, so that for all k with $0 \le k \le K$, we have

$$f(x_k) > f^* + \frac{\alpha C^2 + \epsilon}{2}.$$

By using this relation in Lemma 8.2.1 with α_k is replaced by α, we obtain for all k with $0 \le k \le K$,

$$
\begin{aligned}
\left(d(x_{k+1}, X^*)\right)^2 &\le \left(d(x_k, X^*)\right)^2 - 2\alpha\big(f(x_k) - f^*\big) + \alpha^2 C^2 \\
&\le \left(d(x_k, X^*)\right)^2 - (\alpha^2 C^2 + \alpha\epsilon) + \alpha^2 C^2 \\
&= \left(d(x_k, X^*)\right)^2 - \alpha\epsilon.
\end{aligned}
$$

Summation of the above inequalities over k for $k = 0, \dots, K$, yields

$$\left(d(x_{K+1}, X^*)\right)^2 \le \left(d(x_0, X^*)\right)^2 - (K+1)\alpha\epsilon,$$

so that

$$\left(d(x_0, X^*)\right)^2 - (K+1)\alpha\epsilon \ge 0,$$

which contradicts the definition of K.   **Q.E.D.**

Note that the estimate (8.12) involves only the iterates obtained at the end of cycles. Since every cycle consists of m subiterations, the total number N of component functions that must be processed in order for Eq. (8.12) to hold is given by

$$N = mK = m \left\lfloor \frac{\left(d(x_0, X^*)\right)^2}{\alpha \epsilon} \right\rfloor.$$

Figure 8.2.6 illustrates the convergence and rate of convergence properties established in the preceding two propositions.

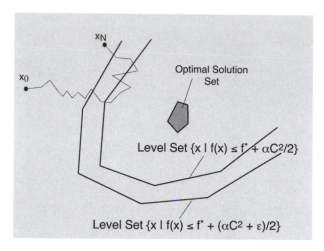

Figure 8.2.6. Illustration of the convergence process of the incremental subgradient method for the case of a constant stepsize α: after no more than

$$K = \left\lfloor \frac{\big(d(x_0, X^*)\big)^2}{\alpha\epsilon} \right\rfloor$$

iterations, the level set

$$\left\{ x \mid f(x) \leq f^* + \frac{\alpha C^2 + \epsilon}{2} \right\}.$$

will be reached.

The Role of the Order of Processing the Components

The error tolerance estimate $\alpha C^2/2$ of Prop. 8.2.2 and the upper bound

$$\liminf_{k\to\infty} f(\psi_{i,k}) \leq f^* + \frac{\alpha C^2}{2}, \qquad \forall\, i = 1,\ldots,m. \tag{8.13}$$

assume the *worst* possible order of processing the components f_i within a cycle. One question that arises is whether this bound is sharp, in the sense that there exists a problem and a processing order, such that for each stepsize α, we can find a starting point for which the sequence $\{\psi_{i,k}\}$ generated by the method satisfies Eq. (8.13). The exercises provide an example where the bound is satisfied within a constant that is independent of the problem data, i.e., for an unfavorable processing order and starting point, the method satisfies

$$\liminf_{k\to\infty} f(\psi_{i,k}) = f^* + \frac{\beta\alpha C^2}{2}, \qquad \forall\, i = 1,\ldots,m,$$

where β is a positive constant that is fairly close to 1. Thus, there is not much room for improvement of the worst-order error tolerance estimate $\alpha C^2/2$.

On the other hand, suppose that we are able to choose the *best* possible order of processing the components f_i within a cycle. Would it then be possible to lower the tolerance estimate $\alpha C^2/2$, and by how much? We claim that with such an optimal choice, it is impossible to lower the tolerance estimate by more than a factor of m. To see this, consider the case where the f_i are the one-dimensional functions $f_i(x) = (C/m)|x|$. Then, because all functions f_i are identical, the order of processing the components is immaterial. If we start at $x_0 = (\alpha C)/2m$, then it can be seen that the method oscillates between x_0 and $-x_0$, and the corresponding function value is

$$f(x_0) = f(-x_0) = \sum_{i=1}^{m} \frac{C}{m} \left| \frac{\alpha C}{2m} \right| = \frac{\alpha C^2}{2m}.$$

Since $f^* = 0$, this example shows that there exists a problem and a starting point such that

$$\liminf_{k \to \infty} f(\psi_{i,k}) = f^* + \frac{\alpha C^2}{2m}, \qquad \forall\, i = 1, \ldots, m. \qquad (8.14)$$

Thus from Eqs. (8.13) and (8.14), we see that for a given stepsize α, the achievable range for the bound

$$\liminf_{k \to \infty} f(\psi_{i,k}) - f^*$$

corresponding to the incremental subgradient method with a fixed processing order is

$$\left[\frac{\alpha C^2}{2m}, \frac{\alpha C^2}{2} \right]. \qquad (8.15)$$

By this we mean that there exists a choice of problem for which we can do no better that the lower end of the above range, even with optimal processing order choice; moreover, for all problems and processing orders, we will do no worse than the upper end of the above range.

From the bound range (8.15), it can be seen that for a given stepsize α, there is significant difference in the performance of the method with the best and the worst processing orders. Unfortunately, it is difficult to find the best processing order for a given problem. In the next section, we will show that, remarkably, *by randomizing the order, we can achieve the lower tolerance error estimate*

$$\frac{\alpha C^2}{2m}$$

in an expected sense.

Diminishing Stepsize Rule

We now consider the case where the stepsize α_k diminishes to zero, but satisfies $\sum_{k=0}^{\infty} \alpha_k = \infty$ (so that the method can "travel" infinitely far if necessary to attain convergence). An example of such a stepsize is $\alpha_k = \beta/(k + \gamma)$, where β and γ are some positive scalars.

Proposition 8.2.4: Assume that the stepsize α_k satisfies

$$\lim_{k \to \infty} \alpha_k = 0, \qquad \sum_{k=0}^{\infty} \alpha_k = \infty.$$

Then, for the sequence $\{x_k\}$ generated by the incremental method (8.9)-(8.11), we have

$$\liminf_{k \to \infty} f(x_k) = f^*.$$

Proof: Assume, to arrive at a contradiction, that the above relation does not hold, so there exists an $\epsilon > 0$ such that

$$\liminf_{k \to \infty} f(x_k) - 2\epsilon > f^*.$$

Then, using the convexity of f and X, there exists a point $\hat{y} \in X$ such that

$$\liminf_{k \to \infty} f(x_k) - 2\epsilon \geq f(\hat{y}) > f^*.$$

Let k_0 be large enough so that for all $k \geq k_0$, we have

$$f(x_k) \geq \liminf_{k \to \infty} f(x_k) - \epsilon.$$

By combining the preceding two relations, we obtain for all $k \geq k_0$,

$$f(x_k) - f(\hat{y}) \geq \liminf_{k \to \infty} f(x_k) - \epsilon - f(\hat{y}) \geq \epsilon.$$

By setting $y = \hat{y}$ in Lemma 8.2.1, and by using the above relation, we have for all $k \geq k_0$,

$$\|x_{k+1} - \hat{y}\|^2 \leq \|x_k - \hat{y}\|^2 - 2\alpha_k \epsilon + \alpha_k^2 C^2 = \|x_k - \hat{y}\|^2 - \alpha_k \left(2\epsilon - \alpha_k C^2\right).$$

Since $\alpha_k \to 0$, without loss of generality, we may assume that k_0 is large enough so that

$$2\epsilon - \alpha_k C^2 \geq \epsilon, \qquad \forall \, k \geq k_0.$$

Therefore for all $k \geq k_0$ we have

$$\|x_{k+1} - \hat{y}\|^2 \leq \|x_k - \hat{y}\|^2 - \alpha_k \epsilon \leq \cdots \leq \|x_{k_0} - \hat{y}\|^2 - \epsilon \sum_{j=k_0}^{k} \alpha_j,$$

which cannot hold for k sufficiently large.   **Q.E.D.**

If we assume in addition that X^* is nonempty and bounded, Prop. 8.2.4 can be strengthened as follows.

Proposition 8.2.5: Assume that X^* is nonempty and bounded, and that the stepsize α_k is such that

$$\lim_{k \to \infty} \alpha_k = 0, \qquad \sum_{k=0}^{\infty} \alpha_k = \infty.$$

Then, for the sequence $\{x_k\}$ generated by the incremental method (8.9)-(8.11), we have

$$\lim_{k \to \infty} d(x_k, X^*) = 0, \qquad \lim_{k \to \infty} f(x_k) = f^*.$$

Proof: The idea is to show that, once x_k enters a certain level set, it cannot get too far away from that set. Fix a $\gamma > 0$, consider the level set

$$L_\gamma = \{y \in X \mid f(y) \leq f^* + \gamma\},$$

which is compact (in view of the boundedness of X^* and the continuity of f), and denote

$$d(\gamma) = \max_{y \in L_\gamma} d(y, X^*).$$

Consider iterations $k \geq k_0$, where k_0 is such that $\gamma \geq \alpha_k C^2$ for all $k \geq k_0$. We distinguish two cases:

Case 1: $x_x \in L_\gamma$. From the iteration (8.9)-(8.11), we have $\|x_{k+1} - x_k\| \leq \alpha_k C$, implying that for all $x^* \in X^*$,

$$\|x_{k+1} - x^*\| \leq \|x_k - x^*\| + \|x_{k+1} - x_k\| \leq \|x_k - x^*\| + \alpha_k C.$$

By taking the minimum above over $x^* \in X^*$, we obtain

$$d(x_{k+1}, X^*) \leq d(\gamma) + \alpha_k C. \tag{8.16}$$

Case 2: $x_k \notin L_\gamma$. From Lemma 8.2.1 we have for all $x^* \in X^*$ and k,

$$\|x_{k+1} - x^*\|^2 \leq \|x_k - x^*\|^2 - 2\alpha_k (f(x_k) - f^*) + \alpha_k^2 C^2. \tag{8.17}$$

Hence, using the fact $\gamma \geq \alpha_k C^2$, we have

$$\begin{aligned}
\|x_{k+1} - x^*\|^2 &< \|x_k - x^*\|^2 - 2\gamma\alpha_k + \alpha_k^2 C^2 \\
&= \|x_k - x^*\|^2 - \alpha_k(2\gamma - \alpha_k C^2) \\
&\leq \|x_k - x^*\|^2 - \alpha_k\gamma,
\end{aligned}$$

implying that

$$d(x_{k+1}, X^*) \leq d(x_k, X^*) - \alpha_k\gamma. \tag{8.18}$$

Thus, for sufficiently large k, when the current iterate x_k is in L_γ, and hence its distance $d(x_k, X^*)$ is within $d(\gamma)$, the distance of the next iterate $d(x_{k+1}, X^*)$ remains within $d(\gamma) + \alpha_k C$ [cf. Eq. (8.16)], and when x_k gets outside L_γ, the distance $d(x_{k+1}, X^*)$ decreases [cf. Eq. (8.18)]. Furthermore, x_k eventually gets within L_γ, in view of Eq. (8.18) and the fact $\sum_{k=0}^{\infty} \alpha_k = \infty$. It follows that for all sufficiently large k, we have

$$d(x_k, X^*) \leq d(\gamma) + \alpha_k C.$$

Therefore, since $\alpha_k \to 0$,

$$\limsup_{k\to\infty} d(x_k, X^*) \leq d(\gamma), \qquad \forall \, \gamma > 0.$$

In view of the compactness of the level sets of f, we have $\lim_{\gamma \to 0} d(\gamma) = 0$, so that $d(x_k, X^*) \to 0$. This relation also implies that $f(x_k) \to f^*$. **Q.E.D.**

Proposition 8.2.5 does not guarantee convergence of the entire sequence $\{x_k\}$. With slightly different assumptions that include an additional mild restriction on α_k, this convergence is guaranteed, as indicated in the following proposition.

Proposition 8.2.6: Assume that X^* is nonempty and that the step-size α_k is such that

$$\sum_{k=0}^{\infty} \alpha_k = \infty, \qquad \sum_{k=0}^{\infty} \alpha_k^2 < \infty.$$

Then the sequence $\{x_k\}$ generated by the incremental method (8.9)-(8.11) converges to some optimal solution.

Proof: From Eq. (8.17), we have for any $x^* \in X^*$ and k,

$$\|x_{k+1} - x^*\|^2 \leq \|x_k - x^*\|^2 - 2\alpha_k \big(f(x_k) - f^*\big) + \alpha_k^2 C^2. \qquad (8.19)$$

Hence for any integer $\overline{k}$, we have

$$\|x_{\overline{k}} - x^*\|^2 \leq \|x_0 - x^*\|^2 + C^2 \sum_{k=0}^{\overline{k}-1} \alpha_k^2,$$

and since $\sum_{k=\overline{k}-1}^{\infty} \alpha_k^2 < \infty$, it follows that the sequence $\{x_k\}$ is bounded. Let $\{x_{k_j}\}$ be a subsequence converging to a limit point $\overline{x}^*$, which in view of Prop. 8.2.5, must belong to X^*. From Eq. (8.19), we have for all j and all integers $\overline{k} > k_j$,

$$\|x_{\overline{k}} - \overline{x}^*\|^2 \leq \|x_{k_j} - \overline{x}^*\|^2 + C^2 \sum_{k=k_j}^{\overline{k}-1} \alpha_k^2,$$

and by taking the limit as $\overline{k} \to \infty$, we obtain for all j,

$$\limsup_{\overline{k} \to \infty} \|x_{\overline{k}} - \overline{x}^*\|^2 \leq \|x_{k_j} - \overline{x}^*\|^2 + C^2 \sum_{k=k_j}^{\infty} \alpha_k^2.$$

The right-hand side tends to 0 as $j \to \infty$, so that $\limsup_{\overline{k} \to \infty} \|x_{\overline{k}} - \overline{x}^*\|^2 = 0$, implying that the entire sequence $\{x_k\}$ converges to $\overline{x}^*$. **Q.E.D.**

Dynamic Stepsize Rule for Known f^*

The preceding results apply to the constant and the diminishing stepsize choices. An interesting alternative for the ordinary subgradient method is the dynamic stepsize rule

$$\alpha_k = \gamma_k \frac{f(x_k) - f^*}{\|g_k\|^2},$$

with $g_k \in \partial f(x_k)$, $0 < \underline{\gamma} \leq \gamma_k \leq \overline{\gamma} < 2$, motivated by Prop. 8.2.1. For the incremental method, we introduce a variant of this stepsize where $\|g_k\|$ is replaced by the upper bound $C = \sum_{i=1}^{m} C_i$ of Assumption 8.2.1:

$$\alpha_k = \gamma_k \frac{f(x_k) - f^*}{C^2}, \qquad 0 < \underline{\gamma} \leq \gamma_k \leq \overline{\gamma} < 2, \qquad (8.20)$$

where

$$C = \sum_{i=1}^{m} C_i \qquad (8.21)$$

and

$$C_i \geq \sup_{k \geq 0}\{\|g\| \mid g \in \partial f_i(x_k) \cup \partial f_i(\psi_{i-1,k})\}, \qquad i = 1, \dots, m. \qquad (8.22)$$

For this choice of stepsize we must be able to calculate suitable upper bounds C_i, which can be done, for example, when the components f_i are polyhedral.

We first consider the case where f^* is known. We later modify the stepsize, so that f^* can be replaced by a dynamically updated estimate.

Proposition 8.2.7: Assume that X^* is nonempty. Then the sequence $\{x_k\}$ generated by the incremental method (8.9)-(8.11) with the dynamic stepsize rule (8.20)–(8.22) converges to some optimal solution.

Proof: From Lemma 8.2.1 with $y = x^* \in X^*$, we have

$$\|x_{k+1} - x^*\|^2 \leq \|x_k - x^*\|^2 - 2\alpha_k \big(f(x_k) - f^*\big) + \alpha_k^2 C^2, \quad \forall \, x^* \in X^*, \ k \geq 0,$$

and by using the definition of α_k [cf. Eq. (8.20)], we obtain

$$\|x_{k+1} - x^*\|^2 \leq \|x_k - x^*\|^2 - \gamma(2 - \overline{\gamma}) \frac{\big(f(x_k) - f^*\big)^2}{C^2}, \qquad \forall \, x^* \in X^*, \ k \geq 0.$$

This implies that $\{x_k\}$ is bounded. Furthermore, $f(x_k) \to f^*$, since otherwise we would have $\|x_{k+1} - x^*\| \leq \|x_k - x^*\| - \epsilon$ for some suitably small $\epsilon > 0$ and infinitely many k. Hence for any limit point $\overline{x}$ of $\{x_k\}$, we have $\overline{x} \in X^*$, and since the sequence $\{\|x_k - x^*\|\}$ is decreasing, it converges to $\|\overline{x} - x^*\|$ for every $x^* \in X^*$. If there are two distinct limit points $\tilde{x}$ and $\overline{x}$ of $\{x_k\}$, we must have $\tilde{x} \in X^*$, $\overline{x} \in X^*$, and $\|\tilde{x} - x^*\| = \|\overline{x} - x^*\|$ for all $x^* \in X^*$, which is possible only if $\tilde{x} = \overline{x}$. **Q.E.D.**

Dynamic Stepsize Rule for Unknown f^*

In most practical problems the value f^* is not known. In this case we may modify the dynamic stepsize (8.20) by replacing f^* with an estimate. This leads to the stepsize rule

$$\alpha_k = \gamma_k \frac{f(x_k) - f_k}{C^2}, \qquad 0 < \underline{\gamma} \leq \gamma_k \leq \overline{\gamma} < 2, \qquad \forall \, k \geq 0, \qquad (8.23)$$

where $C = \sum_{i=1}^m C_i$ is the upper bound of Assumption 8.2.1, and f_k is an estimate of f^*.

We discuss two procedures for updating f_k. In both procedures, f_k is equal to the best function value $\min_{0\leq j\leq k} f(x_j)$ achieved up to the kth iteration minus a positive amount δ_k, which is adjusted based on the algorithm's progress. The first adjustment procedure is simple but is only guaranteed to yield a δ-optimal objective function value, with δ positive and arbitrarily small (unless $f^* = -\infty$ in which case the procedure yields the optimal function value). The second adjustment procedure for f_k is more complex but is guaranteed to yield the optimal value f^* in the limit.

In the first adjustment procedure, f_k is given by

$$f_k = \min_{0\leq j\leq k} f(x_j) - \delta_k, \tag{8.24}$$

and δ_k is updated according to

$$\delta_{k+1} = \begin{cases} \rho\delta_k & \text{if } f(x_{k+1}) \leq f_k, \\ \max\{\beta\delta_k, \delta\} & \text{if } f(x_{k+1}) > f_k, \end{cases} \tag{8.25}$$

where δ, β, and ρ are fixed positive constants with $\beta < 1$ and $\rho \geq 1$. Thus in this procedure, we essentially "aspire" to reach a target level that is smaller by δ_k over the best value achieved thus far. Whenever the target level is achieved, we increase δ_k or we keep it at the same value depending on the choice of ρ. If the target level is not attained at a given iteration, δ_k is reduced up to a threshold δ. This threshold guarantees that the stepsize α_k of Eq. (8.23) is bounded away from zero, since from Eq. (8.24), we have $f(x_k) - f_k \geq \delta$ and hence

$$\alpha_k \geq \gamma \frac{\delta}{C^2}.$$

As a result, the method behaves similar to the one with a constant stepsize (cf. Prop. 8.2.2), as indicated by the following proposition.

Proposition 8.2.8: Let $\{x_k\}$ be the sequence generated by the incremental method (8.9)-(8.11), and the dynamic stepsize rule (8.23) with the adjustment procedure (8.24)–(8.25).

(a) If $f^* = -\infty$, then

$$\inf_{k\geq 0} f(x_k) = f^*.$$

(b) If $f^* > -\infty$, then

$$\inf_{k\geq 0} f(x_k) \leq f^* + \delta.$$

Proof: To arrive at a contradiction, assume that

$$\inf_{k\geq 0} f(x_k) > f^* + \delta. \tag{8.26}$$

Each time the target level is attained [i.e., $f(x_k) \leq f_{k-1}$], the current best function value $\min_{0 \leq j \leq k} f(x_j)$ decreases by at least δ [cf. Eqs. (8.24) and (8.25)], so in view of Eq. (8.26), the target value can be attained only a finite number of times. From Eq. (8.25) it follows that after finitely many iterations, δ_k is decreased to the threshold value and remains at that value for all subsequent iterations, i.e., there is an index $\overline{k}$ such that

$$\delta_k = \delta, \qquad \forall \, k \geq \overline{k}. \tag{8.27}$$

In view of Eq. (8.26), there exists $\overline{y} \in X$ such that $\inf_{k \geq 0} f(x_k) - \delta \geq f(\overline{y})$. From Eqs. (8.24) and (8.27), we have

$$f_k = \min_{0 \leq j \leq k} f(x_j) - \delta \geq \inf_{k \geq 0} f(x_k) - \delta \geq f(\overline{y}), \qquad \forall \, k \geq \overline{k},$$

so that

$$\alpha_k \big(f(x_k) - f(\overline{y})\big) \geq \alpha_k \big(f(x_k) - f_k\big) = \gamma_k \left(\frac{f(x_k) - f_k}{C}\right)^2, \qquad \forall \, k \geq \overline{k}.$$

By using Lemma 8.2.1 with $y = \overline{y}$, we have

$$\|x_{k+1} - \overline{y}\|^2 \leq \|x_k - \overline{y}\|^2 - 2\alpha_k \big(f(x_k) - f(\overline{y})\big) + \alpha_k^2 C^2, \qquad \forall \, k \geq 0.$$

By combining the preceding two relations and the definition of α_k [cf. Eq. (8.23)], we obtain

$$\|x_{k+1} - \overline{y}\|^2 \leq \|x_k - \overline{y}\|^2 - 2\gamma_k \left(\frac{f(x_k) - f_k}{C}\right)^2 + \gamma_k^2 \left(\frac{f(x_k) - f_k}{C}\right)^2$$

$$= \|x_k - \overline{y}\|^2 - \gamma_k(2 - \gamma_k) \left(\frac{f(x_k) - f_k}{C}\right)^2$$

$$\leq \|x_k - \overline{y}\|^2 - \underline{\gamma}(2 - \overline{\gamma})\frac{\delta^2}{C^2}, \qquad \forall \, k \geq \overline{k},$$

where the last inequality follows from the facts $\gamma_k \in [\underline{\gamma}, \overline{\gamma}]$ and $f(x_k) - f_k \geq \delta$ for all k. By summing the above inequalities over k, we have

$$\|x_k - \overline{y}\|^2 \leq \|x_{\overline{k}} - \overline{y}\|^2 - (k - \overline{k})\underline{\gamma}(2 - \overline{\gamma})\frac{\delta^2}{C^2}, \qquad \forall \, k \geq \overline{k},$$

which cannot hold for sufficiently large k – a contradiction.   **Q.E.D.**

We now consider another procedure for adjusting f_k, which guarantees that $f_k \rightarrow f^*$, and that the associated method converges to the optimum. In this procedure we reduce δ_k whenever the method "travels" for a long distance without reaching the corresponding target level.

Path-Based Incremental Target Level Algorithm

Step 0: (*Initialization*) Select x_0, $\delta_0 > 0$, and $B > 0$. Set $\sigma_0 = 0$, $f_{-1} = \infty$. Set $k = 0$, $l = 0$, and $k(l) = 0$ [$k(l)$ will denote the iteration number when the l-th update of f_k occurs].

Step 1: (*Function evaluation*) Calculate $f(x_k)$. If $f(x_k) < f_{k-1}$, then set $f_k = f(x_k)$. Otherwise set $f_k = f_{k-1}$ [so that f_k keeps the record of the smallest value attained by the iterates that are generated so far, i.e., $f_k = \min_{0 \le j \le k} f(x_j)$].

Step 2: (*Sufficient descent*) If $f(x_k) \le f_{k(l)} - \frac{\delta_l}{2}$, then set $k(l+1) = k$, $\sigma_k = 0$, $\delta_{l+1} = \delta_l$, increase l by 1, and go to Step 4.

Step 3: (*Oscillation detection*) If $\sigma_k > B$, then set $k(l+1) = k$, $\sigma_k = 0$, $\delta_{l+1} = \frac{\delta_l}{2}$, and increase l by 1.

Step 4: (*Iterate update*) Set $f_k = f_{k(l)} - \delta_l$. Select $\gamma_k \in [\underline{\gamma}, \overline{\gamma}]$ and compute x_{k+1} via Eqs. (8.9)-(8.11) using the stepsize (8.23).

Step 5: (*Path length update*) Set $\sigma_{k+1} = \sigma_k + \alpha_k C$. Increase k by 1 and go to Step 1.

The algorithm uses the same target level $f_k = f_{k(l)} - \delta_l$ for $k = k(l), k(l) + 1, \ldots, k(l + 1) - 1$. The target level is updated only if sufficient descent or oscillation is detected (Step 2 or Step 3, respectively). It can be shown that the value σ_k is an upper bound on the length of the path traveled by iterates $x_{k(l)}, \ldots, x_k$ for $k < k(l + 1)$. Whenever σ_k exceeds the prescribed upper bound B on the path length, the parameter δ_l is decreased, which (possibly) increases the target level f_k.

We will show that $\inf_{k \ge 0} f(x_k) = f^*$ even if f^* is not finite. First, we give a preliminary result showing that the target values f_k are updated infinitely often (i.e., $l \to \infty$), and that $\inf_{k \ge 0} f(x_k) = -\infty$ if δ_l is nondiminishing.

Lemma 8.2.2: Let $\{x_k\}$ be the sequence generated by the path-based incremental target level algorithm. Then, we have $l \to \infty$, and either $\inf_{k \ge 0} f(x_k) = -\infty$ or $\lim_{l \to \infty} \delta_l = 0$.

Proof: Assume that l takes only a finite number of values, say $l = 0, 1, \ldots, \bar{l}$. In this case we have $\sigma_k + \alpha_k C = \sigma_{k+1} \le B$ for all $k \ge k(\bar{l})$, so that $\lim_{k \to \infty} \alpha_k = 0$. But this is impossible, since for all $k \ge k(\bar{l})$ we have

$$\alpha_k = \gamma_k \frac{f(x_k) - f_k}{C^2} \ge \underline{\gamma} \frac{\delta_{\bar{l}}}{C^2} > 0.$$

Hence we must have $l \to \infty$.

Let $\delta = \lim_{l\to\infty} \delta_l$. If $\delta > 0$, then from Steps 2 and 3 it follows that for all l large enough, we have $\delta_l = \delta$ and

$$f_{k(l+1)} - f_{k(l)} \leq -\frac{\delta}{2},$$

implying that $\inf_{k\geq 0} f(x_k) = -\infty$. **Q.E.D.**

We have the following convergence result.

Proposition 8.2.9: Let $\{x_k\}$ be the sequence generated by the path-based incremental target level algorithm. Then, we have

$$\inf_{k\geq 0} f(x_k) = f^*.$$

Proof: If $\lim_{l\to\infty} \delta_l > 0$, then, according to Lemma 8.2.2, we have

$$\inf_{k\geq 0} f(x_k) = -\infty$$

and we are done, so assume that $\lim_{l\to\infty} \delta_l = 0$. Consider the set L given by

$$L = \left\{ l \in \{1, 2, \ldots\} \ \middle| \ \delta_l = \frac{\delta_{l-1}}{2} \right\},$$

and note that L is an infinite set. Then, from Steps 3 and 5, we obtain

$$\sigma_k = \sigma_{k-1} + \alpha_{k-1}C = \sum_{j=k(l)}^{k-1} C\alpha_j,$$

so that $k(l+1) = k$ and $l+1 \in L$ whenever $\sum_{j=k(l)}^{k-1} \alpha_j C > B$ at Step 3. Hence

$$\sum_{j=k(l-1)}^{k(l)-1} \alpha_j > \frac{B}{C}, \qquad \forall\, l \in L.$$

Since $\delta_l =\to 0$ and L is an infinite set, we have

$$\sum_{k=0}^{\infty} \alpha_k \geq \sum_{l\in L} \sum_{j=k(l-1)}^{k(l)-1} \alpha_j > \sum_{l\in L} \frac{B}{C} = \infty. \tag{8.28}$$

Now, in order to arrive at a contradiction, assume that $\inf_{k\geq 0} f(x_k) > f^*$, so that for some $\hat{y} \in X$ and some $\epsilon > 0$

$$\inf_{k\geq 0} f(x_k) - \epsilon \geq f(\hat{y}). \tag{8.29}$$

Since $\delta_l \to 0$, there is a large enough $\hat{l}$ such that $\delta_l \leq \epsilon$ for all $l \geq \hat{l}$, implying that for all $k \geq k(\hat{l})$

$$f_k = f_{k(l)} - \delta_l \geq \inf_{k \geq 0} f(x_k) - \epsilon \geq f(\hat{y}).$$

Using this relation, Lemma 8.2.1 with $y = \hat{y}$, and the definition of α_k, we obtain

$$\|x_{k+1} - \hat{y}\|^2 \leq \|x_k - \hat{y}\|^2 - 2\alpha_k \big(f(x_k) - f(\hat{y})\big) + \alpha_k^2 C^2.$$
$$\leq \|x_k - \hat{y}\|^2 - 2\alpha_k \big(f(x_k) - f_k\big) + \alpha_k^2 C^2$$
$$= \|x_k - \hat{y}\|^2 - \gamma_k (2 - \gamma_k) \frac{\big(f(x_k) - f_k\big)^2}{C^2}$$
$$\leq \|x_k - \hat{y}\|^2 - \underline{\gamma}(2 - \overline{\gamma}) \frac{\big(f(x_k) - f_k\big)^2}{C^2}.$$

By summing these inequalities over $k \geq k(\hat{l})$, we have

$$\frac{\underline{\gamma}(2 - \overline{\gamma})}{C^2} \sum_{k=k(\hat{l})}^{\infty} \big(f(x_k) - f_k\big)^2 \leq \|x_{k(\hat{l})} - \hat{y}\|^2.$$

By the definition of α_k in Eq. (8.23), it follows that $\sum_{k=k(\hat{l})}^{\infty} \alpha_k^2 < \infty$, implying that $\alpha_k \to 0$. Since $\sum_{k=0}^{\infty} \alpha_k = \infty$ [cf. Eq. (8.28)], according to Prop. 8.2.4, we must have

$$\liminf_{k \to \infty} f(x_k) = f^*.$$

Hence $\inf_{k \geq 0} f(x_k) = f^*$, which contradicts Eq. (8.29). **Q.E.D.**

In an attempt to improve the efficiency of the path-based incremental target level algorithm, one may introduce parameters $\beta, \tau \in (0, 1)$ and $\rho \geq 1$ (whose values will be fixed at Step 0), and modify Steps 2 and 3 as follows:

Step 2′: If $f(x_k) \leq f_{k(l)} - \tau\delta_l$, then set $k(l + 1) = k$, $\sigma_k = 0$, $\delta_{l+1} = \rho\delta_l$, increase l by 1, and go to Step 4.

Step 3′: If $\sigma_k > B$, then set $k(l + 1) = k$, $\sigma_k = 0$, $\delta_{l+1} = \beta\delta_l$, and increase l by 1.

It can be seen that the result of Prop. 8.2.9 still holds for this modified algorithm. If we choose $\rho > 1$ at Step 3′, then in the proofs of Lemma 8.2.2 and Prop. 8.2.9 we have to replace $\lim_{l \to \infty} \delta_l$ with $\limsup_{l \to \infty} \delta_l$.

Let us remark that there is no need to keep the path bound B fixed. Instead, we can decrease B as the method progresses. However, in order to preserve the convergence result of Prop. 8.2.9, we have to ensure that

B remains bounded away from zero for otherwise f_k might converge to a nonoptimal value.

We finally note that all the results presented in this section are valid for the variant of the incremental method that does not use projections within the cycles, but rather employs projections at the end of cycles:

$$\psi_{i,k} = \psi_{i-1,k} - \alpha_k g_{i,k}, \qquad g_{i,k} \in \partial f_i(\psi_{i-1,k}), \quad i = 1, \ldots, m,$$

where $\psi_{0,k} = x_k$ and the iterate x_{k+1} is given by

$$x_{k+1} = P_X(\psi_{m,k}).$$

This variant may be of interest if the projection is a relatively expensive operation.

8.2.2   Subgradient Methods with Randomization

It can be verified that the convergence analysis of the preceding subsection goes through assuming any order for processing the component functions f_i, as long as each component is taken into account exactly once within a cycle. In particular, at the beginning of each cycle, we could reorder the components f_i by either shifting or reshuffling and then proceed with the calculations until the end of the cycle. However, the order used can significantly affect the rate of convergence of the method. Unfortunately, determining the most favorable order may be very difficult in practice. A technique that is popular in practice for incremental methods is to reshuffle randomly the order of the functions f_i at the beginning of each cycle. A variation of this method is to pick randomly a function f_i at each iteration rather than to pick each f_i exactly once in every cycle according to a randomized order. In this section, we analyze this type of method for the constant, diminishing, and dynamic stepsize rules.

We focus on the randomized method given by

$$x_{k+1} = P_X\big(x_k - \alpha_k g(\omega_k, x_k)\big), \tag{8.30}$$

where ω_k is a random variable taking equiprobable values from the set $\{1, \ldots, m\}$, and $g(\omega_k, x_k)$ is a subgradient of the component f_{ω_k} at x_k. This simply means that if the random variable ω_k takes a value j, then the vector $g(\omega_k, x_k)$ is a subgradient of f_j at x_k. Throughout this section we assume the following.

Assumption 8.2.2: For the randomized method (8.30):

(a) $\{\omega_k\}$ is a sequence of independent random variables, each uniformly distributed over the set $\{1, \ldots, m\}$. Furthermore, the sequence $\{\omega_k\}$ is independent of the sequence $\{x_k\}$.

(b) The set of subgradients $\{g(\omega_k, x_k) \mid k = 0, 1, \ldots\}$ is bounded, i.e., there exists a positive constant C_0 such that with probability 1

$$\|g(\omega_k, x_k)\| \le C_0, \qquad \forall\, k \ge 0.$$

Note that if the set X is compact or the components f_i are polyhedral, then Assumption 8.2.2(b) is satisfied. The proofs of several propositions in this section rely on the following slight generalization of the Supermartingale Convergence Theorem given by Neveu ([Neu75], p. 33, see also Bertsekas and Tsitsiklis [BeT96], p. 148).

Proposition 8.2.10: (Supermartingale Convergence Theorem) Let Y_k, Z_k, and W_k, $k = 0, 1, 2, \ldots$, be three sequences of random variables and let $\mathcal{F}_k$, $k = 0, 1, 2, \ldots$, be sets of random variables such that $\mathcal{F}_k \subset \mathcal{F}_{k+1}$ for all k. Suppose that:

(1) The random variables Y_k, Z_k, and W_k are nonnegative, and are functions of the random variables in $\mathcal{F}_k$.

(2) For each k, we have $E\{Y_{k+1} \mid \mathcal{F}_k\} \le Y_k - Z_k + W_k$.

(3) There holds $\sum_{k=0}^{\infty} W_k < \infty$.

Then, we have $\sum_{k=0}^{\infty} Z_k < \infty$, and the sequence Y_k converges to a nonnegative random variable Y, with probability 1.

Constant Stepsize Rule

The following proposition addresses the constant stepsize rule.

Proposition 8.2.11: Let $\{x_k\}$ be the sequence generated by the randomized incremental method (8.30), with the stepsize α_k fixed to some positive constant α.

(a) If $f^* = -\infty$, then with probability 1

$$\inf_{k \geq 0} f(x_k) = f^*.$$

(b) If $f^* > -\infty$, then with probability 1

$$\inf_{k \geq 0} f(x_k) \leq f^* + \frac{\alpha m C_0^2}{2}.$$

Proof: By adapting Lemma 8.2.1 to the case where f is replaced by f_{ω_k}, we have

$$\|x_{k+1} - y\|^2 \leq \|x_k - y\|^2 - 2\alpha\big(f_{\omega_k}(x_k) - f_{\omega_k}(y)\big) + \alpha^2 C_0^2, \quad \forall\, y \in X, \quad k \geq 0.$$

By taking the conditional expectation with respect to $\mathcal{F}_k = \{x_0, \ldots, x_k\}$, the method's history up to x_k, we obtain for all $y \in X$ and k

$$E\{\|x_{k+1} - y\|^2 \mid \mathcal{F}_k\} \leq \|x_k - y\|^2 - 2\alpha E\{f_{\omega_k}(x_k) - f_{\omega_k}(y) \mid \mathcal{F}_k\} + \alpha^2 C_0^2$$

$$= \|x_k - y\|^2 - 2\alpha \sum_{i=1}^{m} \frac{1}{m}\big(f_i(x_k) - f_i(y)\big) + \alpha^2 C_0^2$$

$$= \|x_k - y\|^2 - \frac{2\alpha}{m}\big(f(x_k) - f(y)\big) + \alpha^2 C_0^2,$$

$$(8.31)$$

where the first equality follows since ω_k takes the values $1, \ldots, m$ with equal probability $1/m$.

Now, fix a positive scalar γ, consider the level set L_γ defined by

$$L_\gamma = \begin{cases} \left\{ x \in X \mid f(x) < -\gamma + 1 + \frac{\alpha m C_0^2}{2} \right\} & \text{if } f^* = -\infty, \\ \left\{ x \in X \mid f(x) < f^* + \frac{2}{\gamma} + \frac{\alpha m C_0^2}{2} \right\} & \text{if } f^* > -\infty, \end{cases}$$

and let $y_\gamma \in X$ be such that

$$f(y_\gamma) = \begin{cases} -\gamma & \text{if } f^* = -\infty, \\ f^* + \frac{1}{\gamma} & \text{if } f^* > -\infty. \end{cases}$$

Note that $y_\gamma \in L_\gamma$ by construction. Define a new process $\{\hat{x}_k\}$ as follows

$$\hat{x}_{k+1} = \begin{cases} P_X\big(\hat{x}_k - \alpha g(\omega_k, \hat{x}_k)\big) & \text{if } \hat{x}_k \notin L_\gamma, \\ y_\gamma & \text{otherwise,} \end{cases}$$

where $\hat{x}_0 = x_0$. Thus the process $\{\hat{x}_k\}$ is identical to $\{x_k\}$, except that once x_k enters the level set L_γ, the process terminates with $\hat{x}_k = y_\gamma$ (since $y_\gamma \in L_\gamma$). We will now argue that $\{\hat{x}_k\}$ (and hence also $\{x_k\}$) will eventually enter each of the sets L_γ.

Using Eq. (8.31) with $y = y_\gamma$, we have

$$E\{\|\hat{x}_{k+1} - y_\gamma\|^2 \mid \mathcal{F}_k\} \leq \|\hat{x}_k - y_\gamma\|^2 - \frac{2\alpha}{m}\left(f(\hat{x}_k) - f(y_\gamma)\right) + \alpha^2 C_0^2,$$

or equivalently

$$E\{\|\hat{x}_{k+1} - y_\gamma\|^2 \mid \mathcal{F}_k\} \leq \|\hat{x}_k - y_\gamma\|^2 - z_k, \tag{8.32}$$

where

$$z_k = \begin{cases} \frac{2\alpha}{m}\left(f(\hat{x}_k) - f(y_\gamma)\right) - \alpha^2 C_0^2 & \text{if } \hat{x}_k \notin L_\gamma, \\ 0 & \text{if } \hat{x}_k = y_\gamma. \end{cases}$$

The idea of the subsequent argument is to show that as long as $\hat{x}_k \notin L_\gamma$, the scalar z_k (which is a measure of progress) is strictly positive and bounded away from 0.

(a) Let $f^* = -\infty$. Then if $\hat{x}_k \notin L_\gamma$, we have

$$\begin{aligned} z_k &= \frac{2\alpha}{m}\left(f(\hat{x}_k) - f(y_\gamma)\right) - \alpha^2 C_0^2 \\ &\geq \frac{2\alpha}{m}\left(-\gamma + 1 + \frac{\alpha m C_0^2}{2} + \gamma\right) - \alpha^2 C_0^2 \\ &= \frac{2\alpha}{m}. \end{aligned}$$

Since $z_k = 0$ for $\hat{x}_k \in L_\gamma$, we have $z_k \geq 0$ for all k, and by Eq. (8.32) and the Supermartingale Convergence Theorem, $\sum_{k=0}^{\infty} z_k < \infty$ implying that $\hat{x}_k \in L_\gamma$ for sufficiently large k, with probability 1. Therefore, in the original process we have

$$\inf_{k\geq 0} f(x_k) \leq -\gamma + 1 + \frac{\alpha m C_0^2}{2}$$

with probability 1. Letting $\gamma \to \infty$, we obtain $\inf_{k\geq 0} f(x_k) = -\infty$ with probability 1.

(b) Let $f^* > -\infty$. Then if $\hat{x}_k \notin L_\gamma$, we have

$$\begin{aligned} z_k &= \frac{2\alpha}{m}\left(f(\hat{x}_k) - f(y_\gamma)\right) - \alpha^2 C_0^2 \\ &\geq \frac{2\alpha}{m}\left(f^* + \frac{2}{\gamma} + \frac{\alpha m C_0^2}{2} - f^* - \frac{1}{\gamma}\right) - \alpha^2 C_0^2 \\ &= \frac{2\alpha}{m\gamma}. \end{aligned}$$

Hence, $z_k \geq 0$ for all k, and by the Supermartingale Convergence Theorem, we have $\sum_{k=0}^{\infty} z_k < \infty$ implying that $\hat{x}_k \in L_\gamma$ for sufficiently large k, so that in the original process,

$$\inf_{k \geq 0} f(x_k) \leq f^* + \frac{2}{\gamma} + \frac{\alpha m C_0^2}{2}$$

with probability 1. Letting $\gamma \to \infty$, we obtain $\inf_{k \geq 0} f(x_k) \leq f^* + \alpha m C_0^2 / 2$.
Q.E.D.

From Prop. 8.2.11(b), it can be seen that when $f^* > -\infty$, the randomized method (8.30) with a fixed stepsize has a better error bound (by a factor m, since $C^2 \approx m^2 C_0^2$) than the one of the nonrandomized method (8.9)-(8.11) with the same stepsize (cf. Prop. 8.2.2). In effect, the randomized method achieves in an expected sense the error tolerance of the nonrandomized method with the best processing order [compare with the discussion in the preceding subsection and Eqs. (8.13) and (8.14)]. Thus when randomization is used, one can afford to use a larger stepsize α than in the nonrandomized method. This suggests a rate of convergence advantage in favor of the randomized method.

A related result is provided by the following proposition, which parallels Prop. 8.2.3 for the nonrandomized method.

Proposition 8.2.12: Let Assumption 8.2.2 hold, and let $\{x_k\}$ be the sequence generated by the randomized incremental method (8.30), with the stepsize α_k fixed to some positive constant α. Then, for any positive scalar ϵ, we have with probability 1

$$\min_{0 \leq k \leq N} f(x_k) \leq f^* + \frac{\alpha m C_0^2 + \epsilon}{2},$$

where N is a random variable with

$$E\{N\} \leq \frac{m \big(d(x_0, X^*)\big)^2}{\alpha \epsilon}.$$

Proof: Define a new process $\{\hat{x}_k\}$ by

$$\hat{x}_{k+1} = \begin{cases} P_X\big(\hat{x}_k - \alpha g(\omega_k, \hat{x}_k)\big) & \text{if } \hat{x}_k \notin L_\gamma, \\ y_\gamma & \text{otherwise,} \end{cases}$$

where $\hat{x}_0 = x_0$ and $\hat{y}$ is some fixed vector in X^*. The process $\{\hat{x}_k\}$ is identical to $\{x_k\}$, except that once x_k enters the level set

$$L = \left\{ x \in X \mid f(x) < f^* + \frac{\alpha m C_0^2 + \epsilon}{2} \right\},$$

the process $\{\hat{x}_k\}$ terminates at the point $\hat{y}$. Similar to the proof of Prop. 8.2.11 [cf. Eq. (8.31) with $y \in X^*$], for the process $\{\hat{x}_k\}$ we obtain for all k,

$$E\left\{ \left(d(\hat{x}_{k+1}, X^*)\right)^2 \mid \mathcal{F}_k \right\} \leq \left(d(\hat{x}_k, X^*)\right)^2 - \frac{2\alpha}{m}\left(f(\hat{x}_k) - f^*\right) + \alpha^2 C_0^2$$
$$= \left(d(\hat{x}_k, X^*)\right)^2 - z_k,$$
(8.33)

where $\mathcal{F}_k = \{x_0, \ldots, x_k\}$ and

$$z_k = \begin{cases} \frac{2\alpha}{m}\left(f(\hat{x}_k) - f^*\right) - \alpha^2 C_0^2 & \text{if } \hat{x}_k \notin L, \\ 0 & \text{otherwise.} \end{cases}$$

In the case where $\hat{x}_k \notin L$, we have

$$z_k \geq \frac{2\alpha}{m}\left(f^* + \frac{\alpha m C_0^2 + \epsilon}{2} - f^*\right) - \alpha^2 C_0^2 = \frac{\alpha\epsilon}{m}. \qquad (8.34)$$

By the Supermartingale Convergence Theorem, from Eq. (8.33) we have

$$\sum_{k=0}^{\infty} z_k < \infty$$

with probability 1, so that $z_k = 0$ for all $k \geq N$, where N is a random variable. Hence $\hat{x}_N \in L$ with probability 1, implying that in the original process we have

$$\min_{0 \leq k \leq N} f(x_k) \leq f^* + \frac{\alpha m C_0^2 + \epsilon}{2}$$

with probability 1. Furthermore, by taking the total expectation in Eq. (8.33), we obtain for all k,

$$E\left\{ \left(d(\hat{x}_{k+1}, X^*)\right)^2 \right\} \leq E\left\{ \left(d(\hat{x}_k, X^*)\right)^2 \right\} - E\{z_k\}$$
$$\leq \left(d(x_0, X^*)\right)^2 - E\left\{ \sum_{j=0}^{k} z_j \right\},$$

where in the last inequality we use the facts $\hat{x}_0 = x_0$ and

$$E\left\{ \left(d(x_0, X^*)\right)^2 \right\} = \left(d(x_0, X^*)\right)^2.$$

Therefore

$$\left(d(x_0, X^*)\right)^2 \geq E\left\{ \sum_{k=0}^{\infty} z_k \right\} = E\left\{ \sum_{k=0}^{N-1} z_k \right\} \geq E\left\{ \frac{N\alpha\epsilon}{m} \right\} = \frac{\alpha\epsilon}{m} E\{N\},$$

where the last inequality above follows from Eq. (8.34).   **Q.E.D.**

Comparison of Deterministic and Randomized Methods

Let us now compare the estimate of the above proposition with the corresponding estimate for the deterministic incremental method. We showed in Prop. 8.2.3 that the deterministic method is guaranteed to reach the level set

$$\left\{ x \ \middle| \ f(x) \le f^* + \frac{\alpha C^2 + \epsilon}{2} \right\}$$

after no more than $\big(d(x_0, X^*)\big)^2/(\alpha\epsilon)$ cycles, where $C = \sum_{i=1} C_i$. To compare this estimate with the one for the randomized method, we note that $C_i \le C_0$, so that C can be estimated as mC_0, while each cycle requires the processing of m component functions. Thus, the deterministic method, in order to reach the level set

$$\left\{ x \ \middle| \ f(x) \le f^* + \frac{\alpha m^2 C_0^2 + \epsilon}{2} \right\},$$

it must process a total of

$$N \le \frac{m\big(d(x_0, X^*)\big)^2}{\alpha\epsilon}$$

component functions (this bound is essentially sharp, as shown in the exercises).

If in the randomized method we use the same stepsize α, then according to Prop. 8.2.12, we will reach with probability 1 the (much smaller) level set

$$\left\{ x \ \middle| \ f(x) \le f^* + \frac{\alpha m C_0^2 + \epsilon}{2} \right\}$$

after processing N component functions, where the expected value of N satisfies

$$E\{N\} \le \frac{m\big(d(x_0, X^*)\big)^2}{\alpha\epsilon}.$$

Thus, for the same value of α and ϵ, the bound on the number of component functions that must be processed in the deterministic method is the same as the bound on the expected number of component functions that must be processed in the randomized method. However, the error term $\alpha m^2 C_0^2$ in the deterministic method is m times larger than the corresponding error term in the randomized method.

Similarly, if we choose the stepsize α in the randomized method to achieve the same error level (in cost function value) as in the deterministic method, then the corresponding expected number of iterations becomes m times smaller. Thus it appears that there is a substantial advantage in favor of the randomized methods, particularly when m is large; see Nedić and Bertsekas [NeB01a] for supporting computational results.

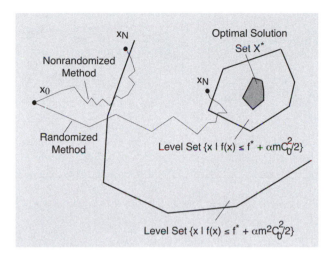

Figure 8.2.7. Comparison of the improvement property of the nonrandomized and randomized incremental subgradient methods. The randomized method is guaranteed to make progress on the average when x_k lies outside a much smaller level set.

The key to understanding the better convergence properties of the randomized method is Eq. (8.31), repeated below for the case where y is equal to an optimal solution x^*:

$$E\big\{\|x_{k+1} - x^*\|^2 \mid \mathcal{F}_k\big\} \le \|x_k - x^*\|^2 - \frac{2\alpha}{m}\big(f(x_k) - f^*\big) + \alpha^2 C_0^2.$$

It shows that the method is guaranteed to make progress on the average when x_k lies outside the level set

$$\left\{x \;\middle|\; f(x) \le f^* + \frac{\alpha m C_0^2}{2}\right\}.$$

By contrast, the nonrandomized method is guaranteed to make progress only when x_k lies outside the much larger level set

$$\left\{x \;\middle|\; f(x) \le f^* + \frac{\alpha m^2 C_0^2}{2}\right\}.$$

Figure 8.2.7 illustrates the preceding discussion.

Diminishing Stepsize Rule

We now consider the case of a diminishing stepsize rule.

Proposition 8.2.13: Assume that the optimal set X^* is nonempty, and that the stepsize α_k in Eq. (8.30) is such that

$$\sum_{k=0}^{\infty} \alpha_k = \infty, \qquad \sum_{k=0}^{\infty} \alpha_k^2 < \infty.$$

Then the sequence $\{x_k\}$ generated by the randomized method (8.30) converges to some optimal solution with probability 1.

Proof: Similar to the proof of Prop. 8.2.11, we obtain for all k and $x^* \in X^*$,

$$E\{\|x_{k+1} - x^*\|^2 \mid \mathcal{F}_k\} \le \|x_k - x^*\|^2 - \frac{2\alpha_k}{m}\left(f(x_k) - f^*\right) + \alpha_k^2 C_0^2$$

[cf. Eq. (8.31) with α and y replaced by α_k and x^*, respectively], where $\mathcal{F}_k = \{x_0, \dots, x_k\}$. By the Supermartingale Convergence Theorem (Prop. 8.2.10), for each $x^* \in X^*$, we have for all sample paths in a set Ω_{x^*} of probability 1

$$\sum_{k=0}^{\infty} \frac{2\alpha_k}{m}\left(f(x_k) - f^*\right) < \infty, \qquad (8.35)$$

and the sequence $\{\|x_k - x^*\|\}$ converges.

Let $\{v_i\}$ be a countable subset of the relative interior $\mathrm{ri}(X^*)$ that is dense in X^* [such a set exists since $\mathrm{ri}(X^*)$ is a relatively open subset of the affine hull of X^*; an example of such a set is the intersection of X^* with the set of vectors of the form $x^* + \sum_{i=1}^{p} r_i \xi_i$, where $\xi_1, \dots, \xi_p$ are basis vectors for the affine hull of X^* and r_i are rational numbers]. The intersection

$$\Omega = \cap_{i=1}^{\infty} \Omega_{v_i}$$

has probability 1, since its complement $\overline{\Omega}$ is equal to the union of the complements $\overline{\Omega_{v_i}}$ of Ω_{v_i}, and we have

$$\mathrm{Prob}\left(\overline{\Omega}\right) = \mathrm{Prob}\left(\cup_{i=1}^{\infty} \overline{\Omega_{v_i}}\right) \le \sum_{i=1}^{\infty} \mathrm{Prob}\left(\overline{\Omega_{v_i}}\right) = 0.$$

For each sample path in Ω, the sequences $\|x_k - v_i\|$ converge so that $\{x_k\}$ is bounded. Furthermore, from Eq. (8.35) we see that

$$\lim_{k \to \infty} \left(f(x_k) - f^*\right) = 0,$$

which by the continuity of f, implies that all the limit points of $\{x_k\}$ belong to X^*. Since $\{v_i\}$ is a dense subset of X^* and the sequences $\|x_k - v_i\|$ converge, it follows that $\{x_k\}$ cannot have more than one limit point, so it must converge to some vector $\overline{x} \in X^*$. **Q.E.D.**

Dynamic Stepsize Rule for Known f^*

One possible version of the dynamic stepsize rule for the method (8.30) has the form

$$\alpha_k = \gamma_k \frac{f(x_k) - f^*}{mC_0^2}, \qquad 0 < \underline{\gamma} \le \gamma_k \le \overline{\gamma} < 2,$$

where $\{\gamma_k\}$ is a deterministic sequence, and requires knowledge of the cost function value $f(x_k)$ at the current iterate x_k. However, it would be inefficient to compute $f(x_k)$ at each iteration since that iteration involves a single component f_i, while the computation of $f(x_k)$ requires all the components. We thus modify the dynamic stepsize rule so that the value of f and the parameter γ_k that are used in the stepsize formula are updated every M iterations, where M is any fixed positive integer, rather than at each iteration. In particular, assuming f^* is known, we use the stepsize

$$\alpha_k = \gamma_p \frac{f(x_{Mp}) - f^*}{mMC_0^2}, \quad 0 < \underline{\gamma} \le \gamma_p \le \overline{\gamma} < 2,$$

$$k = Mp, \dots, M(p+1) - 1, \quad p = 0, 1, \dots,$$

$$(8.36)$$

where $\{\gamma_p\}$ is a deterministic sequence. We can choose M greater than m, if m is relatively small, or we can select M smaller than m, if m is very large.

Proposition 8.2.14: Assume that X^* is nonempty. The sequence $\{x_k\}$ generated by the randomized method (8.30) with the stepsize (8.36) converges to some optimal solution with probability 1.

Proof: By adapting Lemma 8.2.1 to the case where $y = x^* \in X^*$ and f is replaced by f_{ω_k}, we have for all $x^* \in X^*$ and all k,

$$\|x_{k+1} - x^*\|^2 \le \|x_k - x^*\|^2 - 2\alpha_k\big(f_{\omega_k}(x_k) - f_{\omega_k}(x^*)\big) + \alpha_k^2 C_0^2.$$

By summing this inequality over $k = Mp, \dots, M(p+1) - 1$ (i.e., over the M iterations of a cycle), we obtain for all $x^* \in X^*$ and all p,

$$\|x_{M(p+1)} - x^*\|^2 \le \|x_{Mp} - x^*\|^2 - 2\alpha_{Mp} \sum_{k=Mp}^{M(p+1)-1} \big(f_{\omega_k}(x_k) - f_{\omega_k}(x^*)\big)$$

$$+ M\alpha_{Mp}^2 C_0^2,$$

since $\alpha_k = \alpha_{Mp}$ for $k = Mp, \dots, M(p+1) - 1$. By taking the conditional expectation with respect to $\mathcal{G}_p = \{x_0, \dots, x_{M(p+1)-1}\}$, we have for all $x^* \in$

X^* and p,

$$E\{\|x_{M(p+1)} - x^*\|^2 \mid \mathcal{G}_p\} \leq \|x_{Mp} - x^*\|^2$$

$$- 2\alpha_{Mp} \sum_{k=Mp}^{M(p+1)-1} E\{f_{\omega_k}(x_k) - f_{\omega_k}(x^*) \mid x_k\}$$

$$+ M^2 \alpha_{Mp}^2 C_0^2$$

$$\leq \|x_{Mp} - x^*\|^2 - \frac{2\alpha_{Mp}}{m} \sum_{k=Mp}^{M(p+1)-1} \left(f(x_k) - f^*\right)$$

$$+ M^2 \alpha_{Mp}^2 C_0^2.$$

(8.37)

We now relate $f(x_k)$ and $f(x_{Mp})$ for $k = Mp, \ldots, M(p+1) - 1$. We have

$$f(x_k) - f^* = \left(f(x_k) - f(x_{Mp})\right) + \left(f(x_{Mp}) - f^*\right)$$

$$\geq \tilde{g}'_{Mp}(x_k - x_{Mp}) + f(x_{Mp}) - f^* \qquad (8.38)$$

$$\geq f(x_{Mp}) - f^* - mC_0\|x_k - x_{Mp}\|,$$

where $\tilde{g}_{Mp}$ is a subgradient of f at x_{Mp} and in the last inequality we use the fact

$$\|\tilde{g}_{Mp}\| = \left\|\sum_{i=1}^{m} \tilde{g}_{i,Mp}\right\| \leq mC_0$$

[cf. Assumption 8.2.2(b)] with $\tilde{g}_{i,Mp}$ being a subgradient of f_i at x_{Mp}. Furthermore, we have for all p and $k = Mp, \ldots, M(p+1) - 1$,

$$\|x_k - x_{Mp}\| \leq \|x_k - x_{k-1}\| + \|x_{k-1} - x_{Mp}\|$$

$$\leq \alpha_{k-1}\|g(\omega_{k-1}, x_{k-1})\| + \|x_{k-1} - x_{Mp}\|$$

$$\leq \cdots \qquad (8.39)$$

$$\leq \alpha_{Mp} \sum_{l=Mp}^{k-1} \|g(\omega_l, x_l)\|$$

$$\leq (k - Mp)\alpha_{Mp}C_0,$$

which when substituted in Eq. (8.38) yields

$$f(x_k) - f^* \geq f(x_{Mp}) - f^* - (k - Mp)m\alpha_{Mp}C_0^2.$$

From the preceding relation and Eq. (8.37) we have

$$E\{\|x_{M(p+1)} - x^*\|^2 \mid \mathcal{G}_{p+1}\} \leq \|x_{Mp} - x^*\|^2 - \frac{2M\alpha_{Mp}}{m}\left(f(x_{Mp}) - f^*\right)$$

$$+ 2\alpha_{Mp}^2 C_0^2 \sum_{k=Mp}^{M(p+1)-1} (k - Mp) + M\alpha_{Mp}^2 C_0^2.$$

(8.40)

Since

$$2\alpha_{Mp}^2 C_0^2 \sum_{k=Mp}^{M(p+1)-1} (k - Mp) + M\alpha_{Mp}^2 C_0^2$$

$$= 2\alpha_{Mp}^2 C_0^2 \sum_{l=1}^{M-1} l + M\alpha_{Mp}^2 C_0^2$$

$$= M^2 \alpha_{Mp}^2 C_0^2,$$

it follows that for all $x^* \in X^*$ and p

$$E\{\|x_{M(p+1)} - x^*\|^2 \mid \mathcal{G}_p\} \leq \|x_{Mp} - x^*\|^2 - \frac{2M\alpha_{Mp}}{m} \left(f(x_{Mp}) - f^*\right)$$
$$+ M^2 \alpha_{Mp}^2 C_0^2.$$

This relation and the definition of α_k [cf. Eq. (8.36)] yield

$$E\{\|x_{M(p+1)} - x^*\|^2 \mid \mathcal{G}_p\} \leq \|x_{Mp} - x^*\|^2 - \gamma_p(2 - \gamma_p) \left(\frac{f(x_{Mp}) - f^*}{mC_0}\right)^2.$$

By the Supermartingale Convergence Theorem, we have

$$\sum_{k=0}^{\infty} \gamma_p(2 - \gamma_p) \left(\frac{f(x_{Mp}) - f^*}{mC_0}\right)^2 < \infty$$

and for each $x^* \in X^*$ the sequence $\{\|x_{Mp} - x^*\|\}$ is convergent, with probability 1. Because $\gamma_p \in [\underline{\gamma}, \overline{\gamma}] \subset (0, 2)$, it follows that

$$\lim_{p \to \infty} \left(f(x_{Mp}) - f^*\right) = 0$$

with probability 1. Convergence of x_{Mp} to some optimal solution with probability 1 now follows similar to the proof of Prop. 8.2.13. Since with probability 1

$$\|x_k - x_{Mp}\| < M\alpha_{Mp} C_0, \qquad k = Mp, \ldots, M(p+1) - 1,$$

[cf. Eq. (8.39)], we see that the entire sequence $\{x_k\}$ converges to some optimal solution with probability 1.   **Q.E.D.**

Dynamic Stepsize Rule for Unknown f^*

In the case where f^* is not known, we modify the dynamic stepsize (8.36) by replacing f^* with a target level estimate f_p. Thus the stepsize is

$$\alpha_k = \gamma_p \frac{f(x_{Mp}) - f_p}{mMC_0^2}, \quad 0 < \underline{\gamma} \leq \gamma_p \leq \overline{\gamma} < 2,$$
$$k = Mp, \ldots, M(p+1) - 1, \quad p = 0, 1, \ldots.$$
$$(8.41)$$

To update the target values f_p, we may use the adjustment procedures described in Section 8.2.1.

In the first adjustment procedure, f_p is given by

$$f_p = \min_{0 \leq j \leq p} f(x_{Mj}) - \delta_p, \tag{8.42}$$

and δ_p is updated according to

$$\delta_{p+1} = \begin{cases} \delta_p & \text{if } f(x_{M(p+1)}) \leq f_p, \\ \max\{\beta\delta_p, \delta\} & \text{if } f(x_{M(p+1)}) > f_p, \end{cases} \tag{8.43}$$

where δ and β are fixed positive constants with $\beta < 1$. Thus all the parameters of the stepsize are updated every M iterations. Note that here the parameter ρ of Eq. (8.25) has been set to 1. Our proof relies on this (relatively mild) restriction. Since the stepsize is bounded away from zero, the method behaves similar to the one with a constant stepsize (cf. Prop. 8.2.12). More precisely, we have the following result.

Proposition 8.2.15: Let Assumption 8.2.2 hold, and let $\{x_k\}$ be the sequence generated by the randomized method (8.30), and the stepsize rule (8.41) with the adjustment procedure (8.42)–(8.43).

(a) If $f^* = -\infty$, then with probability 1

$$\inf_{k \geq 0} f(x_k) = f^*.$$

(b) If $f^* > -\infty$, then with probability 1

$$\inf_{k \geq 0} f(x_k) \leq f^* + \delta.$$

Proof: (a) Define the events

$$H_1 = \left\{ \lim_{p \to \infty} \delta_p > \delta \right\}, \qquad H_2 = \left\{ \lim_{p \to \infty} \delta_p = \delta \right\}.$$

Given that H_1 occurred there is an integer R such that $\delta_R > \delta$ and

$$\delta_p = \delta_R, \qquad \forall\, p \geq R.$$

We let R be the smallest integer with the above property and we note that R is a discrete random variable taking nonnegative integer values. In view of Eq. (8.43), we have for all $p \geq R$,

$$f(x_{M(p+1)}) \leq f_p.$$

Then from the definition of f_p [cf. Eq. (8.42)], the relation

$$\min_{0 \le j \le p} f(x_{Mj}) \le f(x_{Mp}),$$

and the fact $\delta_p = \delta_R$ for all $p \ge R$, we obtain

$$f(x_{M(p+1)}) \le f(x_{Mp}) - \delta_R, \qquad \forall\, p \ge R.$$

Summation of the above inequalities yields

$$f(x_{Mp}) \le f(x_{MR}) - (p - R)\delta_R, \qquad \forall\, p \ge R.$$

Therefore, given that H_1 occurred, we have

$$\inf_{p \ge 0} f(x_{Mp}) \ge \inf_{p \ge 0} f(x_{Mp}) = -\infty$$

with probability 1, i.e.,

$$P\left\{ \inf_{p \ge 0} f(x_{Mp}) = -\infty \,\Big|\, H_1 \right\} - 1. \tag{8.44}$$

Now assume that H_2 occurred. The event H_2 occurs if and only if, after finitely many iterations, δ_p is decreased to the threshold value δ and remains at that value for all subsequent iterations. Thus H_2 occurs if and only if there is an index S such that

$$\delta_p = \delta, \qquad \forall\, p \ge S. \tag{8.45}$$

Let S be the smallest integer with the above property (a random variable), and note that we have $H_2 = \cup_{s \ge 0} B_s$, where for all integers s, B_s is the event $\{S = s\}$.

Similar to the proof of Prop. 8.2.14 [cf. Eq. (8.40)], we have for all $y \in X$ and p,

$$
\begin{aligned}
E\{\|x_{M(p+1)} - y\|^2 \mid \mathcal{G}_p, B_s\} &= E\{\|x_{M(p+1)} - y\|^2 \mid \mathcal{G}_p\} \\
&\le \|x_{Mp} - y\|^2 - 2\gamma_p \frac{f(x_{Mp}) - f_p}{m^2 C_0^2} \left(f(x_{Mp}) - f(y) \right) \\
&\quad + \gamma_p^2 \frac{\left(f(x_{Mp}) - f_p \right)^2}{m^2 C_0^2},
\end{aligned}
$$
$$\tag{8.46}$$

where $\mathcal{G}_p = \{x_0, \ldots, x_{Mp-1}\}$. Now, fix a nonnegative integer γ and let $y_\gamma \in X$ be such that

$$f(y_\gamma) = -\gamma - \delta.$$

Consider a new process $\{\hat{x}_k\}$ defined by

$$\hat{x}_{k+1} = \begin{cases} P_X\big(\hat{x}_k - \alpha_k g(\omega_k, \hat{x}_k)\big) & \text{if } f(\hat{x}_{Mp}) \geq -\gamma, \\ y_\gamma & \text{otherwise,} \end{cases}$$

for $k = Mp, \ldots, M(p+1)-1$, $p = 0, 1, \ldots$, and $\hat{x}_0 = x_0$. The process $\{\hat{x}_k\}$ is identical to $\{x_k\}$ up to the point when x_{Mp} enters the level set

$$L_\gamma = \big\{x \in X \mid f(x) < -\gamma\big\},$$

in which case the process $\{\hat{x}_k\}$ terminates at the point y_γ. Therefore, given B_s, the process $\{\hat{x}_{Mp}\}$ satisfies Eq. (8.46) for all $p \geq s$ and $y = y_\gamma$, i.e., we have

$$E\big\{\|\hat{x}_{M(p+1)} - y_\gamma\|^2 \mid \mathcal{G}_p\big\} \leq \|\hat{x}_{Mp} - y_\gamma\|^2$$
$$- 2\gamma_p \frac{f(\hat{x}_{Mp}) - f_p}{m^2 C_0^2}\big(f(\hat{x}_{Mp}) - f(y_\gamma)\big) + \gamma_p^2 \frac{\big(f(\hat{x}_{Mp}) - f_p\big)^2}{m^2 C_0^2},$$

or equivalently

$$E\big\{\|\hat{x}_{M(p+1)} - y_\gamma\|^2 \mid \mathcal{G}_p\big\} \leq \|\hat{x}_{Mp} - y_\gamma\|^2 - z_p,$$

where

$$z_p = \begin{cases} 2\gamma_p \frac{f(\hat{x}_{Mp}) - f_p}{m^2 C_0^2}\big(f(\hat{x}_{Mp}) - f(y_\gamma)\big) - \gamma_p^2 \frac{\big(f(\hat{x}_{Mp}) - f_p\big)^2}{m^2 C_0^2} & \text{if } \hat{x}_{Mp} \notin L_\gamma, \\ 0 & \text{if } \hat{x}_{Mp} = y_\gamma. \end{cases}$$

By using the definition of f_p [cf. Eq. (8.42)] and the fact $\delta_p = \delta$ for all $p \geq s$ [cf. Eq. (8.45)], we have for $p \geq s$ and $\hat{x}_{Mp} \notin L_\gamma$,

$$f(y_\gamma) \leq \min_{0 \leq j \leq p} f(\hat{x}_{Mj}) - \delta = f_p,$$

which when substituted in the preceding relation yields for $p \geq s$ and $\hat{x}_{Mp} \notin L_\gamma$,

$$z_p \geq \gamma_p(2 - \gamma_p)\frac{\big(f(\hat{x}_{Mp}) - f_p\big)^2}{m^2 C_0^2} \geq \underline{\gamma}(2 - \overline{\gamma})\frac{\delta^2}{m^2 C_0^2}.$$

The last inequality above follows from the facts $\gamma_p \in [\underline{\gamma}, \overline{\gamma}]$ and $f(\hat{x}_{Mp}) - f_p \geq \delta$ for all p [cf. Eqs. (8.42)–(8.43)]. Hence $z_p \geq 0$ for all k, and by the Supermartingale Convergence Theorem, we obtain $\sum_{p=s}^{\infty} z_p < \infty$ with probability 1. Thus, given B_s and for sufficiently large p, we have $\hat{x}_{Mp} \in L_\gamma$ with probability 1, implying that in the original process

$$P\bigg\{\inf_{p \geq 0} f(x_{Mp}) \leq -\gamma \;\bigg|\; B_s\bigg\} = 1.$$

By letting $\gamma \to \infty$ in the preceding relation, we obtain

$$P\left\{\inf_{p\geq 0} f(x_{Mp}) = -\infty \;\middle|\; B_s\right\} = 1.$$

Since $H_2 = \cup_{s\geq 0} B_s$, it follows that

$$P\left\{\inf_{p\geq 0} f(x_{Mp}) = -\infty \;\middle|\; H_2\right\}$$

$$= \sum_{s=0}^{\infty} P\left\{\inf_{p\geq 0} f(x_{Mp}) = -\infty \;\middle|\; B_s\right\} P(B_s)$$

$$= \sum_{s=0}^{\infty} P(B_s) = 1.$$

Combining this equation with Eq. (8.44), we have with probability 1

$$\inf_{p\geq 0} f(x_{Mp}) = -\infty,$$

so that $\inf_{k\geq 0} f(x_k) = -\infty$ with probability 1.

(b) Using the proof of part (a), we see that if $f^* > -\infty$, then H_2 occurs with probability 1. Thus, as in part (a), we have $H_2 = \cup_{s\geq 0} B_s$, where for all integers $s \geq 0$, B_s is the event $\{S = s\}$ and S is as in Eq. (8.45).

Fix a positive integer γ and let $y_\gamma \in X$ be such that

$$f(y_\gamma) = f^* + \frac{1}{\gamma}.$$

Consider the process $\{\hat{x}_k\}$ defined by

$$\hat{x}_{k+1} = \begin{cases} P_X\big(\hat{x}_k - \alpha_k g(\omega_k, \hat{x}_k)\big) & \text{if } f(\hat{x}_{Mp}) \geq f^* + \delta + \frac{1}{\gamma}, \\ y_\gamma & \text{otherwise,} \end{cases}$$

for $k = Mp, \ldots, M(p+1)-1$, $p = 0, 1, \ldots$, and $\hat{x}_0 = x_0$. The process $\{\hat{x}_k\}$ is the same as the process $\{x_k\}$ up to the point where x_{Mp} enters the level set

$$L_\gamma = \left\{x \in X \;\middle|\; f(x) < f^* + \delta + \frac{1}{\gamma}\right\},$$

in which case the process $\{\hat{x}_k\}$ terminates at the point y_γ. The remainder of the proof is similar to the proof of part (a).   **Q.E.D.**

The target level f_p can also be updated according to the second adjustment procedure discussed in the preceding subsection. In this case, it can be shown that a convergence (with probability 1) result analogous to Prop. 8.2.9 holds. We omit the lengthy details.

8.3 CUTTING PLANE METHODS

In this section, we will discuss the *cutting plane method*, which (like the subgradient method) calculates a single subgradient at each iteration, but (unlike the subgradient method) uses all the subgradients previously calculated. Consider again the dual problem

$$\text{maximize} \quad q(\mu)$$
$$\text{subject to} \quad \mu \in M.$$

We continue to assume that M is closed, and that at each $\mu \in M$, a subgradient of q can be calculated. The cutting plane method consists of solving at the kth iteration the problem

$$\text{maximize} \quad Q^k(\mu)$$
$$\text{subject to} \quad \mu \in M,$$

where the dual function is replaced by a polyhedral approximation Q^k, constructed using the points μ^i generated so far and their subgradients $g(x_{\mu^i})$, which are denoted by g^i. In particular, for $k = 1, 2, \ldots,$

$$Q^k(\mu) = \min\left\{q(\mu^0) + (\mu - \mu^0)'g^0, \ldots, q(\mu^{k-1}) + (\mu - \mu^{k-1})'g^{k-1}\right\} \quad (8.47)$$

and μ^k maximizes $Q^k(\mu)$ over $\mu \in M$,

$$Q^k(\mu^k) = \max_{\mu \in M} Q^k(\mu); \quad (8.48)$$

see Fig. 8.3.1. We assume that the maximum of $Q^k(\mu)$ above is attained for all k. For those k for which this is not guaranteed, artificial bounds may be placed on the components of μ, so that the maximization will be carried out over a compact set and consequently the maximum will be attained.

The following proposition establishes the convergence properties of the cutting plane method. An important special case arises when the primal problem is a linear program, in which case the dual function is polyhedral and can be put in the form

$$q(\mu) = \min_{i \in I}\{a_i'\mu + b_i\}, \quad (8.49)$$

where I is a finite index set, and $a_i \in \Re^r$ and b_i are given vectors and scalars, respectively. Then, the subgradient g^k in the cutting plane method is a vector a_{i^k} for which the minimum in Eq. (8.49) is attained (cf. Prop. 8.1.2). In this case, the following proposition shows that the cutting plane method converges finitely; see also Fig. 8.3.2.

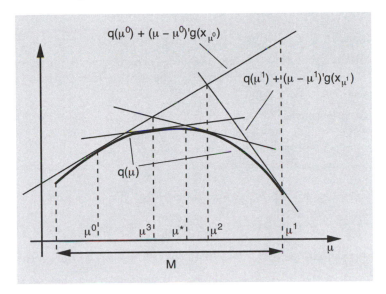

Figure 8.3.1. Illustration of the cutting plane method. With each new iterate μ^i, a new hyperplane $q(\mu^i) + (\mu - \mu^i)'g^i$ is added to the polyhedral approximation of the dual function.

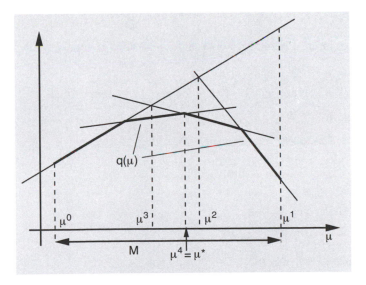

Figure 8.3.2. Illustration of the finite convergence property of the cutting plane method in the case where q is polyhedral. What happens here is that if μ^k is not optimal, a new cutting plane will be added at the corresponding iteration, and there can be only a finite number of cutting planes.

Proposition 8.3.1:

(a) Assume that $\{g^k\}$ is a bounded sequence. Then every limit point of a sequence $\{\mu^k\}$ generated by the cutting plane method is a dual optimal solution.

(b) Assume that the dual function q is polyhedral of the form (8.49). Then the cutting plane method terminates finitely; that is, for some k, μ^k is a dual optimal solution.

Proof: (a) Since for all i, g^i is a subgradient of q at μ^i, we have

$$q(\mu^i) + (\mu - \mu^i)'g^i \geq q(\mu), \qquad \forall\, \mu \in M,$$

so from the definitions (8.47) and (8.48) of Q^k and μ^k, it follows that

$$Q^k(\mu^k) \geq Q^k(\mu) \geq q(\mu), \qquad \forall\, \mu \in M. \tag{8.50}$$

Suppose that a subsequence $\{\mu^k\}_K$ converges to $\overline{\mu}$. Then, since M is closed, we have $\overline{\mu} \in M$, and by using Eq. (8.50) and the definitions (8.47) and (8.48) of Q^k and μ^k, we obtain for all k and all $i < k$,

$$q(\mu^i) + (\mu^k - \mu^i)'g^i \geq Q^k(\mu^k) \geq Q^k(\overline{\mu}) \geq q(\overline{\mu}). \tag{8.51}$$

By taking the limit superior above as $i \to \infty$, $i \in K$, and $k \to \infty$, $k \in K$, we obtain

$$\limsup_{\substack{i\to\infty,\, k\to\infty,\\ i\in K,\, k\in K}} \{q(\mu^i) + (\mu^k - \mu^i)'g^i\} \geq \limsup_{k\to\infty,\, k\in K} Q^k(\mu^k).$$

By using the assumption that the subgradient sequence $\{g^i\}$ is bounded, we have

$$\lim_{\substack{i\to\infty,\, k\to\infty,\\ i\in K,\, k\in K}} (\mu^k - \mu^i)'g^i = 0,$$

implying that

$$\limsup_{i\to\infty,\, i\in K} q(\mu^i) \geq \limsup_{k\to\infty,\, k\in K} Q^k(\mu^k).$$

Furthermore, from Eq. (8.51) it follows that

$$\liminf_{k\to\infty,\, k\in K} Q^k(\mu^k) \geq q(\overline{\mu}).$$

Moreover, by the upper-semicontinuity of q, we have

$$q(\overline{\mu}) \geq \limsup_{i\to\infty,\, i\in K} q(\mu^i).$$

The preceding three relations imply that

$$\lim_{k \to \infty, \, k \in K} Q^k(\mu^k) = q(\overline{\mu}).$$

Combining this equation with Eq. (8.50), we obtain

$$q(\overline{\mu}) \geq q(\mu), \qquad \forall \, \mu \in M,$$

showing that $\overline{\mu}$ is a dual optimal solution.

(b) Let i^k be an index attaining the minimum in the equation

$$q(\mu^k) = \min_{i \in I} \{ a_i' \mu^k + b_i \},$$

so that a_{i^k} is a subgradient at μ^k. From Eq. (8.50), we see that if $Q^k(\mu^k) = q(\mu^k)$, then μ^k is a dual optimal solution. Therefore, if μ^k is not dual optimal, we must have $Q^k(\mu^k) > q(\mu^k) = a_{i^k}' \mu^k + b_{i^k}$. Since

$$Q^k(\mu^k) = \min_{0 \leq m \leq k-1} \{ a_{i^m}' \mu^k + b_{i^m} \},$$

the pair $\left(a_{i^k}, b_{i^k} \right)$ is not equal to any of the pairs $\left(a_{i^0}, b_{i^0} \right), \ldots, \left(a_{i^{k-1}}, b_{i^{k-1}} \right)$. It follows that there can be only a finite number of iterations for which μ^k is not dual optimal.   **Q.E.D.**

The reader may verify that the boundedness assumption in Prop. 8.3.1(a) can be replaced by the assumption that $q(\mu)$ is real-valued for all $\mu \in \Re^r$, which can be ascertained if X is a finite set, or alternatively if f and g_j are continuous, and X is a compact set. Despite the finite convergence property shown in Prop. 8.3.1(b), the cutting plane method often tends to converge slowly, even for problems where the dual function is polyhedral. Indeed, typically one should base termination on the upper and lower bounds

$$Q^k(\mu^k) \geq \max_{\mu \in M} q(\mu) \geq \max_{0 \leq i \leq k-1} q(\mu^i),$$

[cf. Eq. (8.50)], rather that wait for finite termination to occur. Nonetheless, the method is often much better suited for solution of particular types of large problems than its competitors.

We note that a number of variants of the cutting plane method have been proposed, for which we refer to the specialized literature cited at the end of the chapter. We discuss one of these variants, which can be combined with the methodology of interior point methods for linear programming.

Central Cutting Plane Methods

These methods maintain a polyhedral approximation

$$Q^k(\mu) = \min\{q(\mu^0) + (\mu - \mu^0)'g^0, \ldots, q(\mu^{k-1}) + (\mu - \mu^{k-1})'g^{k-1}\}$$

to the dual function q, but they generate the next vector μ^k by using a somewhat different mechanism. In particular, instead of maximizing Q^k as in Eq. (8.48), the methods obtain μ^k by finding a "central pair" (μ^k, z^k) within the subset

$$S^k = \{(\mu, z) \mid \mu \in M, \ \tilde{q}^k \leq q(\mu), \ \tilde{q}^k \leq z \leq Q^k(\mu)\},$$

where $\tilde{q}^k$ is the best lower bound to the optimal dual value that has been found so far,

$$\tilde{q}^k = \max_{i=0,\ldots,k-1} q(\mu^i).$$

The set S^k is illustrated in Fig. 8.3.3.

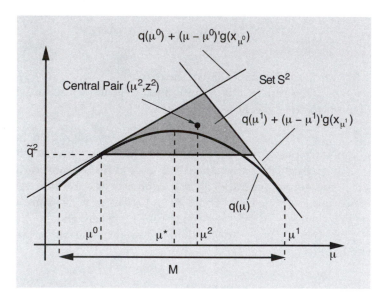

Figure 8.3.3. Illustration of the set

$$S^k = \{(\mu, z) \mid \mu \in M, \ \tilde{q}^k \leq q(\mu), \ \tilde{q}^k \leq z \leq Q^k(\mu)\},$$

in the central cutting plane method.

There are several possible methods for finding the central pair (μ^k, z^k). Roughly, the idea is that the central pair should be "somewhere in the middle" of S. For example, consider the case where S is polyhedral with nonempty interior. Then (μ^k, z^k) could be the *analytic center* of S, where for any polyhedron

$$P = \{y \mid a_p'y \le c_p, \, p = 1, \ldots, m\}$$

with nonempty interior, its analytic center is the unique maximizer of

$$\sum_{p=1}^{m} \ln(c_p - a_p'y)$$

over $y \in P$. Another possibility is the *ball center* of S, i.e., the center of the largest inscribed sphere in S; for the polyhedron P given above, the ball center can be obtained by solving the following problem with optimization variables (y, σ):

> maximize σ
>
> subject to $a_p'(y + d) \le c_p$, $\quad \forall \, d$ with $\|d\| \le \sigma$, $p = 1, \ldots, m$.

It can be seen that this problem is equivalent to the linear program

> maximize σ
>
> subject to $a_p'y + \|a_p\|\sigma \le c_p$, $\quad p = 1, \ldots, m$.

The convergence properties of central cutting plane methods are satisfactory, even though they do not terminate finitely in the case of a polyhedral q. Furthermore, these methods have benefited from advances in the implementation of interior point methods; see the references cited at the end of the chapter.

8.4 ASCENT METHODS

The cutting plane and subgradient methods are not monotonic in the sense that they do not guarantee a dual function improvement at each iteration. We will now briefly discuss some of the difficulties in constructing ascent methods, i.e., methods where the dual function is increased at each iterate. For simplicity, we will restrict ourselves to the case where the dual function is real-valued, and the dual problem is unconstrained (i.e., $M = \Re^r$). Note, however, that constraints may be lumped into the dual function by using an exact penalty.

Steepest Ascent

The steepest ascent direction d_μ of a concave function $q : \Re^n \mapsto \Re$ at a vector μ is obtained by solving the problem

$$\text{maximize} \ \ q'(\mu; d)$$
$$\text{subject to} \ \ \|d\| \leq 1,$$

where

$$q'(\mu; d) = \lim_{\alpha \downarrow 0} \frac{q(\mu + \alpha d) - q(\mu)}{\alpha}$$

is the directional derivative of q at μ in the direction d. Using Prop. 4.2.2 (and making an adjustment for the concavity of q), we have

$$q'(\mu; d) = \min_{g \in \partial q(\mu)} d'g,$$

and by the saddle point result of Prop. 2.6.9 [using also the compactness of $\partial q(\mu)$], we have

$$\max_{\|d\| \leq 1} q'(\mu; d) = \max_{\|d\| \leq 1} \min_{g \in \partial q(\mu)} d'g = \min_{g \in \partial q(\mu)} \max_{\|d\| \leq 1} d'g = \min_{g \in \partial q(\mu)} \|g\|.$$

It follows that the steepest ascent direction is

$$d_\mu = \frac{\bar{g}_\mu}{\|\bar{g}_\mu\|},$$

where $\bar{g}_\mu$ is the unique minimum-norm subgradient at μ,

$$\|\bar{g}_\mu\| = \min_{g \in \partial q(\mu)} \|g\|. \tag{8.52}$$

The steepest ascent method has the form

$$\mu^{k+1} = \mu^k + \alpha^k \bar{g}^k,$$

where $\bar{g}^k$ is the vector of minimum norm in $\partial q(\mu^k)$, and α^k is a positive stepsize that guarantees that

$$q(\mu^{k+1}) > q(\mu^k)$$

if μ^k is not optimal. A potentially important disadvantage of this method is that to calculate the steepest ascent direction, one needs to compute all the subgradients at the current iterate. It is possible, however, to compute an approximate steepest ascent direction by employing approximations to the subdifferential that involve only a finite number of subgradients (see Exercise 4.16).

ϵ-Ascent Methods

The steepest ascent method also suffers from a theoretical disadvantage: if the stepsize α^k is chosen by optimization along the steepest ascent direction, i.e.,

$$\alpha_k \in \arg \max_{\alpha > 0} q(\mu^k + \alpha \overline{g}^k),$$

the method may fail to converge to the optimum (see Wolfe [Wol75] for an example). To correct this flaw, one may use a slightly different ascent direction, obtained by projecting in the manner of Eq. (8.52) on a set that is larger than the subdifferential. Such a set should include not only the subgradients at the current iterate, but also the subgradients of neighboring points. One possibility is to replace the subdifferential $\partial q(\mu)$ in the projection of Eq. (8.52) with the ϵ-subdifferential $\partial_\epsilon q(\mu)$ at μ, where ϵ is a positive scalar. The resulting method, called the ϵ-*ascent method*, has the form

$$\mu^{k+1} = \mu^k + \alpha^k \tilde{g}^k,$$

where $\tilde{g}^k$ is the minimum norm ϵ-subgradient at μ^k,

$$\|\tilde{g}^k\| = \min_{g \in \partial_\epsilon q(\mu^k)} \|g\|. \tag{8.53}$$

This method was discussed in Section 4.3. It can be implemented by either calculating explicitly the ϵ-subdifferential $\partial_\epsilon q(\mu)$ (in problems with favorable structure) or by approximating $\partial_\epsilon q(\mu)$ with a finite number of ϵ-subgradients, generated with a procedure described in Exercise 4.17. This procedure is central in a class of methods, called *bundle methods*, which are closely related to the ϵ-ascent method. We refer to the specialized literature given at the end of the chapter for further discussion.

An interesting variation of the ϵ-ascent method applies to the case where the dual function consists of the sum of several concave functions:

$$q(\mu) = q_1(\mu) + \cdots + q_m(\mu).$$

As we have seen, this is very common, particularly for problems with separable structure. Then, as discussed in Section 4.3, we can approximate the ϵ-subdifferential of q with the vector sum

$$\tilde{\partial}_\epsilon q(\mu) = \partial_\epsilon q_1(\mu) + \cdots + \partial_\epsilon q_m(\mu)$$

in the projection problem (8.53). The method consists of the iteration

$$\mu^{k+1} = \mu^k + \alpha^k \tilde{g}^k,$$

where $\tilde{g}^k$ is the minimum norm vector in $\tilde{\partial}_\epsilon q(\mu^k)$,

$$\|\tilde{g}^k\| = \min_{g \in \tilde{\partial}_\epsilon q(\mu^k)} \|g\|.$$

In many cases, the approximation $\tilde{\partial}_\epsilon q(\mu)$ may be obtained much more easily than the exact ϵ-subdifferential $\partial_\epsilon q(\mu)$, thereby resulting in an algorithm that is much easier to implement.

8.5 NOTES, SOURCES, AND EXERCISES

Subgradient methods were first introduced in the former Soviet Union during the middle 60s by Shor; the works of Ermoliev and Polyak were also particularly influential. Description of these works are found in many sources, including Shor [Sho85], Ermoliev [Erm83], and Polyak [Pol87]. An extensive bibliography for the early period of the subject, together with some important early papers are given in the edited volume by Balinski and Wolfe [BaW75].

Incremental gradient methods for differentiable cost functions have a long history and find extensive application in the training of neural networks, among other areas. Incremental subgradient methods and their analysis are more recent. They were first proposed by Kibardin [Kib79], although his work went largely unnoticed outside the former Soviet Union. They were studied more recently by Solodov and Zavriev [SoZ98], Ben-Tal, Margalit, and Nemirovski [BMN00], and Nedić and Bertsekas [NeB01a]. Incremental subgradient methods that are somewhat different from the ones discussed here have been proposed by Kaskavelis and Caramanis [KaC98], and Zhao, Luh, and Wang [ZLW99]. The analysis given here is due to Nedić and Bertsekas [NeB01a] (also partly reproduced in the textbook by Bertsekas [Ber99a]).

The first dynamic stepsize adjustment procedure of Section 8.2.1 was proposed by Nedić and Bertsekas [NeB01a] (this rule is new even when specialized to the ordinary subgradient method). The second adjustment procedure is based on the ideas and algorithms of Brännlund [Brä93] (see also Goffin and Kiwiel [GoK99]). The idea of randomization of the component selection and the corresponding convergence analysis are due to Nedić and Bertsekas [NeB01a] (also reported in [Ber99a]). The analysis of the convergence rate of incremental subgradient methods is due to Nedić and Bertsekas [NeB01b].

A distributed asynchronous implementation of incremental subgradient methods, with and without randomization, was given by Nedić, Bertsekas, and Borkar [NBB01]. In the randomized distributed version, the multiple processors, whenever they become available, select at random a component f_i, calculate the subgradient g_i, and execute the corresponding incremental subgradient step. The algorithm is asynchronous in that different processors use slightly differing copies of the current iterate x_k; this contributes to the efficiency of the method, because there is no waiting time for processors to synchronize at the end of iterations. Despite the asynchronism, convergence can be shown thanks to the use of a diminishing stepsize, similar to related distributed asynchronous gradient method analyses for differentiable optimization problems (see Tsitsiklis, Bertsekas, and Athans [TBA86], and Bertsekas and Tsitsiklis [BeT89]).

There are several other subgradient-like methods that we have not discussed. For example, in place of a subgradient, one may use an ϵ-

subgradient that is obtained by approximate minimization of the Lagrangian $L(\cdot, \mu^k)$ (see Exercise 8.6). This method also admits an incremental versions (see Nedić [Ned02], which provides an analysis of several variations of subgradient methods).

Another important class of methods, which we have not discussed, are the, so called, subgradient methods with space dilation, which were proposed and developed extensively by Shor; see his textbooks [Sho85] and [Sho98].

Cutting plane methods were introduced by Cheney and Goldstein [ChG59], and by Kelley [Kel60]. For analysis of related methods, see Ruszczynski [Rus89], Lemaréchal and Sagastizábal [LeS93], Mifflin [Mif96], Bonnans et. al. [BGL95], Kiwiel [Kiw97], Burke and Qian [BuQ98], and Mifflin, Sun, and Qi [MSQ98]. Central cutting plane methods were introduced by Elzinga and Moore [ElM75]. More recent proposals, some of which relate to interior point methods, are described in the textbook by Ye [Ye97], and the survey by Goffin and Vial [GoV02].

Ascent methods for minimax problems and the steepest ascent method in particular, were proposed by Demjanov [Dem66], [Dem68], and Grinold [Gri72]. The general forms of the steepest ascent and ϵ-ascent methods, based on projection on the subdifferential and ϵ subdifferential, respectively, were first proposed by Bertsekas and Mitter [BeM71], [BeM73]. Bundle methods, proposed by Lemarechal [Lem74], [Lem75], and Wolfe [Wol75], provided effective implementations of ϵ-ascent ideas (see Exercises 4.16 and 4.17).

EXERCISES

8.1

This exercise shows that nondifferentiabilities of the dual function often tend to arise at the most interesting points and thus cannot be ignored. Consider problem (P) and assume that for all $\mu \geq 0$, the infimum of the Lagrangian $L(x, \mu)$ over X is attained by at least one $x_\mu \in X$. Show that if there is a duality gap, then the dual function $q(\mu) = \inf_{x \in X} L(x, \mu)$ is nondifferentiable at every dual optimal solution. *Hint:* If q is differentiable at a dual optimal solution μ^*, by the theory of Section 4.2, we must have $\partial q(\mu^*)/\partial \mu_j \leq 0$ and $\mu_j^* \partial q(\mu^*)/\partial \mu_j = 0$ for all j. Use this to show that μ^* together with any vector x_{μ^*} that minimizes $L(x, \mu^*)$ over X satisfy the conditions for an optimal solution-geometric multiplier pair.

8.2 (Sharpness of the Error Tolerance Estimate)

Consider the unconstrained optimization of the two-dimensional function

$$f(x_1, x_2) = \sum_{i=1}^{M} C_0 \big(|x_1 + 1| + 2|x_1| + |x_1 - 1| + |x_2 + 1| + 2|x_2| + |x_2 - 1| \big),$$

where C_0 is a positive constant, by using the incremental subgradient method with a constant stepsize α. Show that there exists a component processing order such that when the method starts a cycle at the point $\bar{x} = (\bar{x}_1, \bar{x}_2)$, where $\bar{x}_1 = \bar{x}_2 = \alpha M C_0$ with $\alpha M C_0 \leq 1$, it returns to $\bar{x}$ at the end of the cycle. Use this example to show that starting from $\bar{x}$, we have

$$\liminf_{k \to \infty} f(\psi_{i,k}) \geq f^* + \frac{\beta \alpha C^2}{2}, \qquad \forall\, i = 1, \ldots, m,$$

for some constant β (independent of C_0 and M), where $C = mC_0$ and $m = 8M$ [cf. Eq. (8.13)].

8.3 (A Variation of the Subgradient Method [CFM75])

Consider the dual problem and the following variation of the subgradient method

$$\mu^{k+1} = P_X(\mu^k + s^k d^k),$$

where

$$d^k = \begin{cases} g^k & \text{if } k = 0, \\ g^k + \beta^k d^{k-1} & \text{if } k > 0, \end{cases}$$

g^k is a subgradient of q at μ^k, and s^k and β^k are scalars satisfying

$$0 < s^k \leq \frac{q(\mu^*) - q(\mu^k)}{\|d^k\|^2},$$

$$\beta^k = \begin{cases} -\gamma \dfrac{g^{k\prime} d^{k-1}}{\|d^{k-1}\|^2} & \text{if } g^{k\prime} d^{k-1} < 0, \\ 0 & \text{otherwise}, \end{cases}$$

with $\gamma \in [0, 2]$, and μ^* is an optimal dual solution. Assuming $\mu^k \neq \mu^*$, show that

$$\|\mu^* - \mu^{k+1}\| < \|\mu^* - \mu^k\|.$$

Furthermore,

$$\frac{(\mu^* - \mu^k)' d^k}{\|d^k\|} \geq \frac{(\mu^* - \mu^k)' g^k}{\|g^k\|},$$

i.e., the angle between d^k and $\mu^* - \mu^k$ is no larger than the angle between g^k and $\mu^* - \mu^k$.

8.4 (Subgradient Randomization for Stochastic Programming)

Consider the stochastic programming problem of Example 8.2.2. Derive and justify a randomized incremental subgradient method that samples the different component functions f_i using the probabilities π_i.

8.5

Give an example of a one-dimensional problem where the cutting plane method is started at an optimal solution but does not terminate finitely.

8.6 (Approximate Subgradient Method)

This exercise deals with a variant of the subgradient method whereby we minimize approximately $L(x, \mu^k)$ over $x \in X$, thereby obtaining a vector $x^k \in X$ with

$$L(x^k, \mu^k) \leq \inf_{x \in X} L(x, \mu^k) + \epsilon.$$

(a) Show that the corresponding constraint vector, $g(x^k)$, is an ϵ-subgradient at μ^k.

(b) Consider the iteration

$$\mu^{k+1} = P_M \left(\mu^k + \frac{q(\mu^*) - q(\mu^k)}{\|g^k\|^2} g^k \right),$$

where μ^* is a dual optimal solution, g^k is an ϵ-subgradient at μ^k, and $\epsilon > 0$. Show that as long as μ^k is such that $q(\mu^k) < q(\mu^*) - 2\epsilon$, we have $\|\mu^{k+1} - \mu^*\| < \|\mu^k - \mu^*\|$.

References

[AHU61] Arrow, K. J., Hurwicz, L., and Uzawa, H., 1961. "Constraint Qualifications in Maximization Problems," Naval Research Logistics Quarterly, Vol. 8, pp. 175-191.

[Aba67] Abadie, J., 1967. "On the Kuhn-Tucker Theorem," in Nonlinear Programming, Abadie, J., (Ed.), North Holland, Amsterdam.

[Ash72] Ash, R. B., 1972. Real Analysis and Probability, Academic Press, N. Y.

[AuF90] Aubin, J.-P., and Frankowska, H., 1990. Set-Valued Analysis, Birkhauser, Boston.

[AuT03] Auslender, A., and Teboulle, M., 2003. Asymptotic Cones and Functions in Optimization and Variational Inequalities, Springer, N. Y.

[Aus96] Auslender, A., 1996. "Non Coercive Optimization Problems," Math. of Operations Research, Vol. 21, pp. 769-782.

[Aus97] Auslender, A., 1997. "How to Deal with the Unbounded in Optimization: Theory and Algorithms," Math. Programing, Vol. 79, pp. 3-18.

[Aus00] Auslender, A., 2000. "Existence of Optimal Solutions and Duality Results Under Weak Conditions," Math. Programing, Vol. 88, pp. 45-59.

[BGL95] Bonnans, J. F., Gilbert, J. C., Lemaréchal, C., Sagastizábal, S. C., 1995. "A Family of Variable Metric Proximal Methods," Math. Programming, Vol. 68, pp. 15-47.

[BGS72] Bazaraa, M. S., Goode, J. J., and Shetty, C. M., 1972. "Constraint Qualifications Revisited," Management Science, Vol. 18, pp. 567-573.

[BMN01] Ben-Tal, A., Margalit, T., and Nemirovski, A., 2001. "The Ordered Subsets Mirror Descent Optimization Method and its Use for the Positron Emission Tomography Reconstruction," in Inherently Parallel Algorithms in Feasibility and Optimization and Their Applications, Eds., Butnariu, D., Censor, Y., and Reich, S., Elsevier Science, Amsterdam, Netherlands.

[BNO02] Bertsekas, D. P., Nedić, A., and Ozdaglar, A. E., 2002. "Min Common/Max Crossing Duality: A Simple Geometric Framework for Convex

Optimization and Minimax Theory," Lab. for Information and Decision Systems Report LIDS-P-2536, M.I.T., Cambridge, MA; to appear in J. Opt. Theory and Appl.

[BOT02] Bertsekas, D. P., Ozdaglar, A. E., and Tseng, P., 2002. Unpublished research.

[BaW75] Balinski, M., and Wolfe, P., (Eds.), 1975. Nondifferentiable Optimization, Math. Programming Study 3, North-Holland, Amsterdam.

[BeG62] Berge, C., and Ghouila-Houri, A., 1962. Programmes, Jeux et Reseau de Transport, Dunod, Paris.

[BeM71] Bertsekas, D. P, and Mitter, S. K., 1971. "Steepest Descent for Optimization Problems with Nondifferentiable Cost Functionals," Proc. 5th Annual Princeton Confer. Inform. Sci. Systems, Princeton, N. J., pp. 347-351.

[BeM73] Bertsekas, D. P., and Mitter, S. K., 1973. "A Descent Numerical Method for Optimization Problems with Nondifferentiable Cost Functionals," SIAM J. on Control, Vol. 11, pp. 637-652.

[BeN01] Ben-Tal, A., and Nemirovski, A., 2001. Lectures on Modern Convex Optimization: Analysis, Algorithms, and Engineering Applications, SIAM, Philadelphia.

[BeN02] Bertsekas, D. P., and Nedić, A., 2002. "A Unified Framework for Existence of Solutions and Absence of Duality Gap in Convex Programs, and Minimax Theory," Unpublished Research.

[BeO00a] Bertsekas, D. P, and Ozdaglar, A. E., 2000. "Pseudonormality and a Lagrange Multiplier Theory for Constrained Optimization," Lab. for Information and Decision Systems Report LIDS-P-2489, M.I.T., Cambridge, MA; J. Opt. Theory and Appl., Vol. 114, 2002, pp. 287-343.

[BeO00b] Bertsekas, D. P, and Ozdaglar, A. E., 2000. "Enhanced Optimality Conditions and Exact Penalty Functions," Proceedings of Allerton Conference, Allerton Park, Ill.

[BeT89] Bertsekas, D. P., and Tsitsiklis, J. N., 1989. Parallel and Distributed Computation: Numerical Methods, Prentice-Hall, Englewood Cliffs, N. J; republished by Athena Scientific, Belmont, MA, 1997.

[BeT96] Bertsekas, D. P., and Tsitsiklis, J. N., 1996. Neuro-Dynamic Programming, Athena Scientific, Belmont, MA.

[BeT97] Bertsimas, D., and Tsitsiklis, J. N., 1997. Introduction to Linear Optimization, Athena Scientific, Belmont, MA.

[Ber75] Bertsekas, D. P., 1975. "Necessary and Sufficient Conditions for a Penalty Method to be Exact," Math. Programming, Vol. 9, pp. 87-99.

[Ber82] Bertsekas, D. P., 1982. Constrained Optimization and Lagrange

Multiplier Methods, Academic Press, N. Y; republished by Athena Scientific, Belmont, MA, 1997.

[Ber98] Bertsekas, D. P., 1998. Network Optimization: Continuous and Discrete Models, Athena Scientific, Belmont, MA.

[Ber99a] Bertsekas, D. P., 1999. Nonlinear Programming: 2nd Edition, Athena Scientific, Belmont, MA.

[Ber99b] Bertsekas, D. P., 1999. "A Note on Error Bounds for Convex and Nonconvex Problems", Computational Optimization and Applications, Vol. 12, 1999, pp. 41-51.

[BoF34] Bonnesen, T., and Fenchel, W., 1934. Theorie der Konvexen Korper, Springer, Berlin; republished by Chelsea, N. Y., 1948.

[BoL00] Borwein, J. M., and Lewis, A. S., 2000. Convex Analysis and Nonlinear Optimization, Springer-Verlag, N. Y.

[BoS00] Bonnans, J. F., and Shapiro, A., 2000. Perturbation Analysis of Optimization Problems, Springer-Verlag, N. Y.

[Bou30] Bouligand, G., 1930. "Sur les Surfaces Depourvues de Points Hyperlimites," Annales de la Societe Polonaise de Mathematique, Vol. 9, pp. 32-41.

[Bou32] Bouligand, G., 1932. Introduction a la Geometrie Infinitesimale Directe, Gauthiers-Villars, Paris.

[Brä93] Brännlund, U., 1993. "On Relaxation Methods for Nonsmooth Convex Optimization," Doctoral Thesis, Royal Institute of Technology, Stockholm, Sweden.

[Bro64] Bronsted, A., 1964. "Conjugate Convex Functions in Topological Vector Spaces," Fys. Medd. Dans. Vid. Selsk., Vol. 34, pp. 1-26.

[BuQ98] Burke, J. V., and Qian, M., 1998. "A Variable Metric Proximal Point Algorithm for Monotone Operators," SIAM J. on Control and Optimization, Vol. 37, pp. 353-375.

[CCP98] Cook, W., Cunningham, W., Pulleyblank, W., and Schrijver, A., 1998. Combinatorial Optimization, Wiley, N. Y.

[CFM75] Camerini, P. M., Fratta, L., and Maffioli, F., 1975. "On Improving Relaxation Methods by Modified Gradient Techniques," Math. Programming Studies, Vol. 3, pp. 26-34.

[Cam68] Camion, P., 1968. "Modules Unimodulaires," J. Comb. Theory, Vol. 4, pp. 301-362.

[Car11] Caratheodory, C., 1911. "Uber den Variabilitatsbereich der Fourierschen Konstanten von Positiven Harmonischen Funktionen," Rendiconto del Circolo Matematico di Palermo, Vol. 32, pp. 193-217; reprinted in Constantin Caratheodory, Gesammelte Mathematische Schriften, Band III (H. Tietze, ed.), C. H. Beck'sche Verlagsbuchhandlung, Munchen, 1955, pp. 78-110.

[ChG59] Cheney, E. W., and Goldstein, A. A., 1959. "Newton's Method for Convex Programming and Tchebycheff Approximation," Numer. Math., Vol. I, pp. 253-268.

[Chv83] Chvatal, V., 1983. Linear Programming, W. H. Freeman and Co., N. Y.

[Cla76] Clarke, F. H., 1976. "A New Approach to Lagrange Multipliers," Math. of Operations Research, Vol. 1, pp. 165-174.

[Cla83] Clarke, F. H., 1983. Optimization and Nonsmooth Analysis, J. Wiley, N. Y.

[DGK63] Danzer, L., Grunbaum, B., and Klee, V. L., 1963. "Helly's Theorem and its Relatives," in Convexity, by Klee, V. L., (Ed.), American Math. Society, Providence, R. I., pp. 101-180.

[Dan63] Dantzig, G. B., 1963. Linear Programming and Extensions, Princeton Univ. Press, Princeton, N. J.

[DeV85] Demjanov, V. F., and Vasilév, L. V., 1985. Nondifferentiable Optimization, Optimization Software Inc., N. Y.

[Dem66] Demjanov, V. F., 1966. "The Solution of Several Minimax Problems," Kibernetika, Vol. 2, pp. 58-66.

[Dem68] Demjanov, V. F., 1968. "Algorithms for Some Minimax Problems," J. of Computer and Systems Science, Vol. 2, pp. 342-380.

[Egg58] Eggleston, H. G., 1958. Convexity, Cambridge Univ. Press, Cambridge.

[EkT76] Ekeland, I., and Temam, R., 1976. Convex Analysis and Variational Problems, North-Holland Publ., Amsterdam.

[Eke74] Ekeland, I., 1974. "On the Variational Principle," J. of Math. Analysis and Applications, Vol. 47, pp. 324-353.

[ElM75] Elzinga, J., and Moore, T. G., 1975. "A Central Cutting Plane Algorithm for the Convex Programming Problem," Math. Programming, Vol. 8, pp. 134-145.

[Erm83] Ermoliev, Yu. M., 1983. "Stochastic Quasigradient Methods and Their Application to System Optimization," Stochastics, Vol. 9, pp. 1-36.

[FGH57] Fan, K., Glicksberg, I., and Hoffman, A. J., 1957. "Systems of Inequalities Involving Convex Functions," Proc. Amer. Math. Soc., Vol. 8, pp. 617-622.

[Fan52] Fan, K., 1952. "Fixed-Point and Minimax Theorems in Locally

Convex Topological Linear Spaces," Proc. Nat. Acad. Sci. USA, Vol. 38, pp. 121-126.

[Fen49] Fenchel, W., 1949. "On Conjugate Convex Functions," Canad. J. of Math., Vol. 1, pp. 73-77.

[Fen51] Fenchel, W., 1951. "Convex Cones, Sets, and Functions," Mimeographed Notes, Princeton Univ.

[FrW56] Frank, M., and Wolfe, P., 1956. "An Algorithm for Quadratic Programming," Naval Research Logistics Quarterly, Vol. 3, pp. 95-110.

[GKT51] Gale, D., Kuhn, H. W., and Tucker, A. W., 1951. "Linear Programming and the Theory of Games," in Activity Analysis of Production and Allocation, Koopmans, T. C., (Ed.), Wiley, N. Y.

[GoK99] Goffin, J. L., and Kiwiel, K. C., 1999. "Convergence of a Simple Subgradient Level Method," Math. Programming, Vol. 85, pp. 207-211.

[GoT56] Goldman, A. J., and Tucker, A. W., 1956. "Theory of Linear Programming," in Linear Inequalities and Related Systems, H. W. Kuhn and A. W. Tucker, eds., Princeton University Press, Princeton, N.J., pp. 53-97.

[GoT71] Gould, F. J., and Tolle, J., 1971. "A Necessary and Sufficient Condition for Constrained Optimization," SIAM J. Applied Math., Vol. 20, pp. 164-172.

[GoT72] Gould, F. J., and Tolle, J., 1972. "Geometry of Optimality Conditions and Constraint Qualifications," Math. Programming, Vol. 2, pp. 1-18.

[GoV02] Goffin, J. L., and Vial, J. P., 2002. "Convex Nondifferentiable Optimization: A Survey Focussed on the Analytic Center Cutting Plane Method," Optimization Methods and Software, Vol. 17, pp. 805-867.

[Gor73] Gordan, P., 1873. "Uber die Auflosung Linearer Gleichungen mit Reelen Coefficienten," Mathematische Annalen, Vol. 6, pp. 23-28.

[Gri72] Grinold, R. C., 1972. "Steepest Ascent for Large-Scale Linear Programs," SIAM Review, Vol. 14, pp. 447-464.

[Gru67] Grunbaum, B., 1967. Convex Polytopes, Wiley, N. Y.

[Gui69] Guignard, M., 1969. "Generalized Kuhn-Tucker Conditions for Mathematical Programming Problems in a Banach Space," SIAM J. on Control, Vol. 7, pp. 232-241.

[HaM79] Han, S. P., and Mangasarian, O. L., 1979. "Exact Penalty Functions in Nonlinear Programming," Math. Programming, Vol. 17, pp. 251-269.

[Hel21] Helly, E., 1921. "Uber Systeme Linearer Gleichungen mit Unendlich Vielen Unbekannten," Monatschr. Math. Phys., Vol. 31, pp. 60-91.

[Hes75] Hestenes, M. R., 1975. Optimization Theory: The Finite Dimensional Case, Wiley, N. Y.

[HiL93] Hiriart-Urruty, J.-B., and Lemarechal, C., 1993. Convex Analysis and Minimization Algorithms, Vols. I and II, Springer-Verlag, Berlin and N. Y.

[HoK71] Hoffman, K., and Kunze, R., 1971. Linear Algebra, Prentice-Hall, Englewood Cliffs, N. J.

[Joh48] John, F., 1948. "Extremum Problems with Inequalities as Subsidiary Conditions," in Studies and Essays: Courant Anniversary Volume, Friedrichs, K. O., Neugebauer, O. E., and Stoker, J. J., (Eds.), Wiley-Interscience, N. Y., pp. 187-204.

[KaC98] Kaskavelis, C. A., and Caramanis, M. C., 1998. "Efficient Lagrangian Relaxation Algorithms for Industry Size Job-Shop Scheduling Problems," IIE Transactions on Scheduling and Logistics, Vol. 30, pp. 1085–1097.

[Kak41] Kakutani, S., 1941. "A Generalization of Brouwer's Fixed Point Theorem," Duke Mathematical Journal, Vol. 8, pp. 457-459.

[Kar39] Karush, W., 1939. "Minima of Functions of Several Variables with Inequalities as Side Conditions," M.S. Thesis, Department of Math., University of Chicago.

[Kel60] Kelley, J. E., 1960. "The Cutting-Plane Method for Solving Convex Programs," J. Soc. Indust. Appl. Math., Vol. 8, pp. 703-712.

[Kib79] Kibardin, V. M., 1979. "Decomposition into Functions in the Minimization Problem," Automation and Remote Control, Vol. 40, pp. 1311-1323.

[Kiw97] Kiwiel, K. C., 1997. "Efficiency of the Analytic Center Cutting Plane Method for Convex Minimization," SIAM J. on Optimization, Vol. 7, pp. 336-346.

[Kle63] Klee, V. L., (Ed.), 1963. Convexity, American Math. Society, Providence, R. I.

[Kle68] Klee, V. L., 1968. "Maximal Separation Theorems for Convex Sets," Trans. Amer. Math. Soc., Vol. 134, pp. 133-148.

[KrM40] Krein, M., and Millman, D., 1940. "On Extreme Points of Regular Convex Sets," Studia Math., Vol. 9, pp. 133-138.

[KuT51] Kuhn, H. W., and Tucker, A. W., 1951. "Nonlinear Programming," in Proc. of the Second Berkeley Symposium on Math. Statistics and Probability, Neyman, J., (Ed.), Univ. of California Press, Berkeley,

CA, pp. 481-492.

[Kuh76] Kuhn, H. W., 1976. "Nonlinear Programming: A Historical View," in Nonlinear Programming, Cottle, R. W., and Lemke, C. E., (Eds.), SIAM-AMS Proc., Vol. IX, American Math. Soc., Providence, RI, pp. 1-26.

[LVB98] Lobo, M. S., Vandenberghe, L., Boyd, S., and Lebret, H., 1998. "Applications of Second-Order Cone Programming," Linear Algebra and Applications, Vol. 284, pp. 193-228.

[LaT85] Lancaster, P., and Tismenetsky, M., 1985. The Theory of Matrices, Academic Press, N. Y.

[LeS93] Lemaréchal, C., and Sagastizábal, C., 1993. "An Approach to Variable Metric Bundle Methods," in Systems Modelling and Optimization, Proc. of the 16th IFIP-TC7 Conference, Compiègne, Henry, J., and Yvon, J.-P., (Eds.), Lecture Notes in Control and Information Sciences 197, pp. 144-162.

[Lem74] Lemaréchal, C., 1974. "An Algorithm for Minimizing Convex Functions," in Information Processing '74, Rosenfeld, J. L., (Ed.), pp. 552-556, North-Holland, Amsterdam.

[Lem75] Lemaréchal, C., 1975. "An Extension of Davidon Methods to Nondifferentiable Problems," Math. Programming Study 3, Balinski, M., and Wolfe, P., (Eds.), North-Holland, Amsterdam, pp. 95-109.

[LuZ99] Luo, Z.-Q., and Zhang, S. Z., 1999. "On the Extension of Frank-Wolfe Theorem," Comput. Optim. Appl., Vol. 13, pp. 87-110.

[MSQ98] Mifflin, R., Sun, D., and Qi, L., 1998. "Quasi-Newton Bundle-Type Methods for Nondifferentiable Convex Optimization, SIAM J. on Optimization, Vol. 8, pp. 583-603.

[MaF67] Mangasarian, O. L., and Fromovitz, S., 1967. "The Fritz John Necessary Optimality Conditions in the Presence of Equality and Inequality Constraints," J. Math. Anal. and Appl., Vol. 17, pp. 37-47.

[McS73] McShane, E. J., 1973. "The Lagrange Multiplier Rule," Amer. Mathematical Monthly, Vol. 80, pp. 922-925.

[Man69] Mangasarian, O. L., 1969. Nonlinear Programming, Prentice-Hall, Englewood Cliffs, N. J.; also SIAM, Classics in Applied Mathematics 10, Phila., PA., 1994.

[Mif96] Mifflin, R., 1996. "A Quasi-Second-Order Proximal Bundle Algorithm," Math. Programming, Vol. 73, pp. 51-72.

[Min11] Minkowski, H., 1911. "Theorie der Konvexen Korper, Insbesondere Begrundung Ihres Ober Flachenbegriffs," Gesammelte Abhandlungen, II, Teubner, Leipsig.

[Min60] Minty, G. J., 1960. "Monotone Networks," Proc. Roy. Soc. London, A, Vol. 257, pp. 194-212.

[Mor62] Moreau, J.-J., 1962. "Fonctions Convexes Duales et Points Proximaux dans en Espace Hilbertien," Comptes Rendus de l'Academie des Sciences de Paris, 255, pp. 2897-2899.

[Mor64] Moreau, J.-J., 1964. "Theoremes 'inf-sup'," Comptes Rendus de l'Academie des Sciences de Paris, 258, pp. 2720-2722.

[Mor76] Mordukhovich, B. S., 1976. "Maximum Principle in the Problem of Time Optimal Response with Nonsmooth Constraints," J. of Applied Mathematics and Mechanics, Vol. 40, pp. 960-969.

[Mor88] Mordukhovich, B. S., 1988. Approximation Methods in Problems of Optimization and Control, Nauka, Moscow.

[NBB01] Nedić, A., Bertsekas, D. P., and Borkar, V. S., 2001. "Distributed Asynchronous Incremental Subgradient Methods," in Inherently Parallel Algorithms in Feasibility and Optimization and Their Applications, Butnariu, D., Censor, Y., and Reich, S., (Eds.), Elsevier Science, Amsterdam, Netherlands.

[NeB01a] Nedić, A., and Bertsekas, D. P., 2001. "Incremental Subgradient Methods for Nondifferentiable Optimization," SIAM J. on Optim., Vol. 12, pp. 109-138.

[NeB01b] Nedić, A., and Bertsekas, D. P., 2001. "Convergence Rate of Incremental Subgradient Algorithms," in Stochastic Optimization: Algorithms and Applications, Uryasev, S., and Pardalos, P. M., (Eds.), Kluwer Academic Publishers, Dordrecht, Netherlands, pp. 223-264.

[NeB02] Nedić, A., and Bertsekas, D. P., 2002. "A Unified Framework for Minimax Theory, and for Existence of Solutions and Absence of Duality Gap in Convex Programs," Unpublished Research.

[NeW88] Nemhauser, G. L., and Wolsey, L. A., 1988. Integer and Combinatorial Optimization, Wiley, N. Y.

[Ned02] Nedić, A., 2002. Subgradient Methods for Convex Minimization, Ph.D. Thesis, Mass. Institute of Technology, Cambridge, MA.

[Neu28] Neumann, J. von, 1928. "Zur Theorie der Gesellschaftsspiele," Math. Ann., Vol. 100, pp. 295-320.

[Neu37] Neumann, J. von, 1937. "Ueber ein Okonomisches Gleichungssystem und eine Verallgemeinerung des Brouwerschen Fixpunktsatzes," Engebn. Math. Kolloq. Wien, Vol. 8, pp. 73-83.

[Nev75] Neveu, J., 1975. Discrete Parameter Martingales, North-Holland, Amsterdam.

[Nur77] Nurminskii, E. A., 1977. "The Continuity of ϵ-Subgradient Mappings," (Russian) Kibernetika (Kiev), pp. 148-149.

[OrR70] Ortega, J. M., and Rheinboldt, W. C., 1970. Iterative Solution of Nonlinear Equations in Several Variables, Academic Press, N. Y.

[OzB03] Ozdaglar, A. E., and Bertsekas, D. P., 2003. "The Relation Between Pseudonormality and Quasiregularity in Constrained Optimization," report in preparation.

[Ozd03] Ozdaglar, A. E., 2003. Pseudonormality and a Lagrange Multiplier Theory for Constrained Optimization, Ph.D. Thesis, Mass. Institute of Technology, Cambridge, MA.

[Pie69] Pietrzykowski, T., 1969. "An Exact Potential Method for Constrained Maxima," SIAM J. Numer. Anal., Vol. 6, pp. 294-304.

[Pol87] Polyak, B. T., 1987. Introduction to Optimization, Optimization Software Inc., N.Y.

[RoW98] Rockafellar, R. T., and Wets, R. J.-B., 1998. Variational Analysis, Springer-Verlag, Berlin.

[Roc64] Rockafellar, R. T., 1964. "Minimax Theorems and Conjugate Saddle Functions," Math. Scand., Vol. 14, pp. 151-173.

[Roc65] Rockafellar, R. T., 1965. "Helly's Theorem and Minima of Convex Functions," Duke Math. J., Vol. 33, pp. 81-90.

[Roc69] Rockafellar, R. T., 1969. "The Elementary Vectors of a Subspace of R^N," in Combinatorial Mathematics and its Applications, by Bose, R. C., and Dowling, T. A., (Eds.), University of North Carolina Press, pp. 104-127.

[Roc70] Rockafellar, R. T., 1970. Convex Analysis, Princeton Univ. Press, Princeton, N. J.

[Roc71] Rockafellar, R. T., 1971. "Ordinary Convex Programs Without a Duality Gap," J. Opt. Theory and Appl., Vol. 7, pp. 143-148.

[Roc84] Rockafellar, R. T., 1984. Network Flows and Monotropic Optimization, Wiley, N. Y.; republished by Athena Scientific, Belmont, MA, 1998.

[Roc93] Rockafellar, R. T., 1993. "Lagrange Multipliers and Optimality," SIAM Review, Vol. 35, pp. 183-238.

[Rud76] Rudin, W., 1976. Principles of Mathematical Analysis, McGraw-Hill, N. Y.

[Rus89] Ruszczynski, A., 1989. "An Augmented Lagrangian Decomposition Method for Block Diagonal Linear Programming Problems," Operations Res. Letters, Vol. 8, pp. 287-294.

[Sch86] Schrijver, A., 1986. Theory of Linear and Integer Programming, Wiley, N. Y.

[Sho85] Shor, N. Z., 1985. Minimization Methods for Nondifferentiable Functions, Springer-Verlag, Berlin.

[Sho98] Shor, N. Z., 1998. Nondifferentiable Optimization and Polynomial Problems, Kluwer, Dordrecht, the Netherlands.

[Sla50] Slater, M., 1950. "Lagrange Multipliers Revisited: A Contribution to Non-Linear Programming," Cowles Commission Discussion Paper, Math. 403.

[SoZ98] Solodov, M. V., and Zavriev, S. K., 1998. "Error Stability Properties of Generalized Gradient-Type Algorithms," J. Opt. Theory and Appl., Vol. 98, pp. 663–680.

[Ste13] Steinitz, H., 1913. "Bedingt Konvergente Reihen und Konvexe System, I," J. of Math., Vol. 143, pp. 128-175.

[Ste14] Steinitz, H., 1914. "Bedingt Konvergente Reihen und Konvexe System, II," J. of Math., Vol. 144, pp. 1-40.

[Ste16] Steinitz, H., 1916. "Bedingt Konvergente Reihen und Konvexe System, III," J. of Math., Vol. 146, pp. 1-52.

[StW70] Stoer, J., and Witzgall, C., 1970. Convexity and Optimization in Finite Dimensions, Springer-Verlag, Berlin.

[Str76] Strang, G., 1976. Linear Algebra and its Applications, Academic Press, N. Y.

[TBA86] Tsitsiklis, J. N., Bertsekas, D. P., and Athans, M., 1986. "Distributed Asynchronous Deterministic and Stochastic Gradient Optimization Algorithms," IEEE Trans. on Aut. Control, Vol. AC-31, pp. 803-812.

[TBN82] Tseng, P., Bertsekas, D. P., Nedić, A., and Ozdaglar, A. E., 2002. Unpublished research.

[Ter85] Terlaky, T., 1985. "On l_p Programming," European J. of Operational Research, Vol. 22, pp. 70-100.

[Val63] Valentine, F. A., 1963. "The Dual Cone and Helly Type Theorems," in Convexity, by Klee, V. L., (Ed.), American Math. Society, Providence, R. I, pp. 473-493.

[Val64] Valentine, F. A., 1964. Convex Sets, McGraw-Hill, N.Y.

[Vor08] Voronoi, G., 1908. "Nouvelles Applications des Parametres Continus a la Theorie des Formes Quadratiques - Premier Memoire. Sur Quelques Proprietes des Formes Quadratiques Positives Parfaites," Journal fur die Reine und Angewandte Mathematik, Vol. 133, pp. 97-178.

[Wet90] Wets, R. J., 1990. "Elementary Constructive Proofs of the Theorems of Farkas, Minkowski and Weyl," in Economic Decision Making: Games, Econometrics and Optimization. Contributions in Honour of Jacques Dreze, Gabszewicz, J. J., Richard, J.-F., and Wolsey, L. A., (Eds.), North-Holland, Elsevier-Science, Amsterdam, pp. 427-432.

[Wey35] Weyl, H., 1935. "Elementare Theorie der Konvexen Polyeder," Commentarii Mathematici Helvetici, Vol. 7, pp. 290-306.

[Wol75] Wolfe, P., 1975. "A Method of Conjugate Subgradients for Minimizing Nondifferentiable Functions," Math. Programming Study 3, Balinski, M., and Wolfe, P., (Eds.), North-Holland, Amsterdam, pp. 145-173.

[Wol98] Wolsey, L. A., 1998. Integer Programming, Wiley, N. Y.

[Ye97] Ye, Y., 1997. Interior Point Algorithms: Theory and Analysis, Wiley Interscience, N. Y.

[ZLW99] Zhao, X., Luh, P. B., and Wang, J., 1999. "Surrogate Gradient Algorithm for Lagrangian Relaxation," J. Opt. Theory and Appl., Vol. 100, pp. 699–712.

[Zan67] Zangwill, W. I., 1967. "Nonlinear Programming via Penalty Functions," Management Sci., Vol. 13, pp. 344-358.

INDEX

A

Affine function 7, 29
Affine hull 36, 75
Affine set 6, 36
Analytic center 509
Approximate minima 155
Arithmetic-geometric mean 73
Arrow-Hurwitz-Uzawa constraint qualification 305
Ascent Methods 509, 513

B

Base of a cone 213
Basis 6
Bidirectionally flat functions 106, 156, 162, 381
Bifunction 152
Bolzano-Weierstrass Theorem 11
Boundary of a set 12
Boundary point 12
Bounded sequence 9
Bounded set 11
Branch-and-bound 358
Bundle methods 511

C

Caratheodory's Theorem 37-39, 68, 76
Cartesian product 5, 80, 132, 214
Central cutting plane methods 508
Chain rule 19, 233, 262, 264
Clarke regularity 261
Closed ball 12
Closed function 28-30
Closed halfspace 108, 113
Closed set 11
Closed set intersection 56-64, 70, 82
Closed sphere 12

Closedness under linear transformation 64-68
Closure of a function 79
Closure of a set 11, 43-47
Closure point 11
Cluster point 11
Coercive function 85
Combinatorial Separation Theorem 220
Combinatorial optimization 210
Compact set 11
Complementary slackness 282, 286, 364
Complementary violation 286
Component of a vector 5
Composition of functions 14, 71
Concave function 24, 26
Cone 22, 36, 70, 74
Cone decomposition 211
Cone generated by a set 36, 37, 75
Cone of feasible directions 248, 250
Cone programming 451
Conic duality 439
Conjugacy Theorem 426
Conjugate functions 422–424, 432
Conjugate duality 421-422
Conjugate saddle function 153
Constancy space of a function 97
Constant stepsize rule 470, 473
Constraint qualification 276, 302, 320
Continuity 14, 48, 264
Continuous differentiability 17
Convergent sequence 9
Convex combination 35
Convex function 23-26, 29-31, 34, 48, 71, 72
Convex hull 36-39, 75, 113, 176
Convex programming 208

Convex set 21, 22
Convex system alternatives 162, 218
Convex-affine system alternatives 218
Convexification of a function 75
Coordinate 5
Cutting plane methods 504-509, 513, 514

D

Danskin's Theorem 245
Decomposition of a convex set 56
Derivative 16
Descent direction 265
Determinant 7
Diagonal matrix 7
Differentiability 17, 229
Differentiable convex functions 31, 261
Differentiation theorems 19, 20
Dimension of a convex set 36
Dimension of a subspace 6
Dimension of an affine set 6
Diminishing stepsize rule 464, 470, 478
Direction of recession 50, 92
Directional derivative 17, 222, 225, 245, 261, 262
Directional differentiability 225
Discrete minimax problem 129
Discrete separable problems 465
Distance function 88
Domain 5
Dual derivative 457
Dual function 346, 355, 356, 432
Duality gap 123, 357, 374, 375, 377-381
Dual problem 129, 346, 355, 414-417, 422, 456
Duality theory 129, 355, 362, 415, 416

E

ϵ-ascent methods 241, 511, 513
ϵ-descent algorithm 241

ϵ-descent direction 266
ϵ-subdifferential 235-237, 264
ϵ-subgradient 235
Effective domain 25, 26
Ekeland's variational principle 154
Elementary vector 219
Empty set 4, 21
Enhanced Farkas' Lemma 312, 419
Enhanced Fritz John conditions 281, 325, 326, 390-392
Enhanced dual Fritz John conditions 399
Envelope function 160
Epigraph 25, 27, 28, 93
Exact penalty function 154, 270, 278, 313, 316–318, 320, 343, 441
Existence of geometric multipliers 366-374
Existence of solutions 85, 94-101, 215, 366, 377-381
Extended real number 4
Extended real-valued function 24, 241, 261, 263, 264
Extended representation 313, 319, 344, 415
Extreme point 180, 183, 184, 187, 188
Euclidean norm 8

F

Face 216
Farkas' Lemma 7, 170, 175, 204, 312, 419
Feasible direction 248
Feasible direction methods 252
Feasible solution 84
Fenchel duality 422
Fenchel duality theorems 434-437
Fenchel's Inequality 448
Finitely generated cone 169, 176
Forward image 5
Frechet differentiability 17
Fritz John conditions for linear contraints 387
Fritz John optimality conditions 270,

273, 281, 325, 326, 384, 387, 390-392, 399
Full rank 7
Fundamental Theorem of Linear Programming 188

G

Gateaux differentiability 17
Generated cone 36-38, 75
Geometric multiplier 209, 346, 347, 353, 367
Global minimum 84
Goldman-Tucker Complementarity Theorem 220, 417
Gordan's Theorem 217
Gradient 16, 20
Gradient matrix 18, 20

H

Halfspace 108
Helly's Theorem 70, 77, 82, 152
Hessian matrix 18
Holder Inequality 74
Horizontal hyperplane 117
Hyperplane 107
Hyperplane separation 108

I

Image 5-7
Implicit Function Theorem 20
Improper function 26, 28
Incremental subgradient methods 464, 512
Indicator function 430
Induced norm 14
Infeasible problem 84, 362
Infimum 4
Informative Lagrange multipliers 288-290, 320
Informative dual optimal solutions 410
Informative geometric multipliers 406, 407
Inner limit 13
Inner product 5

Integer programming 189, 210, 358
Interior of a set 12, 39
Interior point 12
Inverse image 5, 7
Inverse matrix 15
Invertible matrix 15
Isomorphic polyhedral sets 216

J

Jacobian matrix 18
Jensen's Inequality 36

K

Karush-Kuhn-Tucker conditions 270, 274
Krein-Milman Theorem 181, 183, 210
Kirchhoff's Laws 449

L

l_1-norm 9, 452
l_∞-norm 8, 452
Lagrange multiplier 270, 281, 299, 318, 320, 325, 332, 353
Lagrangian function 129, 282, 347
Lagrangian Saddle Point Theorem 360
Lagrangian stationarity 282
Left derivative 225
Left continuous function 14
Level set 24-27, 93, 103, 267
Limit 9, 13
Limit point 11
Line segment principle 40
Lineality space 54, 211
Linear equation 7
Linear function 7, 76
Linear independence constraint qualification (LICQ) 271, 304
Linear inequality 7
Linear programming 184, 188, 362
Linear programming duality 362
Linear system alternatives 218
Linear transformation 7, 19, 30, 44, 53, 55, 65, 79, 81, 212, 213, 233,

264
Linearity criterion 333, 334
Linearly independent vectors 6
Lipschitz continuity 154, 263
Local maximum 86
Local minimum 86
Lower limit 10
Lower semicontinuous function 14, 27, 28, 71, 224

M

Mangasarian-Fromowitz constraint qualification 305
Matrix norm 14
Max crossing point problem 120
Max function 245
Maximum norm 8
Maximum point 4, 84
Mean Value Theorem 19, 264
Min common point problem 120
Min common/max crossing duality 120-128, 163, 196-199, 449
Minimal Lagrange multipliers 289, 290
Minimax inequality 131
Minimax problems 77, 343
Minimax theory 128-151, 199-203, 343, 360-362, 437-439
Minimum point 4, 84
Minkowski-Weyl Representation 176
Minkowski-Weyl Theorem 171, 210
Monotonically nondecreasing sequence 9
Monotonically nonincreasing sequence 9
Monotropic programming 69, 220, 241, 450

N

Negative halfspace 108
Neighborhood 12
Network optimization 69, 191, 220, 449
Nonexpansive mapping 88

Nonlinear Farkas' Lemma 204-210, 367
Nonnegative combination 36
Nonsingular matrix 15
Nonsmooth 17
Nonsmooth analysis 69
Nonvertical hyperplane 117, 118
Norm 8, 29, 452
Norm equivalence 12
Normal 252
Normal cone 252, 266, 267
Normal of a hyperplane 108
Nullspace 7, 53, 65

O

Open ball 12
Open halfspace 108
Open set 11
Open sphere 12
Optimality conditions 34, 255-260, 268, 350, 351, 361
Orthogonal complement 6
Orthogonal vectors 5
Outer limit 13

P

Parallel subspace 6
Pareto optimality 419
Partial derivative 16
Partial minimization 101-107, 158, 214
Penalty methods 159, 278
Penalty parameter 278, 314, 446
Pointed cone 213
Polar cone 166, 212
Polar Cone Theorem 167
Polyhedral Proper Separation Theorem 193, 210
Polyhedral cone 168, 213
Polyhedral function 178, 214, 215, 432
Polyhedral programming 420
Polyhedral set 175, 213
Positive combination 36
Positive definite matrix 15, 16, 34

Positive halfspace 108
Positive homogeneity 430
Positive semidefinite matrix 15, 16, 34
Posynomial 73
Primal function 346, 374, 375, 382, 420, 432
Primal problem 346, 365
Projection theorem 88
Proper Separation Theorem 115, 117
Proper function 25
Properly separating hyperplane 114
Proximal mapping 160
Pseudonormality 280, 288, 302, 308, 316, 317, 320, 332-334, 389
Pythagorean Theorem 8

Q

Quadratic function 34
Quadratic penalty function 278
Quadratic programming 91, 100, 215, 365, 415
Quadratic programming duality 365, 415
Quasiconvex function 157
Quasinormality 313
Quasiregularity 268, 276, 313, 320

R

R-multipliers 299, 308, 319, 320
Radon's Theorem 77
Range 5
Range space 7
Rank 7
Recession cone of a function 93
Recession cone of a set 50, 80-82, 93
Recession Cone Theorem 50
Regular 69, 252-254, 261
Relative boundary 40
Relative boundary point 40
Relative interior 40, 43-47, 78, 79
Relative interior point 40
Relatively open 40
Relaxed problem 190

Right derivative 225
Right continuous function 14

S

Saddle Point Theorem 150
Saddle Point Theory 128-151
Saddle point 131, 164
Schwarz Inequality 8, 15
Second order expansions 19
Second order cone programming 451
Semiconvex problem 353
Semidefinite programming 69
Sensitivity 273, 275, 297, 340, 383, 415
Separable problems 465
Separating Hyperplane Theorem 111
Separating hyperplane 108
Sequence 9
Set intersection 56-64, 82
Simplex method 189, 190
Singular matrix 15
Slater condition 208, 307, 333-335, 371
Smooth 17
Span 6
Sphere 12
Steepest ascent methods 510, 513
Steepest descent direction 265, 370
Stochastic programming 466
Strict complementarity 342
Strict local minimum 86
Strictly convex function 24, 26, 31, 34
Strict Separation Theorem 111
Strictly separating hyperplane 111
Strong duality 123, 368, 371, 373, 374
Strong Lagrange multipliers 290
Strong separation 161
Strongly convex function 72
Subdifferential 227, 228, 231-233, 262, 263
Subgradient 227, 241-243, 382
Subgradient methods 460, 512-515

Subgradient methods with random-
ization 488, 494, 512
Subsequence 10
Subspace 6
Sufficient optimality conditions 34,
339, 340
Supermartingale Convergence The-
orem 489
Support 219
Support function 69, 430, 431
Supporting Hyperplane Theorem 109
Supporting hyperplane 109
Supremum 4
Symmetric matrix 7, 16

T

Tangent 248
Tangent cone 248-250, 266, 267
Target level algorithm 485
Theorems of the alternative 217-
219
Total unimodularity 190
Triangle inequality 8
Tucker's Complementarity Theorem
220

U

Unbounded problem 362
Unimodularity 190, 217
Upper limit 10
Upper semicontinuous function 14,
224, 226

V

Variational analysis 69
Vector sum of sets 5, 67, 68, 82
Vertex 78
Visualization Lemma 348

W

Weak Duality Theorem 123, 356
Weierstrass' Theorem 85

Y

Young Inequality 74

Z

Zero-sum games 130, 416

ATHENA SCIENTIFIC BOOKS

1. Convex Analysis and Optimization, by Dimitri P. Bertsekas, with Angelia Nedić and Asuman E. Ozdaglar, 2003, ISBN 1-886529-45-0, 560 pages

2. Introduction to Probability, by Dimitri P. Bertsekas and John N. Tsitsiklis, 2002, ISBN 1-886529-40-X, 430 pages

3. Dynamic Programming and Optimal Control: Second Edition, Vols. I and II, by Dimitri P. Bertsekas, 2001, ISBN 1-886529-08-6, 704 pages

4. Nonlinear Programming, Second Edition, by Dimitri P. Bertsekas, 1999, ISBN 1-886529-00-0, 791 pages

5. Network Optimization: Continuous and Discrete Models by Dimitri P. Bertsekas, 1998, ISBN 1-886529-02-7, 608 pages

6. Network Flows and Monotropic Optimization by R. Tyrrell Rockafellar, 1998, ISBN 1-886529-06-X, 634 pages

7. Introduction to Linear Optimization by Dimitris Bertsimas and John N. Tsitsiklis, 1997, ISBN 1-886529-19-1, 608 pages

8. Parallel and Distributed Computation: Numerical Methods by Dimitri P. Bertsekas and John N. Tsitsiklis, 1997, ISBN 1-886529-01-9, 718 pages

9. Neuro-Dynamic Programming, by Dimitri P. Bertsekas and John N. Tsitsiklis, 1996, ISBN 1-886529-10-8, 512 pages

10. Constrained Optimization and Lagrange Multiplier Methods, by Dimitri P. Bertsekas, 1996, ISBN 1-886529-04-3, 410 pages

11. Stochastic Optimal Control: The Discrete-Time Case, by Dimitri P. Bertsekas and Steven E. Shreve, 1996, ISBN 1-886529-03-5, 330 pages

Nonlinear Programming: 2nd Edition
Dimitri P. Bertsekas
Massachusetts Institute of Technology

This book is an insightful and comprehensive treatment of nonlinear programming. It focuses on constrained and unconstrained iterative algorithms, Lagrange multiplier theory, duality theory, and large scale optimization methods. It includes all the major classical topics as well as important recent research. It presents a more algorithmic view of several of the topics in the present book.

Review:

"This is a beautifully written book by a prolific author ... who has taken painstaking care in making the presentation extremely lucid.... The style is unhurried and intuitive yet mathematically rigorous."

"The numerous figures in the book are extremely well thought out and are used in a very effective way to elucidate the text. The detailed and self-explanatory long captions accompanying each figure are extremely helpful."

"The 80 pages constituting the four appendixes serve as a masterfully written introduction to the field of nonlinear programming that can be used as a self-contained monograph. Teachers using this book could easily assign these appendixes as introductory or remedial material."

Olvi Mangasarian (Optima, March 1997)

Among its special features, the book:

- provides extensive coverage of iterative optimization methods within a unifying framework

- provides a detailed treatment of interior point methods for linear programming

- covers in depth duality theory from both a variational and a geometric/convex analysis point of view

- includes much new material on a number of topics, such as neural network training, discrete-time optimal control, and large-scale optimization

- includes a large number of examples and exercises

- contains many exercises with solutions posted on the internet

- developed through extensive classroom use in first-year graduate courses.

Contents: 1. Unconstrained Optimization. 2. Optimization Over a Convex Set. 3. Lagrange Multiplier Theory. 4. Lagrange Multiplier Algorithms. 5. Duality and Convex Programming. 6. Dual Methods.

ISBN 1-886529-00-0, 780 pp., hardcover, 1999

Network Optimization: Continuous and Discrete Models
Dimitri P. Bertsekas
Massachusetts Institute of Technology

A comprehensive, insightful, and up-to-date treatment of one of the most elegant and applications-rich classes of optimization problems.

Linear, nonlinear, and discrete network optimization problems are discussed extensively, including their theory and algorithms, and their applications in fields such as communication, transportation, manufacturing, logistics, and production planning. A unified approach bridges the gap between continuous (linear/nonlinear) problems, and discrete (integer/combinatorial) problems.

From the review by Panos Pardalos (J. of Opt. and Software, 1998):

"This beautifully written book provides an introductory treatment of linear, nonlinear, and discrete network optimization problems... The textbook is addressed not only to students of optimization but to all scientists in numerous disciplines who need network optimization methods to model and solve problems. This book is an engaging read and it is highly recommended either as a textbook or as a reference on network optimization."

Among its special features, the book:

- provides a comprehensive account of the theory and the practical application of the principal algorithms for linear and nonlinear network flow problems, including simplex, dual ascent, and auction

- describes the main models for discrete network optimization, such as traveling salesman, constrained shortest path, vehicle routing, multidimensional assignment, facility location, network design, etc

- describes the main methods for integer-constrained network problems, such as branch-and-bound, Lagrangian relaxation, genetic algorithms, tabu search, simulated annealing, and rollout algorithms

- discusses extensively auction algorithms, based on the author's extensive research on the subject

- contains many examples, practical applications, and exercises

Contents: 1. Introduction. 2. Shortest Path Problems. 3. The Max-Flow Problem. 4. The Min-Cost Flow Problem. 5. Simplex Methods for Min-Cost Flow. 6. Dual Ascent Methods for Min-Cost Flow. 7.Auction Algorithms for Min-Cost Flow. 8. Nonlinear Network Optimization. 9. Convex Separable Network Problems. 10. Network Problems with Integer Constraints. Appendix: Mathematical Background.

ISBN: 1-886529-02-7, 608 pp., hardcover, 1998

Introduction to Linear Optimization

Dimitris Bertsimas and John N. Tsitsiklis

Massachusetts Institute of Technology

This book provides a unified, insightful, and modern treatment of linear optimization, that is, linear programming, network flow problems, and discrete optimization. It includes classical topics as well as the state of the art, in both theory and practice. It is an excellent supplement to the present book.

From the review by T. V. Marana (Optima, June 1997):

"The true merit of the book, however, lies in its pedagogical qualities which are so impressive that I have decided to adopt it for a course in linear programming ..."

"... the overall writing style is pleasant and to-the-point."

"One reading of this book is sufficient to appreciate the tremendous amount of the quality effort that the authors have put into the writing, and I strongly recommend it ..."

Among its special features, the book:

- develops the major algorithms and duality theory through a geometric perspective

- provides a thorough treatment of the geometry, convergence, and complexity of interior point methods

- covers the main methods for network flow problems

- contains a detailed treatment of integer programming formulations and algorithms

- discusses the art of formulating and solving large scale problems through practical case studies

- includes a large number of examples and exercises. Has been developed through extensive classroom use in graduate courses.

Contents: 1. Introduction. 2. The geometry of linear programming. 3. The simplex method. 4. Duality theory. 5. Sensitivity analysis. 6. Large scale optimization. 7. Network flow problems. 8. Complexity of linear programming and the ellipsoid method. 9. Interior point methods. 10. Integer programming formulations. 11. Integer programming methods. 12. The art in linear optimization.

ISBN 1-886529-19-1, 608 pp., hardcover, 1997

Constrained Optimization and Lagrange Multiplier Methods
Dimitri P. Bertsekas
Massachusetts Institute of Technology

This reference textbook, first published in 1982 by Academic Press, remains the authoritative and comprehensive treatment of some of the most widely used constrained optimization methods, including the augmented Lagrangian/multiplier and sequential quadratic programming methods. It is an excellent supplement to the present book and our Nonlinear Programming book.

From the review by S. Zlobec in SIAM Review:

"This is an excellent reference book. The author has done a great job in at least three directions. First, he expertly, systematically and with ever-present authority guides the reader through complicated areas of numerical optimization. This is achieved by carefully explaining and illustrating (by figures, if necessary) the underlying principles and theory. Second, he provides extensive guidance on the merits of various types of methods. This is expremely useful to practitioners. Finally, this is truly a state of the art book on numerical optimization."

Among its special features, the book:

- treats extensively augmented Lagrangian methods, including an exhaustive analysis of the associated convergence and rate of convergence properties

- develops comprehensively sequential quadratic programming and other Lagrangian methods

- provides a detailed analysis of differentiable and nondifferentiable exact penalty methods

- presents nondifferentiable and minimax optimization methods based on smoothing

- contains much in depth research not found in any other textbook

Contents: 1. Introduction. 2. The method of multipliers for equality constrained problems. 3. The method of multipliers for inequality constrained problems and nondifferentiable optimization. 4. Exact penalty methods and Lagrangian methods. 5. Nonquadratic penalty functions – convex programming .

ISBN 1-886529-04-3, 410 pp., softcover, 1996

Dynamic Programming and Optimal Control
Dimitri P. Bertsekas
Massachusetts Institute of Technology

This two-volume textbook develops in depth dynamic programming, a central algorithmic method for optimal control, sequential decision making under uncertainty, and combinatorial optimization.

The first volume is oriented towards modeling, conceptualization, and finite-horizon problems, but also includes a substantive introduction to infinite horizon problems that is suitable for classroom use. The second volume is oriented towards mathematical analysis and computation, and treats infinite horizon problems extensively. The text contains many illustrations, worked-out examples, and exercises.

Reviews:

"Here is a tour-de-force in its field." David K. Smith, in the Journal of Operational Research Society. "In conclusion, this book is an excellent source of reference ... The main strengths of the book are the clarity of the exposition, the quality and variety of the examples, and its coverage of the most recent advances." Thomas W. Archibald, in IMA Jnl. of Math. Appl. in Business and Industry.

Among its special features, the book:

- provides a unifying framework for sequential decision making

- treats simultaneously deterministic and stochastic control problems popular in control theory and operations research

- develops the theory of deterministic optimal control problems including the Pontryagin Minimum Principle

- describes the recent simulation-based approximation techniques of neuro-dynamic programming/reinforcement learning

- provides a comprehensive treatment of infinite horizon problems

Contents of Volume 1: 1. The Dynamic Programming Algorithm. 2. Deterministic Systems and the Shortest Path Problem. 3. Deterministic Continuous-Time Optimal Control. 4. Problems with Perfect State Information. 5. Problems with Imperfect State Information. 6. Suboptimal and Adaptive Control. 7. Introduction to Infinite Horizon Problems.

Contents of Volume 2: 1. Infinite Horizon - Discounted Problems. 2. Stochastic Shortest Path Problems. 3. Undiscounted Problems. 4. Average Cost per Stage Problems. 5. Continuous-Time Problems.

ISBN: 1-886529-08-6, 848 pp., hardcover, 2001

Neuro-Dynamic Programming
Dimitri P. Bertsekas and John N. Tsitsiklis
Massachusetts Institute of Technology

This is the first textbook that fully explains the neuro-dynamic programming/reinforcement learning methodology, which is a recent breakthrough in the practical application of neural networks and dynamic programming to complex problems of planning, optimal decision making, and intelligent control.

From the review by George Cybenko for IEEE Computational Science and Engineering, May 1998:

"Neurodynamic Programming is a remarkable monograph that integrates a sweeping mathematical and computational landscape into a coherent body of rigorous knowledge. The topics are current, the writing is clear and to the point, the examples are comprehensive and the historical notes and comments are scholarly."

"In this monograph, Bertsekas and Tsitsiklis have performed a Herculean task that will be studied and appreciated by generations to come. I strongly recommend it to scientists and engineers eager to seriously understand the mathematics and computations behind modern behavioral machine learning."

Among its special features, the book:

- Describes and unifies a large number of NDP methods, including several that are new

- Describes new approaches to formulation and solution of important problems in stochastic optimal control, sequential decision making, and discrete optimization

- Rigorously explains the mathematical principles behind NDP

- Illustrates through examples and case studies the practical application of NDP to complex problems from optimal resource allocation, optimal feedback control, data communications, game playing, and combinatorial optimization

- Presents extensive background and new research material on dynamic programming and neural network training

Neuro-Dynamic Programming is the winner of the 1997 INFORMS CSTS prize for research excellence in the interface between Operations Research and Computer Science

ISBN 1-886529-10-8, 512 pp., hardcover, 1996

Parallel and Distributed Computation: Numerical Methods
Dimitri P. Bertsekas and John N. Tsitsiklis
Massachusetts Institute of Technology

This highly acclaimed work, first published by Prentice Hall in 1989, is a comprehensive and theoretically sound treatment of parallel and distributed numerical methods. It focuses on algorithms that are naturally suited for massive parallelization, and it explores the fundamental convergence, rate of convergence, communication, and synchronization issues associated with such algorithms. This is an extensive book, which aside from its focus on parallel and distributed algorithms, contains a wealth of material on a broad variety of computation and optimization topics.

Reviews:

"This major contribution to the literature belongs on the bookshelf of every scientist with an interest in computational science, directly beside Knuth's three volumes and Numerical Recipes..." Anna Nagurney, University of Massachusetts, in the Intern. J. of Supercomputer Applications

"This major work of exceptional scholarship summarizes more than three decades of research into general-purpose algorithms for solving systems of equations and optimization problems." W. Smyth, in Comp. Rev.

Among its special features, the book:

- quantifies the performance of parallel algorithms, including the limitations imposed by the communication and synchronization penalties

- provides a comprehensive convergence analysis of asynchronous methods and a comparison with their asynchronous counterparts

- covers extensively parallel and distributed algorithms for systems of equations, variational inequalities, nonlinear programming, shortest paths, dynamic programming, network flows, and large-scale decomposition

- includes extensive research material on optimization methods, asynchronous algorithm convergence, rollback synchronization, asynchronous communication network protocols, and others

- contains many exercises with solutions posted on the internet

- contains much in depth research not found in any other textbook

ISBN 1-886529-01-9, 718 pp., softcover, 1997

Network Flows and Monotropic Optimization
R. Tyrrell Rockafellar
University of Washington

A rigorous and comprehensive treatment of network flow theory and monotropic optimization by one of the world's most renowned applied mathematicians.

This classic textbook, first published by J. Wiley in 1984, covers extensively the duality theory and the algorithms of linear and nonlinear network optimization optimization, and their significant extensions to monotropic programming (separable convex constrained optimization problems, including linear programs). It is an excellent supplement to the present book.

Monotropic programming problems are characterized by a rich interplay between combinatorial structure and convexity properties. Rockafellar develops, for the first time, algorithms and a remarkably complete duality theory for these problems.

From the review by Dimitri P. Bertsekas in SIAM Review:

"By creating an elegant unifying framework for a broad range of subjects, Rockafellar's book represents an important event in the evolution of optimization theory. Besides its creative aspect, the book is thoughtful, well written, and packed with a wealth of material. The large number of exercises (a total of 479!), most of them extensions and elaborations of the theory, enhance its value as a class textbook, and provide fertile grounds for self-study and inspiration. Every student and practitioner of optimization should take a careful look at this book."

Among its special features, the book:

- treats in-depth the duality theory for linear and nonlinear network optimization

- uses a rigorous step-by-step approach to develop the principal network optimization algorithms

- covers the main algorithms for specialized network problems, such as max-flow, feasibility, assignment, and shortest path

- develops in detail the theory of monotropic programming, based on the author's highly acclaimed research

- contains many examples, illustrations, and exercises

- contains much new material not found in any other textbook

ISBN 1-886529-06-X, 634 pp., hardcover, 1998

Introduction to Probability
Dimitri P. Bertsekas and John N. Tsitsiklis
Massachusetts Institute of Technology

An intuitive, yet precise introduction to probability theory, stochastic processes, and probabilistic models used in science, engineering, economics, and related fields. This is the currently used textbook for "Probabilistic Systems Analysis," an introductory probability course at the Massachusetts Institute of Technology, attended by a large number of undergraduate and graduate students.

The book strikes a balance between simplicity in exposition and sophistication in analytical reasoning. Some of the more mathematically rigorous analysis has been just intuitively explained in the text, so that complex proofs do not stand in the way of an otherwise simple exposition. At the same time, some of this analysis and the necessary mathematical results are developed (at the level of advanced calculus) in theoretical problems, which are included at the end of the corresponding chapter.

Among its special features, the book:

- Develops the basic concepts of probability, random variables, stochastic processes, laws of large numbers, and the central limit theorem

- Illustrates the theory with many examples

- Provides many theoretical problems that extend the book's coverage and enhance its mathematical foundation (solutions are included in the text)

- Provides many exercises that enhance the understanding of the basic material (solutions are posted on the publisher's website)

- Is supplemented by many unsolved practice exercises and other supportive class material posted on the publisher's website

- Has been developed through extensive classroom use and experience at the Masschusetts Institute of Technology

Contents: 1. Sample Space and Probability. 2. Discrete Random Variables. 3. General Random Variables. 4. Further Topics on Random Variables and Expectations. 5. The Bernoulli and Poisson Processes. 6. Markov Chains. 7. Limit Theorems.

ISBN: 1-886529-40-X, 430 pp., hardcover, 2002